Methods in Microbiology
Volume 34

Recent titles in the series

Volume 24 *Techniques for the Study of Mycorrhiza*
JR Norris, DJ Reed and AK Varma

Volume 25 *Immunology of Infection*
SHE Kaufmann and D Kabelitz

Volume 26 *Yeast Gene Analysis*
AJP Brown and MF Tuite

Volume 27 *Bacterial Pathogenesis*
P Williams, J Ketley and GPC Salmond

Volume 28 *Automation*
AG Craig and JD Hoheisel

Volume 29 *Genetic Methods for Diverse Prokaryotes*
MCM Smith and RE Sockett

Volume 30 *Marine Microbiology*
JH Paul

Volume 31 *Molecular Cellular Microbiology*
P Sansonetti and A Zychlinsky

Volume 32 *Immunology of Infection*, 2nd edition
SHE Kaufmann and D Kabelitz

Volume 33 *Functional Microbial Genomics*
B Wren and N Dorrell

Methods in Microbiology

Volume 34
Microbial Imaging

Edited by

Tor Savidge

*Department of Pediatric GI
Massachusetts General Hospital & Harvard Medical School
Charlestown, MA
USA*

*Department of Internal Medicine
The University of Texas Medical Branch
Galveston, TX
USA*

and

Charalabos Pothoulakis

*Division of Gastroenterology
Beth Israel Deaconess Medical Center, Harvard Medical School
Boston, MA
USA*

ELSEVIER
ACADEMIC
PRESS

Amsterdam Boston Heidelberg London New York Oxford Paris
San Diego San Francisco Singapore Sydney Tokyo

ELSEVIER B.V.
Radarweg 29
P.O. Box 211, 1000 AE Amsterdam
The Netherlands

ELSEVIER Inc.
525 B Street, Suite 1900
San Diego, CA 92101-4495
USA

ELSEVIER Ltd
The Boulevard, Langford Lane
Kidlington, Oxford OX5 1GB
UK

ELSEVIER Ltd
84 Theobalds Road
London WC1X 8RR
UK

© 2005 Elsevier Ltd. All rights reserved.

This work is protected under copyright by Elsevier Ltd, and the following terms and conditions apply to its use:

Photocopying
Single photocopies of single chapters may be made for personal use as allowed by national copyright laws. Permission of the Publisher and payment of a fee is required for all other photocopying, including multiple or systematic copying, copying for advertising or promotional purposes, resale, and all forms of document delivery. Special rates are available for educational institutions that wish to make photocopies for non-profit educational classroom use.

Permissions may be sought directly from Elsevier's Rights Department in Oxford, UK: phone (+44) 1865 843830, fax (+44) 1865 853333, e-mail: permissions@elsevier.com. Requests may also be completed on-line via the Elsevier homepage (http://www.elsevier.com/locate/permissions).

In the USA, users may clear permissions and make payments through the Copyright Clearance Center, Inc., 222 Rosewood Drive, Danvers, MA 01923, USA; phone: (+1) (978) 7508400, fax: (+1) (978) 7504744, and in the UK through the Copyright Licensing Agency Rapid Clearance Service (CLARCS), 90 Tottenham Court Road, London W1P 0LP, UK; phone: (+44) 20 7631 5555; fax: (+44) 20 7631 5500. Other countries may have a local reprographic rights agency for payments.

Derivative Works
Tables of contents may be reproduced for internal circulation, but permission of the Publisher is required for external resale or distribution of such material. Permission of the Publisher is required for all other derivative works, including compilations and translations.

Electronic Storage or Usage
Permission of the Publisher is required to store or use electronically any material contained in this work, including any chapter or part of a chapter.

Except as outlined above, no part of this work may be reproduced, stored in a retrieval system or transmitted in any form or by any means, electronic, mechanical, photocopying, recording or otherwise, without prior written permission of the Publisher.
Address permissions requests to: Elsevier's Rights Department, at the fax and e-mail addresses noted above.

Notice
No responsibility is assumed by the Publisher for any injury and/or damage to persons or property as a matter of products liability, negligence or otherwise, or from any use or operation of any methods, products, instructions or ideas contained in the material herein. Because of rapid advances in the medical sciences, in particular, independent verification of diagnoses and drug dosages should be made.

First edition 2005

A catalogue record is available from the British Library

Cover photograph: Adapted from Chapter 4, figure 15D, courtesy of Neu and Lawrence.

ISBN: 0-12-521534-7 (Hardbound)
ISBN: 0-12-521535-5 (Comb-bound)

∞ The paper used in this publication meets the requirements of ANSI/NISO Z39.48-1992 (Permanence of Paper). Printed in Great Britain.

Working together to grow
libraries in developing countries

www.elsevier.com | www.bookaid.org | www.sabre.org

ELSEVIER   BOOK AID International   Sabre Foundation

Contents

Contributors ix

Preface xiii

1. *In Situ* Hybridization Methods to Study Microbial Populations and Their Interactions with Human Host Cells 1
   R Holm

2. Fluorescent Protein Probes in Fungi 27
   KJ Czymmek, TM Bourett and RJ Howard

3. Live-cell Imaging of Filamentous Fungi Using Vital Fluorescent Dyes and Confocal Microscopy 63
   PC Hickey, SR Swift, MG Roca and ND Read

4. One-photon versus Two-photon Laser Scanning Microscopy and Digital Image Analysis of Microbial Biofilms 89
   TR Neu and JR Lawrence

5. Applications of Cryo- and Transmission Electron Microscopy in the Study of Microbial Macromolecular Structure and Bacterial-Host Cell Interactions 137
   MI Fernandez, M-C Prevost, PJ Sansonetti and G Griffiths

6. Microbial Surfaces Investigated Using Atomic Force Microscopy ... 163
   YF Dufrêne and DJ Müller

7. Positron Emission Tomography Imaging of Clinical Infectious Diseases 199
   C Van de Wiele, O De Winter, H Ham and R Dierckx

8. Biosensor Characterization of Structure-Function Relationships in Viral Proteins 213
   L Choulier, D Altschuh, G Zeder-Lutz and MHV Van Regenmortel

9. RT *In Situ* PCR: Protocols and Applications 239
   AF Nicol and GJ Nuovo

10. Real-Time Fluorescent PCR Techniques to Study Microbial-Host Interactions 255
    IM Mackay, KE Arden and A Nitsche

11. Design and Use of Functional Gene Microarrays (FGAs) for the Characterization of Microbial Communities 331
    CW Schadt, J Liebich, SC Chong, TJ Gentry, Z He, H Pan and J Zhou

Index 369

Series Advisors

Gordon Dougan Director of The Centre for Molecular Microbiology and Infection, Department of Biological Sciences, Imperial College of Science, Technology and Medicine, London SW7 2AZ, UK

Graham J Boulnois Schroder Ventures Life Sciences Advisers (UK) Limited, 71 Kingsway, London WC2B 6ST, UK

Jim Prosser Department of Molecular and Cell Biology, University of Aberdeen, Institute of Medical Sciences, Foresterhill, Aberdeen AB25 2ZD, UK

Ian R Booth Professor of Microbiology, Department of Molecular and Cell Biology, University of Aberdeen, Institute of Medical Sciences, Foresterhill, Aberdeen AB25 2ZD, UK

David A Hodgson Reader in Microbiology, Department of Biological Sciences, University of Warwick, Coventry CV4 7AL, UK

David H Boxer Department of Biochemistry, Medical Sciences Institute, Dundee DD1 4HN, UK

Contributors

D Altschuh UMR7100, CNRS, Ecole Supérieure de Biotechnologie de Strasbourg, Boulevard Sébastien Brandt, BP 10413, 67412 Illkirch Cedex, France

Katherine E Arden Clinical Virology Research Unit, Sir Albert Sakzewski Virus Research Centre, Royal Children's Hospital, Brisbane, Qld, Australia

Timothy M Bourett Crop Genetics Research and Development, DuPont Experimental Station, Wilmington, DE 19880-0402, USA

Song C Chong Environmental Sciences Division, Oak Ridge National Laboratory, P.O. Box 2008, Oak Ridge, TN 37831-6038, USA

Laurence Choulier UMR7100, CNRS, Ecole Supérieure de Biotechnologie de Strasbourg, Boulevard Sébastien Brandt, BP 10413, 67412 Illkirch Cedex, France

Kirk J Czymmek Department of Biological Sciences, University of Delaware, Newark, DE 19716, USA

Olivier De Winter Department of Nuclear Medicine, University Hospital Ghent, De Pintelaan 185, 9000 Ghent, Belgium

Rudi Dierckx Department of Nuclear Medicine, University Hospital Ghent, De Pintelaan 185, 9000 Ghent, Belgium

Yves F Dufrêne Unité de chimie des interfaces, Université Catholique de Louvain, Croix du Sud 2/18, B-1348 Louvain-la-Neuve, Belgium

M Isabel Fernandez Department de Biologia Cl·lular i Anatomia Patològica, Facultat de Medicina, Institut d'Investigacions Biomèdiques August Pi i Sunyer (IDIBAPS), University of Barcelona, Casanova, 143, E-08036 Barcelona, Spain

Terry J Gentry Environmental Sciences Division, Oak Ridge National Laboratory, P.O. Box 2008, Oak Ridge, TN 37831-6038, USA

Gareth Griffiths European Molecular Biology Laboratory, Heidelberg, Germany

Hamphrey Ham Department of Nuclear Medicine, University Hospital Ghent, De Pintelaan 185, 9000 Ghent, Belgium

Zhili He Environmental Sciences Division, Oak Ridge National Laboratory, P.O. Box 2008, Oak Ridge, TN 37831-6038, USA

Patrick C Hickey Fungal Cell Biology Group, Institute of Cell Biology, University of Edinburgh, Rutherford Building, Edinburgh EH9 3JH, UK

Ruth Holm Department of Pathology, The Norwegian Radium Hospital, Oslo, Norway

Richard J Howard Crop Genetics Research and Development, DuPont Experimental Station, Wilmington, DE 19880-0402, USA

John R Lawrence NWRI National Water Research Institute, Saskatoon, Sask., Canada

Jost Liebich Environmental Sciences Division, Oak Ridge National Laboratory, P.O. Box 2008, Oak Ridge, TN 37831-6038, USA

Ian M Mackay Clinical Virology Research Unit, Sir Albert Sakzewski Virus Research Centre, Royal Children's Hospital, Brisbane, Qld, Australia; and Clinical Medical Virology Centre, University of Queensland, Brisbane, Qld, Australia

Daniel J Müller BioTechnological Center, University of Technology Dresden, Tatzberg 47-51, D-01307 Dresden, Germany

Thomas R Neu Department of River Ecology, UFZ Centre for Environmental Research Leipzig-Halle, Magdeburg, Germany

Alcina Frederica Nicol Laboratory of Immunology-IPEC/FIOCRUZ, Av. Brasil 4365, Manguinhos-RJ 21045-900, Brazil

Andreas Nitsche Robert Koch Institute, Berlin, Germany

Gerard J Nuovo Ohio State University Medical Center, Department of Pathology, Columbus, OH 43210, USA

Hongbin Pan Environmental Sciences Division, Oak Ridge National Laboratory, P.O. Box 2008, Oak Ridge, TN 37831-6038, USA

Marie-Christine Prevost Plate-Forme de Microscopie Electronique, Institute Pasteur, Paris, France

Nick D Read Fungal Cell Biology Group, Institute of Cell Biology, University of Edinburgh, Rutherford Building, Edinburgh EH9 3JH, UK

M Gabriela Roca Fungal Cell Biology Group, Institute of Cell Biology, University of Edinburgh, Rutherford Building, Edinburgh EH9 3JH, UK

Philippe J Sansonetti Unité de Pathogénie Microbienne Moléculaire, Institute Pasteur, Paris, France

Christopher W Schadt Environmental Sciences Division, Oak Ridge National Laboratory, P.O. Box 2008, Oak Ridge, TN 37831-6038, USA

Samuel R Swift Fungal Cell Biology Group, Institute of Cell Biology, University of Edinburgh, Rutherford Building, Edinburgh EH9 3JH, UK

Christophe Van de Wiele Department of Nuclear Medicine, University Hospital Ghent, De Pintelaan 185, 9000 Ghent, Belgium

Marc H V Van Regenmortel UMR7100, CNRS, Ecole Supérieure de Biotechnologie de Strasbourg, Boulevard Sébastien Brandt, BP 10413, 67412 Illkirch Cedex, France

Gabrielle Zeder-Lutz UMR7100, CNRS, Ecole Supérieure de Biotechnologie de Strasbourg, Boulevard Sébastien Brandt, BP 10413, 67412 Illkirch Cedex, France

Jizhong Zhou Environmental Sciences Division, Oak Ridge National Laboratory, P.O. Box 2008, Oak Ridge, TN 37831-6038, USA

Preface

Using various imaging methods developed and refined since the 1960s it is now possible to characterize the physiology and pharmacology of individual microorganisms. Recent advances in imaging technology and the application of molecular biology to this discipline has provided new microbial imaging tools that not only complement more classical methods, but in many cases significantly enhance the sensitivity and efficiency in which studies may be conducted.

The aim of this volume is to provide a comprehensive reference text for the novice-to-expert that covers important theoretical and technological advances in 'microbial imaging'. Emphasis has been placed on detailing methodology, as well as providing a theoretical background and listing potential applications of specific imaging tools to a broad range of microorganisms. These technologies are now applicable to a wide range of problems in contemporary microbiology, including strain selection, understanding microbial structure, function and pathophysiology, as well as in the development of anti-microbial agents and vaccines.

The compiled volume covers 11 distinct, yet complementary imaging methods that permit the investigator to study microorganisms in their host environments under molecular, ultrastructural or physiological real-time conditions. These technologies range from *in situ* hydridization (Chapter 1) and RT *in situ* PCR (Chapter 9) that localize specific low copy transcripts in histological sections, to more classical methods of studying microbial-host ultrastructure using electron microscopy (Chapter 5). Molecular characterization of microbial communities using real-time fluorescent PCR and functional gene microarrays is considered in Chapters 10 and 11. Comparative fluorescent detection of protein expression in microbial populations using specialized one versus two-photon laser scanning microscopy is considered in Chapter 4. Detailed methodology outlining the use of fluorescent protein probes and vital dyes in fungi is covered in Chapters 2 and 3. Finally, the potential real-time structure-function and clinical applications of more specialized 'microbial imaging' techniques such as atomic force microscopy, positron emission tomography and surface plasma resonance are considered in Chapters 6–8.

Importantly, in preparing *Microbial Imaging* we are deeply indebted to the outstanding authors who contributed their extensive expertise, and in many cases provided previously unpublished data. We hope that this work will prove useful to the scientific community as a whole, and we are especially grateful to the editorial staff at Elsevier for their invaluable role in preparing the volume.

Tor Savidge and Harry Pothoulakis

1 *In Situ* Hybridization Methods to Study Microbial Populations and Their Interactions with Human Host Cells

Ruth Holm
Department of Pathology, The Norwegian Radium Hospital, Oslo, Norway

◆◆◆

CONTENTS

Introduction
Methodology
Microorganisms that interact with human host cells

◆◆◆◆◆◆ INTRODUCTION

In situ hybridization techniques are used to localize specific DNA and RNA sequences within individual cells, either in tissue sections or in whole cell preparations. It was introduced in 1969 (Gall and Pardue, 1969; John *et al.*, 1969) and was used primarily for the localization of cellular DNA sequences. *In situ* hybridization is the only method permitting direct visualization of DNA and RNA sequences in a morphological context. Initially, nucleic acid probes were labeled isotopically and detected by autoradiography. However, the use of radioactively labeled probes limited the application to research and diagnostic areas. Since the introduction of nonisotopic probes, the use of *in situ* hybridization has grown considerably (Komminoth, 1992; Pang and Baum, 1993; DeLellis, 1994; Still *et al.*, 2000; Wang, 2003). These methods offer economic and other advantages over radioactive methods, by providing greater probe stability, avoiding hazardous radiation, improving resolution, and providing shorter turnover periods in the analysis of results. A potential drawback of nonisotopic *in situ* hybridization is that the sensitivity of the techniques have been reported to be less or at comparable levels to radioactive methods (Lurain *et al.*, 1986; Mifflin *et al.*, 1987; Nuovo and Richart *et al.*, 1989; Grubendorf-Conen and Cremer, 1990). However, the sensitivity is dependent on the detection system used (Herrington *et al.*, 1989; Furuta *et al.*, 1990; Morris *et al.*, 1990; Holm *et al.*, 1992; Zehbe *et al.*, 1997; Holm, 2000).

In situ hybridization has become established as an essential method in the study of cell and molecular biology, and is a powerful technique to visualize host–pathogen interactions (Holm *et al.*, 1994; Plummer *et al.*, 1998; Reed *et al.*, 1998; Moter and Göbel, 2000; Jayshree *et al.*, 2003). In many laboratories *in situ* hybridization is a routine procedure, particularly in the diagnosis of viral disease. A number of viruses such as human papillomavirus (HPV) (Syrjanen, 1992; Holm *et al.*, 1994; Herrington *et al.*, 1995), epstein-barr virus (EBV) (MacMahon *et al.*, 1991), cytomegalovirus (CMV) (Wolber and Lloyd, 1988; Rimsza *et al.*, 1996), human sarcoma viruses (HSV) (Reed *et al.*, 1998), polyomaviruses (JC, BK) (Boerman *et al.*, 1989), hepatitis B virus (Blum *et al.*, 1983; Jayshree *et al.*, 2003), hepatitis C virus (van Dekken *et al.*, 2003), parvoviruses (Porter *et al.*, 1988) and human immunodeficiency virus (HIV-1) (Gerrard *et al.*, 1994; Wenig *et al.*, 1996) have been detected by *in situ* hybridization. The purpose of this chapter is to outline *in situ* hybridization techniques, paying particular attention to the different detection systems that are available. Furthermore, the chapter will focus on HPV DNA integration of human chromosomes.

◆◆◆◆◆◆ METHODOLOGY

The sensitivity and outcome of *in situ* hybridization are highly dependent on numerous variables, such as the effects of tissue preparation on the retention and accessibility of target DNA and RNA, the type of probe construction, efficiency of probe labeling, hybridization conditions, and the quality of the detection system applied.

Fixation

Optimal fixation for *in situ* hybridization is to preserve maximum retention of nucleic acid in the tissue, and at the same time maintain morphological detail and allowing access of the probe to its target. In contrast to stable DNA, mRNA is more susceptible to degradation by RNases in the tissue. Therefore, when mRNA is to be hybridized, the tissue should be rapidly fixed or frozen. Delay in fixation results in a loss of signal for mRNA (Pringle *et al.*, 1989). However, DNA and mRNA with a high copy number can be detected in paraffin-embedded archival tissue (Holm *et al.*, 1997). Although some loss of mRNA will occur during paraffin processing, this may be counter-balanced by improved tissue morphology and a clearer resolution of the final results (Chinery *et al.*, 1992). A variety of fixatives have been assessed (Weiss and Chan, 1991) and buffered paraformaldehyde (a cross-linking fixative) has been successfully used in many studies (Brigati *et al.*, 1983; Höfler *et al.*, 1986). Glutaraldehyde (another cross-linking fixative) can also be used, but requires more prolonged unmasking of target sequences (Uehara *et al.*, 1993). Precipitating fixatives (e.g. acetic acid–alchol mixtures or Bouin's fixation) have also been used for *in situ* hybridization (Nuovo, 1991), but the hybridization target retention is lower than with formalin, thus lowering the intensity of signal.

Slide Preparation and Pretreatment of Tissues

Material for *in situ* hybridization is usually mounted on coated glass slides to improve adhesion of cells and sectioned material to the glass surface. Silane (3-aminopropyltriethoxy-silane, APES) (Rentrop *et al.*, 1986) is now more frequently used than poly-L-lysine.

In order to detect DNA and mRNA, probes have to penetrate complex and dense biological structure to reach their targets. For this reason the biological material has to be pretreated to increase the accessibility of these nucleic acid targets. This treatment may involve exposure to diluted acids, detergents and proteolytic enzymes. Protease treatment is particularly needed after cross-linking fixation and improved signal from probes longer than 100 bp. Most protocols prefer proteinase K (Holm *et al.*, 1997), but other enzymes have also been applied successfully including pepsin (Murphy *et al.*, 1995) and protease VIII (Fleming *et al.*, 1992). The optimal concentration and lengths of treatment for the enzymes depends on the tissue and preparation, and needs to be adjusted empirically. After protease treatment, it is essential to arrest digestion and this is carried out by immersion in a glycin solution, followed by a brief fixation step in 4% paraformaldehyde.

Types of Probe

The probe is the fragment of DNA or RNA that is labeled for detection during the *in situ* hybridization experiment. Methods for cloning, isolating, and handling DNA sequences are beyond the scope of this chapter, and are covered in detail elsewhere (Sambrook *et al.*, 1989). Three types of probes are commonly used and include long double-stranded DNA (thousands of bases), short DNA oligonucleotides (15–50 bases) and single-stranded RNA (usually less than 1000 bases).

Single-stranded RNA probes

RNA probes are frequently used for *in situ* hybridization to RNA targets (Cox *et al.*, 1984; Holm *et al.*, 1997). RNA probes are obtained by inserting a specific DNA sequence into an appropriate transcription vector (e.g. Bluescript) containing two different RNA polymerase promoters located on either side of the multiple cloning site. RNA polymerases are then used to produce single-stranded, labeled RNA probes. cDNA insert can be transcribed in either direction and sense (control, mRNA sequence) and antisense (cRNA sequence) strands can thus be obtained. Because these probes are single stranded, reannealing in solution does not occur, so a greater percentage of the probe is available for hybridization. The use of RNA probes also offers other advantages: the affinity of RNA–RNA hybrids is high and the high background binding of single-stranded RNA to tissues can be overcome by use of RNase in the post-hybridization washes, removing only single-stranded RNA and leaving the newly formed hybrids untouched. The major disadvantage of using RNA probes is their lability, which requires special precautions to avoid contamination with the RNase prior to or during hybridization.

DNA oligonucleotide probes

Oligonucleotide probes are of increasing importance for *in situ* hybridization (Crabb *et al.*, 1992; Larsson and Hougaard, 1993) and they are very effective in applications in which there are relatively high copy numbers of the target nucleic acids. These probes are quickly and easily made and are generated on an automated DNA synthesizer. They are end-labeled either by extending the sequence at the 3' end with labeled deoxyuridine triphosphate (dUTP) using terminal deoxynucleotide transferase (Stahl *et al.*, 1993), or by attaching a reporter molecule to the 5' end. Their short length makes oligonucleotide probes able to penetrate tissue sections better (Nunez *et al.*, 1989), but if very short they may bind nonspecifically due to the need to establish less stringent hybridization and washing conditions. A drawback with oligonucleotide probes compared to larger probes, are the low level of sensitivity because the oligonucleotide probe is only looking at 15–50 bases of a target sequence which might be 8000 bases long. Use of cocktails of oligonucleotide probes may increase the sensitivity (Pringle and Ruprai, 1990; Trembleau and Bloom, 1995).

Double-stranded DNA probes

Traditionally, the most frequently used probes were double-stranded DNA. Long DNA probes are frequently used to localize single copy genes to individual chromosomes and to identify translocations or other chromosomal abnormalities in both interphase and metaphase cells (Misra *et al.*, 1995; Rosenthal *et al.*, 2002). Furthermore, these probes are often applied in viral detection (Holm *et al.*, 1994; Berumen *et al.*, 1995). Double-stranded DNA can be labeled by nick translation (Rigby and Dieckmann, 1977) or by random priming (Feinberg and Vogelstein, 1984), but nick translation is probably the most widely employed method. Nick translation employs the enzymes DNase I and DNA polymerase I, and can be performed with specific insert only, or both the specific insert DNA and plasmid vector (see Protocol 1). As the labeling method introduces nicks at random into the double-stranded DNA molecules, labeled fragments of different length are produced. It has been suggested that these fragments anneal together via their overlapping portions to produce a probe network leading to amplification of the signal. These probes have the advantage of being stable because contamination with DNase (which are heat labile) is less of problem than with RNase (which even survive boiling). Furthermore, double-stranded DNA has a high level of specificity because of their length, and sensitivity because of the high number of incorporated reporter molecules. One drawback is that the probe requires denaturation to produce single-stranded DNA before hybridization and strand reannealing of the probes may occur during this process.

Probe Labeling

Initially, nucleic acid probes were labeled isotopically and detected by autoradiography. Several different isotopes have been used to radiolabel probes for *in situ* hybridization, including ^{3}H, ^{35}S, ^{32}P and ^{33}P. The choice of isotope usually reflects a compromise between quality of resolution and the time of exposure. ^{3}H provides the best resolution, but requires long exposure; ^{35}S gives more rapid results and a reasonable resolution; ^{32}P require the shortest exposure time, but offers poor resolution. ^{33}P is a newly developed isotope that has properties compared with ^{35}S, and is technically superior to either ^{32}P or ^{35}S for *in situ* hybridization (Johnston *et al.*, 1998). Problems with probe stability, hazardous radiation, poor resolution, and speed of visualization have directed attention to the development of nonisotopic probes.

Now, nonradioactive labels, where a fluorophore, a hapten, or other chemical group is attached to a nucleotide moiety, are used widely for labeling DNA and RNA. These nonradioactive labels may be incorporated using standard molecular labeling techniques (see Protocol 1) and detection uses a color reaction or fluorescence. Among the choices for nonradioactive labels, biotin is currently favored by many investigators since detection systems for biotin are well established in most laboratories on the basis of its widespread use in immunohistochemistry (Albertson *et al.*, 1988; Bloch, 1993). However, biotinylated probes may in some situations cause trouble due to the presence of endogenous biotin within tissues like kidney, liver, lymph nodes, and tonsils (Wood and Warnke, 1981; Banerjee and Pettit, 1984). The presence of endogenous biotin may be minimized by preincubation of the sections with diluted avidin solution followed by incubation with diluted biotin solution (Miller and Kubier, 1997). However, other investigators have reported a limited success using such blocking methods (Wood and Warnke, 1981). Therefore, in organs with high concentrations of endogenous biotin, other labels such as digoxigenin may be used (Holm *et al.*, 1992; Komminoth, 1992). Digoxigenin has shown good results and high sensitivity when detected using immunohistochemical techniques with antibodies to digoxigenin. Particularly for viral detection by *in situ* hybridization, nonisotopic labeling using either biotin or digoxigenin is the method of choice. Fluorescent markers such as fluorescein may be used for direct visualization (Bauman, 1985), particularly in cytogenetic studies or for immunohistochemical detection.

Protocol 1

1. Probe labeling using nick translation method

1. Mix the following in a 1.5 ml microcentrifuge tube on ice: 5 µl 10× dNTP mixture, X µl (1 µg) HPV16 DNA (double-stranded DNA not denaturated), Y µl distilled water to 45 µl, 5 µl 10× enzyme mixture
2. Mix and centrifuge briefly (15 000g for 5 s).
3. Incubate at 16°C in water bath for 1 h.
4. Add 5 µl 0.5 M EDTA (pH 8.0) to stop the reaction.

2. Separation of unincorporated nucleotides from labeled DNA probe

1. Transfer the labeled DNA probe to a Bio-Gel P-60 gel column. After transferring buffer 1 to the column, collect 20 fractions each of 50 µl in microcentrifuge tubes.
2. Spot 1 µl of each fraction to a membrane and dry in an oven at 80°C for 1.5 h.

3. Detection of labeled DNA probe

1. Prehybridize membrane for 1 h in a box with prehybridization solution.
2. Wash the membrane in buffer 2 for 2×15 min. Shake gently.
3. Incubate the membrane in streptavidin-alkaline phosphatase diluted in buffer 3 for 1 h. Shake gently.
4. Wash the membrane in buffer 2 for 2×15 min. Transfer the membrane to buffer 4 (2 min). Shake gently.
5. Incubate membrane in substrate solution for 1 h in the dark.
6. Wash the membrane in buffer 2 for 5 min.
7. Dry the membrane at room temperature and store in dark.
8. Combine fractions with highest level of labeled DNA probe and concentrate if necessary.

4. Calculate the final concentration of the DNA probe

1. Spot 1 µl of different dilutions of control biotinylated pBR322 and labeled DNA probe to a membrane and dry in an oven at 80°C for 1.5 h.
2. Then repeat steps 1–7 as described in the section "Detection of labeled DNA probe".
3. Compare the staining intensity between labeled DNA probe and control biotinylated pBR322. Calculate the probe concentration.

5. Reagents

1. HPV16 DNA (Dürst et al., 1983) (kindly provided by Prof. Dr Harald Zur Hausen and Dr Ethel-Michele de Villiers, Heidelberg, Germany)
2. BioNick Labeling System (cat. no. 18247-015, Invitrogen Life Technologies)

3. 10×dNTP mixture: 0.2 mM each dCTP, dGTP, dTTP; 0.1 mM dATP; 0.1 mM biotin-14-dATP; 500 mM Tris–HCl (pH 7.8); 50 mM MgCl$_2$; 100 mM β-mercaptoethanol; 100 µg/ml nuclease-free BSA
4. 10×enzyme mixture: 0.5 U/µl DNA polymerase I; 0.0007 U/µl DNase I; 50 mM Tris–HCl (pH 7.5); 5 mM magnesium chloride; 0.1 mM phenylmethylsulfonyl fluoride; 50% (v/v) glycerol; 100 µg/ml nuclease-free BSA
5. Bio-Gel P-60 gel, medium (cat. no. 150-4160, Bio-Rad Laboratories)
6. Membrane Hybon-N (cat. no. RPN 2020 N, Amersham Pharmacia Biotech)
7. Streptavidine-alkaline phosphatase (cat. no. 19542-018, Invitrogen Life Technologies)
8. Prehybridization solution: 0.05% Tween 20, 0.1% Sarkosyl, 0.02% SDS, 1% blocking reagent (cat. no. 1096176, Bohringer), 5×SSC
9. Substrate solution: 66 µl nitroblue tetrazolium (NBT) (cat. no. 14968001, Promega) and 33 µl 5-bromo-4-chloro-3-indolyl phosphate (BCIP) (cat. no. 14104002, Promega) in 10 ml buffer 4
10. Control pBR322 (cat. no. Y01591 in kit cat. no. 18247-015, Invitrogen Life Technologies)
11. Buffer 1: 10 mM Tris pH 7.5, 1 mM EDTA
12. Buffer 2: 0.1 M Tris pH 7.5, 0.15 M NaCl, 0.05% Tween 20
13. Buffer 3: 0.1 M Tris pH 7.5, 0.15 M NaCl, 1% BSA, 0.05% Tween 20
14. Buffer 4: 0.1 M Tris pH 9.5, 0.1 M NaCl, 0.05 M MgCl$_2$
15. Buffer 5: 10 mM Tris pH 8.0, 1 mM EDTA

Hybridization Conditions

One important advantage of *in situ* hybridization is that the degree of specificity of hybridization reactions may be controlled by varying the reaction conditions. The degree of specificity depends on probe concentration and length, stringency of hybridization conditions, kinetics of hybridization and preservation of tissue morphology. The optimal probe concentration is that which produces the greatest signal-to-noise ratio, and needs to be evaluated for each probe. Because background may be linearly related to probe concentration (Cox *et al.*, 1984), the desired probe concentration is, therefore, the lowest required to saturate target nucleic acid. The hybridization solution normally includes one or more hydrated components, such as dextran sulfate, polyvinylpyrrolidone (PVP) and ficoll to increase the effective probe concentration, giving a large enough volume of solution to cover the preparation without using excessive probe. The stringency at which the *in situ* hybridization experiment is carried out determines the approximate percentage of nucleotides that are correctly matched in the probe and target. Stringency increases with temperature and formamide concentration, and is inversely related to salt concentration. Under conditions of high stringency (high temperature, high formamide and low salt concentration) only probes with a high degree of homology to the target sequence will form stable hybrids. Under conditions of low stringency (low temperature, low formamide and high salt concentration) probes may bind to target sequences with lower homology, thus allowing for nonspecific hybridization signal (Lawrence and Singer, 1985; Höfler et al., 1986). To determine optimal stringency conditions of hybridization, a range of temperatures and salt concentrations must be tested. In order to preserve tissue structure and limit section loss from slides, hybridization is generally performed at a relatively low temperature. As this may increase nonspecific interactions, post-hybridization washing of increased stringency is necessary to dissociate nonhomologous hybrids. A series of stringency washes includes buffers of decreasing salt concentration and sometimes formamide is included.

Detection of Hybridized Nonisotopic Probe

A drawback of the nonisotopic *in situ* hybridization method is the low sensitivity (Brigati et al., 1983; Lurain et al., 1986; Mifflin et al., 1987; Albertson et al., 1988; Grubendorf-Conen and Cremer, 1990). However, the sensitivity and outcome of *in situ* hybridization depends to a high degree on the quality of the detection system applied (Herrington et al., 1989; Furuta et al., 1990; Morris et al., 1990; Holm et al., 1992; Zehbe et al., 1997; Holm, 2000). Visualization of nonisotopically labeled hybrids may be achieved by using well-established immunohistochemical procedures. Detection steps that are required depend on the modified nucleotide used to label the probe. If probes are directly labeled with fluorochromes, like fluorescein (FITC), rhodamine, Texas Red and CY3, then no immunohistochemistry is required for visualization (Bentz et al., 1994). However, immunological amplification using anti-fluorochrome antibodies and second antibodies is possible, and provides a greater signal intensity (Weigant et al., 1991; Holm et al., 1997). Biotinylated probes may be detected using avidin, streptavidin or monoclonal anti-biotin (Holm et al., 1992; McQuaid and Allan, 1992; Bloch, 1993; Holm, 2000), whereas probes labeled with digoxigenin require anti-digoxigenin (Furuta et al., 1990; Holm et al., 1992; Terenghi and Polak, 1994). The antibodies, avidin and streptavidin are conjugated to fluorophores, metals or enzymes, such as peroxidase and alkaline phosphatase. Most commonly alkaline phosphatase is used with the NBT/BCIP chromogenic substrate, which catalyzes the reduction of NBT and oxidises BCIP giving an insoluble purple precipitate (Holm et al., 1992). Development of peroxidase systems can be carried out using 3′,3′-diaminobenzidine (DAB). If gold-labeled reagents are used, enhancement with gold or silver ions may provide an *in situ* hybridization method with high sensitivity (Springall et al., 1984; Roth et al., 1992; Zehbe et al., 1997; Tubbs et al., 2002).

Because of the previously reported low sensitivity of nonradioactive *in situ* hybridization methods (Brigati et al., 1983; Lurain et al., 1986; Mifflin et al., 1987; Albertson et al., 1988; Grubendorf-Conen and Cremer, 1990) an extremely sensitive nonisotopic polymerase chain reaction (PCR) *in situ* technique was developed (Haase et al., 1990; Nuovo et al., 1991, 1992). However, false positive reactions due to diffusion of amplicons into cells that do not contain the DNA/RNA sequence before amplification have been reported (Bagasra *et al.*, 1994; Komminoth and Long, 1993; Long *et al.*, 1993). (See Chapter 9.)

In an attempt to obtain a nonisotopic *in situ* hybridization method with high sensitivity we have compared 12 different detection systems using HPV 6 and HPV 16 biotin-, digoxigenin-, and FITC-labeled probes on condylomas, cervical carcinomas, CaSki, and SiHa cells (Table 1.1) (Holm *et al.*, 1992; Holm, 2000). The cells and tissues have been fixed in 4% buffered formalin and paraffin-embedded. Detection system no. 8 (streptavidin-HRP/biotinylated tyramid/streptavidin-nanogold/silver intensification) and biotinylated HPV 16 probe gave the highest sensitivity, but the morphology of the tissue was poor. In CaSki cells the best results were obtained with biotinylated HPV 16 probe and detection systems no. 5 (mouse anti-biotin/anti-mouse/mouse AP-anti-AP/NBT-BCIP) (Figures 1.1a and 1.2a) and no. 7 (streptavidin-HRP/biotinylated tyramide/streptavidin-HRP/DAB) (Figures 1.1b and 1.2b) showing an intense specific hybridization signal in the nuclei. However, demonstration of HPV 16 DNA in SiHa cells (containing one to two copies of HPV 16 DNA) (Callahan *et al.*, 1992) was possible only by use of detection system no. 5 (mouse anti-biotin/anti-mouse/mouse AP-anti-AP/NBT-BCIP) (Figures 1.1a and 1.2c) (see Protocol 2). The success of identifying one to two copies of HPV 16 DNA using biotinylated double-stranded DNA probe was not achieved with digoxigenin- and FITC-labeled probes. In conclusion, our results of a comparative nonisotopic *in situ* hybridization study using biotin-, digoxigenin-, and FITC-labeled probes and 12 different detection systems demonstrated that detection system no. 5 (mouse anti-biotin/anti-mouse/mouse AP-anti-AP/NBT-BCIP) provided the best sensitivity combined with good tissue morphology. This detection method can identify single viral copies in formalin-fixed and paraffin-embedded material.

Protocol 2

1. *In situ* hybridization

1. Cut 4–6 μm thick sections on to slides coated with 3-aminopropyl-trietoxysilane (Rentrop *et al.*, 1986), air dry and bake at 56°C for 1 h.
2. Deparaffinize in xylene (2×5 min).
3. Rinse and rehydrate in degraded ethanols and autoclaved distilled water (2–3 min each).
4. Immerse slides in 0.02 N HCl for 10 min.
5. Wash in phosphate buffer saline (PBS) (2×5 min).
6. Permeabilize with 0.01% Triton X-100 for 1.5 min.
7. Wash in PBS (2×5 min) at room temperature and PBS at 37°C for 5 min.
8. Add 100 μl proteinase K (range 100–1000 μg/ml) diluted in buffer 1 to each slide and incubate at 37°C for 15 min.
9. To stop the proteinase K treatment wash in PBS containing 2 mg glycin/ml (2×5 min).

Table 1.1. Detection of hybridized nonisotopic probe

No.	Label	Primary layer	Secondary layer	Third layer	Substrate
1	Biotin	Streptavidin-AP			NBT/BCIP
2	Biotin	Streptavidin-FITC	AP-anti-FITC		NBT/BCIP
3	Biotin	Mouse anti-biotin	Biotin-anti-mouse	Streptavidin-AP	NBT/BCIP
4	Biotin	Mouse anti-biotin	Rabbit AP-anti-mouse	AP-anti-rabbit	NBT/BCIP
5	Biotin	Mouse anti-biotin	Anti-mouse	Mouse AP-anti-AP	NBT/BCIP
6	Biotin	Mouse anti-biotin	AP-anti-mouse	Mouse AP-anti-AP	NBT/BCIP
7	Biotin	Streptavidin-HRP	Biotinylated tyramide	Streptavidin-HRP	DAB
8	Biotin	Streptavidin-HRP	Biotinylated tyramide	Streptavidin-nanogold	Silver intensification
9	Digoxigenin	AP-anti-digoxigenin			NBT/BCIP
10	Digoxigenin	Sheep AP-anti-digoxigenin	Biotin anti-sheep	Streptavidin-AP	NBT/BCIP
11	FITC	Mouse anti-FITC	Anti-mouse	Mouse AP-anti-AP	NBT/BCIP
12	FITC	Mouse anti-FITC	Dextran-AP-anti-mouse		NBT/BCIP

FITC, fluorescein; AP, alkaline phosphatase; HRP, horseradish peroxidase; NBT, nitroblue tetrazolium; BCIP, 5-bromo-4 chloro-3-indolylphosphate; DAB, 3′,3′-diaminobenzidine.

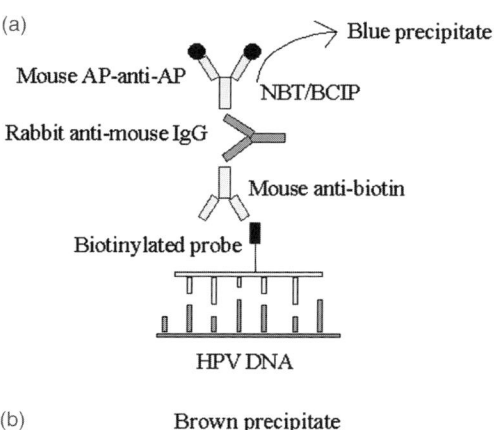

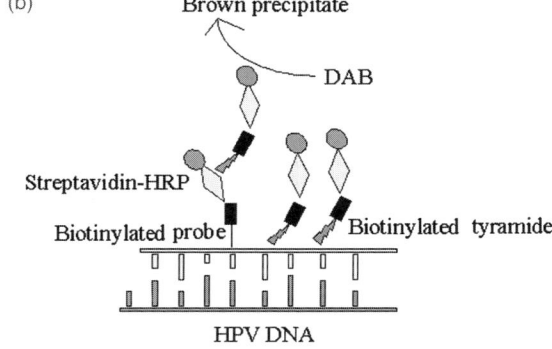

Figure 1.1. *In situ* hybridization methods (a) detection system no. 5 (mouse anti-biotin/anti-mouse/mouse AP-anti-AP/NBT-BCIP), and (b) detection system no. 7 (streptavidin-HRP/biotinylated tyramide/streptavidin-HRP/DAB). AP, alkaline phosphatase; HRP, horseradish peroxidase; NBT, nitroblue tetrazolium; BCIP, 5-bromo-4 chloro-3-indolylphosphate; DAB, 3′,3′-diaminobenzidine.

10. Immerse slides in 4% paraformaldehyde fixative in PBS for 5 min at room temperature.
11. Rinse slides in PBS (2×5 min).
12. Dehydrate slides in 70, 96, and 100% ethanol (1 min each) and air dry.
13. Add biotinylated DNA probe (0.1–2 ng/μl) in 10 μl hybridization buffer on the section and cover with a siliconized cover slip (18×18 mm). Seal cover slip with rubbercement.
14. Denature probe at 95°C for 10 min and cool on ice.
15. Hybridize in a moist chamber overnight at 37°C.
16. Float off cover slips by incubating slides in 2×SSC. Remove the rubbercement with forceps. After removing the cover slips wash the slides in 2×SSC for 20 min at room temperature, 0.1×SSC for 20 min at 56°C (preheated solution) and 2×SSC for 5 min at room temperature.
17. Wash in buffer 2 (2×5 min).
18. Nonspecific antibody binding is blocked with normal rabbit serum diluted 1/25 in buffer 3 for 20 min.
19. Incubate sections with mouse anti-biotin in buffer 3 for 30 min.

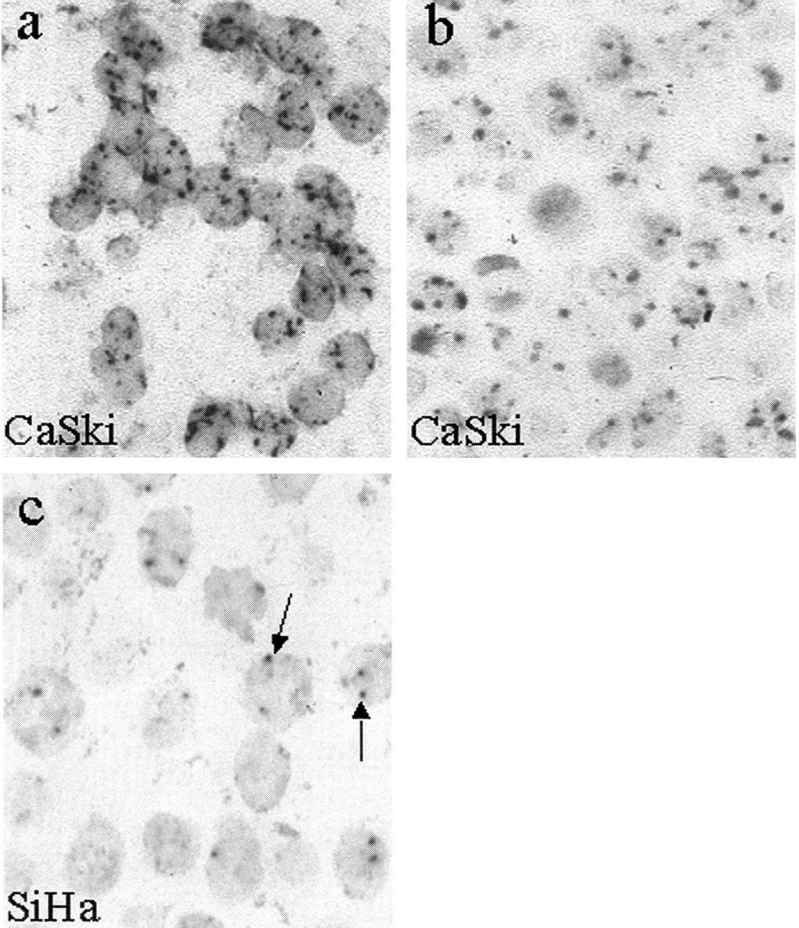

Figure 1.2. Positive HPV 16 hybridization signal in CaSki cells using biotinylated probe and detection systems (a) no. 5 (mouse anti-biotin/anti-mouse/mouse AP-anti-AP/NBT-BCIP), (b) no. 7 (streptavidin-HRP/biotinylated tyramide/streptavidin-HRP/DAB), and (c) positive HPV 16 hybridization signal (arrows) in SiHa cells using biotinylated probe and detection system no. 5 (mouse anti biotin/anti-mouse/mouse AP-anti-AP/NBT-BCIP). AP, alkaline phosphatase; HRP, horseradish peroxidase; NBT, nitroblue tetrazolium; BCIP, 5-bromo-4 chloro-3-indolylphosphate; DAB, 3′,3′-diaminobenzidine. Reprinted with permission from Holm, "A highly sensitive nonisotopic detection method for *in situ* hybridization" in Appl. Immunohistochem. Mol. Morphol. 8(2) 162–165, 2000. Copyright Lippincott Williams & Wilkins. (See colour plate 1.)

20. Wash in buffer 2 (2×5 min).
21. Incubate sections in rabbit anti-mouse IgG diluted in buffer 3 for 30 min.
22. Rinse in buffer 2 (2×5 min).
23. Incubate sections in alkaline phosphatase mouse anti-alkaline phosphatase in buffer 3 in the dark for 30 min.
24. Wash in buffer 2 (2×5 min).

25. Sections are placed in buffer 4 for 5 min prior to incubation in substrate solution.
26. The slides are the incubated in substrate solution for 45 min in the dark.
27. Wash in buffer 5 (2×5 min) and distilled water for 5 min.
28. Add two or three drops water based glycerin jelly and cover with cover slip.

2. Reagents

- 3-aminopropyltrietoxysilane (cat. no. A-3648, Sigma)
- Triton-X 100 (Sigma)
- Proteinase K (cat. no. P-2308, Sigma)
- Glycin (cat. no. 161-0718, BioRad)
- Paraformaldehyde (cat. no. P-6148, Sigma)
- Hybridization buffer: 2×SSC, 1×Denhardt's [0.02% Ficoll (cat. no. F-2637, Sigma), 0.02% BSA (cat. no B-2518, Sigma), 0.02% PVP (cat. no. P-5288, Sigma)], 50% deionized formamide (cat. no. F-7503, Sigma), 10% dextran sulfate (cat. no. D-8906, Sigma), 400 µg/ml herring sperm DNA (cat. no. D-6898, Sigma)
- Sigmacote (cat. no. SL-2, Sigma)
- Rubbercement (cat. no. 95306500, Royal Talens)
- Normal rabbit serum (cat. no. X-0902, Dako)
- Mouse anti-biotin (cat. no. M-0743, Dako)
- Rabbit anti- mouse IgG (cat. no. Z-0259, Dako)
- Alkaline phosphatase mouse anti-alkaline phosphatase (cat. no. D-0651, Dako)
- Substrate solution: 66 µl NBT (cat. no. 14968001, Promega), 33 µl BCIP (cat. no. 14104002, Promega) and 1 ml 0.01 M levamisol (cat. no. L-9756, Sigma) in 9 ml buffer 4
- Buffer 1: 0.1 M Tris pH 8.0, 50 mM EDTA
- Buffer 2: 0.1 M Tris pH 7.5, 0.15 M NaCl
- Buffer 3: 0.1 M Tris pH 7.5, 0.15 M NaCl, 1% BSA
- Buffer 4: 0.1 M Tris pH 9.5, 0.1 M NaCl, 0.05 M $MgCl_2$
- Buffer 5: 10 mM Tris pH 8.0, 1 mM EDTA

Controls

In order to prove specificity of hybridization signals, controls must be included when detecting mRNA and DNA in tissue sections and whole mounts. Specific hybridization may be easily confused with unwanted hybrid formation between probe and weakly homologous sequences or with nonspecific interaction between probe and non-nucleic acid tissue components. Depending on the probe used and the target nucleic acid, several control methods are available.

Positive controls, a component of every *in situ* hybridization experiment, is important to check that the hybridization and detection have worked. Tissue or cell lines always known to work with the probe should be included as positive controls.

When using double-stranded DNA probes the most frequently used negative controls are nonhomologous probes, such as labeled pBR322 DNA. For DNA oligonucleotide probes, the ideal control is a scrambled oligonucleotide of the same length and nucleotide content, but with three or four nucleotides shuffled around such that it represents a nonsense probe. Furthermore, a corresponding sense probe of the same length can be used. When testing RNA probes the use of sense transcripts allows control probes to be of identical specific activity and fragment length. This allows the hybridization observed with antisense probes to be assigned only to specific hybridization arising from the probe sequence and not to any other properties of the probes. Other negative controls include pretreatment of sections with RNase or DNase, which should, in theory, abolish or reduce the specific signal. However, RNase may enhance the signal for a specific mRNA, presumably by unmasking the target sequence rather than degrading it (McNicol and Farquharson, 1997).

When identifying mRNA it is important to confirm that mRNA is intact in tissues. Different probes have been used, such as polydT probes which hybridized with the polyA tail which characterizes mRNA (Montone and Tomaszewiski, 1993) and β-actin (Holm *et al.*, 1997).

◆◆◆◆◆◆ MICROORGANISMS THAT INTERACT WITH HUMAN HOST CELLS

In situ hybridization methods have been widely used for the identification of infectious agents in tissue (Syrjanen, 1992; Moter and Göbel, 2000). Ribosomal RNA (rRNA) are abundant RNA sequences, which are highly conserved, and can be used for classification of bacteria (Montone, 1994). Tuberculosis is caused by infection with species of the *Mycobacterium tuberculosis* complex (MTC) and is globally the most important cause of death from a single pathogen (Reichman, 1997). The final diagnosis of tuberculosis requires differentiation of species of MTC from nontuberculous mycobacteria (NTB). Recently, Stender and co-workers (1999) have used a FISH method and PNA probes target selected regions of mycobacterial 16S rRNA sequences which allows distinction between MTC and NTB. Using a immunoenzymatic detection method for *in situ* hybridization, *Mycobacterium leprae* has been identified in frozen sections of skin biopsies obtained from patients suffering from lepromatous leprosy (Arnoldi *et al.*, 1992). *Helicobacter pylori* is established as the major aetiological agent in gastritis and peptic ulcer. Bashir *et al.* (1994) used a probe labeled with biotin to visualize *H. pylori* in gastric biopsy specimens. Recently, a set of oligonucleotide probes and a FISH method was used to identify bacteria (*Pseudomonas aeruginosa*, *Stenotrophomonas maltophilia*, *Burkholderia cepacia*, *Haemophilus influenzae*, *Streptococcus pyogenes*, *Staphyloccus aereus*, and *Candida albicans*) in clinical samples obtained from cystic fibrosis patients (Hogardt *et al.*, 2000).

Viruses such as HPV (Syrjanen, 1992; Holm *et al.*, 1994; Herrington *et al.*, 1995), EBV (MacMahon *et al.*, 1991), CMV (Wolber and Lloyd, 1988;

Rimsza et al., 1996), HSV (Reed et al., 1998), polyomaviruses (JC, BK) (Boerman et al., 1989), hepatitis B virus (Blum et al., 1983; Jayshree et al., 2003), hepatitis C virus (van Dekken et al., 2003), parvoviruses (Porter et al., 1988) and HIV-1 (Gerrard et al., 1994; Wenig et al., 1996) have been detected by in situ hybridization.

EBV is associated with a number of human malignancies (Anagnostopoulos and Hummel, 1996). Due to the low copy number of EBV genomes in latent infected cells, the use of in situ hybridization to detect EBV DNA is unsatisfactory (see Chapter 9). However, the two EBV-encoded nuclear RNAs (EBER1 and EBER2), which are present in high copy numbers in latent EBV-infected cells, can be identified with the technique (Wu et al., 1991). EBV infection has been detected by in situ hybridization in Hodgkin's disease (Beck et al., 2001) and non-Hodgkin's lymphoma (Hummel et al., 1995; Natatsuka et al., 2002). In individuals with HIV-associated lymphomas EBV has been identified in 40–60% of the cases (Flaitz et al., 2002), whereas in 18 cases of AIDS-related primary central nervous system lymphoma a strong positive signal in tumor cells indicate abundant expression of the EBV-EBR1 transcript (MacMahon et al., 1991). In situ hybridization methods have also been used to confirm infectious mononucleosis in atypical cases (Shin et al., 1991).

CMV has been identified in viral encephalitis in patients with AIDS using in situ hybridization techniques (Musiani et al., 1994). CMV is also associated with lower respiratory tract illness (Malcolm et al., 2001) and intestinal tissue sections from patients with ulcerative colitis and Crohn's disease (Rahbar et al., 2003). Recently, Harkins and co-workers (2002) found CMV localized to neoplastic cells in human colorectal polyps and adenocarcinomas, whereas Samanta et al. (2003) identify CMV in preneoplastic and neoplastic prostatic lesions. Hepatitis B and C virus infections seem to be an important factor in the development of hepatocellular carcinoma. These two virus types have been identified in dysplastic liver lesions and in hepatocellular carcinomas using in situ hybridization (Jayshree et al., 2003; van Dekken et al., 2003).

Viruses that have received the most attention include HPV; more than 80 HPV types have been characterized (Carr and Gyorfi, 2000), and they are divided into two major categories referred to as cutaneous and mucosal. Cutanous types are generally responsible for warts on hands and feets (Uezato et al., 1999), whereas mucosal types are identified in the anogenital tract (Dürst et al., 1983; Weaver et al., 1989; Holm et al., 1994; Beutner and Tyring, 1997; Carr and Gyorfi, 2000), respiratory tract, oral cavity and the conjunctiva (McDonnell et al., 1987; Odrich et al., 1991). To date there are at least 15 HPV types which are associated with high cervical intraepithelial neoplasia (CIN) and cervical carcinomas (Lorincz et al., 1992). HPV types 6, 11, 42, 43, and 44 were found to be low-risk viruses, HPV types 31, 33, 35, 51, 52, and 58 were identified as intermediate-risk viruses, whereas HPV types 16, 18, 45, and 56 were found to be high-risk viruses in the development of cervical cancer (Lorincz et al., 1992).

Chromosomal integration of HPV 16 DNA and HPV 18 DNA has been suggested to be an important step for initiation and progression of cervical cancers (Dürst et al., 1985; Fukushima et al., 1990; Duensing and Münger, 2002). HPV 16 DNA and HPV 18 DNA have been found to be integrated near *myc* genes in invasive genital cancers, which supports the hypothesis that integration plays a role in tumor initiation and progression via activation of cellular oncogenes (Dürst et al., 1987; Couturier et al., 1991). Furthermore, integration of HPV DNA into the human chromosome disrupts the expression of virus E1-E2 genes (Choo et al., 1987; Kalantari et al., 1998). The gene product of an undisturbed E2 region suppresses the expression of E6 and E7, and inactivation of the E2 gene is thought to result in over expression of E6 and E7 oncoproteins. The gene products from high-risk HPV E6 and E7 oncoproteins are frequently sufficient for transformation through their interaction with cell cycle proteins. Whereas, high-risk HPV E6 oncoproteins induces accelerated degradation of tumor suppressor protein p53 (Scheffner et al., 1990), high-risk HPV E7 inactivates retinoblastoma (Rb) tumor suppressor protein (Dyson et al., 1989; Boyer et al., 1996) and interacts with the cyclin-dependent kinase inhibitors p21$^{Waf/Cip1}$ (Funk et al., 1997; Jones et al., 1997) and p27^{Kip1} (Zerfass-Thome et al., 1996). Furthermore, high-risk HPV E7-dependent activation of cyclin E and cyclin A gene expression have been found (Zerfass et al., 1995), whereas high-risk HPV E6 increases the expression of cyclin B (Cho et al., 2003).

Integration of HPV 16 DNA into cervical carcinomas has been studied using Southern blot restriction analysis and two-dimensional gel electrophoresis (Dürst et al., 1985; Fukushima et al., 1990; Cooper et al., 1991; Kristiansen et al., 1994). These methods are time consuming, and fresh tissue is needed. Previously, Cooper and co-workers (1991) and Kristiansen et al. (1994) have investigated the physical state of HPV DNA in cervical cancers using *in situ* hybridization methods, which are less time consuming and can be performed on formalin-fixed paraffin-

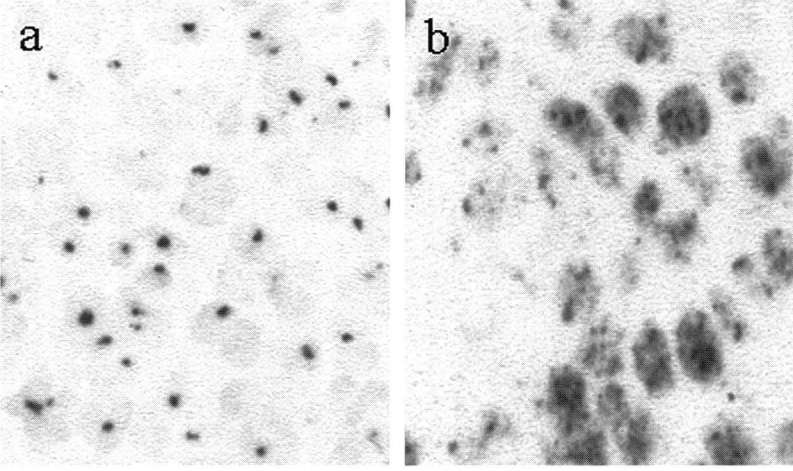

Figure 1.3. Positive HPV 16 hybridization signal in two cervical carcinomas. A punctuate signal (a) and a punctuate and diffuse signal (b). (See colour plate 2.)

Table 1.2. *In situ* hybridization signal and physical state of HPV

In situ hybridization signal	Physical state of HPV DNA
Diffuse	Episomal
Punctate	Integrated
Diffuse/punctuate	Episomal/integrated

embedded tissue. *In situ* hybridization analysis on cervical carcinomas showed a diffuse, a punctuate (Figure 1.3a) or both diffuse/punctuate (Figure 1.3b) signal indicating episomal, integrated or combined episomal/integrated HPV 16 DNA, respectively (Cooper *et al.*, 1991; Kristiansen *et al.*, 1994) (Table 1.2).

In a previous study, Unger and co-workers (1995) identified HPV 16, 18, 31, 33, and 35 DNA in 26 of 47 cervical carcinomas using *in situ* hybridization. The 22 patients with only integrated HPV DNA have a shorter disease-free life expectancy than patients with episomal or episomal/integrated HPV DNA. An explanation may be that the presence of episomal HPV DNA in the tumor is that virus E1 and E2 would be retained and could act on host and viral genes.

References

Albertson, D. G., Fishpool, R., Sherrington, P., Nacheva, E. and Milstein, M. (1988). Sensitive and high resolution in situ hybridization to human chromosomes using biotin labelled probes: assignment of the human thymocyte CD1 antigen genes to chromosome 1. *EMBO J.* **7**, 2801–2806.

Alonso, M. J., Gomez, F., Munoz, E., Abad, M. M., Roldan, M., Curiel, I., Paz, J. I., Bullon, A. and Lopez-Bravo, A. (1992). Comparative study of in situ hybridization and immunohistochemical techniques for the detection of human papillomavirus in lesions of the uterine cervix. *Eur. J. Histochem.* **36**, 271–278.

Anagnostopoulos, I. and Hummel, M. (1996). Epstein-Barr virus in tumours. *Histopathology* **29**, 297–315.

Arnoldi, J., Schluter, C., Duchrow, M., Hubner, L., Ernst, M., Teske, A., Flad, H. D., Gerdes, J. and Bottger, E. C. (1992). Species-specific assessment of *Mycobacterium leprae* in skin biopsies by in situ hybridization and polymerase chain reaction. *Lab. Investig.* **66**, 618–623.

Bagasra, O., Seshamma, T., Hansen, J., Bobroski, L., Saikumari, P., Pestaner, J. P. and Pomerantz, R. J. (1994). Application of *in situ* PCR methods in molecular biology. I. Details of methodology for general use. *Cell Vision* **1**, 324–335.

Banerjee, D. and Pettit, S. (1984). Endogenous avidin-binding activity in human lymphoid tissue. *J. Clin. Pathol.* **37**, 223–225.

Bashir, M. S., Lewis, F. A., Quirke, P., Lee, A. and Dixon, M. F. (1994). In situ hybridization for identification of *Helicobacter pylori* in paraffin wax embedded tissue. *J. Clin. Pathol.* **47**, 862–866.

Bauman, J. G. J. (1985). Fluoresence microscopical hybridocytochemistry. *Acta Histochem.* **31**, 9–18.

Bedell, M. A., Jones, K. H. and Laimins, L. A. (1986). The E6-E7 region of human papillomavirus type 18 is sufficient for transformation of NIH 343 and rat-1 cells. *J. Virol.* **61**, 3635–3640.

Beck, A., Päzolt, D., Grabenbauer, G. G., Nicholls, J. M., Herbst, H., Young, L. Y. and Niedobitek, G. (2001). Expression of cytokine and chemokine genes in Epstein-Barr virus-associated nasopharyngeal carcinoma: comparison with Hodgkin's disease. *J. Pathol.* **194**, 145–151.

Bentz, M., Döhner, H., Cabot, G. and Lichter, P. (1994). Fluorescence in situ hybridization in leukemias: "the FISH are spawning". *Leukemia* **8**, 1447–1452.

Berumen, J., Unger, E. R., Casas, L. and Figueroa, P. (1995). Amplification of human papillomavirus types 16 and 18 in invasive cervical cancer. *Hum. Pathol.* **26**, 676–681.

Beutner, K. R. and Tyring, S. (1997). Human papillomavirus and human disease. *Am. J. Med.* **102**, 9–15.

Bloch, B. (1993). Biotinylated probes for *in situ* hybridization histochemistry – use for messenger RNA detection. *J. Histochem. Cytochem.* **41**, 1751–1754.

Blum, H. E., Stowring, L., Figus, A., Montgomery, C. K., Haase, A. T. and Vyas, G. N. (1983). Detection of hepatitis B virus DNA in hepatocytes, bile duct epithelium and vascular elements by in situ hybridization. *Proc. Natl Acad. Sci. USA* **80**, 6685–6688.

Boerman, R. H., Arnoldous, E. P. J., Raap, A. K., Peters, A. C. B., Ter Schegget, J. and Van Der Ploeg, M. (1989). Diagnosis of progressive multifocal leucencephalopathy by in situ hybridization techniques. *J. Clin. Pathol.* **42**, 153–161.

Boyer, S. N., Wazer, D. E. and Band, V. (1996). E7 protein of human papilloma virus-16 induces degradation of retinoblastoma protein through the ubiquitin–proteasome pathway. *Cancer Res.* **56**, 4620–4624.

Brigati, D. J., Myerson, D., Leary, J. J., Spalholz, B., Travis, S. Z., Fong, C. K., Hsiung, G. D. and Ward, D. C. (1983). Detection of viral genomes in cultured cells and paraffin-embedded tissues using biotin-labelled hybridization probes. *Virology* **126**, 32–50.

Callahan, D. E., Karim, A., Zheng, G., Tso, P. O. P. and Lesko, S. A. (1992). Quantitation and mapping of integrated human papillomavirus on human metaphase chromosomes using a fluoresence microscope image system. *Cytometry* **13**, 453–461.

Carr, J. and Gyorfi, T. (2000). Human papillomavirus. Epidemiology, transmission, and pathogenesis. *Clin. Lab. Med.* **20**, 235–255.

Chinery, R., Poulsom, R., Rogers, L., Jeffery, R. E., Longcroft, J. M., Hanby, A. M. and Wright, N. A. (1992). Localization of intestinal trefoil-factor mRNA in rat stomach and intestine by hybridization in situ. *Biochem. J.* **285**, 5–8.

Cho, N. H., Lim, S. Y., Kim, Y. T., Kim, D., Kim, Y. S. and Kim, J. W. (2003). G2 chechpoint in uterine cervical cancer with HPV 16 E6 according to p53 polymorphism and its screening value. *Gynecol. Oncol.* **90**, 15–22.

Choo, K. B., Pan, C. C. and Han, S. (1987). Integration of human papillomavirus type 16 into cellular DNA of cervical carcinoma: preferential deletion of the E2 gene and invariable retention of the long control region and the E6/E7 open reading frames. *Virology* **161**, 259–261.

Cooper, K., Herrington, C. S., Stickland, J. E., Evans, M. F. and McGee, J. O. (1991). Episomal and integrated human papillomavirus in cervical neoplasis shown by non-isotopic in situ hybridization. *J. Clin. Pathol.* **44**, 990–996.

Cordon, S. A., Sant-Cassia, L. J., Easton, A. J. and Morris, A. G. (1999). The integration of HPV-18 DNA in cervical carcinoma. *J. Clin. Pathol.: Mol. Pathol.* **52**, 275–282.

Couturier, J., Sastra-Garau, S., Schneider-Maunoury, S., Labib, A. and Orth, G. (1991). Integration of papillomavirus DNA near *myc* genes in genital carcinomas and its consequences for proto-oncogene expression. *J. Virol.* **65**, 4534–4538.

Cox, K. H., DeLeon, D. V., Angerer, L. M. and Angerer, R. C. (1984). Detection of mRNA in sea urchin embryos by in situ hybridization using asymmetric RNA probe. *Dev. Biol.* **101**, 485–502.

Crabb, I. D., Hughes, S. S., Hicks, D. G., Puzas, J. E., Tsao, G. J. Y. and Rosier, R. N. (1992). Nonradioactive *in situ* hybridization using digoxygenin-labeled oligo-deoxynucleotides. application to musculoskeletal tissues. *Am. J. Pathol.* **141**, 579–589.

DeLellis, R. A. (1994). In situ hybridization techniques for the analysis of gene expression: application in tumor pathology. *Hum. Pathol.* **25**, 580–585.

Duensing, S. and Münger, K. (2002). The human papillomavirus type 16 E6 and E7 oncoproteins independently induce numerical and structural chromosome instability. *Cancer Res.* **62**, 7075–7082.

Dürst, M., Gissmann, L., Ikenberg, H. and Zur Hausen, H. (1983). A papillomavirus DNA from a cervical carcinoma and its prevalence in cancer biopsy samples from different geographic regions. *Proc. Natl Acad. Sci. USA* **80**, 3812–3815.

Dürst, M., Kleinheinz, A., Hotz, M. and Gissmann, L. (1985). The physical state of human papillomavirus type 16 DNA in benign and malignant genital tumours. *J. Gen. Virol.* **66**, 1515–1522.

Dürst, M., Croce, C. M., Gissmann, L., Schwarz, E. and Huebner, K. (1987). Papillomavirus sequences integrate near cellular oncogene in some cervical carcinomas. *Proc. Natl Acad. Sci. USA* **84**, 1070–1074.

Dyson, N., Howley, P. M., Munger, K. and Harlow, E. (1989). The human papillomavirus-16 E7 oncoprotein is able to bind to the retinoblastoma gene product. *Science* **243**, 934–937.

Feinberg, A. P. and Vogelstein, B. (1984). A technique for radiolabeling DNA restriction endonuclease fragments to high specific activity. *Anal. Biochem.* **137**, 266–267.

Flaitz, C. M., Nichols, C. M., Walling, D. M. and Hicks, M. J. (2002). Plasmablastic lymphoma: an HIV-associated entity with primary oral manifestations. *Oral Oncol.* **38**, 96–102.

Fleming, K. A., Evans, M., Ryley, K. C., Franklin, D., Lovell-Badge, R. H. and Morey, A. L. (1992). Optimization of non-isotopic *in situ* hybridization on formalin-fixed, paraffin embedded material using digoxigenine-labelled probes and transgenic tissue. *J. Pathol.* **167**, 9–17.

Fukushima, M., Yamakawa, Y., Shimano, S., Hashimoto, M., Sawada, Y. and Fujinaga, K. (1990). The physical state of human papillomavirus 16 DNA in cervical carcinoma and cervical intraepithelial neoplasia. *Cancer* **66**, 2155–2161.

Funk, J. O., Waga, S., Harry, J. B., Espling, E., Stillman, B. and Galloway, D. A. (1997). Inhibition of CDK activity and PCNA-dependent DNA replication by p21 is blocked by interaction with the HPV-16 E7 oncoprotein. *Genes Dev.* **11**, 2090–2100.

Furuta, Y., Sano, K., Meguro, M. and Nagashima, K. (1990). In situ hybridization with digoxigenin-labeled DNA probes for detection of viral genomes. *J. Clin. Pathol.* **43**, 806–809.

Gall, J. G. and Pardue, M. L. (1969). Formation and detection of RNA–DNA hybrid molecules in cytological preparations. *Proc. Natl Acad. Sci. USA* **63**, 378–383.

Gerrard, J. G., McGahan, S. L., Milliken, J. S., Mathys, J. M. J. and Wills, E. J. (1994). Australia first case of AIDS-*Pneumocystis carinii* pneumonia and HIV in 1981. *Med. J. Aust.* **160**, 247–250.

Grubendorf-Conen, E.-I. and Cremer, S. (1990). The demonstration of human papillomavirus 16 genomes in the nuclei of genital cancers using two different methods of in situ hybridization. *Cancer* **65**, 238–243.

Haase, A. T., Retzel, E. F. and Staskus, K. A. (1990). Amplification and detection of lentiviral DNA inside cells. *Proc. Natl Acad. Sci. USA* **87**, 4971–4975.

Harkins, L., Volk, A. L., Samanta, M., Mikolaenko, I., Britt, W. J., Bland, K. I. and Cobbs, C. S. (2002). Specific localisation of human cytomegalovirus nucleic acids and proteins in human colorectal cancer. *Lancet* **360**, 1557–1563.

Herrington, C. S., Burns, J., Graham, A. K., Evans, M. and McGee, J. O. (1989). Interphase cytogenetics using biotin and digoxigenin labeled probes I: relative sensitivity of both reporter molecules for detection of HPV 16 in CaSki cells. *J. Clin. Pathol.* **42**, 592–600.

Herrington, C. S., Anderson, S. M., Troncone, G., de Angelis, M. L., Noell, H., Chimera, J., Van Eyck, S. L. and McGee, J. O. D. (1995). Comparative analysis of human papillomavirus detection by polymerase chain reaction and nonisotopic in situ hybridization (NISH). *J. Clin. Pathol.* **48**, 415–419.

Hogardt, M., Trebesius, K., Geiger, A. M., Hornef, M., Rosenecker, J. and Heesemann, J. (2000). Specific and rapid detection by fluorecent in situ hybridization of bacteria in clinical sample obtained from cystic fibrosis patients. *J. Clin. Microbiol.* **38**, 818–825.

Höfler, H., Childers, H., Montminy, M. R., Lechan, R. M., Goodman, R. H. and Wolfe, H. J. (1986). *In situ* hybridization methods for detection of somatostatin mRNA in tissue sections using antisense RNA probes. *Histochem. J.* **18**, 597–604.

Holm, R. (2000). A highly sensitive nonisotopic detection method for in situ hybridization. *Appl. Immunohistochem. Mol. Morphol.* **8**, 162–165.

Holm, R., Karlsen, F. and Nesland, J. M. (1992). In situ hybridization with nonisotopic probes using different detection systems. *Mod. Pathol.* **5**, 315–319.

Holm, R., Tanum, G., Karlsen, F. and Nesland, J. M. (1994). Prevalence and physical state of human papillomavirus DNA in anal carcinomas. *Mod. Pathol.* **7**, 449–453.

Holm, R., Flørenes, V. A., Erikstein, B. and Nesland, J. M. (1997). Expression of stromelysin-3 in medullary carcinoma of the breast. *Anticancer Res.* **17**, 3725–3728.

Hummel, M., Anagnostopoulos, J., Korbjuhn, P. and Stein, H. (1995). Epstein-Barr virus in B cell non-Hodgkin's lymphomas: unexpected infection patterns and different infection incidence in low- and high-types. *J. Pathol.* **175**, 263–271.

Jayshree, R. S., Sridhar, H. and Devi, G. M. (2003). Surface, core, and X genes of hepatitis B virus in hepatocelular carcinoma: an in situ hybridization study. *Cancer* **99**, 63–67.

John, H. A., Birnstiel, M. L. and Jones, K. W. (1969). RNA–DNA hybrids at the cytological level. *Nature* **223**, 582–587.

Johnston, D., Hatzis, D. and Sunday, M. E. (1998). Expression of v-Ha-*ras* driven by the calcitonin/calcitonin gene-peptide promoter: a novel transgenic murine model for medullary carcinoma. *Oncogene* **16**, 167–177.

Jones, D. L., Alani, R. M. and Munger, K. (1997). The human papillomavirus E7 oncoprotein can uncouple cellular differentiation and proliferation in human keratinocytes by abrogating $p21^{Cip1}$-mediated inhibition of cdk2. *Genes Dev.* **11**, 2101–2111.

Kalantari, M., Karlsen, F., Kristensen, G., Holm, R., Hagmar, B. and Johansson, B. (1998). Disruption of the E1 and E2 reading frames of HPV 16 in cervical carcinoma is associated with poor prognosis. *Int. J. Gynecol. Pathol.* **17**, 146–153.

Kerstens, H. M. J., Poddighe, P. J. and Hanselaar, A. G. J. M. (1995). A novel in situ hybridization signal amplification method based on the deposition of biotinylated tyramide. *J. Histochem. Cytochem.* **43**, 347–352.

Komminoth, P. (1992). Digoxigenin as an alternative probe labeling for in situ hybridization. *Diagn. Mol. Pathol.* **2**, 142–150.

Komminoth, P. and Long, A. A. (1993). In situ polymerase chain reaction. An overview of methods, applications and limitations of a new molecular technique. *Virchows Arch. B Cell Pathol.* **64**, 67–73.

Kristiansen, E., Jenkins, A. and Holm, R. (1994). Coexistence of episomal and integrated HPV 16 DNA in squamous cell carcinoma of the cervix. *J. Clin. Pathol.* **47**, 253–256.

Larsson, L. I. and Hougaard, D. M. (1993). Sensitive detection of rat gastrin mRNA by *in situ* hybridization with chemically biotinylated oligodeoxynucleotides: validation, quantitation and double-staining studies. *J. Histochem. Cytochem.* **41**, 157–163.

Lawrence, J. B. and Singer, R. H. (1985). Quantitative analysis of *in situ* hybridization methods for detection of actin gene expression. *Nucleic Acids Res.* **13**, 1777–1799.

Long, A. A., Komminoth, P., Lee, E. and Wolfe, H. J. (1993). Comparison of indirect and direct *in-situ* polymerase, chain reaction in cell preparations and tissue sections. *Histochemistry* **99**, 151–162.

Lorincz, A. T., Reid, R., Jenson, A. B., Greenberg, M. D., Lancaster, W. and Kurman, R. J. (1992). Human papillomavirus infection of the cervix: relative risk associations of 15 common anogenital types. *Obstet. Gynecol.* **79**, 328–337.

Lurain, N. S., Thompson, K. D. and Farrand, S. K. (1986). Rapid detection of cytomegalovirus in clinical specimens by using biotinylated DNA probes and analysis of cross-reactivity with herpesimplex virus. *J. Clin. Microbiol.* **24**, 724–730.

MacMahon, E. M., Glass, J. D., Hayward, S. D., Mann, R. B., Becker, P. S., Charache, P., McArthur, J. C. and Ambinder, R. F. (1991). Epstein-Barr virus in AIDS-related primary central nervous system lymphoma. *Lancet* **338**, 969–973.

Malcolm, E., Arruda, E., Hayden, F. G. and Kaiser, L. (2001). Clinical features of patients with acute respiratory illness and rhinovirus in their bronchoalveolar lavages. *J. Clin. Virol.* **21**, 9–16.

McDonnell, P. J., McDonnell, J. M., Kessis, T., Green, W. R. and Shah, K. V. (1987). Detection of human papillomavirus type 6/11 DNA in conjunctival papillomas by in situ hybridization with radioactive probes. *Hum. Pathol.* **18**, 1115–1119.

McNicol, A. M. and Farquharson, M. A. (1997). *In situ* hybridization and its diagnostic applications in pathology. *J. Pathol.* **182**, 250–261.

McQuaid, S. and Allan, G. M. (1992). Detection protocols for biotinylated probes – optimization using multistep techniques. *J. Histochem. Cytochem.* **40**, 569–574.

Mifflin, T. E., Bowden, J., Lovell, M. A., Bruns, D. E., Hayden, F. G., Gröschel, D. H. M. and Savory, J. C. (1987). Comparison of radioactive and biotinylated probes for detection of cytomegalovirus. *Clin. Biochem.* **20**, 231–235.

Miller, R. T. and Kubier, P. (1997). Blocking of endogenous avidin-binding activity in immunohistochemistry. *Appl. Immunohistochem.* **5**, 63–66.

Misra, D. N., Dickman, P. S. and Yunis, E. J. (1995). Fluoresence in situ hybridization (FISH) detection of MYC oncogene amplification in neuroblastoma using paraffin-embedded tissue. *Diagn. Mol. Pathol.* **4**, 128–135.

Montone, K. T. (1994). *In situ* hybridization for ribosomal RNA sequences: a rapid sensitive method for diagnosis of infectious pathogens in anatomic pathology substrates. *Acta Histochem. Cytochem.* **27**, 601–606.

Montone, K. T. and Tomaszewski, J. E. (1993). *In situ* hybridization protocol for overall preservation of messenger RNA in fixed tissues with polyd(T) oligonucleotide probe. *J. Histotechnol.* **16**, 315–322.

Morris, R. G., Arends, M. J., Bishop, P. E., Sizer, K., Duvall, E. and Bird, C. C. (1990). Sensitivity of digoxigenin and biotin labeled probes for detection of human papillomavirus by in situ hybridization. *J. Clin. Pathol.* **43**, 800–805.

Moter, A. and Göbel, U. B. (2000). Fluoresence in situ hybridization (FISH) for direct visualization of microorganisms. *J. Microbiol. Meth.* **41**, 85–112.

Murphy, D. S., Hoare, S. F., Going, J. J., Mallon, E. E., George, W. D., Kaye, S. B., Brown, R., Black, D. M. and Keith, W. N. (1995). Characterization of extensive genetic alteration in ductal carcinoma in situ by fluorescence in situ hybridization and molecular analysis. *J. Natl Cancer Inst.* **87**, 1694–1704.

Musiani, M., Zerbini, M., Venturoli, S., Gentilomi, G., Borghi, V., Pietrosemoli, P., Pecora, M. and La Placa, M. (1994). Rapid diagnosis of cytomegalovirus encephalitis in patients with AIDS using in situ hybridization. *J. Clin. Pathol.* **47**, 886–891.

Nakatsuka, S., Yao, M., Hoshida, Y., Yamamoto, S., Luchi, K. and Aozasa, K. (2002). Pyothorax-associated lymphoma: a review of 106 cases. *J. Clin. Oncol.* **20**, 4255–4260.

Nunez, D. J., Davenport, A. P., Emson, P. C. and Brown, M. J. (1989). A quantitative in situ hybridization method using computer-assisted image analysis. Validation and measurement of atrial natriuretic factor (ANF) messenger RNA in rat heart. *Biochem. J.* **263**, 121–127.

Nuovo, G. J. (1991). Comparison of Bouin solution and buffered formalin fixation on the detection rate by in situ hybridization of human papillomavirus DNA in genital tract lesions. *J. Histotechnol.* **14**, 13–17.

Nuovo, G. J. and Richart, R. M. (1989). A comparison of biotin and ^{35}S-based in situ hybridization methodologies for detection of human papillomasirus DNA. *Lab. Investig.* **61**, 471–476.

Nuovo, G. J., Gallery, F., MacConnell, P., Becker, J. and Bloch, W. (1991). An improved technique for the in situ detection of DNA after polymerase chain reaction amplification. *Am. J. Pathol.* **139**, 1239–1244.

Nuovo, G. J., Gallery, F. and MacConnell, P. (1992). Detection of amplified HPV6 and 11 DNA in vulva lesions by hot start PCR *in situ* hybridization. *Mod. Pathol.* **5**, 444–448.

Odrich, M. G., Jakobiec, F. A., Lancaster, W. D., Kenyon, K. R., Kelly, L. D., Kornmehl, E. W., Steinert, R. F., Grove, Jr., A. S., Shore, J. W. and Gregoire, L. (1991). A spectrum of bilateral squamous conjunctival tumors associated with human papillomavirus type 16. *Ophthalmology* **98**, 628–635.

Pang, M. and Baum, L. G. (1993). Application of nonisotopic RNA in situ hybridization to detect endogenous gene expression in pathologic specimens. *Diagn. Mol. Pathol.* **2**, 277–282.

Plummer, T. B., Sperry, A. C., Xu, H. S. and Lloyd, R. V. (1998). In situ hybridization detection of low copy nucleic acid sequences using catalyzed reporter deposition and its usefulness in clinical human papillomavirus typing. *Diagn. Mol. Pathol.* **7**, 76–84.

Pollanen, R., Vuopala, S. and Lehto, V. P. (1993). Detection of human papillomavirus infection by nonisotopic in situ hybridization in condylomatous and CIN lesions. *J. Clin. Pathol.* **46**, 936–939.

Porter, H., Quantrill, A. M. and Fleming, K. A. (1988). B19 parvovirus infection of myocardial cells. *Lancet* **5**, 535–536.

Pringle, J. H. and Ruprai, A. K. (1990). *In situ* hybridization of immunoglobulin light chain mRNA in paraffin sections using biotinylated or hapten-labeled oligonucleotid probes. *J. Pathol.* **162**, 197–207.

Pringle, J. H., Primrose, L., Kind, C. N., Talbot, I. C. and Lauder, I. (1989). *In situ* hybridization demonstration of poly-adenylated RNA sequence in formalin-fixed paraffin sections using a biotinylated oligonucleotide poly d(t) probe. *J. Pathol.* **158**, 279–286.

Rahbar, A., Bostrom, L., Lagerstedt, U., Magnusson, I., Soderberg-Naucler, C. and Sundqvist, V. A. (2003). Evidence of active cytomegalovirus infection and increased production of IL-6 in tissue specimens obtained from patients with inflammatory bowel disease. *Inflamm. Bowel Dis.* **9**, 154–161.

Reed, J. A., Nador, R. G., Spaulding, D., Tani, Y., Cesarman, E. and Knowles, D. M. (1998). Demonstration of Kaposi's sarcoma-associated herpes virus cyclin D homolog in cutane Kaposi's sarcoma by colorimetric in situ hybridization using a catalyzed signal amplification system. *Blood* **91**, 3825–3832.

Reichman, L. B. (1997). Tuberculosis elimination-what's to stop us? *Int. J. Tubercul. Lung Dis.* **1**, 3–11.

Rentrop, M., Knapp, B., Winter, H. and Schweizer, J. (1986). Aminoalkylsilane-treated glass slides as support for in situ hybridization of keratin cDNA to frozen tissue sections under varying fixation and pretreatment conditions. *Histochem. J.* **18**, 271–276.

Rigby, P. W. J. and Dieckmann, M. (1977). Labelling deoxyribonucleic acid to high specific activity *in vivo* by nick translation with DNA polymerase I. *J. Mol. Biol.* **113**, 237–251.

Rimsza, L. M., Vela, E. E., Frutiger, Y. M., Rangel, C. S., Solano, M., Richter, L. C., Grogan, T. M. and Bellamy, W. T. (1996). Rapid automated combined in situ hybridization and immunohistochemistry for sensitive detection of cytomegalovirus in paraffin-embedded tissue biopsies. *Am. J. Clin. Pathol.* **106**, 544–548.

Rosenthal, S. I., Depowski, P. L., Sheehan, C. E. and Ross, J. S. (2002). Comparison of HER-2/*neu* oncogene amplification detected by fluorescence in situ hybridization in lobular and ductal breast cancer. *Appl. Immunohistochem. Mol. Morphol.* **10**, 40–46.

Roth, J., Saremasland, P., Warhol, M. J. and Heitz, P. U. (1992). Improved accuracy in diagnostic immunohistochemistry, lectin histochemistry and *in situ* hybridization using a gold-labeled horseradish peroxidase antibody and silver intensification. *Lab. Investig.* **67**, 263–269.

Samanta, M., Harkins, L., Klemm, K., Britt, W. J. and Cobbs, C. S. (2003). High prevalence of human cytomegalovirus in prostatic intraepithelial neoplasia and prostatic carcinoma. *J. Urol.* **170**, 998–1002.

Sambrook, J., Fritsch, E. F. and Maniatis, T. (1989). *Molecular Cloning: A Laboratory Manual.* Cold Spring Harbor Laboratory Press, New York.

Scheffner, M., Werness, B. A., Huibregtse, J. M., Levine, A. J. and Howley, P. M. (1990). The E6 oncoprotein encoded by human papillomavirus types 16 and 18 promotes the degradation of p53. *Cell* **63**, 1129–1136.

Shin, S. S., Berry, G. J. and Weiss, L. M. (1991). Infectious mononucleosis – diagnosis by *in situ* hybridization in two cases with atypical features. *Am. J. Surg. Pathol.* **15**, 625–631.

Springall, D. R., Hacker, G. W., Grimelius, G. W. and Polak, J. M. (1984). The potential of the immunogold-silver staining method for paraffin sections. *Histochemistry* **81**, 603–608.

Stahl, W. L., Eakin, T. J. and Baskin, D. G. (1993). Selection of oligonucleotide probes for detection of mRNA isoforms. *J. Histochem. Cytochem.* **41**, 1735–1740.

Stender, H., Lund, K., Petersen, K. H., Rasmussen, O. F., Hongmanee, P., Miörner, H. and Godtfredsen, S. E. (1999). Fluoresence in situ hybridization assay using peptide nucleic acid probes for differentiation between tuberculous and nontuberculous mycobacterium species in smears of mycobacterium cultures. *J. Clin. Microbiol.* **37**, 2760–2765.

Still, K., Robson, C. N., Autzen, P., Robinson, M. C. and Hamdy, F. C. (2000). Localization and quantification of mRNA for matrix metalloproteinase-2 (MMP-2) and tissue inhibitor of matrix metalloproteinase-2 (TIMP-2) in human benign and malignant prostatic tissue. *Prostate* **42**, 18–25.

Syrjanen, S. (1992). Viral gene detection by *in situ* hybridization. In *Diagnostic Molecular Pathology. A Practical Approach* (C. S. Herrington and J. O. D. McGee, eds), vol. 1, pp. 103–140. Oxford University Press, Oxford.

Syrjanen, S., Partanen, P., Mantyjarvi, R. and Syrjanen, K. (1988). Sensitivity of in situ hybridization techniques using biotin and ^{35}S labeled human papillomavirus (HPV) DNA probes. *J. Virol. Meth.* **19**, 225–238.

Terenghi, G. and Polak, J. M. (1994). Detecting mRNA in tissue sections with digoxigenin-labeled probes. *Meth. Mol. Biol.* **28**, 193–199.

Trembleau, A. and Bloom, F. E. (1995). Enhanced sensitivity for light and electron microscopic *in situ* hybridization with multiple simultaneous nonradioactive oligodeoxynucleotide probes. *J. Histochem. Cytochem.* **43**, 829–841.

Tubbs, R., Pettay, J., Skacel, M., Powell, R., Stoler, M., Roche, P. and Hainfeld, J. (2002). Gold-facilitated in situ hybridization. A bright-field autometallographic alternative to fluorescence in situ hybridization for detection of *HER-2/neu* gene amplification. *Am. J. Pathol.* **160**, 1589–1595.

Uehara, F., Ohba, N., Nakashima, Y., Yanagita, T., Ozawa, M. and Muramatsu, T. (1993). A fixative suitable for *in situ* hybridization histochemistry. *J. Histochem. Cytochem.* **41**, 947–953.

Uezato, H., Hagiwara, K., Ramuzi, S. T., Khaskhely, N. M., Nagata, T., Nagamine, Y., Nonal, S., Asato, T. and Oshiro, M. (1999). Detection of human papilloma virus type 56 in extragenital Bowen's disease. *Acta Derm. Venereol.* **79**, 311–313.

Unger, E. R., Vernon, S. D., Thoms, W. W., Nisenbaum, R., Spann, C. O., Horowitz, I. R., Icenogle, J. P. and Reeves, W. C. (1995). Human papillomavirus and disease-free survival in Figo stage 1b cervical cancer. *J. Infect. Dis.* **172**, 1184–1190.

van Dekken, H., Wink, J., Alers, J. C., de Man, R. A., Ijzermans, J. N. and Zondervan, P. E. (2003). Genetic evaluation of the dysplasia-carcinoma sequence in chronic viral liver disease: a detailed analysis of two cases and a review of the literature. *Acta Histochem.* **105**, 29–41.

Wang, J. Q. (2003). Analysis of mRNA expression using double in situ hybridization labeling with isotopic and nonisotopic probes. *Meth. Mol. Med.* **79**, 153–159.

Weaver, M. G., Abdul-Karim, F. W., Dale, G., Sorensen, K. and Huang, Y. T. (1989). Detection and localization of human papillomavirus in penil condylomas and squamous cell carcinoma using *in situ* hybridization with biotinylated DNA viral probes. *Mod. Pathol.* **2**, 94–100.

Weigant, J., Ried, T., Nederlof, P., van de Ploeg, M., Tanke, H. J. and Raap, A. K. (1991). *In situ* hybridization with fluoresceinated DNA. *Nucleic Acids Res.* **19**, 3237–3241.

Weiss, L. M. and Chan, Y. Y. (1991). Effects of different fixatives on detection of nucleic acids from paraffin-embedded tissues by *in situ* hybridization using oligonucleotide probes. *J. Histochem. Cytochem.* **39**, 1237–1241.

Wenig, B. M., Thompson, L. D., Frankel, S. S., Burke, A. P., Abbondanzo, S. L., Sesterhenn, L. and Heffner, D. K. (1996). Lymphoide changes of the nasopharyngeal and palatine tonsils that are indicative of human immunodeficiency virus infection. A clinicopathologic study of 12 cases. *Am. J. Surg. Pathol.* **20**, 572–587.

Wolber, R. A. and Lloyd, R. V. (1988). Cytomegalovirus detection by nonisotopic *in situ* DNA hybridization and viral antigen immunostaining using a two-colour technique. *Hum. Pathol.* **19**, 736–741.

Wood, G. S. and Warnke, R. (1981). Suppression of endogenous avidin-binding activity in tissues and its relevance to biotin-avidin detection system. *J. Histochem. Cytochem.* **29**, 1196–1204.

Wu, T. C., Mann, R. B., Epstein, J. I., MacMahon, E., Lee, W. A., Charach, P., Hayward, S. D., Kurman, R. J., Hayward, G. S. and Ambinder, R. F. (1991). Abundant expression of EBER1 small nuclear RNA in nasopharyngeal carcinoma. A morphological distinctive target for detection of Epstein-Barr virus in formalin-fixed paraffin-embedded carcinoma specimens. *Am. J. Pathol.* **138**, 1461–1469.

Zehbe, I., Hacker, G. W., Su, H., Hauser-Kronberger, C., Hainfeld, J. F. and Tubbs, R. (1997). Sensitive *in situ* hybridization with catalyzed reporter deposition, streptavidin-nanogold, and silver acetate autometallography. *Am. J. Pathol.* **150**, 1553–1561.

Zerfass, K., Schulze, A., Spitkovsky, D., Friedman, V., Henglein, B. and Jansen-Dürr, P. (1995). Sequential activation of cyclin E and cyclin A gene expression by human papillomavirus type 16 E7 through sequences necessary for transformation. *J. Virol.* **69**, 6389–6399.

Zerfass-Thome, K., Zwerschke, W., Mannhardt, B., Tindle, R., Botz, J. W. and Jansen-Durr, P. (1996). Inactivation of the cdk inhibitor p27^{KIP1} by the human papillomavirus type 16 E7 oncoprotein. *Oncogene* **13**, 2323–2330.

2 Fluorescent Protein Probes in Fungi

Kirk J Czymmek[1], Timothy M Bourett[2] and Richard J Howard[2]

[1] Department of Biological Sciences, University of Delaware, Newark, DE 19716, USA;
[2] Crop Genetics Research and Development, DuPont Experimental Station, Wilmington, DE 19880-0402, USA

◆◆

CONTENTS

Introduction
Advancements in live-cell imaging
Applications
New developments and future perspectives
Conclusions

◆◆◆◆◆◆ INTRODUCTION

Live imaging of cellular events is fundamental for understanding many aspects of cellular function. As probes in cell biology, fluorescent proteins repeatedly have demonstrated their utility as genetically encoded intrinsically fluorescent reporters. In heterologous organisms (Chalfie et al., 1994), these radiant molecules have cultivated a paradigm shift in the way that modern genomics is applied to solve a multitude of questions in diverse disciplines such as microbiology, cell biology, developmental and neurobiology. Arguably one of the greatest virtues of this technology is the ability to genetically append a target protein, with a fluorescent protein, creating a vital fluorescent fusion that can be expressed and monitored in living cells. As a result, fluorescent proteins have fundamentally transformed the spatio-temporal analysis of protein function. In addition, new fluorescent protein technologies allow for much more than simple observations of fusion protein distribution. An ever-growing family of fluorescent protein variants is enabling far more sophisticated studies of protein function and illuminating wide-ranging processes from gene expression to protein kinetics and intercellular signaling. With concomitant advances in microscopy-related instrumentation new fluorescent protein-based experimental approaches are generating novel findings in all areas of biology.

In this chapter, we will describe the array of fluorescent proteins that are currently available and how they have been employed for investigating fungal cell biology and phytopathology. This includes a discussion of fluorescence-based imaging techniques such as confocal microscopy, multiphoton microscopy, and spectral imaging. Finally, we explore applications of more sophisticated fluorescent protein-based technologies and reflect on what these techniques will bring to these disciplines.

◆◆◆◆◆◆ ADVANCEMENTS IN LIVE-CELL IMAGING

Since the inception of the first compound light microscope nearly 400 years ago, optical microscopy has played significant roles in our understanding of fungal biology. Many general contrast-enhancing cellular stains (e.g. Congo red, trypan blue, fast green) traditionally were used to provide a limited means of distinguishing various cellular components such as the cell wall, cytoplasm, and fungal mucilage. Some of these dyes were applied to differentiate fungal entities from the host during pathogenesis, but many were not compatible with live-cell imaging. During the past half-century, incremental enhancements in optical technology have shown their value for visualization of a multitude of fungal and plant structures. The introduction of contrast-enhancing optical methods such as differential interference contrast (DIC) and phase contrast allowed, for the first time, imaging of live cells with little inherent contrast. Although these methods brought a greater appreciation of dynamic cellular and subcellular events, in many cases the number of cellular components that could be discriminated unequivocally by these imaging advancements remained low.

The next major advance in live-cell imaging came with the introduction of vital fluorescent dyes to identify specific intracellular regions or compartments. The catalog of potential targets of these dyes has expanded rapidly and for the foreseeable future will continue to provide very valuable and versatile means for visualizing certain cellular components. The breadth of utility of these fluorescent probes ranges from measuring intracellular pH and calcium, to staining vacuoles, nuclei, mitochondria and the endomembrane system (see Chapter 3). However, in the majority of cases, these dyes lack the specificity to localize individual molecules (Rieder and Khodjakov, 2003). Other limitations include variability of uptake, partitioning within the cell, and chemical and/or photo-induced toxicity. Many dyes used routinely to image subcellular compartments in higher plants and animals are not taken up into live fungal cells. For example, in our hands the commonly used vital nuclear dyes DAPI and Hoescht 33258 are excluded from all but moribund or dead cells of filamentous fungi.

Microinjection has been one means for introducing dyes into animal and plant cells but, historically, has been difficult to use with filamentous fungi because of the small size of hyphae and high intracellular turgor pressures. However, the galistan expansion femtosyringe – with a tip diameter of less than 100 nm – recently used to microinject individual plant organelles (Knoblauch *et al.*, 1999), and the demonstration of tip-less laser induced dye loading (Greulich, 2001; Tirlapur and König, 2002), may warrant revisiting microinjection and laser methods for dye loading of fungal cells.

Further subcellular exploration was realized in the early 1980s with the advent of video-enhanced light microscopy (see Allen *et al.*, 1981a,b; Inoué, 1981, 1986). By using the contrast sensitivity of the video camera, this technology permitted the resolution of cellular entities smaller than the theoretical limits of light-based optical systems. This technology was exploited to image individual microtubules to form the basis of *in vitro* motility assays that subsequently led to the identification of microtubule-associated motor proteins (Vale *et al.*, 1985). More recently video-enhanced microscopy was used to generate striking images of satellite Spitzenkörper in growing hyphal tip cells of several fungi (López-Franco *et al.*, 1995). As with other transmitted light techniques, such as DIC and phase contrast imaging, it remained difficult to discriminate among the plethora of cellular structures that are resolvable in living cells (Reider and Khodjakov, 2003).

Reporter proteins such as β-glucuronidase (GUS), luciferase and fluorescent proteins offer some advantages over dye-based imaging. As an enzyme reporter, GUS is highly sensitive with a signal that can be amplified to produce abundant enzyme-catalyzed product. However, both GUS and luciferase are undesirable for live cell studies because they require addition of a substrate that may be differentially efficient at reaching the enzyme expressed within cells or tissues. As a result, destructive fixation and permeabilization steps generally are required, with the potential for introducing artifacts, and/or allowing diffusion of the reaction product away from the original site of expression.

Fluorescent proteins are excellent molecular markers since they are encoded genetically, require no substrates or co-factors for activity and can be employed in any genetically tractable organism. In general, fungi are obligingly amenable to transformation protocols (Gold *et al.*, 2000), with the exception of obligate pathogens and mycorrhizae (Harrier and Millam, 2001). An advantage of fluorescent proteins over dyes is that cells can be more easily observed for extended periods. Over time many fluorescent dyes do not remain confined within the target compartment, and photo-induced toxicity – always a concern – may be exacerbated by prolonged light exposure. Fluorescent proteins have the added potential benefit of being targeted efficiently to subcellular compartments with the appropriate targeting signals and have been used for a variety of purposes including (1) tagging of specific organisms, (2) protein localization, (3) targeting/compartmentalization studies and organelle dynamics, (4) tracking and quantification of specific protein populations, e.g. using photoactivatable fluorescent proteins, (5) assessment of protein–protein interactions, e.g. using fluorescence resonance energy transfer (FRET), and (6) biosensors to investigate biological events and signals. Many of these capabilities have already been employed successfully for the study of topics in fungal cell biology (Cormack, 1998) and include examples such as Spitzenkörper dynamics (Knechtle *et al.*, 2003), septum formation (Wendland and Philippsen, 2002), autophagic bodies (Pinan-Lucarre *et al.*, 2003), secretion (Gordon *et al.*, 2000), and others described throughout this chapter.

Fluorescent Proteins

While the existence of fluorescent proteins in marine organisms has been known for some time, it was not until 1992 that the green fluorescent protein (GFP) from the jellyfish *Aequorea victoria* was cloned (Prasher *et al.*, 1992), and subsequently introduced into heterologous organisms (Chalfie *et al.*, 1994) and fungi (Cormack, 1998). Wild type GFP is an auto-catalyzed 238 amino acid protein that requires only oxygen to form a unique 11-stranded β-barrel structure with a fluorophore-containing α-helix located within the center (Tsien, 1998). While both the folding of the protein and a subsequent cyclization reaction (the chromophore consists of Ser65, Tyr66, and Gly67) are relatively rapid events, the oxidation that follows is the rate-limiting step in formation of the mature fluorescent molecule.

In order to enhance its fluorescence properties and utility for use with a broad range of organisms, wild type GFP has been modified in several ways. Researchers found it necessary to modify the encoding nucleotide sequence to ensure efficient translation due to inter-species differences in codon usage. In addition, it was discovered that the wild type GFP sequence had a cryptic intron that prevented GFP formation in plants (Haseloff *et al.*, 1997). Wild type GFP folds efficiently near or below room temperature but poorly at 37°C. Most GFP mutants contain modifications that improve folding at higher temperatures making them more useful for expression in mammalian cells and *E. coli*. Although not required, there is evidence that chaperonins may facilitate folding (Wang *et al.*, 2002).

Wild type GFP has two excitation maxima, a major peak at 398 nm and a minor at 475 nm. Since illumination of cells with UV radiation is potentially damaging, the molecule was altered successfully to "red-shift" the excitation to a single peak at 488 nm. Additional efforts were devoted to modify the spectral properties of the wild type molecule, including excitation and emission characteristics, to provide a greater choice of colors as well as increased brightness through improved extinction coefficients and quantum yields (Heim and Tsien, 1996). A shortlist of commonly used fluorescent proteins and some of their characteristics can be found in Table 2.1.

Motivated by the desire to generate fluorescent proteins to serve with GFP as the complementary half of a donor/acceptor FRET pair, color variants with altered spectral characteristics were engineered through a series of amino acid substitutions (Haseloff, 1999; Tsien, 1998). These color variants have proven to be valuable for multi-spectral imaging applications as well as for FRET. The best known of these may be the commercially available "enhanced fluorescent proteins:" viz. blue (EBFP), cyan (ECFP), green (EGFP), and yellow (EYFP). EBFP has been used little since it was shown to exhibit low fluorescence intensity (Cubitt *et al.*, 1999), photobleach more easily (Ellenberg *et al.*, 1998), and require harmful UV excitation. While EGFP has been the most used of these enhanced fluorescent proteins the spectral characteristics of ECFP and EYPF make them the preferred choice for FRET applications. More recently a new version of a GFP-based yellow fluorescent protein, named "Venus", was shown to exhibit improved brightness and increased speed of fluorophore maturation and holds great promise as an alternative to EYFP (Nagai et al., 2002).

Unfortunately, researchers have been unable to mutate A. victoria-derived GFP to yield variants with emission in the red region of the visible spectrum, and so additional marine organisms were scrutinized for the presence of homologous proteins. The first non-jellyfish fluorescent protein to be cloned was isolated from an Anthozoan relative of jellyfish (Matz et al., 1999; Fradkov et al., 2002). This red-emitting fluorescent protein, available commercially as DsRed, was of particular interest since it offered significant spectral separation from GFP and may therefore be used for multi-spectral imaging applications. In addition, the green excitation of DsRed potentially avoids the background autofluorescence often prevalent during excitation with shorter wavelengths. However, the utility of DsRed was not realized due to its tendency to form tetramers and lengthy maturation period – as long as 48 h (Baird et al., 2000). In addition, the transient green fluorescent intermediate of DsRed that forms during maturation can interfere with the emission of other green fluorophores during multi-labeling experiments.

Several groups have worked to improve the characteristics of DsRed. Bevis and Glick (2002) engineered a mutant with a reduced maturation time – a ca. 45 min half time for maturation – while Campbell et al. (2002) derived a monomeric version of DsRed through a total of 17 amino acid changes. This monomeric DsRed has the potential to function as the acceptor in a green/red FRET pair but the reduction in brightness compared with its tetrameric progenitor molecule DsRed may be a limiting factor (Zhang et al., 2002). The tetrameric nature of DsRed can also be problematic when used in fusion protein experiments. Hetero-oligomeric tagging was described recently as a method to reduce this problem and entailed the expression of an excess of free non-fluorescent fluorescent protein along with the fusion protein to generate a complex containing a single target protein and fluorescent tag (Bulina et al., 2003). Although oligomerization has been less of an issue with GFP, it also has a weak tendency to dimerize and hence a monomeric version of GFP was engineered (Zacharias et al., 2002). In the jellyfish, dimerization is thought to play a role during association of GFP with its companion bioluminescent protein aequorin. Such dimerization in vivo has no apparent impact on fluorescence. However, when GFP is used in fusion proteins, the larger size of dimeric complexes can introduce steric hindrance effects and may lead to aggregation.

Several years after the introduction of DsRed, the cloning of a number of other reef coral fluorescent proteins (RCFPs) was reported by Matz et al. (1999). Unlike GFP, the RCFPs do not represent companion proteins to a bioluminescent molecule and so their natural physiological role in vivo remains obscure. In addition to DsRed, this group consisted of AmCyan (from Anemonia majano), ZsGreen and ZsYellow (from Zoanthus sp.),

Table 2.1. Properties of select fluorescent proteins

Protein	Excitation max.	Emission max.	Extinction coefficient	Quantum yield	Origin
AcGFP	475	505	32 500	0.82	*Aequorea coerulescens*; Clontech Laboratories, 2002, Living Colors User Manual (Becton Dickinson, Palo Alto, CA) and Gurskaya et al. (2003)
AmCyan	458	486	40 000	0.24	*Anemonia majano*; Matz et al. (1999)
AsRed 2	576	592	56 200	0.05	*Anemonia sulcata*; Clontech Laboratories, 2002, Living Colors User Manual (Becton Dickinson, Palo Alto, CA) and Lukyanov et al. (2000)
DsRed	558	583	57 000	0.79	*Discosoma sp.*; Campbell et al. (2002) and Matz et al. (1999)
DsRed 2	561	587	43 800	0.55	*Discosoma sp.*; Bevis and Glick (2002)
DsRed-Express	557	579	31 000	0.42	*Discosoma sp.*; Matz et al. (1999)
DsRed, monomeric	584	607	44 000	0.25	*Discosoma sp.*; Campbell et al. (2002)
DsRed T4	555	586	30 300	0.44	*Discosoma sp.*; Bevis and Glick (2002)

Name	Ex	Em	Extinction coefficient	QY	Source; Reference
EBFP	382	445	26–31 000	0.17–0.26	*Aequorea victoria*; Tsien (1998)
ECFP	439	476	32 500	0.4	*A. victoria*; Tsien (1998)
EGFP	484	510	55–57 000	0.60	*A. victoria*; Tsien (1998)
EYFP	512	529	83 400	0.61	*A. victoria*; Nagai et al. (2002) and Tsien (1998)
GFP, WT	398/475	504	25–30 000	0.79	*A. victoria*; Tsien (1998)
Gold or GdFP	466	574	–	–	*A. victoria*; Bae et al. (2003)
HcRed	588	618	20 000	0.15	*Heteractis crispa*; Clontech Laboratories, 2002, Living Colors User Manual (Becton Dickinson, Palo Alto, CA) and Gurskaya et al. (2001)
KFP1	580	600	59 000	0.13	*Asulcata*; Chudakov et al. (2003)
Venus	515	528	92 200	0.57	*A. victoria*; Nagai et al. (2002)
ZsGreen	493	505	43 000	0.91	*Zoanthus sp*.; Clontech Laboratories, 2002, Living Colors User Manual (Becton Dickinson, Palo Alto, CA) and Matz et al. (1999)
ZsYellow	528	538	20 200	0.42	*Zoanthus sp*.; Gurskaya et al. (2003)

AsRed (from *Anemonia sulcata*), and HcRed (from *Heteractis crispa*). Each of these RCFPs exhibited distinct excitation and emission spectra (Matz *et al.*, 1999; Lukyanov *et al.*, 2000; Gurskaya *et al.*, 2001) and when expressed in the fungal cytosol have been very useful as reporters of fungal ingress during plant pathogenesis (Bourett *et al.*, 2002; Czymmek *et al.*, 2002). It was discovered later that RCFPs, like DsRed, form oligomers and high molecular weight aggregates. When RCFPs were expressed in the cytosol, the oligomeric nature of these proteins was manifest by nuclear exclusion, fluorescent aggregates (Figure 2.1), and elevated levels of fluorescence in vacuoles. In addition, targeting of the RCFPs to organelles such as the endoplasmic reticulum and mitochondria was fraught with difficulties (Figure 2.2).

HcRed is a novel fluorescent protein in that it was first isolated as a non-fluorescent chromophore that required subsequent mutagenesis to yield a fluorescent molecule (Gurskaya *et al.*, 2001). Its far-red emission makes it a potentially useful fluorescent protein in multi-labeling experiments. However, it is not an especially bright fluorophore (Table 2.1; Campbell *et al.*, 2002), and though its propensity to form oligomers has been mitigated by site-directed mutagenesis (Gurskaya *et al.*, 2001), dimerization still excludes HcRed from FRET applications.

For researchers who sought alternatives to *A. victoria*-derived GFP, the report of a GFP fluorescent protein generated by mutagenesis of a colorless non-fluorescent protein from *Aequorea coerulescens* (Gurskaya *et al.*, 2003),

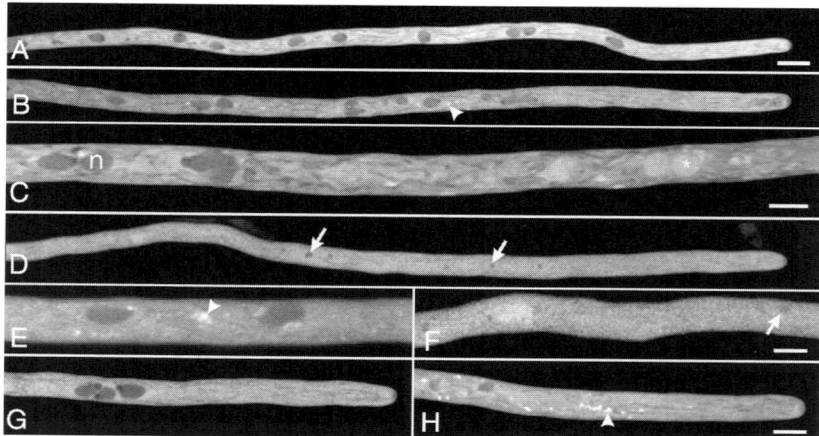

Figure 2.1. Fluorescent proteins expressed in the fungal cytosol. Laser scanning confocal images of fluorescence from transformants of the multinucleate fungus *Fusarium verticillioides* expressing AsRed (A), ZsGreen (B, C), AcGFP (D), DsRed (E), monomeric DsRed (F) and ZsYellow (G, H). Note that fluorescence was excluded from nuclei during interphase and the presence of fluorescent aggregates (arrowheads) within hyphae that expressed the tetrameric reef coral fluorescent proteins (RCFPs) ZsGreen, DsRed, and ZsYellow. At the asynchronous onset of mitosis (C) the RCFPs entered each nucleus in a wave across the cell as evidenced here by both fluorescence-excluding nuclei (n) and brightly fluorescent nuclei (asterisk). Both AcGFP and monomeric DsRed entered the nucleus but were excluded by nucleoli (arrows). Bar for A, B and D, 10 μm. Bar for C and E, F, and G and H, 5 μm.

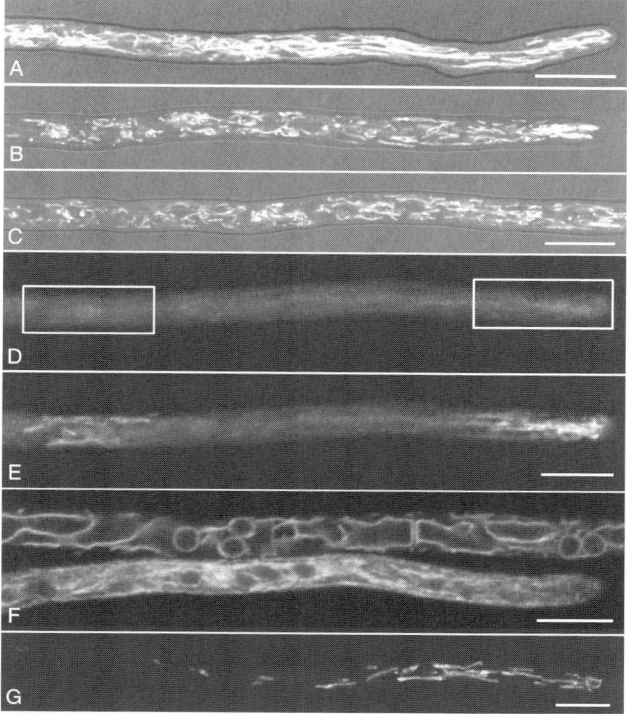

Figure 2.2. Fluorescent proteins targeted to intracellular compartments. Using the initial 64 amino acids of malate dehydrogenase as a targeting sequence, the fluorescent proteins EGFP (A), ZsGreen (B, C), and PA-GFP (D, E) were directed to mitochondria in hyphae of *Magnaporthe grisea* (A) and *Fusarium verticillioides* (B–E). ZsGreen formed fluorescent aggregates within mitochondria. There was minimal fluorescence of PA-GFP (D), until activated (activated regions outlined) by pulsed 810 nm excitation from a titanium:sapphire laser (E). Transformant expression of AcGFP with a C-terminal signal peptide and N-terminal KDEL ER-retention sequence showed fluorescence within a highly dynamic tubular network (F). The chimeric protein in identical constructs with ZsGreen substituted for AcGFP was targeted to a different subcellular compartment, probably of vacuolar origin (G). The clarity of these structures in a slowly acquired z-series suggests that they were not very mobile. Images A–F are single optical sections acquired with a confocal microscope. Images A–C are shown as an overlay of the fluorescence and transmitted light channels. Image G is a maximum intensity projection of 12 confocal optical sections. Bar, 7 μm.

and its commercial availability as AcGFP (BD Biosciences Clontech), were significant. Because of the tetrameric nature of ZsGreen, an RCFP, there had been no monomeric GFP other than *A. victoria*-derived GFP. Fungal expression of AcGFP has been reported (Fuchs, Czymmek, and Sweigard, unpublished) and it has performed well in our trials with fungi including both cytosolic expression (Figure 2.1) and targeting to specific organelles (Figure 2.2).

A red-emitting "gold" fluorescent protein derived from *A. victoria* GFP was created recently by substitution of a synthetic amino acid analogue for tryptophan within the α-helix of the fluorophore using selective pressure incorporation techniques (Bae *et al.*, 2003). While holding much promise for imaging cells grown *in vitro*, where the required non-canonical amino acids could be added as supplements to the growth medium, it is less clear how this fluorescent protein could be exploited for *in planta* observations of host–pathogen or mycorrhizal interactions.

Limitations of Fluorescent Proteins

In selecting the appropriate fluorescent protein for a specific application, one must consider many different properties of these molecules, as they exhibit significant variability. The greatest shortcoming of fluorescent proteins may be related to the time required for complete maturation of the fluorophore. It is for this reason that the utility of all fluorescent proteins is limited when assessing temporally sensitive aspects of transient expression. While progress has been made to hasten molecular folding (Heim and Tsien, 1996), under ideal conditions a minimum of 30 min has normally been required between induction of gene expression and generation of a fluorescence-competent molecule. Thus, for promoter activation studies there will be at least a 30 min lag following induction before the reporter can be visualized. More dire, however, is the >48 h required for maturation of the RCFP DsRed (Baird *et al.*, 2000), thus representing an additional significant impediment for the use of these wild type RCFPs.

Although imaging of live cells over time is a tremendous advantage provided by the use of fluorescent proteins it is sometimes useful to fix samples. For example, with immunological methods fixation is required to localize fluorescent proteins at high spatial resolution via electron microscopy. Chalfie *et al.* (1994) reported that wild type GFP fluorescence survived both glutaraldehyde and formaldehyde fixation. Similarly, we

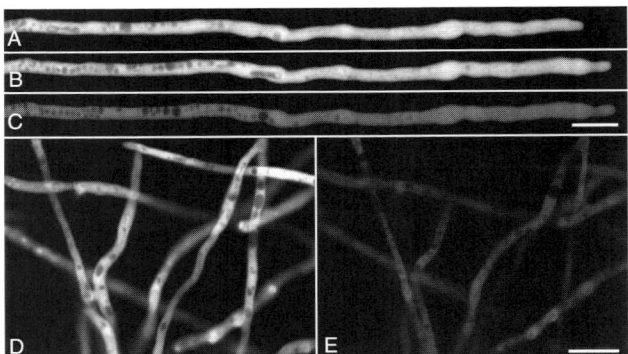

Figure 2.3. Changes in fluorescent protein intensity following fixation. Fluorescence intensity of hyphae of *M. grisea* expressing EYFP in the cytoplasm (A) was unaffected by buffer treatment (B) but was reduced by treatment with 3% formaldehyde in the same buffer (C). Fluorescence intensity was also reduced when EGFP expressing hyphae (D) were fixed with 2% buffered glutaraldehyde (E). Despite this reduction in brightness, the fluorescence intensity was more than sufficient for generation of useful data (not illustrated). A, B, and C shown at the same magnification. D and E are the same magnification. Bars, 7 μm.

have observed that both formaldehyde and glutaraldehyde diminished but did not destroy the fluorescence of EGFP or EYFP (Figure 2.3). We also found that glutaraldehyde but not formaldehyde destroyed the fluorescence of AmCyan and ZsGreen (unpublished). Differential effects of solvents have been reported, but Brock *et al.* (1999) showed that while buffered formaldehyde preserved fluorescence, a combination of formaldehyde and methanol was required to prevent spurious redistribution of a plasma membrane-targeted GFP fusion.

In their review of fluorescent protein targeting to plant compartments, Hawes *et al.* (2001) cautioned researchers about potential misinterpretations of fluorescent protein-derived data, and argued for the inclusion of independent immunology-based cytological controls. They suggested further that localization of fusion proteins be confirmed at electron microscope resolution to ensure that the tagged protein was delivered with fidelity to the same subcellular site as the endogenous protein. Although sensible, such verification would depend upon the availability and specificity of antibodies and, thus, would not always be possible. Several antibodies raised against wild type and enhanced fluorescent proteins are available commercially. In choosing an appropriate antibody the processing method selected for electron microscopy should be considered. For example, antibodies useful for material prepared by conventional chemical fixation (Gordon *et al.*, 2000) may not be suitable for samples prepared by freeze substitution. Successful immunolocalization of an EGFP γ-tubulin fusion, following cryo-fixation and freeze substitution, has been reported in *Saccharomyces cerevisiae* (Chial *et al.*, 1998). Epitope tagging the fluorescent protein would be an alternative strategy (Polishchuk *et al.*, 2000). A fluorescent protein-based solution may be available in "fluorobodies". These are reagents that combine fluorescent proteins with the complementarity-determining regions (CDRs) normally found in antibody variable domains (Zeytun *et al.*, 2003). In fluorobodies the exposed loops of the strands at one end of the β-barrel are replaced, by CDRs from a human library. Following selection using phage display methodologies, genetically encoded fluorescent probes for specific antigens can be rapidly identified.

FlAsH/ReAsH technology is another method for correlating fluorescence signal with electron microscope observations (Adams *et al.*, 2002). Available commercially as "Lumio" in-cell labeling kits (Invitrogen Life Technologies, Carlsbad, CA), the fluorescence signal is generated when a cell-permeant bi-arsenic reagent binds to a tetracystein-containing peptide engineered into the target protein. Introduction of either FlAsH or ReAsH reagent generates green or red emission, respectively, making visible the location of the tagged protein. Through sequential application of FlAsH and then ReAsH reagent the relative age of tagged proteins can be determined (Gaietta *et al.*, 2002) in much the same manner as fluorescent timer (see discussion regarding measurement of protein kinetics). Although not as bright as FlAsH, the ReAsH reagent has the capacity to photo-oxidize diaminobenzidine yielding an electron scattering reaction product. In this way, the fluorescence patterns observed with light microscopy can be correlated with structures resolved at high resolution in the electron microscope (Gaietta *et al.*, 2002).

Imaging Modalities

The ability to observe interactions within or between cells, in real time, adds insight to our understanding of biological processes. Since proper protein functioning requires that molecules be at the appropriate place, and time, the advantages of time-resolved live-cell imaging cannot be overstated. Fluorescent proteins can be imaged using several different types of microscopes – wide field equipped with sensitive CCD cameras, laser scanning, confocal and multiphoton – with advantages and disadvantages to each. Transmitted light techniques (viz. brightfield, darkfield, phase contrast, DIC) can be combined with their complementary fluorescent signals to provide a spatial reference and additional sample information. For example, it is very useful to image growing cells with transmitted light especially during drug treatments that may alter morphology or affect the normal distribution of organelles.

The quality of traditional fluorescence microscope imaging is limited by contaminating fluorescence signal from regions of the sample above or below the plane of focus. Contributions to the image from these regions will reduce resolution and contrast and prevent optical sectioning of the sample. Confocal and multiphoton microscopy, and deconvolution, address these problems and have pushed the limits of photon-based microscopy to remarkable levels. The non-invasive three-dimensional (3D) quality, high contrast and spatial resolution afforded by these technologies are extremely useful for imaging fluorescent proteins.

Confocal microscopy remains the preferred method for many fungal applications especially with thick samples such as hyphal colonies, fungal–host interactions, and fungal associated biofilms. Most confocal microscopes employ a raster scanning laser and pinhole aperture to reject out-of-focus and/or scattered light. In this way "confocality" is achieved and the signal-to-noise ratio is increased. Alternatively, a rapidly spinning disk covered with a patterned array of small apertures, equipped with micro-lens technology, covers the sample with multiple micro-beams as the disk spins. Signal is captured simultaneously as fluorescence emission with a CCD camera (Nakano, 2002). The main advantage of these spinning disk instruments is speed of acquisition (in practice ca. 20 frames per second) suitable for recording dynamic events including fluorescent proteins in fungi (Ovechkina *et al.*, 2003) but, in general, confocality is reduced significantly, especially in thick samples. In addition, these systems are not compatible with more sophisticated types of analysis such as fluorescence loss in photobleaching (FLIP), fluorescence localization after photobleaching (FLAP), and fluorescence recovery after photobleaching (FRAP), and use with photoactivatable fluorescent proteins.

Multiphoton microscopy extends the potential for optical sectioning deeper into samples with a pulsed near infra-red (NIR) laser light source that can be less injurious to living cells. The multiphoton effect occurs when a fluorescent molecule absorbs simultaneously two or more photons producing an electronic transition from the ground to excited state as the summed energy of the incident photons. For example, the simultaneous absorption of two NIR photons (each 710 nm) can yield an electronic transition equivalent to a single UV photon (355 nm). The multiphoton effect has a quadratic dependency on illumination intensity hence the likelihood of such an event occurring outside the objective focal spot decreases rapidly along the beam path (z-axis). Ultimately, an inherently confocal image is generated, without the use of a pinhole aperture, by raster scanning the laser spot across the specimen. Reviews on confocal and/or multiphoton microscopy are available that address specific applications in mycological research including basic theoretical considerations (Czymmek, 2004; Czymmek *et al.*, 1994; Kwon *et al.*, 1993).

With the introduction of commercially available spectral imaging instruments employing linear unmixing algorithms (Dickinson *et al.*, 2001; Hiraoka *et al.*, 2002; Zimmermann *et al.*, 2003), the ability to differentiate fluorophores with overlapping emission spectra has become readily accessible (Bourett *et al.*, 2002). This approach increases the potential number of fluorescent colors that can be differentiated simultaneously and should lead to improved FRET analysis (Zimmermann *et al.*, 2002). Since many fluorescent proteins have broad multiphoton cross-sections (i.e. excitation spectra), making them difficult to excite independently, spectral imaging may be important especially when conducting multi-label experiments as a strategy for minimizing cross talk.

Total internal reflection fluorescence microscopy (TIRFM) is another optical sectioning technique that has gained recent popularity (Sako and Uyemura, 2002). This technique employs a laser light directed through glass or optically compatible plastic at a high incident angle to illuminate cells adhering at the glass–medium interface. Due to differences in refractive index between the glass/plastic and the cell medium the incident laser is reflected back into the glass. But in addition, a very narrow region (typically less than 200 nm from the coverslip) is illuminated by an evanescent wave of the same frequency as the excitation energy, and may be used to excite fluorophores such as fluorescent proteins. TIRFM is desirable for its significantly improved axial resolution, ability to detect single fluorophores, and a demonstrated reduction in photodamage and photobleaching of living samples. Because imaging is limited to the proximity of the coverslip this technique will be restricted largely to imaging cells grown *in vitro*, but may be quite useful for growth and development studies of contact-dependent fungal structures such as appressoria (Apoga *et al.*, 2004).

Following image acquisition, regardless of the technology employed, images may be enhanced using deconvolution algorithms (Swedlow and Platani, 2002). Deconvolution is a computational alternative to the physical methods of confocal and multiphoton optical sectioning and is applied most often to conventional fluorescence images, but provides additional benefit for confocal and multiphoton data, as well. The optical transfer function of the whole microscope system influences the image quality of a given sample, regardless of whether viewed through the oculars or digitally acquired. Deconvolution seeks to reverse these effects mathematically, thereby enhancing contrast and resolution as well as imparting an optical sectioning property to individual optical sections and 3D stacks of images.

◆◆◆◆◆◆ APPLICATIONS

Reporters of Gene Expression

By fusing the coding sequence of a fluorescent protein to a particular promoter sequence, time-resolved gene expression information can be acquired. This approach has been used widely with the β-glucuronidase reporter (GUS). The main advantage of promoter fusions over full-length chimeric proteins is that concern about the reporter altering protein function is reduced. This type of expression data can be extremely valuable in building an understanding of host–pathogen interactions, when the relative timing of expression of fungal pathogenesis genes and corresponding plant resistance genes are thought to influence the outcome of interactions. In this way, stage-specific changes during pathogenesis in expression of calmodulin (Liu and Kolattukudy, 1999) and an endopolygalacturonase (Dumas et al., 1999) have been noted.

As for any fluorescent protein application, potential adverse effects of expression must be addressed. Several fluorescent proteins have been reported toxic when expressed in certain cell types (Haseloff et al., 1997). Reducing the level of expression can mitigate toxicity, but for promoter fusions the best solution may be to simply choose a different fluorescent protein. The targeted subcellular compartment can also influence toxicity, though the mechanisms for this are not fully understood. The fact that hydrogen peroxide is thought to form during the oxidation step of fluorophore maturation has been suggested to explain cytotoxic effects of GFP when expressed at very high levels and why certain subcellular compartments (e.g. mitochondria) may be more tolerant (Tsien, 1998).

Reporters of Protein Distribution

Perhaps the most common application of fluorescent proteins is their use as fusions with specific proteins. Ultimately, the goal is to produce a fluorescent chimeric protein where protein function is unaltered. Ideally, functionality should be confirmed, by showing that expression of the tagged gene product rescues a null mutant of the endogenous gene. For example, a fluorescent protein-tagged avirulence gene should complement a defective copy of that gene in a virulent mutant to restore the avirulence phenotype. This practice is used commonly in S. cerevisiae where homologous integration and gene replacement experiments are routine. However, for most filamentous fungi, the tagged protein is often expressed as a second molecular species where verification of chimeric protein function may be more difficult. Minimally, the fluorescent protein tag should not adversely affect the normal growth and differentiation of the transformant.

The fluorescent protein is usually fused to either the C- or N-terminus of the target protein, however it has been possible to insert into an external loop (Siegel and Isacoff, 1997). Since it is difficult to predict which fusion strategy will be best, it is advisable to construct both C- and N-terminal fusions. An exception to this rule would occur if the C- or N-terminus contained targeting information. If both C- and N-terminal constructs are unsuccessful then engineering a flexible spacer between the fluorescent protein and target protein may be beneficial (e.g. Doyle and Botstein, 1996).

Different strategies can be used to compensate for a loss of target protein function in fusion protein constructs. For example, in cells that do not tolerate the complete replacement of an endogenous cytoskeletal element, (e.g. actin) with a tagged gene (Doyle and Botstein, 1996), one can utilize multi-copy genes by tagging the gene product expressed at lower levels. This strategy was used for α-tubulin in *S. cerevisiae* (Carminati and Stearns, 1999). With single copy genes an alternative approach might be expressing a tagged version as a second molecular species (Han *et al.*, 2001; Takano *et al.*, 2001). In this case, it would be important to express the fusion protein at a lower rate than the endogenous gene. Several options are available to achieve the proper expression level including judicious choice of constitutive promoter, or use of an inducible promoter that can be modulated (Carminati and Stearns, 1999), or a non-native promoter (Figures 2.4 and 2.5). Others have chosen indirect methods to label the cytoskeleton by using microtubule- (Marc *et al.*, 1998) or actin-binding proteins (Kost *et al.*, 1998).

There are additional important considerations when utilizing fluorescent proteins for these methods. As mentioned previously, it is desirable to show genetically that there has been no compromise in the functionality

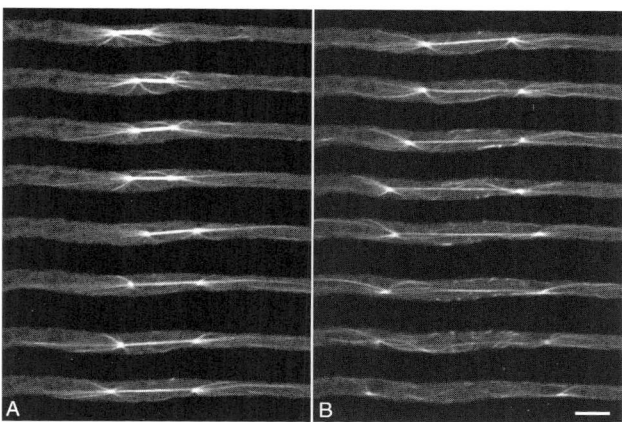

Figure 2.4. Mitosis in a hyphal tip cell of *M. grisea* expressing a β-tubulin-EYFP chimeric protein. Labeling microtubules in this fashion enabled time-resolved imaging of the mitotic spindle. Frames 13, 29, 33, 38, 57, 71, 75, and 81 (A, top to bottom), and 85, 99, 109, 122, 141, 149, 168, 206 (B, top to bottom) with a continual capture rate of 5.2 frames per second. Bar, 5 μm.

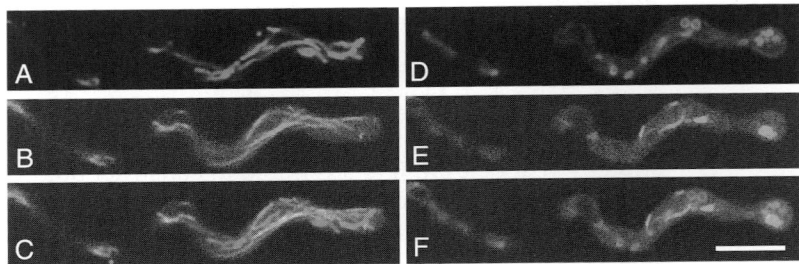

Figure 2.5. Multi-color imaging combining fluorescent chimeric proteins with a conventional fluorescent dye. Rhodamine B staining (A and D) and expression of a β-tubulin-EYFP chimeric protein (B and E) resulted in the visualization of both microtubules and mitochondria (overlays shown in C and F) in a living *M. grisea* germ tube treated for 0 (A–C) or 16.5 min (D–F) with the anti-microtubule agent benomyl at 20 ppm. Bar, 7 μm. (See colour plate 3.)

of the fusion protein. In addition, it is important that fusion proteins are targeted properly and not over- or under-expressed within the cell. Under-expression can result in a weak signal that is difficult or impossible to detect. Fluorescence speckle microscopy has developed as a clever way to observe protein kinetics in such situations, as was the case for tubulin in the spindle of *S. cerevisiae* (Maddox *et al.*, 2000). When under-expressed at the proper ratio, relative to endogenous tubulin, GFP-tagged tubulin molecules inserted infrequently enough into the microtubule polymer to create a speckled appearance that could be tracked during spindle development or astral microtubule formation. Over-expression of a fluorescent protein or fusion protein may lead to mistargeting within the cell, a decline in overall cell health, or lethality. Proteins that are non-functional or improperly targeted are more likely to perturb cells (Hawes *et al.*, 2001). Perhaps the best advice regarding an optimal expression level of a fluorescent fusion protein is that it should be as low as possible to generate a signal that is sufficient to be imaged easily (Sawin, 1999).

The choice of promoter to drive expression is probably the most important factor in controlling the level of expression of a fusion protein. For an overview of commonly used fungal promoters, see Lorang *et al.*, 2001; Pöggeler *et al.*, 2003; Spellig *et al.*, 1996. Maor *et al.* (1998) noted that basidiomycete promoters used by Spellig *et al.* (1996) for *Ustilago maydis* were not suitable for GFP expression in the ascomycetes *Cochliobolus heterostrophus*, *Trichoderma harzianum*, or *Neurospora crassa*. The choice of integration strategy and the number of copies of inserted genes also influence the ultimate level of cellular expression. Depending upon the limitations imposed by a given organism, plasmids can be designed to either integrate into the genome or exist as extra-nuclear self-replicating entities. Integration into the genome can be homologous, leading to gene replacement, or ectopic where both the wild type and fluorescent protein-tagged protein are expressed. Since the site of integration within the genome can affect gene expression, when feasible, directed ectopic integration into specific loci is recommended (Sawin, 1999).

In the realm of filamentous fungi, fluorescent protein fusions have been exploited extensively to study basic fungal cell biology in a select group of model organisms including *Aspergillus nidulans*, *N. crassa*, and *U. maydis*. For example, the roles of microtubules, motor molecules, and other microtubule-associated proteins in various aspects of cell growth and homeostasis have been investigated (Aist and Morris, 1999; Lee and Plamann, 2001; Straube *et al.*, 2003; Steinberg *et al.*, 2001), often facilitated by the identification of genes with *S. cerevisiae* homologs. Thorough coverage of these topics is beyond the scope of this review.

Tags of Subcellular Compartments

Fluorescent proteins have been targeted to a variety of subcellular compartments. Transformants expressing organelle-targeted fluorescent proteins can be extremely useful in monitoring the distribution and dynamics of these organelles during cellular events such as mitosis and cytokinesis, and during plant pathogenesis. In the absence of a signal peptide, and specific targeting signals, fluorescent proteins are synthesized in the fungal cytosol where they are generally excluded from most cellular organelles with the exception of the nucleus. Nuclear pores allowed for bi-directional diffusion of monomeric fluorescent proteins but the tetrameric nature of the RCFPs prevented their entry into the fungal nucleus during interphase (Figure 2.1).

Using a signal peptide for introduction of a fluorescent protein into the endoplasmic reticulum (ER), and an N-terminal KDEL ER-retention sequence, the fungal ER compartment has been successfully labeled (Fernández-Ábalos *et al.*, 1998; Suelmann *et al.*, 1997) and has been described as a network of branched tubules. We were unsuccessful in using such constructs to target the RCFPs ZsGreen and AmCyan to the ER (Figure 2.2). The linear fluorescence entities that we observed showed little mobility and were restricted to a region just behind the hyphal apex (Figure 2.2), suggesting that the fluorescent proteins were being degraded, possibly within the tubular vacuolar network (Cole *et al.*, 1998). We saw a similar aberrant pattern when the RCFPs were targeted for secretion with a signal peptide (unpublished). Substitution of AcGFP for the RCFPs in the ER-targeted construct resulted in a fluorescence pattern more representative of ER distribution (Figure 2.2).

Targeting sequences such as the 5-end of citrate synthase (Suelmann and Fischer, 2000) or the presequence-coding region of a mitochondrial ATP synthase (Fuchs *et al.*, 2002) have been used to direct fluorescent proteins to mitochondria. Similarly, we used the first 64 amino acids – including nine positively charged amino acids – of the nuclear-encoded malate dehydrogenase gene to tag mitochondria in both *Magnaporthe grisea* and *Fusarium verticillioides* (Figure 2.2). While we were able to label and track mitochondria using both the RCFP ZsGreen and AcGFP, the use of ZsGreen resulted in the formation of numerous intra-mitochondrial fluorescent aggregates and somewhat abnormal mitochondrial morphology in subapical cells (Figure 2.2). These aggregates were not observed when the same AcGFP- or EGFP-containing constructs were expressed and the morphology of mitochondria was typical of non-transformed living cells (Figure 2.2).

Fluorescent chimeric proteins destined for secretion have been observed in several fungi (Gordon *et al.*, 2000; Masai *et al.*, 2003). In filamentous fungi secretion appears to be the default pathway when proteins exit the Golgi in the absence of any vacuolar targeting sequence, i.e. when expression constructs contain a signal peptide but no additional targeting sequence (Conesa *et al.*, 2001). As mentioned previously we had difficulty in using RCFPs in this capacity. When *A. victoria*-derived fluorescent proteins were successfully secreted, there was an accumulation of fluorescence at the hyphal tip and in association with incipient septa (Gordon *et al.*, 2000).

A strategy used to target fluorescent proteins to the nucleus utilized a nuclear localization signal (NLS, Fernández-Ábalos *et al.*, 1998; Suelmann *et al.*, 1997; Maruyama *et al.*, 2002). The NLS was required for active transport into the nucleus via the nuclear pore complex. The NLS-containing fluorescent protein must be fused to another protein so that its molecular mass is large enough (>50 kDa) to prevent diffusion back out of the nucleus (Grebenok *et al.*, 1997).

Tools for Studying Protein Kinetics

Time-lapse imaging of steady state emission by a uniform population of fluorescing molecules cannot typically be used to follow their movement, or turnover, within cells. Fluorescence correlation microscopy (FCM), FRAP, and FLIP are time-lapse imaging methods that can characterize the mobility of tagged molecules within a cell (see Lippincott-Schwartz and Patterson, 2003). Measurement of the diffusion of fluorescent molecules and application of the appropriate algorithms can report on the molecular associations of these molecules within the cell. For example, quantification of the mobility dynamics might indicate whether a fluorescent molecule is being actively transported, merely diffusing, or a component of a multimeric protein complex.

The maturation kinetics and high stability of fluorescent proteins can be problematic since at any given point in time fluorescent proteins representing a whole range of "molecular ages" exist within a cell (e.g. newly synthesized unfolded, folded, early mature, late mature). Both destabilized fluorescent proteins (Li *et al.*, 1998) and a fluorescent protein mutant known as "fluorescent timer" (Terskikh *et al.*, 2000) were developed to study gene expression kinetics. Unfortunately, destabilized fluorescent proteins have proven to be very dim. Based on the RCFP DsRed, the spectral emission of fluorescent timer changes over time, showing green fluorescence initially and, over ca. 16 h, the fluorescence emission shifts to red. Thus, changes in expression are indicated by the ratio of green to red fluorescence: regions of cells that are green have recently initiated expression, regions that are yellow or orange have shown continuous expression and those that are red have ceased expression.

A promising approach for studying protein kinetics is the use of photoactivatable fluorescent proteins such as photoactivatable GFP (PA-GFP; Patterson and Lippincott-Schwartz, 2002), Kaede (Ando *et al.*, 2002), and kindling fluorescent protein (KFP1; Chudakov *et al.*, 2003). These fluorescent proteins have little inherent fluorescence, but become fluorescence competent when activated by exposure to specific wavelengths of light (Figure 2.2). Photoactivation produces a population of fluorescent molecules that can be monitored over time: independent of *de novo* synthesis, proteins that were not mature at the time of irradiation, and mature proteins that were not irradiated.

While fluorescent proteins tend to be very stable molecules, their fluorescence intensity may be influenced by the subcellular environment, in which they are expressed. For example, targeting fluorescent proteins to highly acidic compartments such as lysosomes can reduce or abolish fluorescence. Proteolytic effects – that often differ between compartments, e.g. secreted proteins may be subjected to high levels of proteolysis – folding efficiencies, and maturation kinetics must also be considered. Such influences probably accounted for the reduction in fluorescence when ER-targeted chimeras resided within the lumen of ER, as compared to strong emission from the cytosol when the chimeras were tethered to the ER by a transmembrane protein (Jennifer Lippincott-Schwartz, personal communication). Depending upon the application, the stability of fluorescent proteins can be considered desirable or a liability. For example, destabilized fluorescent proteins with rapid turnover have been engineered to allow meaningful measurements of protein kinetics (Li *et al.*, 1998).

Tools for Plant Pathology

Fluorescent proteins have vast potential as a tool for revealing many of the unknown and often complex aspects of plant pathogenic interactions. Plant pathologists traditionally have cultured organisms from infected regions of plant tissue and characterized the isolates for verification of the disease-causing organism and/or strain. This approach has been necessary because of the ubiquitous presence of a variety of fungal entities. When fungal-infected plants are examined under the microscope it remains often equivocal whether the observed fungus was truly the fungus applied by the researcher. The use of fungal transformants expressing cytoplasmic fluorescent proteins has enhanced greatly our ability to identify the experimental organism and monitor fungal ingress during plant pathogenesis or growth of microbes associated with biofilms (Davey and O'Toole, 2000) or soil (Bloemberg *et al.*, 2000; Egener *et al.*, 1998). For example, fluorescent protein-tagged infection hyphae of *M. grisea* in barley were noted during *in planta* imaging of fixed and living leaf tissue (Figure 2.6; Czymmek *et al.*, 2002). A similar approach was used successfully for *in vivo* evaluation of vascular wilt caused by *Fusarium oxysporum* root infection in *Arabidopsis* (Figure 2.6) or tomato (Lagopodi *et al.*, 2002). In *Cochliobolus heterostrophus*, the pathogen associated with Southern corn blight, GFP was used for *in planta* evaluation of fungal

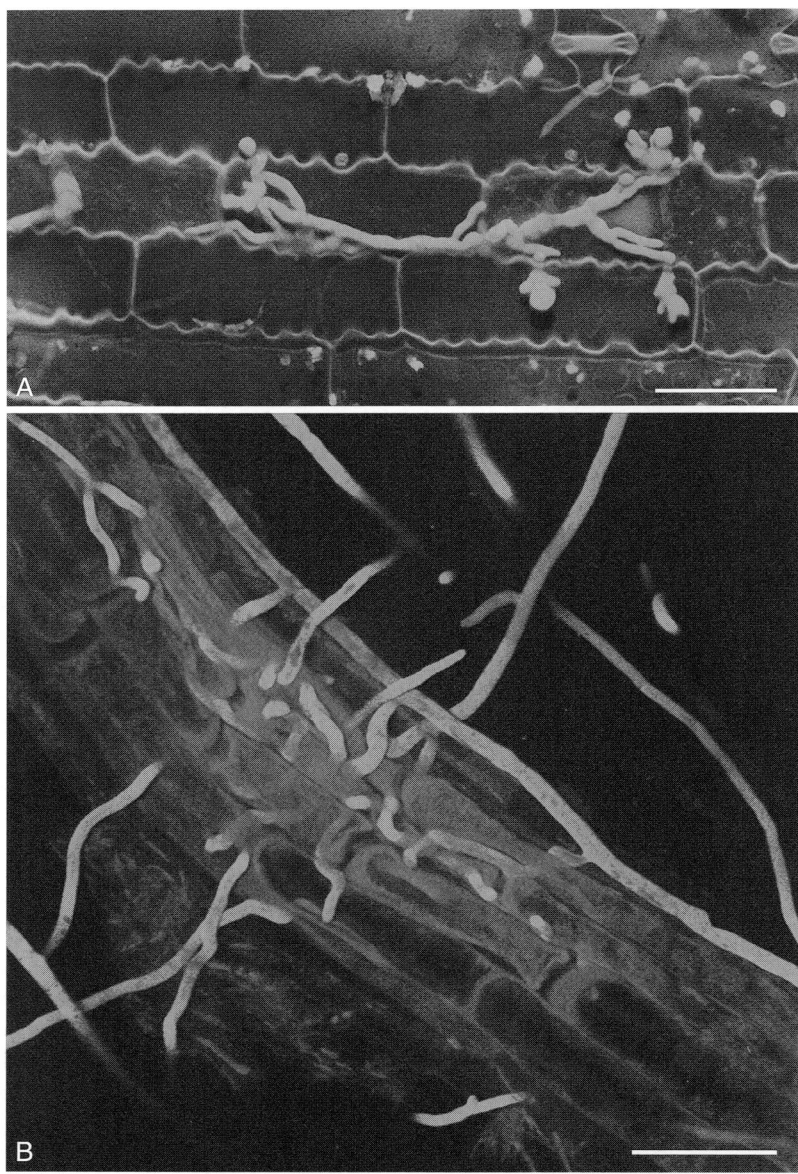

Figure 2.6. Two-channel multiphoton/confocal imaging of fluorescence-tagged pathogens *in planta*. Cytosolic expression of either AmCyan in *Colletotrichum graminicola* (A) or ZsGreen in *Fusarium oxysporum* (B) facilitated imaging of host–pathogen interactions within a maize leaf (A) or *Arabidopsis* root (B). Single photon excitation with either a 458 or 488 nm laser was sufficient to image AmCyan and ZsGreen, respectively. Significant autofluorescence of plant cell walls was generated in a separate channel by multiphoton excitation at 730 nm. Images A and B were based on a series of optical sections rendered as shadow projections using Zeiss LSM Image Visart 3D rendering software. Bar for A, 70 μm; B, 37 μm. (See colour plate 4.)

biomass in corn leaves (Maor *et al.*, 1998). Chen *et al.* (2003) demonstrated the feasibility of a GFP-based assay for quantification of *Colletotrichum* biomass from infected tobacco leaf extracts, correlating data with Northern blot analysis of actin expression. In fact, the success and versatility of GFP as a non-invasive vital cell marker for fungi has spurred many researchers to address aspects of pathogenesis with renewed vigor. GFP use in this capacity has facilitated the visualization of *Venturia inaequalis* during apple scab pathogenesis (Fitzgerald *et al.*, 2003) and *Mycosphaerella* sp. infection of banana leaves (Balint-Kurti *et al.*, 2001). Ecological studies of the biocontrol fungus *Clonostachys rosea* imaged in soil, vermiculite, carrot, and barley (Lübeck *et al.*, 2002), and fungal endophyte/host plant interactions with *Neotyphodium lolii* (Mikkelsen *et al.*, 2001), have benefited as well. Moreover, GFP expression in *F. oxysporum* allowed for the demonstration that a MAP kinase was required for root penetration and subsequent pathogenesis in tomato (Di Pietro *et al.*, 2001, 2003). Where fluorescent proteins have been exploited to visualize fungal pathogens *in planta*, the stability of the fluorescent proteins was advantageous especially because it allowed imaging of both metabolically active and comparatively inactive cells.

Clearly the use of a bright fluorescent cytoplasmic fungal tag has utility, however using this technology to study the expression of pathogenesis-related genes will provide additional and significant biological insight. For example, expression of the hydrophobin gene MPG1 in *M. grisea* was explored by creating a fusion protein with a destabilized GFP – as a means to monitor transient expression phenomena – and shown to be highly expressed during early infection stages (Soanes *et al.*, 2002). The isocitrate lyase promoter from *N. crassa* was used to monitor carbon metabolism during *Tapesia yallundae* pathogenesis of wheat (Bowyer *et al.*, 2000). Interestingly, the expression of this promoter was shown to occur only on the host surface, an observation that was confirmed with the same promoters for *Mycosphaerella graminicola* infection of wheat (Rohel *et al.*, 2001). Isshiki *et al.* (2003) utilized a GFP fusion to an endopolygalacturonase, or its promoter, to investigate the post-harvest citrus pathogen *Alternaria citri*. It was found that expression of this enzyme was induced by pectin in the peel but absent in the juice sac area and, significantly, disruption mutants failed to penetrate the peel. In *M. grisea* a fluorescent protein fusion in the form of a pathogenesis-related protein, Pth11-GFP, localized to the plasma membrane in a stage-specific manner (DeZwaan *et al.*, 1999). Finally, the use of biolistics for transient vacuolar peroxidase expression in leaves showed increased susceptibility to *Blumeria graminis* (Kristensen *et al.*, 2001) and proved valuable for monitoring the influence of antifungal proteins on powdery mildew infected barley leaves (Nielsen *et al.*, 1999).

Studies of fungal allies can benefit from these approaches in much the same way as filamentous fungi. Reported examples include GFP expressing *Phytophthora* species during invasion of *Arabidopsis*, to compare effectiveness of salicylic acid analogues in disease resistance (Si-Ammour *et al.*, 2003), and *P. palmivora* zoospores to study wound responses of ryegrass roots (van West *et al.*, 2003). In some instances, investigators used fluorescent protein technology to address the host-side of interactions only. For example, the subcellular re-organization of GFP-tagged tubulin, actin, endoplasmic reticulum, and Golgi was studied in *Arabidopsis* leaf cells during reaction to the oömycete pathogen *Peronospora parasitica* and non-pathogenic *Phytophthora sojae* (Takemoto et al., 2003).

Creating fluorescent protein fusions with genes that are thought to play key roles during microbe–plant interactions will reveal spatio-temporal expression of these genes at a resolution unattainable by molecular extraction methods. Undoubtedly, the application of fluorescent protein color variants to address multiple targets, e.g. ECFP and EYFP for *Rhizobium*–root interactions (Stuurman et al., 2000), will permit not only the exploration of interrelationships between heterogeneous populations of cells, and protein interactions within individual cells, but also *in vivo* gene expression and protein localization patterns from both the host and pathogen.

Another area with significant potential is the use of fluorescent proteins as reporters for enhancer-trapping (Gonzalez and Bejarano, 2000) protocols for gene discovery. For example, random genome insertion of GFP under control of a basal promoter was used recently to identify *U. maydis* genes expressed only during host interactions (Aichinger et al., 2003). Other innovative genetic approaches are sure to develop in this rapidly evolving discipline.

Measuring Protein–Protein Interactions Using FRET

Protein–protein interactions play a key role in the structural and functional organization of living cells. Identification of interactions and characterization of their physiological significance is a prime directive in multidisciplinary biological research (Mendelsohn and Brent, 1999) as we transcend structural genomics and proceed toward the next logical step of functional genomics (Sweigard and Ebole, 2001). FRET, a technique used to measure the relative distance between molecules based on the non-radiative transfer of energy between an excited donor fluorophore and an acceptor molecule, has proven to be an important tool for exploring protein–protein interactions (Sekar and Periasamy, 2003) and has been underutilized with filamentous fungi. The efficiency of FRET is a function of the distance of separation between the donor and acceptor, occurring only if the proximity is less than 10 nm, and can be measured using either fluorescence intensity or fluorescence lifetime (Hanley et al., 2001). The latter requires dedicated instrumentation that has only recently become commercially available, and both types of measurements require specific corrections and numerous controls to generate meaningful data (Xia and Liu, 2001). FRET is appropriate for assessing inter- or intra-molecular proximity changes that occur during binding, association, conformational shifts, diffusion and catalysis within cells (Jares-Erijman and Jovin, 2003).

Early FRET studies used exogenous proteins covalently labeled with fluorophore tags. The advent of genetically encoded fluorescent protein tags has simplified the introduction of tagged proteins into cells, and facilitated the engineering of sophisticated physiological indicators commonly referred to as FRET-based biosensors (Meyer and Teruel, 2003; Zaccolo and Pozzan, 2000). The first such biosensor molecule was "chameleon", a calcium ion indicator utilizing calmodulin (Miyawaki *et al.*, 1997; Allen *et al.*, 1999). This molecule was designed so that the conformational change that followed calcium ion binding was sufficient to bring the component, intramolecular fluorescent proteins of a FRET pair close enough for an energy transfer from donor to acceptor. Thus, an increase in calcium ion concentration was indicated by an increase in FRET efficiency.

◆◆◆◆◆◆ NEW DEVELOPMENTS AND FUTURE PERSPECTIVES

FRET-based Biosensors

Analysis of the spatio-temporal dynamics of biochemical reactions in single cells is possible now using single gene FRET-based biosensors. They are of particular interest in the area of cellular communication and signal transduction (Miyawaki, 2003). Protein phosphorylation by specific intracellular kinases has been shown to play critical roles during signal transduction (Hunter, 2000). Towards that end FRET-based biosensors, called "phocuses", have been developed to monitor phosphorylation in real time (Sato *et al.*, 2002). These phocus indicators can be tailor-made to detect the activity of specific kinases, and are engineered as a tandem fusion protein with four domains – fluorescent protein/FRET-donor, substrate domain, phosphorylation recognition domain, and fluorescent protein/FRET-acceptor – with a flexible amino acid linker between the substrate and recognition domains. Phosphorylation of the substrate domain by the proper kinase facilitates its binding to the recognition domain. This produces a conformational change in the phocus molecule, bringing the FRET donor and acceptor into molecular proximity, allowing FRET to occur. Other biosensors have been developed to measure the second messengers cAMP (Zaccolo *et al.*, 2000) and cGMP (Honda *et al.*, 2001; Sato *et al.*, 2000).

Development of FRET-based indicators for plant pathology will require a substantial increase in our fundamental understanding of the biochemistry of these interactions, but the possibilities are intriguing. Gene-for-gene interactions involve a highly specific interplay between pathogen and host, interactions where the spatio-temporal details of gene product deployment will likely determine the outcome of the interaction. FRET has been used to investigate protein cleavage by specific proteases, including proteases that are active during programmed cell death (Xu *et al.*, 1998; Luo *et al.*, 2001). Though not a regular phenomenon associated with plant cells, programmed cell death does occur in plants (Pennell and Lamb, 1997) and its manifestation in the hypersensitive response during pathogen attack is thought to play an essential role in plant defense

(Heath, 2000). Thus, biosensors could be designed as indicators of different phases of the hypersensitive response.

Bimolecular Fluorescence Complementation

Complementary fragments of fluorescent proteins, that intrinsically lack fluorescence, have been engineered and used successfully as fusion proteins that fluoresce when brought together to form dimers (Hu and Kerppola, 2003). This approach, known as bimolecular fluorescence complementation (BiFC), has been expanded to "multicolor BiFC" by utilizing fragments from three different fluorescent proteins to generate an assortment of complementation complexes. Each complex has distinct spectral characteristics (i.e. excitation and emission maxima) and can be differentiated in the microscope. Thus, by tagging each of a set of proteins with a unique fluorescent protein fragment one can examine a range of protein–protein interactions in live cells in a time-resolved fashion. The number of complexes that could be imaged simultaneously would be determined by the spectral resolution of the imaging system and should be enhanced by recent advances in spectral imaging (Dickinson *et al.*, 2001; Zimmermann *et al.*, 2002). This ability to monitor multiple interactions may be crucial for understanding a biological system since cellular responses are controlled by a complicated web of protein interactions. Even for monitoring single protein–protein interactions the BiFC approach has a major practical advantage over FRET – any fluorescence above background would be significant.

Laser Capture Microdissection

Laser capture microdissection is a powerful tool that enables researchers to select carefully and remove individual or groups of cells, even subcellular compartments, for genetic and protein analysis. The number of applications using this technology is increasing rapidly, especially in biomedical research where the precise dissection of target cells from within a complex and heterogeneous population of cells is difficult (Michener *et al.*, 2002). A similarly complex situation exists in fungal-infected plant tissue during pathogenesis where traditional biochemical extraction methods are of limited utility. If sufficient quantities of specific cells are collected – typically, ranging from several hundred to several thousand (Kerk *et al.*, 2003) – significant genetic information can be derived *without* contamination from undesired cellular material.

Commercially available LCM instruments vary with regard to the means of specimen microdissection. A system developed by the National Institutes of Health uses NIR light to melt an overlaid plastic polymer, selectively coating underlying cellular contents that subsequently are removed when the polymer is pulled away from the sample. A second LCM system uses a pulsed nitrogen laser (337 nm) focused to a spot as small as 1 μm in diameter, to ablate the perimeter of selected regions by cold laser irradiation. The laser light is then defocused, directly below the selected region, and a pulse of light catapults the tissue into a microfuge tube cap for collection. Both systems have a software interface to allow users to employ region-of-interest tools, with a computer-controlled robotic stage, to define the areas to be removed for analysis. For plant material, samples for microdissection are typically either cryo-sections (Asano *et al.*, 2002; Nakazono *et al.*, 2003) or de-embedded paraffin sections (Kerk *et al.*, 2003). To facilitate laser catapulting the sections can be air-dried onto a thin membrane support surface. Microaspiration is an alternative method to LCM for removal of selected regions and has, for example, been used to sample the cytoplasm from esophageal gland cells of the soybean cyst nematode *Heterodera glycines* during root infection (Wang *et al.*, 2001).

We have used a fluorescent protein-expressing strain of *M. grisea* to facilitate the selection of infection hyphae from infected host plant material (Figure 2.7). Laser ablation was capable of cutting through tenacious plant cell walls to allow for selection of infection hyphae within

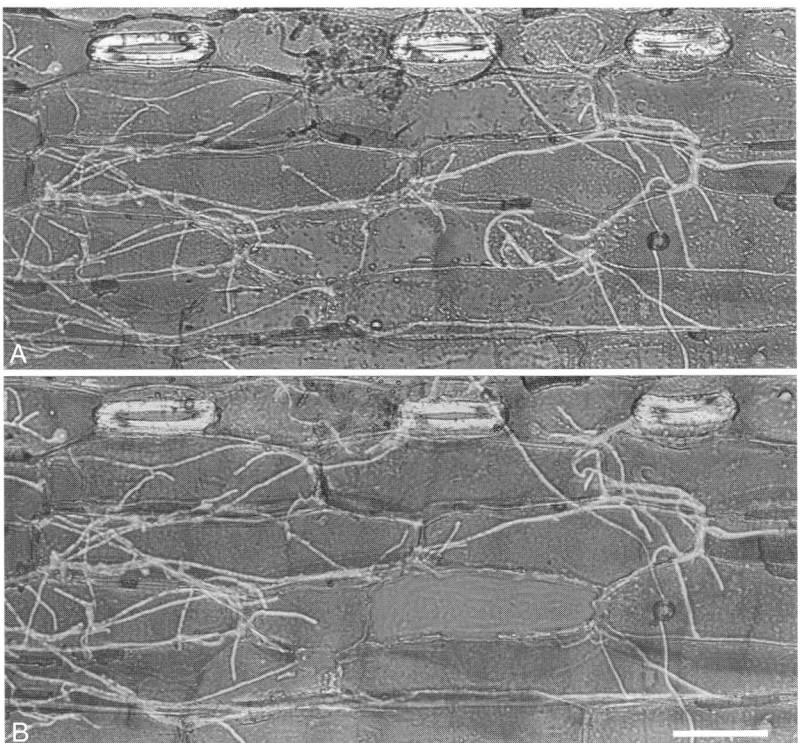

Figure 2.7. Fluorescent protein-assisted laser microdissection. Expression of ZsGreen in the cytoplasm of *M. grisea* facilitated the identification of infection hyphae within barley leaf epidermal cells (A). One capture event recovered the contents of a single infected epidermal cell in a two-step procedure combining laser ablation and separation (B, outlined by red dots). A second ablation, lower and to the left (also outlined by red dots), showed the location where a tip of a single infection hypha was precisely excised and captured. Bar, 45 µm. (See colour plate 5.)

a single epidermal cell at a precise stage during the host–pathogen interaction. This approach eliminated "contaminating" material from surrounding plant and fungal cells. To evaluate the relevant gene expression of host cells, plant material could also be captured. For a comparative gene expression study, LCM was used to selectively capture hyphae from both the sheath and Hartig net of an ectomycorrhizal association between *Amanita muscaria* and *Populus* roots. As measured by RT-PCR, expression of a hexose transporter was elevated in the Hartig net while phenylalanine ammonium lyase was highly expressed in the sheath (Nehls et al., 2001).

Four-dimensional Analysis

The acquisition of 3D data sets over time, often referred to as 4D imaging, provides an enormous new avenue for exploration and documentation of fluorescent protein localization. Three-dimensional data collection has been an effective way for evaluating thick or even relatively thin samples since all cells are inherently 3D. A tremendous amount of information can be obtained from even a single 3D "snapshot". Two-dimensional *in vivo* time course documentation of fluorescent proteins has become routine and potent for recording interactions between targeted proteins, especially when using co-transformation with spectral fluorescent protein mutants. Four-dimensional imaging has provided a wealth of information from which relevant data can be extracted with the appropriate software (Thomas et al., 1996). Fluorescent protein labeled structures such as nuclei can be tracked and measured in 4D with relative ease, as was done in *Arabidopsis* root hairs (van Bruaene et al., 2003). Similarly, fungal hyphae expressing fluorescent proteins in the cytosol were imaged in 4D during invasion of roots (Fogg and Czymmek, unpublished) and leaves (Czymmek et al., 2002). Such images have provided an extraordinary glimpse into individual host–pathogen encounter sites and a sense of the potential of 4D microscopy in exploring underlying molecular mechanisms of resistance and susceptibility.

◆◆◆◆◆◆ CONCLUSIONS

Though many of the fluorescent protein-based approaches mentioned here have not been used with filamentous fungi, we feel that exploitation of the full repertoire of techniques will, eventually, advance the science all the more. The potential growth in use of these technologies for fungi is reflected in the increase in the number of citations in the literature over the last few years. Fluorescent proteins are being used in a variety of fungal research areas from basic cell biology to biotechnology (Siedenberg et al., 1999; Wang et al., 2003) to ecological studies of plant symbionts (Bergero et al., 2003) and soil-inhabiting fungi. In plant pathology fluorescent proteins are proving their worth as reporters for identifying and visualizing pathogens *in planta*, and by facilitating the analysis of gene expression of both pathogen and host during interactions. In the near future, one could easily envision their use with high-throughput technologies for rapid screening of mutants (Avila *et al.*, 2003) or prospective fungicides.

Advances in genome sequencing and bioinformatics will have a profound impact and lead to a dramatic increase in the use of fluorescent proteins. With the elucidation of the complete genome sequence for a number of filamentous fungi, with several additional imminent (see Whitehead Institute Fungal Genome Initiative website: http://www.broad.mit.edu/annotation/fungi/fgi/), the stage is now set for fungi in the functional genomics era (Bennett and Arnold, 2001). In combination with the myriad of fluorescent protein-based techniques currently available, the next decade should be a very exciting time for fungal cell biology and plant pathology. This potential marriage is exemplified by a recent publication in which the subcellular distribution of the majority of the *S. cerevisiae* proteome had been mapped systematically using fluorescent protein chimeras (Ghaemmaghami *et al.*, 2003). It is likely that such global studies will point researchers in the right direction, but the verification of putative molecular mechanisms will require diligent effort with an emphasis on *biology*.

Acknowledgements

We thank James A. Sweigard (DuPont Crop Genetics Research and Development) for donating transformants of *Magnaporthe grisea*, *Fusarium verticillioides*, and *F. oxysporum*, and Dr. Todd DeZwaan (currently at Paradigm Genetics, Inc.) and Jonathon Anobile (University of Delaware) for construction of β-tubulin::EYFP and mitochondrial-targeted PA-GFP constructs, respectively. Melissa Fogg prepared transformants of *F. oxysporum* generated in the laboratory of James A. Sweigard and assisted with the imaging of infected root samples. We also express our gratitude to Dr. Seogchan Kang (Pennsylvania State University) for providing samples of *F. oxysporum* and *Arabidopsis thaliana*, and Dr. Alexandre Conceicao (DuPont Crop Genetics Research and Development) for engineering fluorescent protein-expressing transformants of *Colletorichum graminicola* and supplying infected maize leaves.

References

Adams, S. R., Campbell, R. E., Gross, L. A., Martin, B. R., Walkup, G. K., Yao, Y., Llopis, J. and Tsien, R. Y. (2002). New biarsenical ligands and tetracysteine motifs for protein labeling in vitro and in vivo: synthesis and biological applications. *J. Am. Chem. Soc.* **124**, 6063–6076.

Aichinger, C., Hansson, K., Eichhorn, H., Lessing, F., Mannhaupt, G., Mewes, W. and Kahmann, R. (2003). Identification of plant-regulated genes in *Ustilago maydis* by enhancer-trapping mutagenesis. *Mol. Gen. Genomics* **270**, 303–314.

Aist, J. R. and Morris, N. R. (1999). Mitosis in filamentous fungi: how we got where we are. *Fungal Genet. Biol.* **27**, 1–25.

Allen, R. D., Travis, J. L., Allen, N. S. and Yilmaz, H. (1981a). Video-enhanced contrast polarization (AVEC-POL) microscopy: a new method applied to the detection of birefringence in the motile reticulopodial network of *Allogromia laticollaris*. *Cell Motil.* **1**, 275–289.

Allen, R. D., Allen, N. S. and Travis, J. L. (1981b). Video-enhanced contrast, differential interference contrast (AVEC-DIC) microscopy: a new method capable of analyzing microtubule-related motility in the reticulopodial network of *Allogromia laticollaris*. *Cell Motil.* **1**, 291–302.

Allen, G. J., Kwak, J. M., Chu, S. P., Llopis, J., Tsien, R. Y., Harper, J. F. and Schroeder, J. I. (1999). Cameleon calcium indicator reports cytoplasmic calcium dynamics in *Arabidopsis* guard cells. *Plant J.* **19**, 735–747.

Ando, R., Hama, H., Yamamoto-Hino, M., Mizuno, H. and Miyawaki, A. (2002). An optical marker based on the UV-induced green-to-red photoconversion of a fluorescent protein. *Proc. Natl. Acad. Sci. USA* **99**, 12651–12656.

Apoga, D., Barnard, J., Craighead, H. G. and Hoch, H. C. (2004). Quantification of substratum contact required for initiation of *Colletotrichum graminicola* appressoria. *Fungal Genet. Biol.* **41**, 1–12.

Asano, T., Masumura, T., Kusano, H., Kikuchi, S., Kurita, A., Shimada, H. and Kadowaki, K. (2002). Construction of a specialized cDNA library from plant cells isolated by laser capture microdissection: toward comprehensive analysis of the genes expressed in the rice phloem. *Plant J.* **32**, 401–408.

Avila, E. L., Zouhar, J., Agee, A. E., Carter, D. G., Chary, S. N. and Raikhel, N. V. (2003). Tools to study plant organelle biogenesis. Point mutation lines with disrupted vacuoles and high-speed confocal screening of green fluorescent protein-tagged organelles. *Plant Physiol.* **133**, 1673–1676.

Bae, J. H., Rubini, M., Jung, G., Wiegand, G., Seifert, M. H. J., Azim, M. K., Kim, J.-S., Zumbusch, A., Holak, T. A., Moroder, L., Huber, R. and Budisa, N. (2003). Expansion of the genetic code enables design of a novel "gold" class of green fluorescent proteins. *J. Mol. Biol.* **328**, 1071–1081.

Baird, G. S., Zacharias, D. A. and Tsien, R. Y. (2000). Biochemistry, mutagenesis, and oligomerization of DsRed, a red fluorescent protein from coral. *Proc. Natl. Acad. Sci. USA* **97**, 11984–11989.

Balint-Kurti, P. J., May, G. D. and Churchill, A. C. L. (2001). Development of a transformation system for *Mycosphaerella* pathogens of banana: a tool for the study of host/pathogen interactions. *FEMS Microbiol. Lett.* **195**, 9–15.

Bennet, J. W. and Arnold, J. (2001). Genomics for fungi. In *The Mycota* (R. J. Howard and N. A. R. Gow, eds), vol. VIII, pp. 267–297. Springer, New York.

Bergero, R., Harrier, L. A. and Franken, P. (2003). Reporter genes: applications to the study of arbuscular mycorrhizal (AM) fungi and their symbiotic interactions with plant roots. *Plant Soil* **255**, 143–155.

Bevis, B. J. and Glick, B. S. (2002). Rapidly maturing variants of the *Discosoma* red fluorescent protein (DsRed). *Nat. Biotechnol.* **20**, 83–87.

Bloemberg, G. V., Wijfjes, A. H. M., Lamers, G. E. M., Stuurman, N. and Lugtenberg, B. J. J. (2000). Simultaneous imaging of *Pseudomonas fluorescens* WCS365 populations expressing three different autofluorescent proteins in the rhizosphere: new perspectives for studying microbial communities. *Mol. Plant-Microbe Interact.* **13**, 1170–1176.

Bourett, T. M., Sweigard, J. A., Czymmek, K. J., Carroll, A. and Howard, R. J. (2002). Reef coral fluorescent proteins for visualizing fungal pathogens. *Fungal Genet. Biol.* **37**, 211–220.

Bowyer, P., Mueller, E. and Lucas, J. (2000). Use of an isocitrate lyase promoter-GFP fusion to monitor carbon metabolism of the plant pathogen *Tapesia yallundae* during infection of wheat. *Mol. Plant Pathol.* **1**, 253–262.

Brock, R., Hamelers, I. H. and Jovin, T. M. (1999). Comparison of fixation protocols for adherent cultured cells applied to a GFP fusion protein of the epidermal growth factor receptor. *Cytometry* **35**, 353–362.

Bulina, M. E., Verkhusha, V. V., Staroverov, D. B., Chudakov, D. M. and Lukyanov, K. A. (2003). Hetero-oligomeric tagging diminishes non-specific aggregation of target proteins fused with Anthozoa fluorescent proteins. *Biochem. J.* **371**, 109–114.

Campbell, R. E., Tour, O., Palmer, A. E., Steinbach, P. A., Baird, G. S., Zacharias, D. A. and Tsien, R. Y. (2002). A monomeric red fluorescent protein. *Proc. Natl. Acad. Sci. USA* **99**, 7877–7882.

Carminati, J. L. and Stearns, T. (1999). Cytoskeletal dynamics in yeast. In *Green Fluorescent Proteins. Methods in Cell Biology* (K. F. Sullivan and S. A. Kay, eds), vol. 58, pp. 87–105. Academic Press, New York.

Chalfie, M., Tu, Y., Euskirchen, G., Ward, W. W. and Prasher, D. C. (1994). Green fluorescent protein as a marker for gene expression. *Science* **263**, 802–805.

Chen, N., Hsiang, T. and Goodwin, P. H. (2003). Use of green fluorescent protein to quantify the growth of *Colletotrichum* during infection of tobacco. *J. Microbiol. Meth.* **53**, 113–122.

Chial, H. J., Rout, M. P., Giddings, Jr. T. H. and Winey, M. (1998). *Saccharomyces cerevisiae* Ndc1p is a shared component of nuclear pore complexes and spindle pole bodies. *J. Cell Biol.* **143**, 1789–1800.

Chudakov, D. M., Belousov, V. V., Zaraisky, A. G., Novoselov, V. V., Staroverov, D. B., Zorov, D. B., Lukyanov, S. and Lukyanov, K. A. (2003). Kindling fluorescent proteins for precise in vivo photolabeling. *Nat. Biotechnol.* **21**, 191–194.

Cole, L., Orlovich, D. A. and Ashford, A. E. (1998). Structure, function and motility of vacuoles in filamentous fungi. *Fungal Genet. Biol.* **24**, 86–100.

Conesa, A., Punt, P. J., van Luijk, N. and van den Hondel, C. A. M. J. J. (2001). The secretion pathway in filamentous fungi: a biotechnological view. *Fungal Genet. Biol.* **33**, 155–171.

Cormack, B. (1998). Green fluorescent protein as a reporter of transcription and protein localization in fungi. *Curr. Opin. Microbiol.* **1**, 406–410.

Cubitt, A. B., Woollenweber, L. A. and Heim, R. (1999). Understanding structure–function relationships in the *Aequorea victoria* green fluorescent protein. In *Green Fluorescent Proteins. Methods in Cell Biology* (K. F. Sullivan and S. A. Kay, eds), vol. 58, pp. 19–30. Academic Press, New York.

Czymmek, K. J. (2005). Exploring fungal activity with confocal and multiphoton microscopy. In *The Fungal Community: Its Organization and Role in the Community* (J. Dighton, P. Oudemans and J. White, eds). Marcel Dekker, New York, in press.

Czymmek, K. J., Whallon, J. W. and Klomparens, K. L. (1994). Confocal microscopy in mycological research. *Exp. Mycol.* **18**, 275–293.

Czymmek, K. J., Bourett, T. M., Sweigard, J. A., Carroll, A. and Howard, R. J. (2002). Utility of cytoplasmic fluorescent proteins for live-cell imaging of *Magnaporthe grisea* in planta. *Mycologia* **94**, 280–289.

Davey, M. E. and O'Toole, G. A. (2000). Microbial biofilms: from ecology to molecular genetics. *Microbiol. Mol. Biol. Rev.* **64**, 847–867.

DeZwaan, T. M., Carroll, A. M., Valent, B. and Sweigard, J. A. (1999). *Magnaporthe grisea* Pth11p is a novel plasma membrane protein that mediates appressorium differentiation in response to inductive substrate cues. *Plant Cell* **11**, 2013–2030.

Dickinson, M. E., Bearman, G., Tille, S., Lansford, R. and Fraser, S. E. (2001). Multi-spectral imaging and linear unmixing add a whole new dimension to laser scanning fluorescence microscopy. *BioTechniques* **31**, 1272–1278.

Di Pietro, A., Garcia-MacEira, F. I., Meglecz, E. and Roncero, M. I. G. (2001). A MAP kinase of the vascular wilt fungus *Fusarium oxysporum* is essential for root penetration and pathogenesis. *Mol. Microbiol.* **39**, 1140–1152.

Di Pietro, A., Madrid, M. P., Caracuel, Z., Delgado-Jarana, J. and Roncero, M. I. G. (2003). *Fusarium oxysporum*: exploring the molecular arsenal of a vascular wilt fungus. *Mol. Plant Pathol.* **4**, 315–325.

Doyle, T. and Botstein, D. (1996). Movement of yeast cortical actin cytoskeleton visualized *in vivo*. *Proc. Natl. Acad. Sci. USA* **93**, 3886–3891.

Dumas, B., Centis, S., Sarrazin, N. and Esquerre-Tugaye, M. T. (1999). Use of green fluorescent protein to detect expression of an endopolygalacturonase gene of *Colletotrichum lindemuthianum* during bean infection. *Appl. Environ. Microbiol.* **65**, 1769–1771.

Egener, T., Hurek, T. and Reinhold-Hurek, B. (1998). Use of green fluorescent protein to detect expression of *nif* genes of *Azoarcus* sp. BH72, a grass-associated diazotroph, on rice roots. *Mol. Plant-Microbe Interact.* **11**, 71–75.

Ellenberg, J., Lippincott-Schwartz, J. and Presley, J. F. (1998). Two-color green fluorescent protein time-lapse imaging. *BioTechniques* **25**, 838–846.

Fernández-Ábalos, J. M., Fox, H., Pitt, C., Wells, B. and Doonan, J. H. (1998). Plant-adapted green fluorescent protein is a versatile vital reporter for gene expression, protein localization and mitosis in the filamentous fungus, *Aspergillus nidulans*. *Mol. Microbiol.* **27**, 121–130.

Fitzgerald, A. M., Mudge, A. M., Gleave, A. P. and Plummer, K. M. (2003). *Agrobacterium* and PEG-mediated transformation of the phytopathogen *Venturia inaequalis*. *Mycol. Res.* **107**, 803–810.

Fradkov, A. F., Verkhusha, V. V., Staroverov, D. B., Bulina, M. E., Yanushevich, Y. G., Martynov, V. I., Lukyanov, S. and Lukyanov, K. A. (2002). Far-red fluorescent tag for protein labelling. *Biochem. J.* **368**, 17–21.

Fuchs, F., Prokisch, H., Neupert, W. and Westermann, B. (2002). Interaction of mitochondria with microtubules in the filamentous fungus *Neurospora crassa*. *J. Cell Sci.* **115**, 1931–1937.

Gaietta, G., Deerinck, T. J., Adams, S. R., Bouwer, J., Tour, O., Lair, D. W., Sosinsky, G. E., Tsien, R. Y. and Ellisman, M. H. (2002). Multicolor and electron microscopic imaging of connexin trafficking. *Science* **296**, 503–507.

Ghaemmaghami, S., Huh, W.-K., Bower, K., Howson, R. W., Belle, A., Dephoure, N., O'Shea, E. K. and Weissman, J. S. (2003). Global analysis of protein expression in yeast. *Nature* **425**, 737–741.

Gold, S. E., Duick, J. W., Redman, R. S. and Rodriquez, R. J. (2000). Molecular transformation, gene cloning, and gene expression systems for filamentous fungi. (G. G. Khachatourians and D. P. Arora, eds), *Applied Mycology and Technology*, vol. 1, pp. 199–238. Elsevier Science, Oxford.

Gonzalez, C. and Bejarano, L. A. (2000). Protein traps: using intracellular localization for cloning. *Trends Cell Biol.* **10**, 162–165.

Gordon, C. L., Archer, D. B., Jeenes, D. J., Doonan, J. H., Wells, B., Trinci, A. P. J. and Robson, G. D. (2000). A glucoamylase:GFP gene fusion to study protein secretion by individual hyphae of *Aspergillus niger*. *J. Microbiol. Meth.* **42**, 39–48.

Grebenok, R. J., Pierson, E., Lambert, G. M., Gong, F.-C., Afonso, C. L., Haldeman-Cahill, R., Carrington, J. C. and Galbraith, D. W. (1997). Green-fluorescent protein fusions for efficient characterization of nuclear targeting. *Plant J.* **11**, 573–586.

Greulich, K. O. (2001). Micromanipulation by laser microbeam and optical tweezers. In *Plant Cell Biology* (C. Hawes and B. Satiat-Jeunemaitre, eds), pp. 87–105. Academic Press, New York.

Gurskaya, N. G., Fradkov, A. F., Terskikh, A., Matz, M. V., Labas, Y. A., Martynov, V. I., Yanushevich, Y. G., Lukyanov, K. A. and Lukyanov, S. A. (2001). GFP-like chromoproteins as a source of far-red fluorescent proteins. *FEBS Lett.* **507**, 16–20.

Gurskaya, N. G., Fradkov, A. F., Pounkova, N. I., Staroverov, D. B., Bulina, M. E., Yanushevich, Y. G., Labas, Y. A., Lukyanov, S. and Lukyanov, K. A. (2003). A colourless green fluorescent protein homologue from the non-fluorescent hydromedusa *Aequorea coerulescens* and its fluorescent mutants. *Biochem. J.* **373**, 403–408.

Han, G., Liu, B., Zhang, J., Zuo, W., Morris, N. R. and Xiang, X. (2001). The *Aspergillus* cytoplasmic dynein heavy chain and NUDF localize to microtubule ends and affect microtubule dynamics. *Curr. Biol.* **11**, 719–724.

Hanley, Q. S., Subramaniam, V., Arndt-Jovin, D. J. and Jovin, T. M. (2001). Fluorescence lifetime imaging: multi-point calibration, minimum resolvable differences, and artifact suppression. *Cytometry* **43**, 248–260.

Harrier, L. A. and Millam, S. (2001). Biolistic transformation of arbuscular mycorrhizal fungi – progress and perspectives. *Mol. Biotechnol.* **18**, 25–33.

Haseloff, J. (1999). GFP variants for multispectral imaging of living cells. In *Green Fluorescent Proteins. Methods in Cell Biology* (K. F. Sullivan and S. A. Kay, eds), vol. 58, pp. 139–151. Academic Press, New York.

Haseloff, J., Siemering, K. R., Prasher, D. C. and Hodge, S. (1997). Removal of a cryptic intron and subcellular localization of green fluorescent protein are required to mark transgenic *Arabidopsis* plants brightly. *Proc. Natl. Acad. Sci. USA* **94**, 2122–2127.

Hawes, C., Saint-Jore, C. M., Brandizzi, F., Zheng, H., Andreeva, A. and Boevink, P. (2001). Cytoplasmic illuminations: in planta targeting of fluorescent proteins to cellular organelles. *Protoplasma* **215**, 77–88.

Heath, M. C. (2000). Hypersensitive response-related death. *Plant Mol. Biol.* **44**, 321–334.

Heim, R. and Tsien, R. Y. (1996). Engineering green fluorescent protein for improved brightness, longer wavelengths and fluorescence resonance energy transfer. *Curr. Biol.* **6**, 178–182.

Hiraoka, Y., Shimi, T. and Haraguchi, T. (2002). Multispectral imaging fluorescence microscopy for living cells. *Cell Struct. Funct.* **27**, 367–374.

Honda, A., Adams, S. R., Sawyer, C. L., Lev-Ram, V., Tsien, R. Y. and Dostmann, W. R. G. (2001). Spatiotemporal dynamics of guanosine 3′,5′-cyclic monophosphate revealed by genetically encoded, fluorescent indicator. *Proc. Natl. Acad. Sci. USA* **98**, 2437–2442.

Hu, C.-D. and Kerppola, T. K. (2003). Simultaneous visualization of multiple protein interactions in living cells using multicolor fluorescence complementation analysis. *Nat. Biotechnol.* **21**, 539–545.

Hunter, T. (2000). Signaling – 2000 and beyond. *Cell* **100**, 113–127.

Inoué, S. (1981). Video image processing greatly enhances contrast, quality, and speed in polarization-based microscopy. *J. Cell Biol.* **89**, 346–356.

Inoué, S. (1986). *Video Microscopy*. Plenum Press, New York, 584 pp.

Isshiki, A., Ohtani, K., Kyo, M., Yamamoto, H. and Akimitsu, K. (2003). Green fluorescent detection of fungal colonization and endopolygalacturonase gene expression in the interaction of *Alternaria citri* with citrus. *Phytopathology* **93**, 768–773.

Jares-Erijman, E. A. and Jovin, T. M. (2003). FRET imaging. *Nat. Biotechnol.* **21**, 1387–1394.

Kerk, N. M., Ceserani, T., Tausta, S. L., Sussex, I. M. and Nelson, T. M. (2003). Laser capture microdissection of cells from plant tissues. *Plant Physiol.* **132**, 27–35.

Knechtle, P., Dietrich, F. and Philippsen, P. (2003). Maximal polar growth potential depends on the polarisome component AgSpa2 in the filamentous fungus *Ashbya gossypii*. *Mol. Biol. Cell* **14**, 4140–4154.

Knoblauch, M., Hibberd, J. M., Gray, J. C. and van Bel, A. J. E. (1999). A galistan expansion femtosyringe for microinjection of eukaryotic organelles and prokaryotes. *Nat. Biotechnol.* **17**, 906–909.

Kost, B., Spielhofer, P. and Chua, N.-H. (1998). A GFP-mouse talin fusion protein labels plant actin filaments *in vivo* and visualizes the actin cytoskeleton in growing pollen tubes. *Plant J.* **16**, 393–401.

Kristensen, B. K., Ammitzbøll, H., Rasmussen, S. K. and Nielsen, K. A. (2001). Transient expression of a vacuolar peroxidase increases susceptibility of epidermal barley cells to powdery mildew. *Mol. Plant Pathol.* **2**, 311–317.

Kwon, Y. H., Wells, K. S. and Hoch, H. C. (1993). Fluorescence confocal microscopy: applications in fungal cytology. *Mycologia* **85**, 721–733.

Lagopodi, A. L., Ram, A. F. J., Lamers, G. E. M., Punt, P. J., van den Hondel, C. A. M. J. J., Lugtenberg, B. J. J. and Bloemberg, G. V. (2002). Novel aspects of tomato root colonization and infection by *Fusarium oxysporum* f. sp. *Radicis-lycopersici* revealed by confocal laser scanning microscopic analysis using the green fluorescent protein as a marker. *Mol. Plant-Microbe Interact.* **15**, 172–179.

Lee, H. and Plamann, M. (2001). Microtubules and molecular motors. In *The Mycota* (R. J. Howard and N. A. R. Gow, eds), vol. VIII, pp. 225–241. Springer, Wein, New York.

Li, X., Zhao, X., Fang, Y., Jiang, X., Duong, T., Fan, C., Huang, C.-C. and Kain, S. R. (1998). Generation of destabilized green fluorescent protein as a transcription reporter. *J. Biol. Chem.* **273**, 34970–34975.

Lippincott-Schwartz, J. and Patterson, G. H. (2003). Development and use of fluorescent protein markers in living cells. *Science* **300**, 87–91.

Liu, Z. M. and Kolattukudy, P. E. (1999). Early expression of the calmodulin gene, which precedes appressorium formation in *Magnaporthe grisea*, is inhibited by self-inhibitors and requires surface attachment. *J. Bacteriol.* **181**, 3571–3577.

López-Franco, R., Howard, R. J. and Bracker, C. E. (1995). Satellite Spitzenkörper in growing hyphal tips. *Protoplasma* **188**, 85–103.

Lorang, J. M., Tuori, R. P., Martinez, J. P., Sawyer, T. L., Redman, R. S., Rollins, J. A., Wolpert, T. J., Johnson, K. B., Rodriguez, R. J., Dickman, M. B. and Ciuffetti, L. M. (2001). Green fluorescent protein is lighting up fungal biology. *Appl. Environ. Microbiol.* **67**, 1987–1994.

Lübeck, M., Knudsen, I. M. B., Jensen, B., Thrane, U., Janvier, C. and Jensen, D. F. (2002). GUS and GFP transformation of the biocontrol strain *Clonostachys rosea* IK726 and the use of these marker genes in ecological studies. *Mycol. Res.* **106**, 815–826.

Lukyanov, K. A., Fradkov, A. F., Gurskaya, N. G., Matz, M. V., Labas, Y. A., Savitsky, A. P., Markelov, M. L., Zaraisky, A. G., Zhao, X., Fang, Y., Tan, W. and Lukyanov, S. A. (2000). Natural animal coloration can be determined by a nonfluorescent green fluorescent protein homolog. *J. Biol. Chem.* **275**, 25879–25882.

Luo, K. Q., Yu, V. C., Pu, Y. and Chang, D. C. (2001). Application of the fluorescence resonance energy transfer method for studying the dynamics of caspase-3 activation during UV-induced apoptosis in living HeLa cells. *Biochem. Biophys. Res. Commun.* **283**, 1054–1060.

Maddox, P. S., Bloom, K. S. and Salmon, E. D. (2000). The polarity and dynamics of microtubule assembly in budding yeast *Saccharomyces cerevisiae*. *J. Cell Biol.* **2**, 36–41.

Maor, R., Puyesky, M., Horwitz, B. A. and Sharon, A. (1998). Use of green fluorescent protein (GFP) for studying development and fungal–plant interaction in *Cochliobolus heterostrophus*. *Mycol. Res.* **102**, 491–496.

Marc, J., Granger, C. L., Brincat, J., Fisher, D. D., Kao, T., McCubbin, A. G. and Cyr, R. J. (1998). A GFP-MAP4 reporter gene for visualizing cortical microtubule rearrangements in living epidermal cells. *Plant Cell* **10**, 1927–1940.

Maruyama, J., Nakajima, H. and Kitamoto, K. (2002). Observation of EGFP-visualized nuclei and distribution of vacuoles in *Aspergillus oryzae* arpA null mutant. *FEMS Microbiol. Lett.* **206**, 57–61.

Masai, K., Maruyama, J., Nakajima, H. and Kitamoto, K. (2003). In vivo visualization of the distribution of a secretory protein in *Aspergillus oryzae* hyphae using the RntA-EGFP fusion protein. *Biosci. Biotechnol. Biochem.* **67**, 455–459.

Matz, M. V., Fradkov, A. F., Labas, Y. A., Savitsky, A. P., Zaraisky, A. G., Markelov, M. L. and Lukyanov, S. A. (1999). Fluorescent proteins from nonbioluminescent Anthozoa species. *Nat. Biotechnol.* **17**, 969–973.

Mendelsohn, A. R. and Brent, R. (1999). Protein interaction methods – toward an endgame. *Science* **284**, 1948–1950.

Meyer, T. and Teruel, M. N. (2003). Fluorescence imaging of signaling networks. *Trends Cell Biol.* **13**, 101–106.

Michener, C. M., Ardekani, A. M., Petricoin, E. F., Liotta, L. A. and Kohn, E. C. (2002). Genomics and proteomics: application of novel technology to early detection and prevention of cancer. *Cancer Detect. Prev.* **26**, 249–255.

Mikkelsen, L., Roulund, N., Lübeck, M. and Jensen, D. F. (2001). The perennial ryegrass endophyte *Neotyphodium lolii* genetically transformed with the green fluorescent protein (*gfp*) and visualization in the host plant. *Mycol. Res.* **105**, 644–650.

Miyawaki, A. (2003). Visualization of the spatial and temporal dynamics of intracellular signaling. *Dev. Cell* **4**, 295–305.

Miyawaki, A., Llopis, J., Heim, R., McCaffery, J. M., Adams, J. A., Ikura, M. and Tsien, R. Y. (1997). Fluorescent indicators for Ca2+ based on green fluorescent proteins and calmodulin. *Nature* **388**, 882–887.

Nagai, T., Ibata, K., Park, E. S., Kubota, M., Mikoshiba, K. and Miyawaki, A. (2002). A variant of yellow fluorescent protein with fast and efficient maturation for cell-biological applications. *Nat. Biotechnol.* **20**, 87–90.

Nakano, A. (2002). Spinning-disk confocal microscopy – a cutting-edge tool for imaging of membrane traffic. *Cell Struct. Funct.* **27**, 349–355.

Nakazono, M., Qiu, F., Borsuk, L. A. and Schnable, P. S. (2003). Laser-capture microdissection, a tool for the global analysis of gene expression in specific plant cell types: identification of genes expressed differentially in epidermal cells or vascular tissues of maize. *Plant Cell* **15**, 583–596.

Nehls, U., Bock, A., Ecke, M. and Hampp, R. (2001). Differential expression of the hexose-regulated fungal genes *AmPAL* and *AmMst1* within *Amanita/Populus* ectomycorrhizas. *New Phytol.* **150**, 583–589.

Nielsen, K., Olsen, O. and Oliver, R. (1999). A transient expression system to assay putative antifungal genes on powdery mildew infected barley leaves. *Physiol. Mol. Plant Pathol.* **54**, 1–12.

Ovechkina, Y., Maddox, P., Oakley, C. E., Xiang, X., Osmani, S. A., Salmon, E. D. and Oakley, B. R. (2003). Spindle formation in *Aspergillus* is coupled to tubulin movement into the nucleus. *Mol. Biol. Cell* **14**, 2192–2200.

Patterson, G. H. and Lippincott-Schwartz, J. (2002). A photoactivatable GFP for selective photolabeling of proteins and cells. *Science* **297**, 1873–1877.

Pennell, R. I. and Lamb, C. (1997). Programmed cell death in plants. *Plant Cell* **9**, 1157–1168.

Pinan-Lucarre, B., Paoletti, M., Dementhon, K., Coulary-Salin, B. and Clave, C. (2003). Autophagy is induced during cell death by incompatibility and is essential for differentiation in the filamentous fungus *Podospora anserina*. *Mol. Microbiol.* **47**, 321–333.

Pöggeler, S., Masloff, S., Hoff, B., Mayrhofer, S. and Kück, U. (2003). Versatile EGFP reporter plasmids for cellular localization of recombinant gene products in filamentous fungi. *Curr. Genet.* **43**, 54–61.

Polishchuk, R. S., Polishchuk, E. V., Marra, P., Alberti, S., Buccione, R., Luini, A. and Mironov, A. A. (2000). Correlative light-electron microscopy reveals the tubular-saccular ultrastructure of carriers operating between Golgi apparatus and plasma membrane. *J. Cell Biol.* **148**, 45–58.

Prasher, D. C., Eckenrode, V. K., Ward, W. W., Prendergast, F. G. and Cormier, M. J. (1992). Primary structure of the *Aequorea victoria* green-fluorescent protein. *Gene* **111**, 229–233.

Reider, C. L. and Khodjakov, A. (2003). Mitosis through the microscope: advances in seeing inside live dividing cells. *Science* **300**, 91–96.

Rohel, E. A., Payne, A. C., Fraaije, B. A. and Hollomon, D. W. (2001). Exploring infection of wheat and carbohydrate metabolism in *Mycosphaerella graminicola* transformants with differentially regulated green fluorescent protein expression. *Mol. Plant-Microbe Interact.* **14**, 156–163.

Sako, Y. and Uyemura, T. (2002). Total internal reflection fluorescence microscopy for single-molecule imaging in living cells. *Cell Struct. Funct.* **27**, 357–365.

Sato, M., Hida, N., Ozawa, T. and Umezawa, Y. (2000). Fluorescent indicators for cyclic GMP based on cyclic GMP-dependent protein kinase Iα and green fluorescent proteins. *Anal. Chem.* **72**, 5918–5924.

Sato, M., Ozawa, T., Inukai, K., Asano, T. and Umezawa, Y. (2002). Fluorescent indicators for imaging protein phosphorylation in single living cells. *Nat. Biotechnol.* **20**, 287–294.

Sawin, K. E. (1999). GFP fusion proteins as probes for cytology in fission yeast. In *Green Fluorescent Proteins. Methods in Cell Biology* (K. F. Sullivan and S. A. Kay, eds), vol. 58, pp. 123–137. Academic Press, New York.

Sekar, R. B. and Periasamy, A. (2003). Fluorescence resonance energy transfer (FRET) microscopy imaging of live cell protein localizations. *J. Cell Biol.* **160**, 629–633.

Si-Ammour, A., Mauch-Mani, B. and Mauch, F. (2003). Quantification of induced resistance against *Phytophthora* species expressing GFP as a vital marker: β-aminobutyric acid but not BTH protects potato and *Arabidopsis* from infection. *Mol. Plant Pathol.* **4**, 237–248.

Siedenberg, D., Mestric, S., Ganzlin, M., Schmidt, M., Punt, P. J., van den Hondel, C. A. M. J. J. and Rinas, U. (1999). *GlaA* promoter controlled production of a mutant green fluorescent protein (S65T) by recombinant *Aspergillus niger* during growth on defined medium in batch and fed-batch cultures. *Biotechnol. Prog.* **15**, 43–50.

Siegel, M. S. and Isacoff, E. Y. (1997). A genetically encoded optical probe of membrane voltage. *Neuron* **19**, 735–741.

Soanes, D. M., Kershaw, M. J., Cooley, R. N. and Talbot, N. J. (2002). Regulation of the *MPG1* hydrophobin gene in the rice blast fungus *Magnaporthe grisea*. *Mol. Plant-Microbe Interact.* **15**, 1253–1267.

Spellig, T., Bottin, A. and Kahmann, R. (1996). Green fluorescent protein (GFP) as a new vital marker in the phytopathogenic fungus *Ustilago maydis*. *Mol. Gen. Genet.* **252**, 503–509.

Steinberg, G., Wedlich-Söldner, R., Brill, M. and Schulz, I. (2001). Microtubules in the filamentous fungal pathogen *Ustilago maydis* are highly dynamic and determine cell polarity. *J. Cell Sci.* **114**, 609–622.

Straube, A., Brill, M., Oakley, B. R., Horio, T. and Steinberg, G. (2003). Microtubule organization requires cell cycle-dependent nucleation at dispersed cytoplasmic sites: polar and perinuclear microtubule organizing centers in the plant pathogen *Ustilago maydis*. *Mol. Biol. Cell* **14**, 642–657.

Stuurman, N., Bras, C. P., Schlaman, H. R. M., Wijfjes, A. H. M., Bloemberg, G. and Spaink, H. P. (2000). Use of green fluorescent protein color variants expressed on stable broad-host-range vectors to visualize *Rhizobia* interacting with plants. *Mol. Plant-Microbe Interact.* **13**, 1163–1169.

Suelmann, R. and Fischer, R. (2000). Mitochondrial movement and morphology depend on an intact actin cytoskeleton in *Aspergillus nidulans*. *Cell Motil. Cytoskeleton* **45**, 42–50.

Suelmann, R., Sievers, N. and Fischer, R. (1997). Nuclear traffic in fungal hyphae: in vivo study of nuclear migration and positioning in *Aspergillus nidulans*. *Mol. Microbiol.* **25**, 757–769.

Swedlow, J. R. and Platani, M. (2002). Live cell imaging using wide-field microscopy and deconvolution. *Cell Struct. Funct.* **27**, 335–341.

Sweigard, J. A. and Ebbole, D. J. (2001). Functional analysis of pathogenicity genes in a genomics world. *Curr. Opin. Microbiol.* **4**, 387–392.

Takano, Y., Oshiro, E. and Okuno, T. (2001). Microtubule dynamics during infection-related morphogenesis of *Colletotrichum lagenarium*. *Fungal Genet. Biol.* **34**, 107–121.

Takemoto, D., Jones, D. A. and Hardham, A. R. (2003). GFP-tagging of cell components reveals the dynamics of subcellular re-organization in response to infection of *Arabidopsis* by oomycete pathogens. *Plant J.* **33**, 775–792.

Terskikh, A., Fradkov, A., Ermakova, G., Zaraisky, A., Tan, P., Kajava, A. V., Zhao, X., Lukyanov, S., Matz, M., Kim, S., Weissman, I. and Siebert, P. (2000). "Fluorescent timer": protein that changes color with time. *Science* **290**, 1585–1588.

Thomas, C., DeVries, P., Hardin, J. and White, J. (1996). Four-dimensional imaging: computer visualization of 3D movements in living specimens. *Science* **273**, 603–607.

Tirlapur, U. K. and König, K. (2002). Femtosecond near-infrared laser pulses as a versatile non-invasive tool for intra-tissue nanoprocessing in plants without compromising viability. *Plant J.* **31**, 365–374.

Tsien, R. Y. (1998). The green fluorescent protein. *Annu. Rev. Biochem.* **67**, 509–544.

Vale, R. D., Reese, T. S. and Sheetz, M. P. (1985). Identification of a novel force-generating protein, kinesin, involved in microtubule-based motility. *Cell* **42**, 39–50.

van Bruaene, N., Joss, G., Thas, O. and van Oostveldt, P. (2003). Four-dimensional imaging and computer-assisted track analysis of nuclear migration in root hairs of *Arabidopsis thaliana*. *J. Microsc.* **211**, 167–178.

van West, P., Appiah, A. A. and Gow, N. A. R. (2003). Advances in research on oomycete root pathogens. *Physiol. Mol. Plant Pathol.* **62**, 99–113.

Wang, X., Allen, R., Ding, X., Goellner, M., Maier, T., de Boer, J. M., Baum, T. J., Hussey, R. S. and Davis, E. L. (2001). Signal peptide-selection of cDNA cloned directly from the esophageal gland cells of the soybean cyst nematode *Heterodera glycine*. *Mol. Plant-Microbe Interact.* **14**, 536–544.

Wang, J. D., Herman, C., Tipton, K. A., Gross, C. A. and Weissman, J. S. (2002). Directed evolution of substrate-optimized GroEL/S chaperonins. *Cell* **111**, 1027–1039.

Wang, L., Ridgway, D., Gu, T. and Moo-Young, M. (2003). Effects of process parameters on heterologous protein production in *Aspergillus niger* fermentation. *J. Chem. Technol. Biotechnol.* **78**, 1259–1266.

Wendland, J. and Philippsen, P. (2002). An IQGAP-related protein, encoded by AgCYK1, is required for septation in the filamentous fungus *Ashbya gossypii*. *Fungal Genet. Biol.* **37**, 81–88.

Xia, Z. and Liu, Y. (2001). Reliable and global measurement of fluorescence resonance energy transfer using fluorescence microscopes. *Biophys. J.* **81**, 2395–2402.

Xu, X., Gerard, A. L. V., Huang, B. C. B., Anderson, D. C., Payan, D. G. and Luo, Y. (1998). Detection of programmed cell death using fluorescence energy transfer. *Nucleic Acids Res.* **26**, 2034–2035.

Zaccolo, M. and Pozzan, T. (2000). Imaging signal transduction in living cells with GFP-based probes. *IUBMB Life* **49**, 375–379.

Zaccolo, M., De Giorgi, F., Cho, C. Y., Feng, L., Knapp, T., Negulescu, P. A., Taylor, S. S., Tsien, R. Y. and Pozzan, T. (2000). A genetically encoded, fluorescent indicator for cyclic AMP in living cells. *Nat. Cell Biol.* **2**, 25–29.

Zacharias, D. A., Violin, J. D., Newton, A. C. and Tsien, R. Y. (2002). Partitioning of lipid-modified monomeric GFPs into membrane microdomains of live cells. *Science* **296**, 913–916.

Zeytun, A., Jeromin, A., Scalettar, B. A., Waldo, G. S. and Bradbury, A. R. M. (2003). Fluorobodies combine GFP fluorescence with the binding characteristics of antibodies. *Nat. Biotechnol.* **21**, 1473–1479.

Zhang, J., Campbell, R. E., Ting, A. Y. and Tsien, R. Y. (2002). Creating new fluorescent probes for cell biology. *Nat. Rev. Mol. Cell Biol.* **3**, 906–918.

Zimmermann, T., Rietdorf, J., Girod, A., Georget, V. and Pepperkok, R. (2002). Spectral imaging and linear un-mixing enables improved FRET efficiency with a novel GFP2-YFP FRET pair. *FEBS Lett.* **531**, 245–249.

Zimmermann, T., Rietdorf, J. and Pepperkok, R. (2003). Spectral imaging and its applications in live cell microscopy. *FEBS Lett.* **546**, 87–92.

3 Live-cell Imaging of Filamentous Fungi Using Vital Fluorescent Dyes and Confocal Microscopy

Patrick C Hickey, Samuel R Swift, M Gabriela Roca and Nick D Read

Fungal Cell Biology Group, Institute of Cell Biology, University of Edinburgh, Rutherford Building, Edinburgh EH9 3JH, UK

CONTENTS

Introduction
Microscope technologies for imaging living fungal cells at high spatial resolution
Vital fluorescent dyes for imaging filamentous fungi
Practical aspects of live-cell imaging
Future directions for live-cell imaging of filamentous fungi

◆◆◆◆◆◆ INTRODUCTION

A revolutionary new perspective of the cell biology of fungal hyphae is arising as a result of using live-cell imaging techniques to analyse organelle and molecular dynamics at high spatial resolution. This has become possible because of the development of a wide range of fluorescent probes (vital dyes and fluorescent proteins) that can be used to non-invasively interrogate living cells, new microscope technologies (e.g. confocal and two-photon microscopy), and powerful computer software and hardware for digital image processing and analysis. These innovations are having a profound impact on the experimental analysis of living fungal hyphae at the single cell level.

The aims of this review are to: (a) provide a brief comparative overview of microscope technologies for imaging living fungal cells at high spatial resolution; (b) review the vital fluorescent dyes which are proving useful for analysing the cell biology of filamentous fungi with the confocal laser scanning microscope (CLSM); (c) define important practical aspects that need to be taken into account for optimal live-cell imaging at high spatial resolution with the CLSM; and (d) briefly indicate future directions for live-cell imaging of filamentous fungi.

◆◆◆◆◆◆ MICROSCOPE TECHNOLOGIES FOR IMAGING LIVING FUNGAL CELLS AT HIGH SPATIAL RESOLUTION

A range of microscope technologies have been developed over the last 20 years which allow one to image living cells at high spatial resolution using optical sectioning. These technologies include: confocal laser scanning microscopy (Czymmek et al., 1994; Pawley, 1995; Sheppard and Shotton, 1997; Diaspro, 2001), spinning disk confocal microscopy (Ichiwara et al., 1999), two-photon microscopy (König, 2000; Diaspro, 2001; Feijo and Moreno, 2004), and deconvolution microscopy (Swedlow, 2003). The advantages and disadvantages of these optical sectioning methods for imaging living fungal cells are summarized in Table 3.1. Of these types of microscopy, confocal laser scanning microscopy has been the most popular, and most of this review is focused on results obtained using this extremely powerful imaging technique.

◆◆◆◆◆◆ VITAL FLUORESCENT DYES FOR IMAGING FILAMENTOUS FUNGI

In Table 3.2 we have compiled a list of vital dyes with their selectivities and associated parameters that we have found to be most useful for live-cell imaging. Below we discuss the use of these dyes for studies on filamentous fungi. We have restricted this discussion mostly to dyes which we have personal experience of.

General Membrane-selective Dyes

The FM-dyes (FM4-64 and FM1-43) are membrane-selective probes which stain the plasma membrane and most organelle membranes in fungal cells (Fischer-Parton et al., 2000; Read and Hickey, 2001). They belong to a class of amphiphilic styryl dyes developed by Betz and colleagues (Betz et al., 1992; 1996; Bolte et al., 2004). The FM-dyes only fluoresce strongly when located in a hydrophobic environment (e.g. a membrane) and thus, advantageously, do not need to be washed out from the medium in which the fungi are growing in.

The FM-dyes enter fungal cells primarily by endocytosis and thus have been much used as endocytosis markers (see Endocytosis Marker Dyes). Once internalized, the dyes become distributed to different organelle membranes and different organelles stain up in a time-dependent manner (Fischer-Parton et al., 2000; Read and Hickey, 2001; Atkinson et al., 2002; Dijksterhuis, 2003). Much of this intracellular movement of dye seems to be via the vesicle trafficking network, and through physical continuity between different membranes of the endomembrane system, but other mechanisms of dye distribution may also possibly exist (Fischer-Parton et al., 2000). The precise pattern of organelle staining with FM4-64 is different to that obtained with FM1-43. Both dyes can be used as general

Table 3.1. Comparison of confocal laser scanning microscopy, spinning disc confocal microscopy and two-photon laser scanning microscopy for live-cell imaging of filamentous fungi at high spatial resolution

Point of comparison	Confocal laser scanning microscopy	Spinning disc confocal microscopy	Two-photon laser scanning microscopy	Deconvolution microscopy
Spatial resolution	Very high	Very high	High	Very high
Temporal resolution	Typically 1 image/s	Typically <8 image/s	Typically 1 image/s	Typically ~10 image/s
Phototoxicity	Usually most phototoxic	Reduced	Often reduced	Reduced
Photobleaching	Usually causes most photobleaching	Reduced	Often reduced	Often reduced
Depth of imaging	Typically <20 nm	Typically <20 nm	Typically <100 nm	Typically <20 nm
Excitation wavelengths	405–647 nm	405–647 nm	700–1050 nm (two-photon excitation)	350–650 nm
Ease of use	More difficult	Relatively easy	Most difficult because of complex laser system	Relatively easy
Size	Medium	Medium	Large	Medium
Cost	Expensive	Expensive	Most expensive	Expensive

Table 3.2. Useful vital dyes for live-cell imaging in fungal hyphae using confocal laser scanning microscopy

Dye	Selectivity	Recommended excitation wavelength (nm)	Maximum or recommended emission wavelength(s) (nm)	Recommended dye concentration (μM)
Calcofluor White M2R	Cell walls	405	500	0.1–25
Carboxy DFFDA	Vacuoles	488	529	10
Carboxy SNARF-1-AM	pH	514	580	2–5
FITC-dextran (10 kDa)	Fluid phase endocytosis	488	514	400 mM
Carboxy SNARF-1-dextran (10 kDa)	pH	514	580	10 mg/ml
DASPMI	Mitochondria	488	605	10–50
DIOC$_6$(3)	Membranes	488	501	0.5–50
Fluorescein diacetate/propidium iodide	Live/dead cells	488	517/617	1.5 mM/0.15 mM
FM1-43	Membranes, endocytosis	514	550	2.5–25
FM4-64	Membranes, endocytosis	514	670	2.5–25
Lucifer yellow carbohydrazide	Fluid phase endocytosis	457	536	440
MDY-64	Vacuole membranes	488	497	10
Mitotracker Green	Mitochondria	488	516	10
Rhodamine 123	Mitochondria	514	529	10–65
Oregon Green-1 dextran (10 kDa)/rhodamine B dextran (10 kDa)	Free calcium	488/568	530/578	1 mM/1 mM

membrane-selective stains, especially for imaging cells at low magnification (Figures 3.1, 3.2 and 3.3A; Hickey *et al.*, 2002; Hickey and Read, 2003). For higher magnification work, FM4-64 has proved superior for staining vacuolar membranes (Figure 3.7B,C) and vesicles within the Spitzenkörper (Figures 3.3B, 3.4, 3.9C, and 3.12), whilst FM1-43 is best for staining mitochondria (Figure 3.5B; Fischer-Parton *et al.*, 2000; see Mitochondrial Dyes). Both dyes tend to strongly stain the plasma membrane (Figures 3.IB, 3.3A, 3.5, 3.7C and 3.9), which can be useful for imaging septum development (Figures 3.1 and 3.5), but neither dye stains the nuclear envelope (Figures 3.3B and 3.4B).

Although many organelles become labelled with FM4-64 and FM1-43, the identity of only a few are known. Furthermore, the time after which different organelles become stained depends particularly on the cell type and growth conditions used. Generally the staining of wide, fast growing

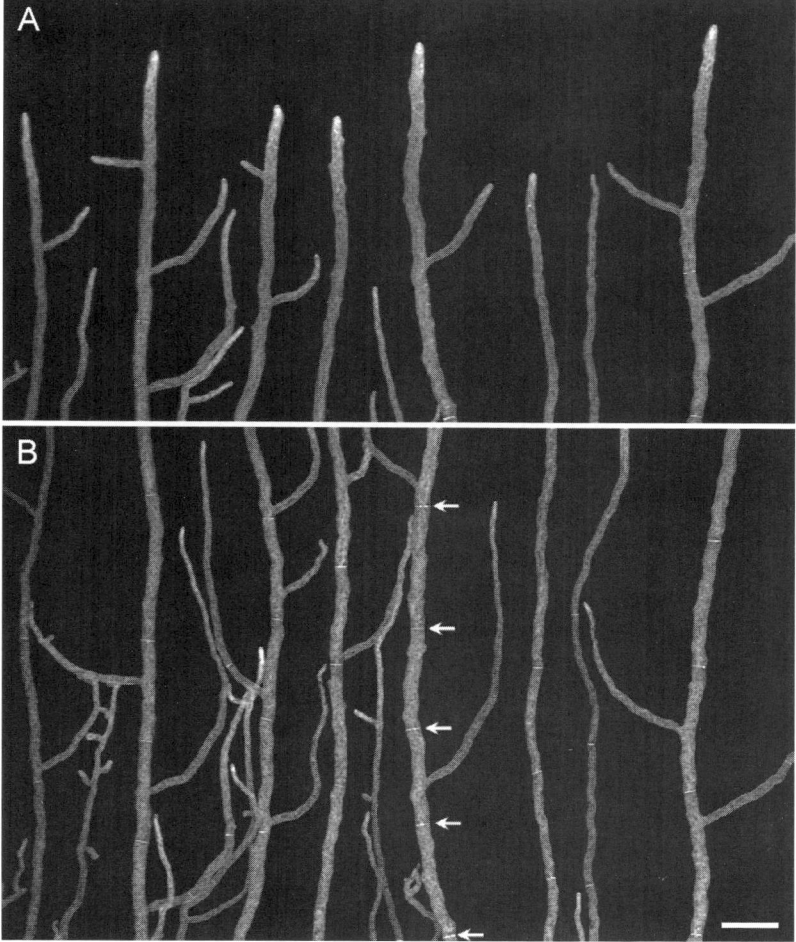

Figure 3.1. *Neurospora crassa* hyphae in the peripheral region of the colony stained with FM1-43. (B) is of the same region as (A) but was imaged 28 min later. Note pronounced staining of septa (arrows in B). Bar = 50 μm.

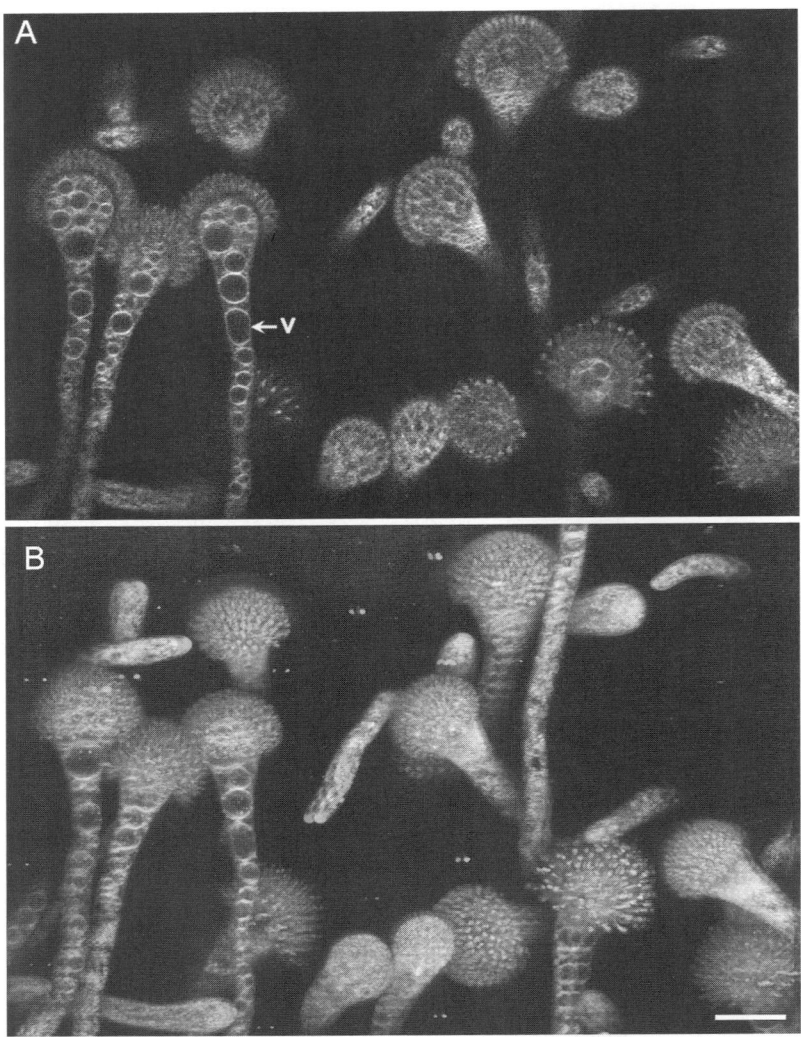

Figure 3.2. *Aspergillus fumigatus* stained with FM4-64. Different stages of conidiophore and conidium development. (A) Single optical section. Note large vacuoles in conidiophores. (B) 3D projection of 50 × 0.5 μm optical sections (requires viewing with green/red stereo glasses). Bar = 20 μm. (See colour plate 6.)

hyphae (e.g. vegetative hyphae at the colony periphery) is quicker than that of narrow, slow growing hyphae (e.g. germ tubes) (Table 3.3). The initial faint staining of the cytoplasm, especially close to the plasma membrane, is assumed to result from the labelling of a "cloud" of endocytic vesicles (Figure 3.9A). The first fluorescent organelles appear as punctate, roughly spherical organelles and have been assumed to be endosomes (Figure 3.9B). Thereafter vesicles, presumed to be primarily secretory in nature, stain up within the Spitzenkörper/apical vesicle cluster (Figure 3.9C) indicating that endocytic pathway(s) are connected to the secretory pathway(s). Vacuole membranes are obviously labelled

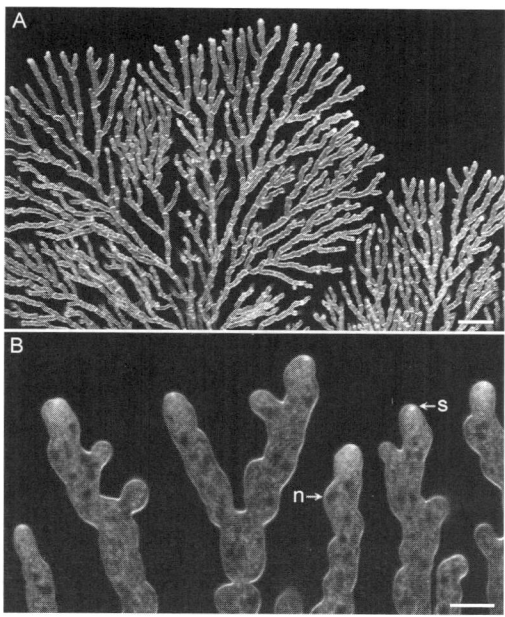

Figure 3.3. Hyperbranching *spray* mutant of *Neurospora crassa* stained with FM4-64. (A) Low-magnification image. Note multiple stained septa (arrows). (B) High-magnification image showing irregular growth of hyphae, multiple branches (many of which abort), small Spitzenkörper (s) and negatively stained nuclei (n). Bar = 10 μm.

and in *N. crassa*, the large spherical vacuoles (Figure 3.7B,C) become stained before those of the tubular vacuolar network (Figure 3.7A; Read and Hickey, 2001; Table 3.3). Interestingly, mitochondrial membranes also stain up with FM4-64 and FM1-43, but most markedly with FM1-43 (see Mitochondrial Dyes) (Fischer-Parton *et al.*, 2000).

In future, it will be very important to identify the different organelles stained with the FM-dyes. One approach will be to double label cells with a FM-dye and a fluorescent protein targeted to a specific organelle. This approach has proved particularly useful in demonstrating that Golgi, but not ER, are stained by FM4-64 in plant cells (Bolte *et al.*, 2004). Another approach will be to localize the FM-dyes at the ultrastructural level, and methods for this have been developed (Henkel *et al.*, 1996).

After prolonged staining (>30 min), FM4-64 and FM1-43 have proven to be the best dyes for providing general staining of living hyphae and sporulating structures, especially when viewed at low magnifications (e.g. with a 20 × objective, Figures 3.1, 3.2 and 3.3A). These dyes are particularly excellent for characterizing the morphological phenotypes of mutants in the living state (Figures 3.3 and 3.11).

Another membrane-selective dye which has been used in filamentous fungi is $DIOC_6(3)$. This is a cationic, cyanine dye which accumulates in the mitochondria, endoplasmic reticulum (ER) and nuclear envelope in a potential-dependent manner. In plants, low dye concentrations favour mitochondrial staining whilst high concentrations will stain ER and the

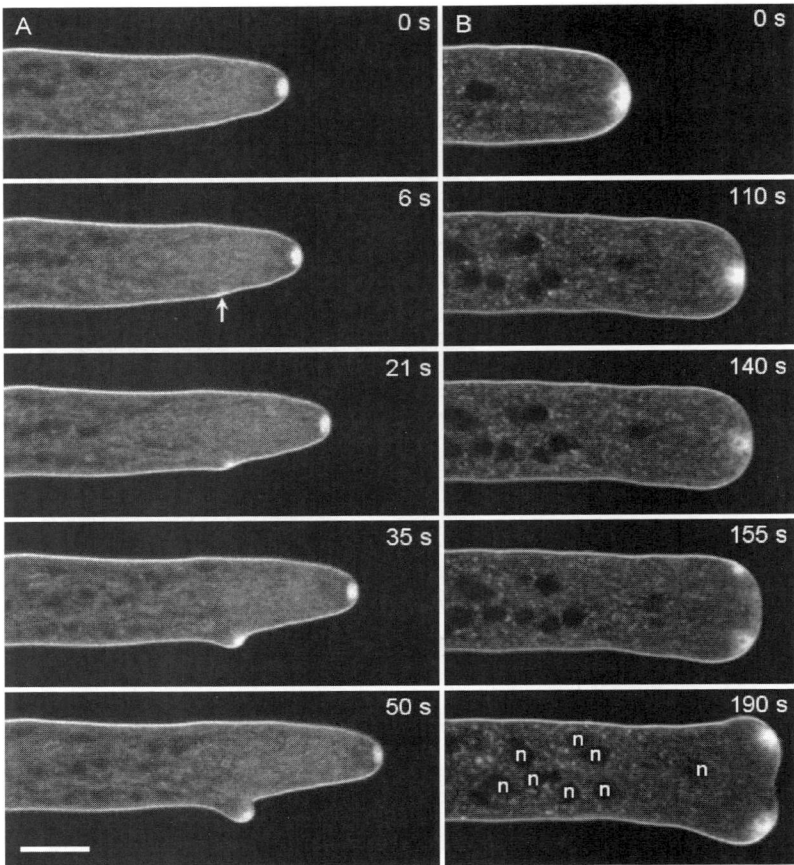

Figure 3.4. (A) *Neurospora crassa* hypha stained with FM4-64 showing sub-apical branch formation. Note the initiation of the Spitzenkörper beneath the plasma membrane (arrow) that has appeared just before the branch emerged. (B) *Sclerotinia sclerotiorum* hyphae stained with FM4-64 showing apical branching. Note the negatively stained nuclei (n) and less stained core region within the Spitzenkörper. Bars = 10 μm.

nuclear envelope (Oparka and Read, 1994). $DIOC_6(3)$ has also been used to stain mitochondria, ER and the nuclear envelope in yeast cells (e.g. Koning *et al.*, 1993) and filamentous fungi (e.g. Koll *et al.*, 2001).

Nuclear Dyes

We have not found any vital nuclear dye to be particularly good for live-cell imaging of fungal hyphae using confocal microscopy.

DAPI is usually used to stained AT-rich regions of DNA within fixed fungal cells. However, it can sometimes be used as a vital probe if the cell is permeable to the dye. Often we find that vegetative hyphae will not take up DAPI whilst germinating spores will. DAPI is normally excited at UV wavelengths which are not very compatible with live-cell imaging. However, it is also excited at 405 nm which is now available for confocal

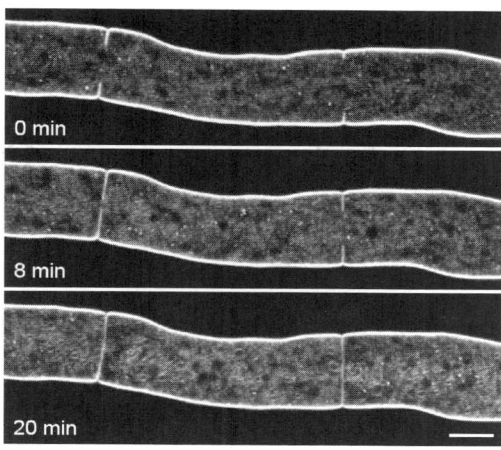

Figure 3.5. Septum formation in *Neurospora crassa* stained with FM4-64. Only the plasma membrane on either side of the septum is stained. Bar = 10 μm.

microscopy in the form of a relatively inexpensive blue diode laser. However, so far we have not been able to perform good live-cell nuclear imaging with DAPI using the CLSM because of the staining of other cellular components (e.g. mitochondrial DNA and possibly polyphosphates) (Roca, M. G. and Read, N. D., unpublished results).

A wide range of cyanine Syto dyes, which stain both DNA and RNA, are available from Molecular Probes Inc. (Haugland, 2002). We have tested a wide range of green fluorescing Syto dyes (Syto 11, 13, 14, 15 and 16) and have found them to be readily cell permeant and to stain nuclei in vegetative hyphae to varying extents. However, these dyes have not proven useful in our live-cell imaging studies because all are very phototoxic and cause organelle disruption when scanned with a laser beam under the CLSM (Hickey, 2001).

As indicated in General Membrane-selective Dyes, $DIOC_6(3)$ can stain up the nuclear envelope, along with mitochondria and ER, if used at high

Table 3.3. Summary of the time course of FM4-64 staining at 20–25°C of vegetative hyphae of *Neurospora crassa* at the colony periphery and conidial germ tubes of *Magnaporthe grisea* (data from: Read and Hickey, 2001; Atkinson et al., 2002)

Cell component stained	Approximate times after adding FM4-64	
	Vegetative hyphae of N. crassa	Germ tubes of M. grisea
Plasma membrane	Immediate	Immediate
Putative endocytic vesicles	<10 s	1–2 min
Putative endosomes	30 s	2 min
Vesicles within Spitzenkörper	1.5 min	2 min
Large spherical vacuoles	10 min	45 min
Tubular vacuoles	15 min	45 min

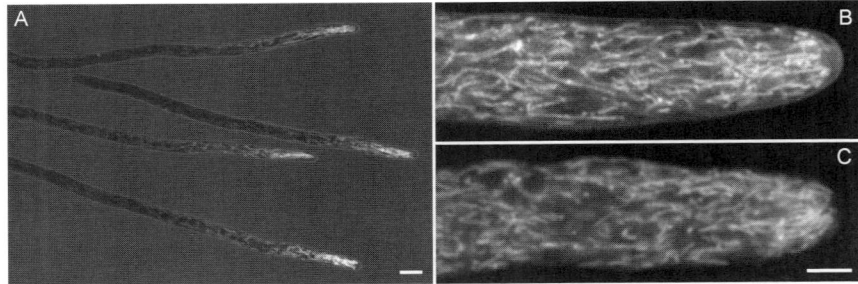

Figure 3.6. Mitochondrial staining. (A) *Aspergillus nidulans* stained with Rhodamine-123. Note that the fluorescence is higher in the hyphal tip regions where the mitochondria are apparently most active. Bar = 20 μm. (B) *Neurospora crassa* hyphal tip stained with FM1-43. (C) *Sclerotinia sclerotiorum* hyphal tip stained with DASPMI. Bars = 10 μm (for B and C).

concentrations (e.g. >25–50 μM). Nuclei can also be often imaged as negatively stained structures in cells which have undergone prolonged staining with FM4-64 (Figures 3.3B and 3.4B). By far the best way of visualizing nuclei in fungi is using a nuclear-targeted fluorescent protein (e.g. GFP) (e.g. Freitag *et al.*, 2004a,b; Figure 3.12). For a detailed description of the use of fluorescent proteins in live-cell imaging see Chapter 2.

Mitochondrial Dyes

Within apical hyphal compartments, mitochondria are typically tubular and long although the length and thickness of individual mitochondria varies between species (Figure 3.6B,C; Fischer-Parton *et al.*, 2000). In actively growing vegetative hyphae the mitochondria are typically tightly packed and numerous (Figure 3.5A–C). In sub-apical regions, mitochondria are shorter and more diffusely distributed (Fischer-Parton *et al.*, 2000). In slow growing germ tubes mitochondria are fewer and shorter (Swift, S., Hickey, P. C. and Read, N. D., unpublished). It is not clear to what extent, if any, that mitochondria in filamentous fungal cells are branched as occurs, for example, in budding yeast cells (Hermann and Shaw, 1998). We have found the following dyes to be useful for staining mitochondria in filamentous fungal cells: Rhodamine 123, DASPMI, FM1-43 and Mitotracker green.

Rhodamine 123 (Johnson *et al.*, 1980) is a cell-permeant, cationic, fluorescent dye that is sequestered by active mitochondria of hyphae within a few minutes. It is a potentiometric dye, the fluorescence of which is directly dependent on the electrochemical gradient across the mitochondrial membrane (the greater the membrane potential, the greater the fluorescence). Generally the mitochondria close to growing tips fluoresce more brightly than those further back indicating that those at the tip are most active (Figure 3.6A). This may be due to the need for high levels of ATP which are required to maintain tip growth. In *Botrytis cinerea* we found that Rhodamine 123 fluorescence could be quenched by collapsing the electrochemical gradient of mitochondrial membranes by treatment with the fungicide azoxystrobin which inhibits mitochondrial membrane transport at the cytochrome bc_1 complex (Swift *et al.*, 2001). Rhodamine 123 also stains the hyphal plasma membrane but often gives background staining of the medium (Figure 3.5C; Fischer-Parton *et al.*, 2000).

DASPMI is a styryl dye (Bereiter-Hahn, 1976) and takes much longer to stain mitochondria (~20 min) than does Rhodamine 123. It is also a potentiometric dye but to a much lesser extent than Rhodamine 123. It does not stain the plasma membrane or give background staining of the medium. A useful feature of DASPMI is that it exhibits a large Stoke's shift with blue excitation and red emission (Table 3.2). It is thus well suited in dual labelling studies (e.g. with GFP which is excited with blue light and emits green fluorescence, Freitag *et al.*, 2004b).

The FM-dyes were developed from DASPMI (Bolte *et al.*, 2004). FM1-43 stains mitochondria of *N. crassa* more rapidly (after ~20 min) compared with FM4-64 (after ~45 min). FM1-43 is also much more selective for mitochondria and stains them more intensely than does FM4-64 (Figure 3.5A; Fischer-Parton *et al.*, 2000) and thus is the preferred mitochondrial stain of the two dyes. FM1-43 also stains the plasma membrane (Figure 3.5A) but to a lesser extent than does Rhodamine 123 (Figure 3.5C). It is not clear how mitochondria are stained by these dyes (i.e. by stained vesicles fusing with mitochondria, by stained membrane being in contact with mitochondria, or possibly by some other mechanism (Fischer-Parton *et al.*, 2000)).

MitoTracker probes are cell-permeant mitochondrion-selective dyes that passively diffuse across the plasma membrane and accumulate in active mitochondria. Various Mitotracker probes are commercially available from Molecular Probes Inc. (Haugland, 2002). We have found that Mitotracker Green stains mitochondria in hyphae well (data not shown).

$DIOC_6(3)$ will stain mitochondria in hyphae and spores when used at low concentrations (typically ~0.5 μM) but we have been unable to obtain mitochondrial staining as good as we routinely achieve with the other mitochondrial dyes.

Vacuolar Dyes

The vacuolar system of filamentous fungi is very complex being composed of a network of very dynamic tubular elements in young hyphal compartments (Figure 3.7A) and large spherical organelles in older compartments (Figures 3.2, 3.7B–D), as well as many small spherical organelles (Figures 3.7A,B) (Cole *et al.*, 1998; Ashford *et al.*, 2001). Three dyes which are useful for staining the vacuolar system are FM4-64, MDY-64 and DFFDA.

FM4-64 (see General Membrane-selective Dyes and Endocytosis Marker Dyes) stains vacuolar membranes as the end point of the endocytic pathway (Figures 3.2 and 3.7C). In *Neurospora* vegetative hyphae FM4-64 starts to stain the large spherical vacuoles after ~10 min

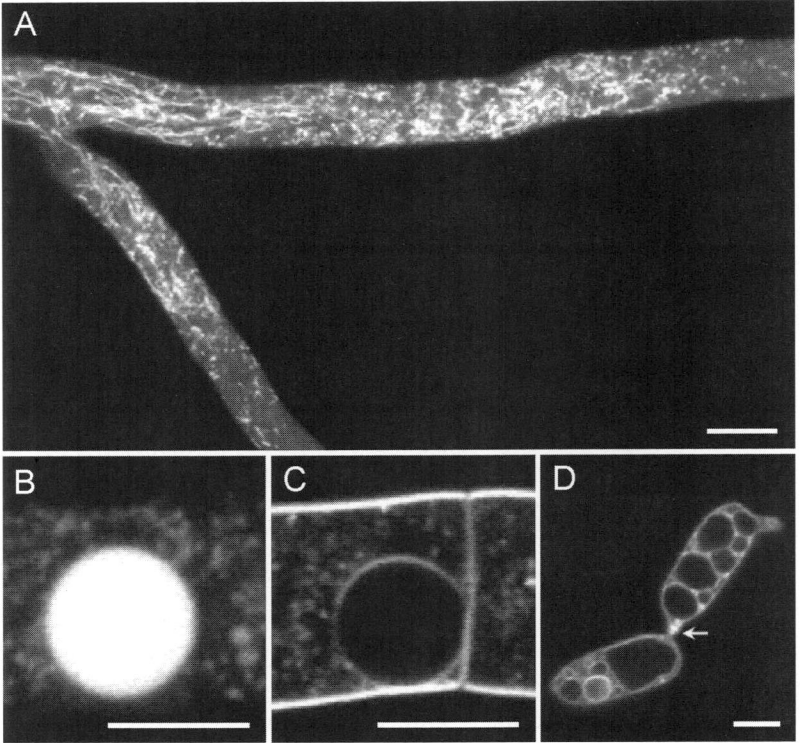

Figure 3.7. Vacuole staining in *Neurospora crassa*. (A) Tubular vacuolar network in apical hyphal compartment and branch stained with carboxy DFFDA. (B) Large and small spherical vacuoles in sub-apical hyphal compartment stained with carboxy-DFFDA. (C) Vacuole membrane of large spherical vacuole in sub-apical hyphal compartment stained with FM4-64. (D) Vacuole membranes within conidia of *Colletotrichum lindemuthianum* stained with MDY-64. Note that the two conidia are fusing via conidial anastomosis tubes (arrow). All bars = 10 μm.

and tubular elements after ~20 min (Table 3.3; Read and Hickey, 2001). In *Magnaporthe* germ tubes, vacuolar compartments stain up later (~45 min, Table 3.3; Atkinson *et al.*, 2002).

MDY-64 is also a yeast vacuolar dye which has been found to also stain up the membranes of vacuoles, and possibly other membranes in spores and hyphae (Figure 3.7D; Cole *et al.*, 1998; Roca *et al.*, 2003), although the precise mechanism of staining is not known.

Oregon Green 488 carboxylic acid (carboxy-DFFDA) is a derivative of 6-carboxyfluorescein diacetate (CFDA) but more photostable than it. Cole *et al.* (1997) provided evidence that both of these lipophilic probes cross the plasma membrane of hyphae of *Pisolithus tinctorius*, and are hydrolysed in the cytoplasm to Oregon Green 488 carboxylic acid and 6-carboxyfluorescein, respectively. They then seem to be rapidly transported across the vacuolar membrane into vacuoles by an organic anion transporter as this step can be inhibited by the anion transport inhibitor probenecid. Carboxy-DFFDA is an excellent dye for staining the lumens of both tubular and spherical vacuoles within hyphae (Figure 3.6A,B).

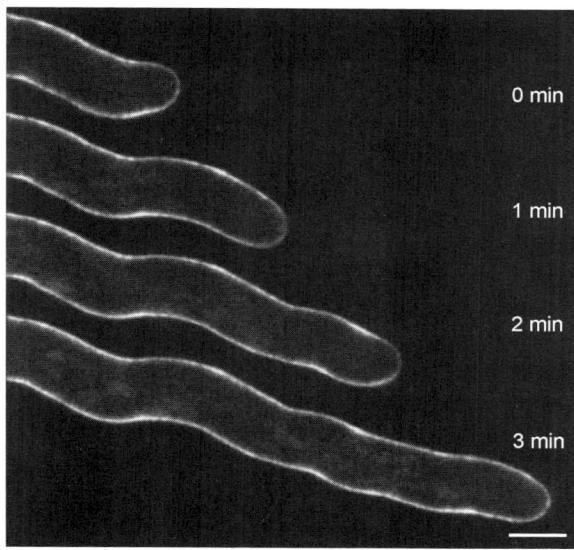

Figure 3.8. Growth of a hyphal branch of *Neurospora crassa* stained with 0.1 μM Calcofluor White M2R. Bar = 5 μm.

Cell Wall Dyes

Calcofluor White M2R is a cell wall stain that binds to chitin by intercalating into nascent chitin chains, and this prevents chitin microfibril assembly (Elorza *et al.*, 1983). It is a widely used cell wall stain for yeast and hyphal cells (both fixed and living) but is normally used at concentrations (typically >25 μM) that increase chitin synthesis and interfere with cell growth (Roncero and Duran, 1985). Nevertheless, at very low concentrations (~1.5 μM), Calcofluor can be useful as a live-cell stain (Figure 3.8).

Endocytosis Marker Dyes

It is generally believed that the FM-dyes are excellent endocytosis markers and reporters of vesicle trafficking in animal, plant, yeast and filamentous fungal cells (Vida and Emr, 1995; Cochilla *et al.*, 1999; Fischer-Parton *et al.*, 2000; Bolte *et al.*, 2004) although there has been some debate about whether these dyes might also be internalized and transported around cells by other mechanisms as well (Fischer-Parton *et al.*, 2000; Torralba and Heath, 2002; Bolte *et al.*, 2004).

The FM-dyes are amphiphilic molecules which are believed to get inserted in the outer leaflet of the plasma membrane bilayer. Because of their amphiphilic nature they are thought to be unable to directly cross the plasma membrane. The primary mechanism of FM-dye entry into cells is probably via endocytic vesicles which are invaginated from the plasma membrane. As indicated in General Membrane-Selective Dyes, these vesicles are then believed to be transported to endosomes, and

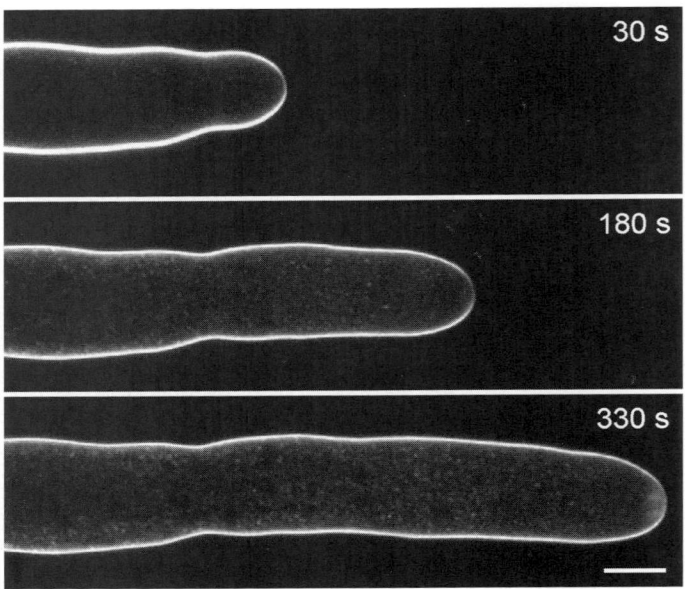

Figure 3.9. Time course of FM4-64 staining of a hyphal tip of *Sclerotinia sclerotiorum*. Bar = 10 μm.

subsequently to other organelles in the vesicle trafficking network resulting in the staining of different organelles in a time-dependent manner (Figures 3.9A–C) (Fischer-Parton *et al.*, 2000; Read and Hickey, 2001; Dijksterhuis, 2003). The endpoint of the endocytic pathway is lytic vacuoles and as described in Vacuolar Dyes, the FM-dyes are good probes for staining vacuolar membranes (Figure 3.2, 3.7C). Consistent with FM-dye uptake being an active process involving endocytosis is the finding that this internalization is inhibited in filamentous fungi by treatments (low temperature or addition of sodium azide) which inhibit ATP production (Hoffman and Mendgen, 1998; Fischer-Parton *et al.*, 2000; Atkinson *et al.*, 2002).

Other mechanisms of FM4 dye internalization (e.g. by flippases) or transport between different organelles (e.g. by lipid transport proteins) have been suggested but we are as yet unaware of any convincing evidence to support the involvement of these alternative pathways (Fischer-Parton *et al.*, 2000; Read and Kalkman, 2003).

Another class of dyes have been used as markers of fluid-phase endocytosis because they are commonly believed to be unable to cross the plasma membrane by passive diffusion but can become trapped within the lumen of endocytic vesicles invaginated from the plasma membrane (Dulic *et al.*, 1991; Hawes *et al.*, 1995). These dyes include Lucifer Yellow carbohydrazide (LYCH) and FITC-dextran. Only very limited success with the uptake of these probes into the cells of filamentous fungi has been achieved (Steinberg *et al.*, 1998; Atkinson *et al.*, 2002) and there are several reports of these probes not being taken up by hyphae (Cole *et al.*, 1997; Steinberg *et al.*, 1998; Fischer-Parton *et al.*, 2000; Torralba and Heath, 2002). The significance of these results is unclear (Read and Kalkman, 2003).

Ca^{2+} and pH Dyes

A very large range of pH- and Ca^{2+}-selective dyes are commercially available (Parton and Read, 1999; Haugland, 2002) for providing quantitative, spatially and temporally resolved measurements of intracellular H^+ or Ca^{2+} concentrations as they undergo dynamic changes within living cells. Most of these probes are designed to measure pH and free Ca^{2+} in the cytosol.

The use of pH- and Ca^{2+}-dyes for imaging and measuring ion concentrations in filamentous fungi is fraught with problems. The first problem is often associated with introducing the dye into the cell, and this can be particularly difficult with Ca^{2+}-dyes (Read et al., 1992; Knight et al., 1993; Parton and Read, 1999). The second problem is that the dyes frequently become sequestered within organelles, and when this occurs in sub-200 nm vesicles it can often be extremely difficult to detect. This can potentially have a profound effect in artifactually generating what appear to be cytosolic ionic gradients. An important control to verify that data obtained using free dyes are reliable is to check whether one obtains similar results with either a dextran-conjugated form of the dye (Parton et al., 1997) or a cytoplasmically targeted ion-sensitive recombinant probe (Miyawaki, 2003), neither of which should be taken up by organelles. Pressure injection of hyphae with dextran-conjugated dyes is not easy but is facilitated by using a pressure probe (Parton et al., 1997). A third problem is quantitation. In hyphae we have found that the only satisfactory quantitative method is to use a ratiometric approach in order to overcome problems of variations in fluorescence brightness due to variations in dye concentration within a cell, cell thickness, dye leakage or dye photobleaching (Parton and Read, 1999).

The only pH-sensitive dye we would recommend for pH imaging with the CLSM is carboxy-SNARF-1 which is a dual emission ratiometric dye. We have successfully used this in both its free form (loaded as its AM-ester) and 10 kDa dextran-conjugated form (Parton et al., 1997).

Calcium ratio imaging with the CLSM is much more difficult to perform because there are no ratiometric dyes which are very suitable for excitation with visible light wavelengths. An alternative approach is to simultaneously load cells with two dyes and ratio one against the other. In our lab, Fischer-Parton (1999) used this approach successfully by pressure injecting both the 10 kDa dextran of Oregon Green (which is Ca^{2+} sensitive) with the 10 kDa dextran of Rhodamine B (which is Ca^{2+} insensitive). Silverman-Gavrila and Lew (2000, 2001, 2003) have ratio imaged Ca^{2+} in hyphae dual loaded with the Ca^{2+}-dyes Fluo3-AM and Fura-Red-AM, although they did not perform appropriate controls with a dextran-conjugated dye or recombinant probe as mentioned above. The methods and problems of imaging and measuring free Ca^{2+} and pH are discussed in detail by Parton and Read (1999). Hopefully, many of the problems will be solved in the future with recombinant probes which have been optimized for Ca^{2+} and pH measurement in filamentous fungi.

Dyes to Test Cell Viability

It is often very useful to test the viability of fungal cells in a cell population. We have found that combining the stains fluorescein diacetate (FDA) and propidium iodide works extremely well for this purpose (Oparka and Read, 1994). Both FDA and propidium iodide can be excited at 488 nm. FDA is readily taken up by living cells and is hydrolysed to fluoroscein which fluoresces green. Propidium iodide, however, is only taken up by cells which are dead or have had their plasma membrane damaged. Only dead or damaged cells, therefore, take up propidium iodide and these fluoresce red. By simultaneously imaging at the appropriate emission wavelengths (Table 3.3) it is thus possible to determine which cells are living (green) and which are dead (red). Because fluorescein is poorly retained by cells, the live/dead cell assay should be performed immediately after adding the dyes. An alternative to using FDA is to use BCECF which is better retained by cells (Haugland, 2002).

◆◆◆◆◆◆ PRACTICAL ASPECTS OF LIVE-CELL IMAGING

Optimal live-cell imaging with the CLSM requires the use of "low-dose" techniques (e.g. using dyes at low non-cytotoxic concentrations with levels of laser irradiation at non-harmful visible wavelengths). Live-cell imaging thus requires more compromises than "dead-cell imaging" (i.e. of fixed cells). It is important to treat one's cells with the respect that they deserve in order that they remain as healthy as possible. The following are important practical aspects which one needs to pay close attention whilst imaging living fungal cells.

Sample Preparation

We have found that the simplest method for preparing samples in order that hyphae stay in the same plane of focus (especially important when imaging at low magnification, Figures 3.1, 3.2, 3.3A) and to reduce spherical aberration to a minimum (and thus have sharply in focus images) is to use the "inverted agar block method" (Figure 3.10). This involves taking a ~ 20 mm^2, 5 mm thick block of agar from whichever region of the colony needs to be imaged, and inverting it onto a droplet of liquid medium (containing dye) upon a glass cover slip (Hickey and Read, 2003). Excess medium can be removed with a filter paper wick. If the hyphae tend to grow into the agar, this can sometimes be inhibited by using a higher agar concentration (e.g. 3% w/v). If using a confocal system mounted on an inverted microscope then it is also useful to place a Petri dish containing water-soaked filter paper over the sample to reduce desiccation to a minimum. Most of the figures shown in the paper were from samples prepared in this way. Using the inverted agar block method we are usually able to image samples for 1 h or more without apparent deleterious effects on growing hyphae (e.g. from O_2 deprivation).

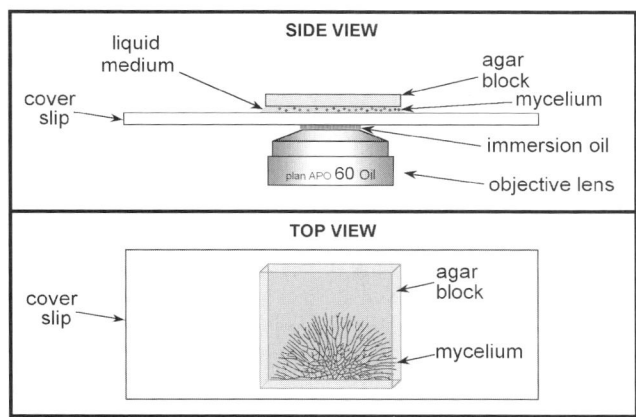

Figure 3.10. Inverted agar block culture method.

Temperature-sensitive mutants can be readily imaged at their restrictive temperatures (normally between 35 and 42°C), or during temperature shift down or shift up, by using a temperature-controlled stage or temperature-controlled box around the confocal microscope. However, it should be noted that rates of evaporation from the sample will be increased at elevated temperatures and thus extra liquid medium may have to be added more frequently during imaging. In Figure 3.11 we show an example of 4D imaging (x, y, z and time) with the *cot-1* mutant of *Neurospora* during a temperature shift down from the non-permissive temperature (37°C) to the permissive temperature (25°C). This temperature decrease results in the hyperbranching phenotype of the mutant (shown in yellow) reverting to the normal hyphal morphology of the wild type (shown in red).

Multiple Labelling Living Cells

Multiple labelling is often possible with living fungal cells. We routinely combine the red fluorescing FM4-64 as a general membrane stain with GFP targeted to different organelles or fused to different proteins (Figure 3.12; Freitag *et al.*, 2004b). The red fluorescing mitochondrial stain DASPMI can also be used in combination with green fluorescing probes such as GFP (Freitag *et al.*, 2004b). Triple labelling is also possible (e.g. with Calcofluor, GFP and FM4-64. Certain combinations of dyes, however, are sometimes incompatible and can result in increased phototoxicity when imaged with the CLSM.

The Signal-to-noise Ratio and Spatial versus Temporal Resolution

In the context of this chapter, the signal-to-noise ratio (S/N) is the ratio of the fluorescent signal to the noise from the surrounding background

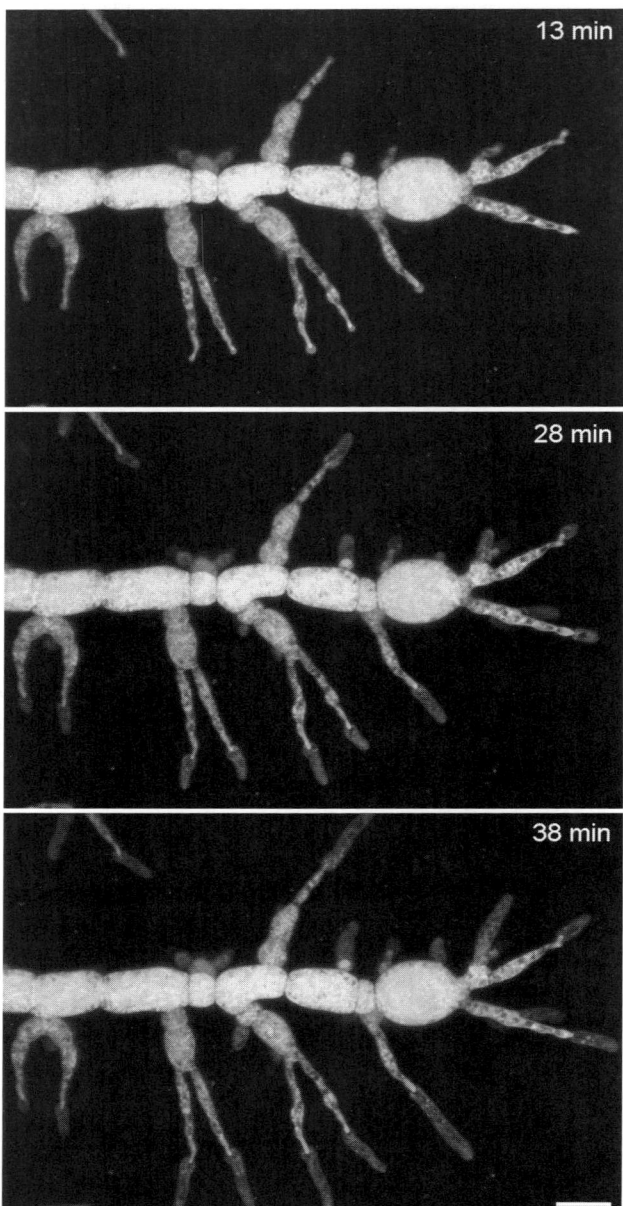

Figure 3.11. Hyperbranching *cot-1* mutant of *Neurospora crassa* stained with FM4-64 grown at 37°C for 3 h and shifted back to 25°C. Each image represents a projection of 30 × 1.0 μm optical sections through the hypha. The times represent the periods after which the hyphae were shifted to 25°C. Note the bulbous compartments of the main hypha and the finely tapered branches (yellow-green) which become wider during recovery of the wild type hyphal phenotype (red). Bar = 20 μm. (See colour plate 7.)

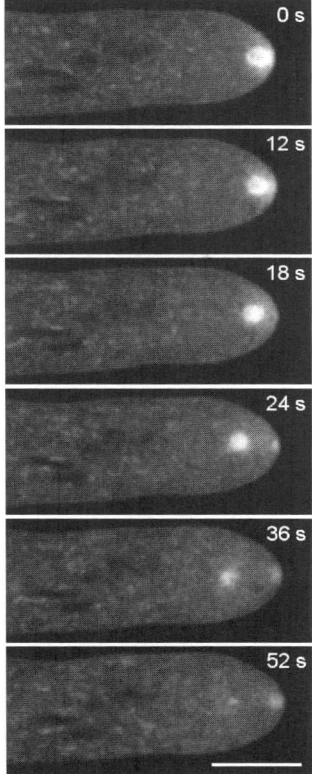

Figure 3.12. *Morchella esculenta*. Hyphal tip stained with FM4-64. Note retraction of the Spitzenkörper and formation of a new one after 24 s. Bar = 10 μm.

(e.g. autofluorescence and electronic noise from the photomultiplier detectors) from which the signal must be distinguished (for a detailed account of noise the reader is referred to Murphy (2001)). Ways of improving the S/N ratio with accompanying problems are indicated in Table 3.4.

Another problem is the compromise that needs to be made between spatial and temporal resolution. Generally increasing the temporal resolution (and thus the time period over which each image is captured) reduces the S/N and the spatial resolution achieved. The S/N ratio can be increased in various ways (see Table 3.4) but this usually results in increased cytotoxicity, phototoxicity, and/or photobleaching. Therefore, the live-cell imager needs to pay much closer attention to the microscope parameters that he/she is using than does the dead-cell imager.

Refractive Index Problems

Multiple changes in the refractive index between the objective and cell and within the cell will significantly reduce the spatial resolution and

Table 3.4. Ways of improving the signal-to-noise ratio for confocal laser scanning microscopy, and the potential problems caused

Ways of increase S/N ratio	Potential problems
Use dyes with a high quantum yield, which are not phototoxic and which are resistant to photobleaching	–
Optimize excitation and emission wavelengths	–
Use high NA objectives	Increased phototoxicity Increased photobleaching
Increase dye concentration	Increased cytotoxicity Increased phototoxicity
Increase laser power	Increased phototoxicity Increased photobleaching
Increase electronic zoom during imaging	Increased phototoxicity Increased photobleaching
Decrease rate of laser scanning	Increased phototoxicity Increased photobleaching
Integrate or average successive images	Increased phototoxicity Increased photobleaching
Open pinhole	Reduced spatial resolution

cause spherical aberration (Murphy, 2001). Some relevant refractive indices here are: air (1.0), immersion oil (1.52), cover slip glass (1.52), growth medium (>1.33), and cytoplasm (>1.33 but varies in different parts of a cell). These refractive index problems (and thus spherical aberration) are accentuated the further the cell is away from the cover slip. Ways of reducing these problems are to: reduce the number of changes of refractive index between the objective and the cell, use oil immersion, water immersion or dipping objectives (without a cover slip), and keep cells as close as possible to the cover slip.

Indicators of Cell Perturbation During Live-cell Imaging

We have found that hyphae of different species loaded with dyes vary in their sensitivities to being scanned with lasers of the CLSM, and varying degrees of cytotoxicity and phototoxicity are experienced with different fluorescent probes and at different excitation wavelengths. It is, therefore, important to pay close attention to the health of one's sample. Various indicators of cell perturbation which can be readily assessed are: slowing down of hyphal growth; narrowing down of the hyphal tip, retraction of the Spitzenkörper from the hyphal tip (Figure 3.12), and changes in organelle morphology.

Movie-making

One of the major advantages of live-cell imaging is that one is able to perform time-lapse experiments and make movies of time-lapse imaging. This can provide an extraordinary insight into the dynamics of labelled organelles and molecules in living fungal cells. Typically we capture 50–500 images at 2–30 s intervals in a time course. These images may then be processed (e.g. contrast and brightness adjusted, cropped and merged), and imported into a movie-making software package (e.g. Adobe Premiere) in which they can be compressed, edited and labelled. We normally save the final movie files in an avi or mpg format which can be easily inserted into Microsoft Powerpoint for presentation purposes. Examples of movies of filamentous fungi imaged with the CLSM after staining with many of the dyes described in this chapter can be found on a CD-ROM (Hickey and Read, 2003) which we have produced for educational purposes, and can be obtained from www.fungalcell.org.

◆◆◆◆◆◆ FUTURE DIRECTIONS FOR LIVE-CELL IMAGING OF FILAMENTOUS FUNGI

In this chapter, we have restricted ourselves to discussing those dyes which are well tried and tested in our lab. Many other commercially available dyes will undoubtedly prove extremely useful for live-cell imaging of filamentous fungi, and the experimenter is strongly encouraged to experiment with them. A good starting point is to leaf through the Molecular Probes catalogue (Haugland, 2002) and explore the Molecular Probes website (www.probes.com). We can expect the number of dyes for live-cell imaging to continually increase and these will compliment the rapidly increasing range of recombinant fluorescent probes which are also being developed (Zhang et al., 2002).

In this chapter, we have also restricted ourselves to discussing the imaging of filamentous fungi with the CLSM because this is the optical sectioning technology which is most widely used and is the one with which we have most experience. However, other optical sectioning techniques (Table 3.1) have different advantages (and disadvantages) to confocal laser scanning microscopy and for certain live-cell imaging applications will provide superior or complimentary results. Furthermore, low-light wide-field fluorescence imaging should also be routinely employed, especially for detecting very low levels of fluorescence. In addition, there are other important technological developments which are beginning to have an impact on live-cell imaging and analysis in filamentous fungi. These include: fluorescence recovery after photobleaching (FRAP) (Lippincott-Schwartz et al., 2001), spectral imaging (Berg, 2004), fluorescence resonance energy transfer (FRET) imaging (Periasamy and Day, 1999; Kenworthy, 2001), fluorescence lifetime imaging microscopy (FLIM) (Lacowicz, 1997; Gadella, 1999) and fluorescence correlation spectroscopy (FCS) (Medina and Schwille, 2002; Levin and Carson, 2004).

Acknowledgements

Thanks are due to the Biological and Biotechnological Sciences Research Council for a CASE Studentship to SRS, a Research Studentship to PCH and a grant (15/P18594) to NDR. Part of the research involved the use of CLSM equipment in the COSMIC (Collaborative, Optical, Spectroscopy, Micromanipulation and Imaging Centre) facility which is a Nikon Partners-in-Research laboratory at the University of Edinburgh.

References

Ashford, A. E., Cole, L. and Hyde, G. J. (2001). Motile tubular vacuole systems. In *Biology of the Fungal Cell.*, *The Mycota* (R. J. Howard and N. A. R. Gow, eds), vol. VIII, pp. 243–265. Springer, Berlin.

Atkinson, H. A., Daniels, A. and Read, N. D. (2002). Live-cell imaging of endocytosis during conidial germination in the rice blast fungus, *Magnaporthe grisea*. *Fungal Genet. Biol.* **37**, 233–244.

Bereiter-Hahn, J. (1976). Dimethylaminostyrylmethylpyridiniumiodine (daspmi) as a fluorescent probe for mitochondria *in situ*. *Biochim. Biophys. Acta* **423**, 1–14.

Berg, R. H. (2004). Evaluation of spectral imaging for plant cell analysis. *J. Microsc.* **214**, 174–181.

Betz, W. J., Mao, F. and Bewick, G. S. (1992). Activity-dependent fluorescent staining and destaining of living vertebrate motor nerve terminals. *J. Neurosci.* **12**, 363–375.

Betz, W. J., Mao, F. and Smith, C. B. (1996). Imaging exocytosis and endocytosis. *Curr. Opin. Neurobiol.* **6**, 365–371.

Bolte, S., Talbot, C., Boutte, Y., Catrice, O., Read, N. D. and Satiat-Jeunemaitre, B. (2004). FM-dyes as experimental probes for dissecting vesicle trafficking in living plant cells. *J. Microsc.* **214**, 159–173.

Chen, L. B. (1989). Fluorescent labeling of mitochondria. *Meth. Cell Biol.* **29**, 103–123.

Cochilla, A. J., Angleson, J. K. and Betz, W. J. (1999). Monitoring secretory membrane with FM1-43 fluorescence. *Annu. Rev. Neurosci.* **22**, 1–10.

Cole, L., Hyde, G. J. and Ashford, A. E. (1997). Uptake and compartmentalisation of fluorescent probes by *Pisolithus tinctorius* hyphae: evidence for an anion transport mechanism at the tonoplast but not for fluid-phase endocytosis. *Protoplasma* **199**, 18–29.

Cole, L., Orlovich, D. A. and Ashford, A. E. (1998). Structure, function and motility of vacuoles in filamentous fungi. *Fungal Genet. Biol.* **24**, 86–100.

Czymmek, K. J., Whallon, J. H. and Klomparens, A. (1994). Confocal microscopy in mycological research. *Exp. Mycol.* **18**, 275–293.

Czymmek, K. J., Bourett, T. M., Sweigard, J. A., Carrol, A. and Howard, R. J. (2002). Utility of cytoplasmic fluorescent proteins for live-cell imaging of *Magnaporthe grisea* in planta. *Mycologia* **94**, 280–289.

Diaspro, A. (2001). *Confocal and Two-Photon Microscopy: Foundations, Applications, and Advances*. Wiley-Liss, New York.

Dijksterhuis, J. (2003). Confocal microscopy of Spitzenkörper dynamics during growth and differentiation of rust fungi. *Protoplasma* **222**, 53–59.

Dulic, V., Egerton, M., Elgundi, I., Raths, S., Singer, B. and Riezman, H. (1991). Yeast endocytosis assays. *Meth. Enzymol.* **194**, 697–710.

Elorza, M. V., Rico, H. and Sentandreu, R. (1983). Calcofluor white alters the assembly of chitin fibrils in *Saccharomyces cerevisiae* and *Candida albicans* cells. *J. Gen. Microbiol.* **129**, 1577–1582.

Feijo, J. A. and Moreno, N. (2004). Imaging plant cells by two-photon excitation. *Protoplasma* **223**, 1–32.

Fischer-Parton, S. (1999). Role of pH, calcium and vesicle trafficking in regulating tip growth of *Neurospora crassa*. PhD Thesis. University of Edinburgh, Edinburgh, UK.

Fischer-Parton, S., Parton, R. M., Hickey, P. C., Dijksterhuis, J., Atkinson, H. A. and Read, N. D. (2000). Confocal microscopy of FM4-64 as a tool for analysing endocytosis and vesicle trafficking in living fungal hyphae. *J. Microsc.* **198**, 246–259.

Freitag, M., Hickey, P. C., Khlafallah, T. K., Read, N. D. and Selker, E. U. (2004a). HP1 is essential for DNA methylation in *Neurospora*. *Mol. Cell.* **13**, 427–434.

Freitag, M., Hickey, P. C., Raju, N. B., Selker, E. U. and Read, N. D. (2004b). GFP as a tool to analyze the organization, dynamics and function of nuclei and microtubles in *Neurospora crassa*. In Press.

Gadella, T. W. J. (1999). Fluorescence lifetime imaging microscopy (FLIM): instrumentation and applications. In *Fluorescent and Luminescent Probes for Biological Activity* (W. T. Mason, ed.), 2nd edn., pp. 467–479. Academic Press, London.

Haugland, R. P. (2002). *Handbook of Fluorescent Probes and Research Products*, 9th edn. Molecular Probes Inc., Eugene, OR.

Hawes, C., Crooks, K., Coeman, J. and Satiat-Jeunemaitre, B. (1995). Endocytosis in plants: fact or artefact? *Plant Cell Environ.* **18**, 1245–1252.

Henkel, A., Lubke, J. and Betz, W. (1996). FM1-43 dyes ultrastructural localization in and release from frog motor nerve terminals. *Proc. Natl Acad. Sci. USA* **93**, 1918–1923.

Hermann, G. J. and Shaw, J. M. (1998). Mitochondrial dynamics in yeast. *Annu. Rev. Cell Dev. Biol.* **14**, 265–303.

Hickey, P. C. (2001). Imaging vesicle trafficking and organelle dynamics in living fungal hyphae. PhD Thesis. University of Edinburgh, Edinburgh, UK.

Hickey, P. C. and Read, N. D. (2003). *Biology of Living Fungi*. British Mycological Society, Stevenage, UK, CD-ROM.

Hickey, P. C., Jacobson, D. J., Read, N. D. and Glass, N. L. (2002). Live-cell imaging of vegetative hyphal fusion in *Neurospora crassa*. *Fungal Genet. Biol.* **37**, 109–119.

Hoffman, J. and Mendgen, K. (1998). Endocytosis and membrane turnover in the germ tube of *Uromyces fabae*. *Fungal Genet. Biol.* **24**, 77–85.

Ichiwara, A., Tanaami, T., Ishida, H. and Shimizu, M. (1999). Confocal fluorescent microscopy using a Nipkow scanner. In *Fluorescent and Luminescent Probes for Biological Activity* (W. T. Mason, ed.), 2nd edn., pp. 344–349. Academic Press, San Diego.

Johnson, L. V., Walsh, M. L. and Chen, L. B. (1980). Localization of mitochondria in living cells with rhodamine 123. *Proc. Natl Acad. Sci. USA* **77**, 990–994.

Kenworthy, A. K. (2001). Imaging protein–protein interactions using fluorescence resonance energy transfer microscopy. *Methods* **24**, 289–296.

Knight, H., Trewavas, A. J. and Read, N. D. (1993). Confocal microscopy of living fungal hyphae microinjected with Ca^{2+}-sensitive fluorescent dyes. *Mycol. Res.* **97**, 1505–1515.

Koll, F., Sidoti, C., Rincheval, V. and Lecellier, G. (2001). Mitochondrial membrane potential and ageing in *Podospora anserina*. *Mech. Ageing Dev.* **122**, 205–217.

König, K. (2000). Multiphoton microscopy in life sciences. *J. Microsc.* **200**, 83–104.

Koning, A. J., Lum, P. Y., Williams, J. M. and Wright, R. (1993). DIOC6 staining reveals organelle structure and dynamics in living yeast cells. *Cell Motil. Cytoskel.* **25**, 111–128.

Lacowicz, J. R. (1997). *Topics in Fluorescence Spectroscopy: Nonlinear and Two-Photon-Induced Fluorescence*, vol. 5. Plenum Press, New York.

Levin, M. K. and Carson, J. H. (2004). Fluorescence correlation spectroscopy and quantitative cell biology. *Differentiation* **72**, 1–10.

Lippincott-Schwartz, J., Snapp, E. and Kenworthy, A. (2001). Studying protein dynamics in living cells. *Nat. Rev. Mol. Cell Biol.* **2**, 444–456.

Medina, M. A. and Schwille, P. (2002). Fluorescence correlation spectroscopy for the detection and study of single molecules in biology. *BioEssays* **24**, 758–764.

Miyawaki, A. (2003). Fluorescence imaging of physiological activity in complex systems using GFP-based probes. *Curr. Opin. Neurobiol.* **13**, 591–596.

Murphy, D. B. (2001). *Fundamentals of Light Microscopy and Electronic Imaging*. Wiley-Liss, New York.

Oparka, K. J. and Read, N. D. (1994). The use of fluorescent probes for studies on living plant cells. In *Plant Cell Biology. A Practical Approach* (N. Harris and K. J. Oparka, eds), pp. 27–50. IRL Press, Oxford.

Parton, R. M. and Read, N. D. (1999). Calcium and pH imaging in living cells. In *Light Microscopy in Biology* (A. J. Lacey, ed.), pp. 211–264. Oxford University Press, Oxford, UK.

Parton, R. M., Fischer, S., Malhó, R., Papasouliotis, O., Jelitto, T. C., Leonard, T. and Read, N. D. (1997). Pronounced cytoplasmic pH gradients are not required for tip growth in plant and fungal cells. *J. Cell Sci.* **110**, 1187–1198.

Pawley, J. B. (1995). *Handbook of Biological Confocal Microscopy*. Plenum Press, New York.

Periasamy, A. and Day, R. N. (1999). Visualizing protein interactions in living cells using digitized GFP imaging and FRET microscopy. *Meth. Cell Biol.* **58**, 293–313.

Read, N. D. and Hickey, P. J. (2001). The vesicle trafficking network and tip growth in fungal hyphae. In *Cell Biology of Plant and Fungal Tip Growth* (A. Geitmann, M. Cresti and I. B. Heath, eds), pp. 137–148. IOS Press, Amsterdam.

Read, N. D. and Kalkman, E. R. (2003). Does endocytosis occur in fungal hyphae? *Fungal Genet. Biol.* **39**, 199–203.

Read, N. D., Allan, W. T. G., Knight, H., Knight, M. R., Knight, R., Malhó, R., Russell, A., Shacklock, P. S. and Trewavas, A. J. (1992). Imaging and measurement of cytosolic free calcium in plant and fungal cells. *J. Microsc.* **166**, 57–86.

Roca, M. G., Davide, L. C., Mendes-Costa, M. C. and Wheals, A. (2003). Conidial anastomosis tubes in Colletotrichum. *Fungal Genet. Biol.* **40**, 138–145.

Roncero, C. and Duran, A. (1985). Effect of calcofluor white and congo red on fungal wall morphogenesis: *in vivo* activation of chitin polymerization. *J. Bacteriol.* **170**, 1950–1954.

Sheppard, C. J. R. and Shotton, D. M. (1997). *Confocal Laser Scanning Microscopy*. IOS Scientific Publishers, Oxford.

Silverman-Gavrila, L. B. and Lew, R. R. (2000). Calcium and tip growth in *Neurospora crassa*. *Protoplasma* **213**, 203–214.

Silverman-Gavrila, L. B. and Lew, R. R. (2001). Regulation of the tip-high [Ca^{2+}] gradient in growing hyphae of the fungus *Neurospora crassa*. *Eur. J. Cell Biol.* **80**, 379–390.

Silverman-Gavrila, L. B. and Lew, R. R. (2003). Calcium gradient dependence of *Neurospora crassa* hyphal growth. *Microbiology* **149**, 2475–2485.

Steinberg, G., Schliwa, M., Lehmler, C., Bölker, M., Kahmann, R. and McIntosh, J. R. (1998). Kinesin from the plant pathogenic fungus *Ustilago maydis* is involved in vacuole formation and cytoplasmic migration. *J. Cell Sci.* **111**, 2235–2246.

Swedlow, J. R. (2003). Quantitative fluorescence microscopy and image deconvolution. *Meth. Cell Biol.* **72**, 346–367.

Swift, S. R., Hart, C. A., Bartlett, D. W. and Read, N. D. (2001). Interactions between azoxystrobin and *Puccinia recondita*, *Erysiphe graminis*, and *Botrytis cinerea* on the microscale. *Scanning* **23**, 153–154.

Torralba, S. and Heath, I. B. (2002). Analysis of three separate probes suggests the absence of endocytosis in *Neurospora crassa* hyphae. *Fungal Genet. Biol.* **37**, 221–232.

Vida, T. A. and Emr, S. D. (1995). A new vital stain for visualizing vacuolar membrane dynamics and endocytosis in yeast. *J. Cell Biol.* **128**, 779–792.

Zhang, J., Campbell, R. E., Ting, A. Y. and Tsien, R. Y. (2002). Creating fluorescent probes for cell biology. *Nat. Rev. Mol. Cell Biol.* **3**, 906–918.

4 One-photon versus Two-photon Laser Scanning Mic roscopy and Digital Image Analysis of Microbial Biofilms

Thomas R Neu[1] and John R Lawrence[2]

[1] Department of River Ecology, UFZ Centre for Environmental Research Leipzig-Halle, Magdeburg, Germany; [2] NWRI National Water Research Institute, Saskatoon, Sask., Canada

◆◆◆

CONTENTS

Introduction
History and evolution of laser scanning microscopy (LSM)
LSM with one-photon excitation
LSM with two-photon excitation
Biofilm samples and mounting
Biofilm LSM imaging
Three-dimensional digital image analysis
New microscopic techniques using laser excitation
Instead of conclusions

◆◆◆◆◆◆ **INTRODUCTION**

Laser scanning microscopy (LSM) nowadays represents a standard optical technique in the life sciences. LSM can be employed in many different modes including reflection and fluorescence, single and multichannel, one-photon and two-photon, with data collection via intensity and lifetime imaging. To provide the reader with a starting point in the field, books and review articles on LSM theory and technical details as well as fluorescent probes and microbiological applications have been detailed in Table 4.1. The interested reader is referred to these publications in order to pick up various topics such as fluorescence basics, fluorescence microscopy, laser microscopy, fluorescent probes, biofilms, intensity imaging, lifetime imaging, basic digital image analysis, future developments, etc. The focus of this chapter is on one-photon and two-photon imaging of microbiological samples and digital image analysis of the resulting LSM data sets.

Table 4.1. Literature on fluorescence, laser microscopy, fluorochromes, image analysis and applications to microbiological samples

Topic	References
Fluorescence basics	Lakowicz (1999), Valeur (2002), Jameson et al. (2003)
Introduction to laser microscopy	Sheppard and Shotton (1997), Müller (2002)
Confocal microscopy	Pawley (1995), Conn (1999)
Advanced laser microscopy	Marriott and Parker (2003a,b)
Fluorescent probes	Haugland (2002), Mason (1999)
Microbial biofilm methods	Doyle and Ofek (1998), Doyle (1999, 2001a,b)
LSM of microbial biofilms	Lawrence et al. (1998b, 2002), Lawrence and Neu (1999), Neu and Lawrence (2002)
LSM of biofilm glyco-conjugates and exopolymers	Neu and Lawrence (1999a,b), Neu et al. (2001), Staudt et al. (2003)
General image analysis and digital image analysis of microbiological data sets	Russ (2002), Wilkinson and Shut (1998)

◆◆◆◆◆◆ HISTORY AND EVOLUTION OF LASER SCANNING MICROSCOPY

The principle of confocal laser scanning microscopy (CLSM) was already suggested by Marvin Minsky in the year 1957. However, at that time the technical requirements, e.g. lasers and computers, to build such an instrument were not available. At a later stage he wrote an article in which he remembers the invention of the confocal microscope (Minsky, 1988). The first instruments were built in 1982 mainly for the semiconductor industry and they were used in the reflection mode only. Later in 1987, when the first instruments became commercially available with reflection and fluorescence options, they were mostly employed in medical and cell biological research. However, it took until 1991 when LSM was first demonstrated to be a valuable tool in microbiology and especially in biofilm research (Lawrence et al., 1991). In the meantime LSM has become an indispensable technique for studying structures and processes in microbial biofilms (Lawrence et al., 2002; Neu and Lawrence, 2002). In terms of excitation LSM can be divided into one-photon and two-photon LSM. In addition, the fluorescence may be recorded as an intensity or lifetime signal. Mostly the fluorescence intensity is recorded after one-photon excitation and the technique is usually called CLSM or one-photon LSM (1P-LSM). If the intensity is recorded after two-photon excitation the technique is called two-photon LSM (2P-LSM) or multiphoton LSM. It has to be pointed out that in the literature various abbreviations are used. Apart from fluorescence intensity the fluorescence signal carries another piece of information, the fluorescence lifetime. If the lifetime of a fluorochrome is recorded the technique is called fluorescence lifetime imaging or fluorescence lifetime imaging microscopy (FLIM).

The history of two-photon LSM also covers several decades until it became a useful microscopic tool in biology. Already in 1931 Maria Göppert-Meyer from Germany who later became a Nobel prize winner theoretically predicted two-photon excitation processes (Göppert-Mayer, 1931). Thirty years later the experimental proof of two-photon excitation was published (Kaiser and Garrett, 1961). But it took until 1990 when two-photon excitation was used for the first time in microscopy (Denk *et al.*, 1990). In comparison to 1P-LSM where visible and ultraviolet continuous wave lasers are used for excitation of fluorochromes, in 2P-LSM a pulsed titanium–sapphire laser producing infrared light is employed. In fact the system usually consists of three lasers, a laser diode bar, a pump laser and an infrared laser. Two-photon excitation has several major advantages, e.g. excitation in the focal plane only, deeper penetration of infrared light into scattering samples, less light scattering, higher resolution in deep areas of the sample, no need for UV excitation and UV optics. As a consequence two-photon systems are suitable for examination of thick, scattering biological samples such as microbial communities. The various aspects of two-photon excitation in microscopy have been covered in several review articles (Cox and Sheppard, 1999; Denk *et al.*, 1995; Williams *et al.*, 1994).

◆◆◆◆◆◆ LSM WITH ONE-PHOTON EXCITATION

As indicated in the introduction and in Table 4.1 there are already several reviews on CLSM of microbiological samples. The basics of analytical microscopy and general technical requirements of a suitable LSM system as well as fluorescence staining techniques have been described by Lawrence *et al.* (2002) and Lawrence and Neu (1999) in detail. These overviews describe how to get started, possible applications in microbiology and the basics for analysis of 3D image data. Furthermore, a structured approach to stepwise examine a new, unknown biofilm sample has been outlined (Neu and Lawrence, 2002). First, in this approach (see Figure 4.1) advantage is taken of the intrinsic sample properties such as reflection and autofluorescence. Second, simple but specific staining for cellular and polymeric biofilm constituents and their distribution is applied. Third, probes for cell identity, cell viability as well as cell and enzyme activity are discussed. Finally the authors suggest various probes for the microenvironment (Neu and Lawrence, 2002). With the exception of reflection imaging, this structured approach may also be applicable if 2P-LSM is used.

Figures 4.2–4.4 demonstrate the first steps of this structured approach through a combination of reflection, autofluorescence, nucleic acid staining and glycoconjugate staining. In Figure 4.2 a lotic aggregate as an example

LSM approach		information
reflection	-	mineral content, cell surface reflection, cell inclusions
autofluorescence	-	phototrophic organisms, cyanobacteria/algae differentiation
nucleic acid staining	-	bacterial cell distribution and biomass
lectin binding analysis	-	exopolymer, glycoconjugate distribution and identity
in situ hybridization	-	cell identity and interactions
viability	-	live and compromised bacteria
cell activity	-	active and inactive bacteria
enzyme activity	-	intracellular and extracellular, distribution
microenvironment	-	permeability, diffusion, pH, ion distribution, heavy metal distribution

Figure 4.1. Structured approach for LSM examination of an unknown biofilm sample after Neu and Lawrence, 2002 (by permission of John Wiley & Sons, Inc.).

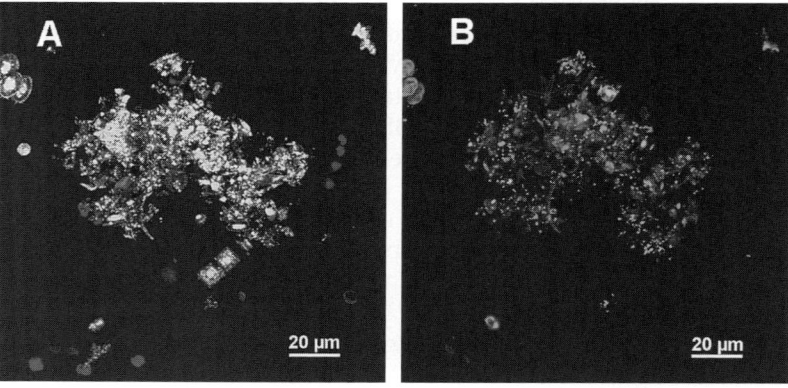

Figure 4.2. 1P-LSM maximum intensity projection of dual channels from the same lotic aggregate. (A) shows the reflection (= white) and autofluorescence (= blue). (B) shows the nucleic acid stained bacteria (= green) and the lectin stained glycoconjugates (= red). (See colour plate 8.)

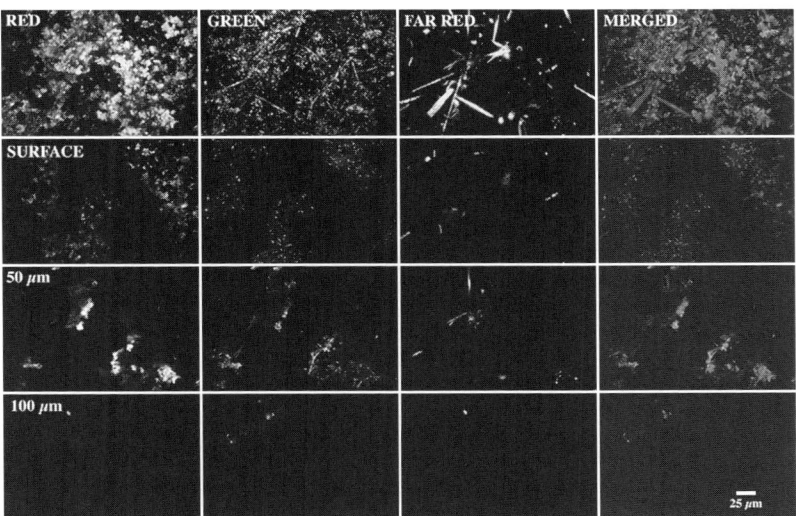

Figure 4.3. 1P-LSM images showing the distribution of bacteria, photosynthetic organisms (algae and cyanobacteria) and exopolymer in a river biofilm. The series illustrates the use of a combination of fluorescent probes (nucleic acid and lectin) in combination with autofluorescence to determine the vertical distribution of bacteria, total photosynthetic organisms and exopolymer in a river biofilm community. (See colour plate 9.)

for a mobile biofilm is shown by imaging dual channels. The second example illustrates a stationary biofilm by imaging of single channels at various depths and as an overlay (Figure 4.3) including quantification of the data (Figure 4.4). Other approaches for using CLSM monitoring in combination with various reporter systems have been carried out. An important application is microautoradiography in combination with fluorescent *in situ* hybridization to determine physiological activity as well as phylogenetic identity (Gray *et al.*, 2000; Lee *et al.*, 1999; Nielsen *et al.*, 2002, 2003; Ouverney and Fuhrman, 1999).

Since the first report in 1991, CLSM has been established as the key technique for studying the structure and function of microbial biofilms. For further detailed information the reader may be referred to the literature in Table 4.1. Nevertheless, there are certain limitations to the applications of these approaches which are listed in Table 4.2. Some of these limitations may be overcome simply by employing two-photon excitation. However, these instruments are not yet readily available in most institutes. As a consequence it is important that the operator is aware of the limitations and is able to collect the data in the most appropriate way to minimize these effects.

◆◆◆◆◆◆ LSM WITH TWO-PHOTON EXCITATION

As already mentioned, the application of two-photon excitation in microscopy starts with the pioneering work of Denk *et al.* (1990).

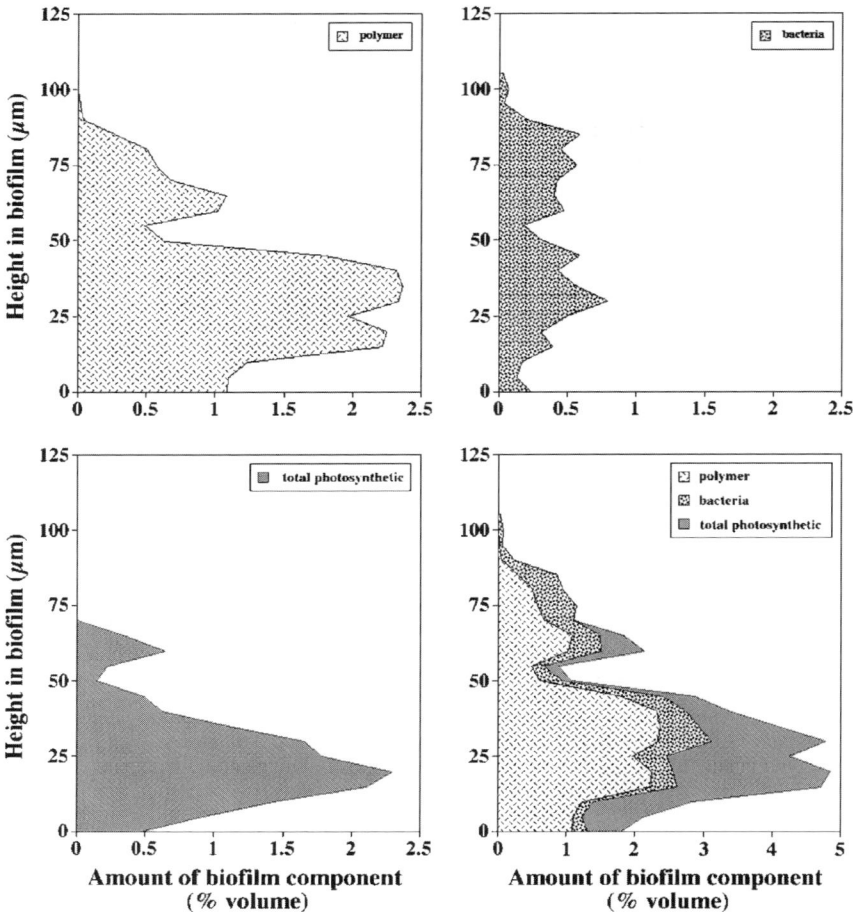

Figure 4.4. The graphs show the results of image analyses of the LSM data sets from Figure 4.3. It illustrates the signals collected in different channels and the vertical distribution of biofilm components.

The literature available since may be divided into papers on technical progress and those dealing with applications in biology. Several authors have published a direct comparison of one-photon versus two-photon microscopy (Centonze and White, 1998; Gu and Sheppard, 1995; Periasamy *et al.*, 1999). Published biological applications of 2P-LSM have been compiled in Table 4.3. Some of the approaches may also be applied for imaging structures and processes in microbiological samples. A number of authors have tested fluorochromes for their general applicability in two-photon excitation (Albota *et al.*, 1998; Bestvater *et al.*, 2002; Neu *et al.*, 2002; Xu *et al.*, 1996). In general, most of the fluorochromes can be used for two-photon excitation although their excitation/emission response will be different from that in 1P-LSM. Whether or not they will show an emission will depend on the laser wavelengths available. For example, if the wavelengths of 700–1050 nm are available they should

Table 4.2. Limitations of confocal or one-photon LSM and possible solutions by using two-photon LSM

One-photon limitations	Comments – solution
Bleaching in–out of focus areas	May cause problems if sampling is done at high axial resolution and if many optical sections are recorded – 2P-LSM
Cell damage in–out of focus areas	Important if the same sample is studied over time – 2P-LSM
Depth of laser penetration	This depends strongly on the density and scattering properties of the sample – 2P-LSM
Fluorescence of background	Examples: biofilms on certain plastic material, some type of rocks, concrete – 2P-LSM
Low resolution in axial direction	In optical microscopy the XZ resolution (low) is unequal to the XY resolution (high)
Large differences in fluorescence emission intensity	The photomultiplier can only be optimized to one intensity in a selected field to be scanned

Table 4.3. Current biological applications of two-photon excitation

Biological topic studied	References
Ion channels, calcium dynamics	Denk (1994), Yuste and Denk (1995)
Quantitative GFP imaging	Drummond *et al.* (2002), Niswender *et al.* (1995)
NAD(P)H redox imaging	Bennett *et al.* (1996), Piston *et al.* (1994, 1999)
Metabolism and transport in fungi	Bago *et al.* (2002)
Plant cells	Cheng *et al.* (2001)
Lymphocyte behaviour	Bousso *et al.* (2002), Miller *et al.* (2002)
Tissue imaging	Masters *et al.* (1999), Ragan *et al.* (2003)
General biological application	Diaspro and Robello (2000), König (2000)

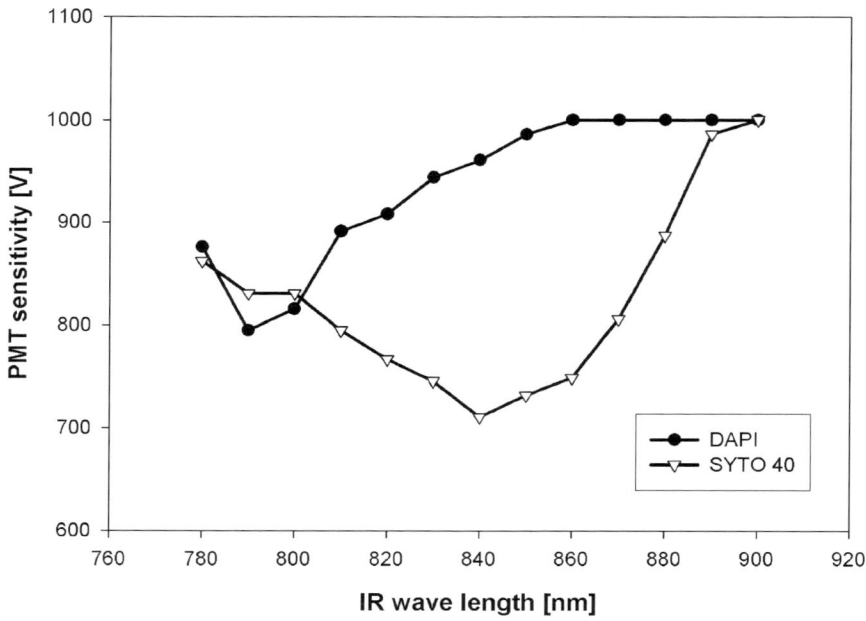

Figure 4.5. Two-photon emission signals of the nucleic acid specific fluorochromes DAPI and SYTO 40 measured in living biofilms. Note the specific excitation wavelengths and the broad nature of the emission spectra.

allow two-photon excitation of fluorochromes having a corresponding 350–525 nm one-photon excitation range. As the fluorochromes in two-photon excitation have a broader excitation cross-section this range may be even wider. Another option in 2P-LSM is that several of the fluorochromes may be combined and excited with one two-photon wavelength only and the resulting emission signals individually collected in separate channels (Neu et al., 2002). Figure 4.5 illustrates the emission signals for two different nucleic acid specific fluorochromes (DAPI, SYTO 40) in 2P-LSM applications. Both fluorochromes may be excited by one photon using a UV laser or a blue laser diode.

Biological samples usually have a high degree of light scattering. Due to this fact it is important to mention a series of comparisons where one-photon versus two-photon imaging in turbid media and scattering samples were measured (Gan and Gu, 2000a,b; Gu et al., 2000a,b; Schilders and Gu, 2000). The results for turbid and scattering media may be summarized as follows: (1) the one-photon signal has a high intensity but a low resolution in deep areas, (2) the two-photon signal has a lower intensity in deep areas, but a higher resolution in these areas where it is still detectable. In other words, the signal intensity in two-photon excitation drops faster than in one-photon excitation. However, resolution is higher in two-photon excitation. These findings may be important for microbiological samples from the environment where a lot of scattering material (clays, colloids, organic matter) can be expected to be present. Figure 4.6 illustrates the direct comparison of 1P-LSM and

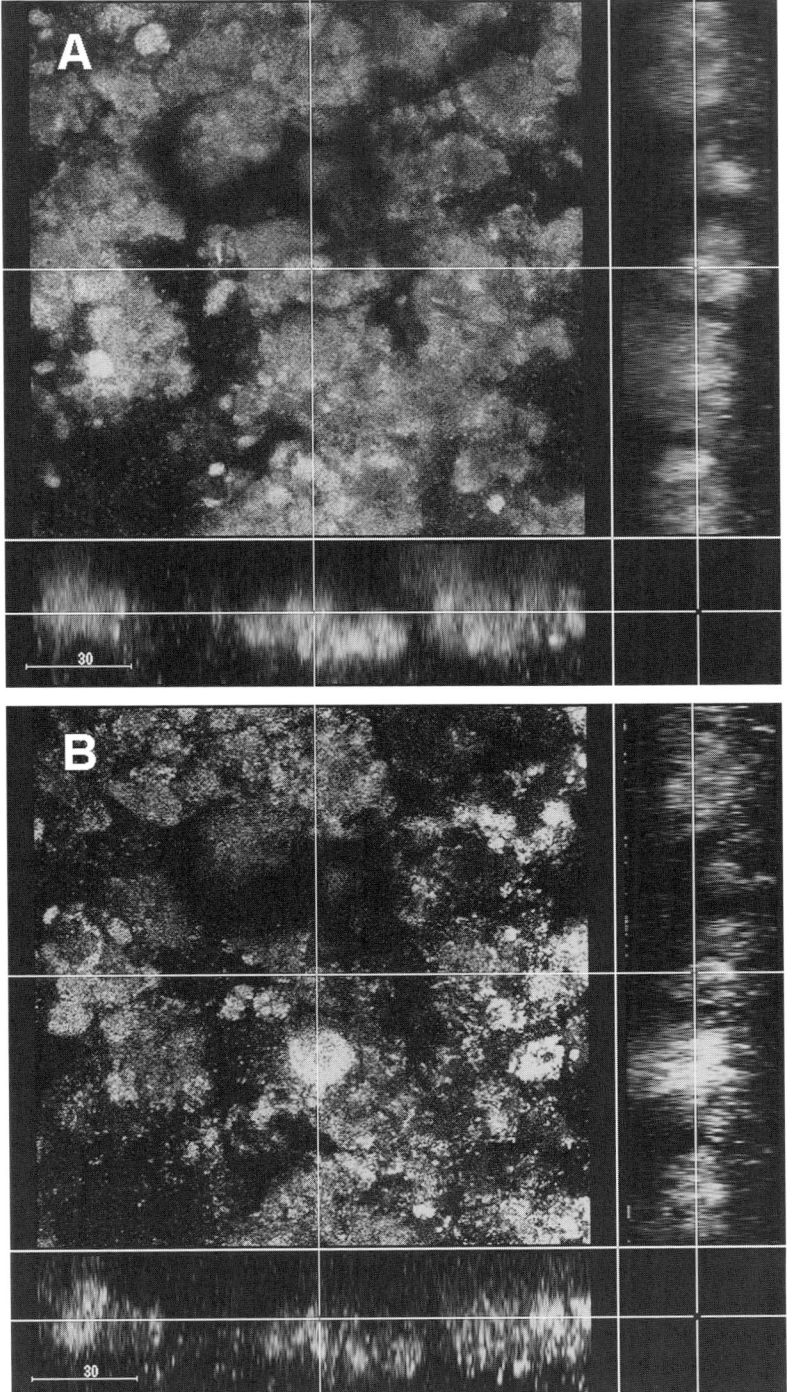

Figure 4.6. 1P-LSM (A) and 2P-LSM (B) of the same location in XY and XZ of laboratory biofilms grown on pumice with EDTA medium. Biofilms were stained with the nucleic acid specific stain SYTO 40. (See colour plate 10.)

2P-LSM in microbiological materials, i.e. a biofilm community growing on a porous pumice support which degrades EDTA.

Two-photon excitation was employed for the first time in microbiology by Sytsma *et al.*, they used acridine orange (AO) to image dense oral biofilms (Sytsma *et al.*, 1998). Later the group demonstrated high resolution in deep oral biofilm locations by means of two-photon excitation after staining with rhodamine B. If compared with one-photon excitation, it could be shown that with two-photon excitation the penetration depth was increased approximately fourfold (Gerritsen and Grauw, 1999; Vroom *et al.*, 1999). A combination of one-photon and two-photon LSM was also employed to study the microbial community of stromatolites. For this purpose the samples were embedded in Nanoplast. After staining, DAPI was imaged in combination with 2P-LSM whereas fluorescein-ConA was imaged in combination with 1P-LSM (Decho and Kawaguchi, 1999). In a second investigation the *in situ* microspatial arrangements of microbial cells and their extracellular polymeric secretions within marine stromatolites including imaging of endolithic cyanobacterial cells within carbonate sand grains were studied using the same approach (Kawaguchi and Decho, 2002).

Another application of two-photon excitation is the measurement of diffusion coefficients in deep biofilm regions. A technique called fluorescence recovery after photobleaching (FRAP) was employed for determination of diffusion coefficients of fluorescein, dextran and BSA *in situ* (Bryers, 2001).

In terms of fluorochromes useful for two-photon excitation their suitability for two-photon excitation has to be established. An assessment of fluorochromes useful for the investigation of microbial biofilms has already been published (Neu *et al.*, 2002). The fluorochromes tested included nucleic acid stains as well as probes for the polymer matrix of biofilms. The results showed that simple doubling of the one-photon wavelength is not an absolute "rule" for two-photon excitation. Consequently, there is flexibility in using one infrared laser line for excitation of several fluorochromes. The possibility of two-photon excitation using a single wavelength and multichannel recording was demonstrated for aquatic biofilm samples where the autofluorescence of phototrophic organisms and a nucleic acid stain was excited with 800 nm. The emission signal was then separated and recorded in two different channels (Neu *et al.*, 2002). Recently, two-photon excitation was employed in studying different types of phototrophic biofilm microorganisms. It was shown that the autofluorescence of algae, based on chlorophyll A only and of cyanobacteria, based on chlorophyll A and phycobilins can be separated by means of 2P-LSM (Neu *et al.*, 2004). An example of two-photon excitation with imaging of emission signals in three channels, 2 × autofluorescence and nucleic acid staining, is shown for a lotic biofilm community (Figure 4.7).

Despite its many advantages, two-photon excitation also has some limitations (Table 4.4). These are related to financial constraints and sample properties. For example, if the full range of possible IR wavelengths (e.g. 700–1010 nm) are to be available for the user there

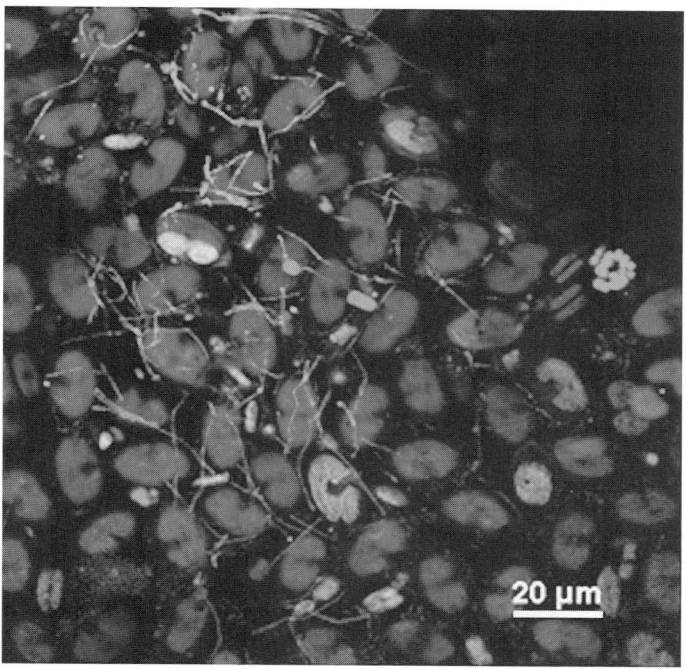

Figure 4.7. 2P-LSM of a SYTO 9 stained lotic biofilm community with excitation at 800 nm. Colour allocation: nucleic acid stain (= green), autofluorescence of cyanobacteria (= pink), autofluorescence of algae (= blue). (See colour plate 11.)

Table 4.4. Limitations of two-photon LSM

Two-photon limitations	Comments
Available infrared wavelengths	To have the full range available (700–1010 nm) several mirror sets (3) are usually necessary
Two-photon signal intensity	Usually lower compared to one photon, therefore the PMT has to be set more sensitive resulting in more noise, thus more averaging is necessary and as a consequence the sample is more bleached and damaged although in the focal region only
Infrared absorbing samples	This may cause problems due to cellular pigments but also with all kind of plastic materials as substratum for biofilm growth
Low resolution in axial direction	In optical microscopy the XZ resolution (low) is unequal to the XY resolution (high)
Large differences in fluorescence emission intensity	The photomultiplier can only be optimized to one intensity in a selected field to be scanned

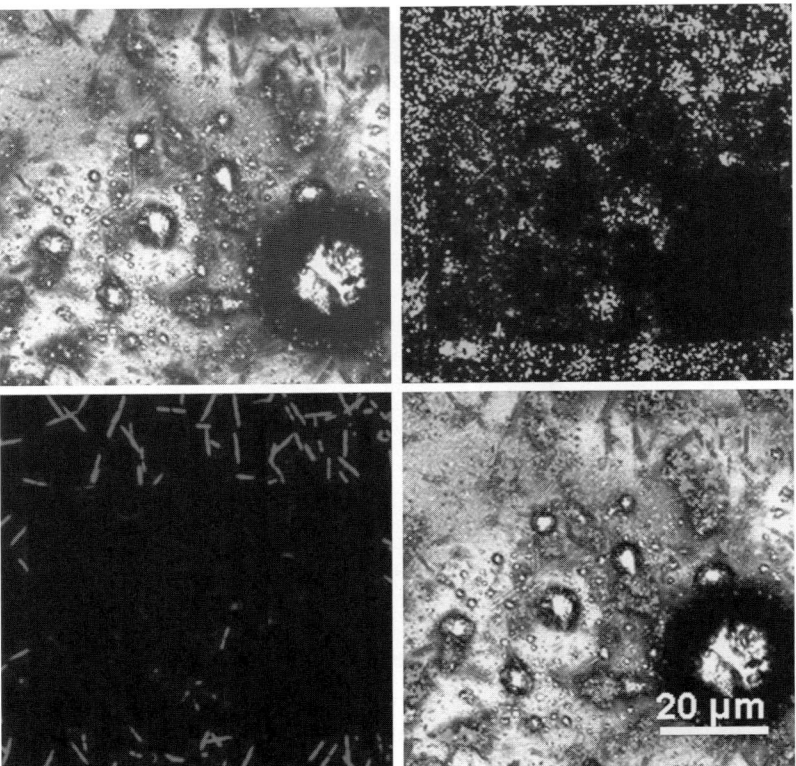

Figure 4.8. Effect of wrong settings in two-photon excitation on biofilm and substratum surface. The images show the bleaching and damaging effect in different channels and as an overlay. Take notice of the bubbles created by melting of the polycarbonate substratum. Colour allocation: reflection (= white), nucleic acid stain (= green), autofluorescence of algae (= blue). See also text for further details. (See colour plate 12.)

is a need for several mirror sets (2 or 3) making the system expensive. Furthermore, the exchange and alignment of mirror sets usually has to be done by the company technician adding to the complexity of use. Another problem can be IR absorption by the sample which results in local heating, thereby the cells will be damaged or in some cases the material (plastics) to which cells are attached will be melted. This is demonstrated in Figure 4.8 where a small region of a biofilm sample was damaged by two-photon excitation using a 63 × lens. Then the same location was scanned by using a 20 × lens. As a result the damaged and intact biofilm and substratum can be imaged.

◆◆◆◆◆◆ BIOFILM SAMPLES AND MOUNTING

The biofilm samples for LSM studies may have their origin in any habitat. We have applied LSM to microbiological samples from very different environments. These included biofilms from moderate habitats such as rivers and lakes but also biofilms from extreme habitats such as pH 1 volcanic creeks, Antarctica or the deep sea. Basically one can distinguish between natural samples directly from the environment and biofilm samples from designed and exposed devices. The latter may originate from the environment, from any technical system or from the medical field.

The only limitation in terms of the biofilm sample is the size and weight of a sample. In general for wide field microscopy using a slide and a cover slip as a mounting option, the sample size is restricted to the few micrometers in between the two glass surfaces. However, for LSM the dimensions of samples may be much larger, from several hundred μm to mm and even cm. Somehow, the sample just has to fit on the microscope between the objective lens and the microscope table. Usually there is a free space of about 2–3.5 cm on an upright standard microscope. Due to the sensitivity of the galvanometer stage the sample should not be too heavy (maximal 150 g). If the scanning is done with the microscope table the sample can be even heavier. How a sample is then mounted depends on the type of microscope available, upright or inverted. Several options are listed in Table 4.5.

With respect to LSM there may be another limitation which is determined by the sample properties. Both types of lasers for one-photon or two-photon excitation cannot indefinitely penetrate a turbid sample. The penetration depth depends on the density and scattering properties of the biofilm sample. As a consequence, if the samples become too thick they have to be embedded and physically sectioned. Embedding may be done using Nanoplast, Epon, Paraffin or a so-called tissue freezing medium with subsequent sectioning using a normal microtome or a cryotome, respectively. The slices cut may be rather thick (e.g. 50–100 μm or more) as they finally will be scanned from the side. In this case, fluorochrome staining of deep samples locations and laser penetration represent no limitation. For example, Nanoplast resin was employed for on-site embedding of stromatolite samples. For this purpose the samples were fixed in formaldehyde and stained prior to embedding. This technique is most suitable if samples have to be transported (Decho and Kawaguchi, 1999). Rocheleau *et al.* used paraffin embedding of USAB granules and sectioning to create 10 μm serial sections. These were then hybridized and imaged to determine the numbers and distribution of methanogens in the granules (Rocheleau *et al.*, 1999). Figure 4.9 illustrates a USAB granule cross-section as observed using 1P-LSM and the results of digital analyses showing the distribution of *Methanosarcina* sp. identified using fluorescent *in situ* hybridization within the granules. Cryosectioning of pure culture and environmental biofilms has been demonstrated in combination with DAPI and CTC staining (Huang *et al.*, 1996; Yu *et al.*, 1994). Sectioning was also employed for a variety of thick biofilm systems. Examples presented in Figure 4.10 are fungal pellets from a bioreactor, biofilms on carrier material from a fluidized bed reactor and phototrophic

Table 4.5. Procedures for biofilm sample mounting and subsequent LSM

Sample	Mounting	LSM, lens
Upright microscope		
Living, hydrated biofilms on flat surfaces of any material (e.g. slides)	Glue surface in 5 cm Petri dish, staining on surface in Petri dish, later flood Petri dish with water	Water immersible lenses, e.g. 40 × 0.8 NA, 63 × 0.9 NA, 63 × 1.2 NA
Living, hydrated biofilms on uneven surface of any material (e.g. rock)	Mount a small piece in Petri dish, staining on surface in Petri dish, later flood Petri dish with water	Water immersible lenses, e.g. 40 × 0.8 NA, 63 × 0.9 NA, 63 × 1.2 NA
Living, hydrated biofilms in form of aggregates, granules or on carrier material (e.g. fluidized bed reactors)	Put a few aggregates or biofilm carriers in cover well chamber, staining in chamber, flood with water, cover with cover slip	Water immersion lenses, e.g. 63 × 1.2 NA
Living, hydrated thick and dense biofilms	Embedding of biofilm necessary, sectioning of thick slices for microscopy	Examination from the side, lens depends on method, mounting with or without cover slip
Thick and dense biofilms fixed and embedded in the field	Sectioning of thick slices for microscopy	Examination from the side, lens depends on method, mounting with or without cover slip

Inverted microscope		
Living, hydrated biofilms on flat surfaces on any material (e.g. slides)	Use chambers with cover slip bottom, mount upside down using spacers, stain inside chamber and later fill chamber with water	Water immersion lenses, e.g. 63 × 1.2 NA. Limitations due to working distance of lens (220 nm)!
Living, hydrated biofilms on uneven surfaces of any material (e.g. leaves, rock)	Mount in chamber with cover slip bottom, stain in chamber and later fill chamber with water	Water immersion lenses, e.g. 63 × 1.2 NA. Limitations due to working distance of lens (220 nm)!
Living, hydrated biofilms in form of aggregates, granules, on carrier material (e.g. fluidized bed reactors)	Put a few aggregates or biofilm carriers in chamber with cover slip bottom, stain in chamber, later fill chamber with water	Water immersion lenses, e.g. 63 × 1.2 NA. Limitations due to working distance of the lens (220 nm)!
Living, hydrated thick and dense biofilms	Embedding of biofilm necessary, sectioning of thick slices for microscopy; transfer slices in chambers with cover slip bottom	Water immersion lenses, e.g. 63 × 1.2 NA. Limitations due to working distance of the lens (220 nm)!
Biofilms fixed and embedded in the field	Sectioning of thick slices for microscopy, transfer slices in chambers with cover slip bottom	Water immersion lenses, e.g. 63 × 1.2 NA. Limitations due to working distance of lens (220 nm)!

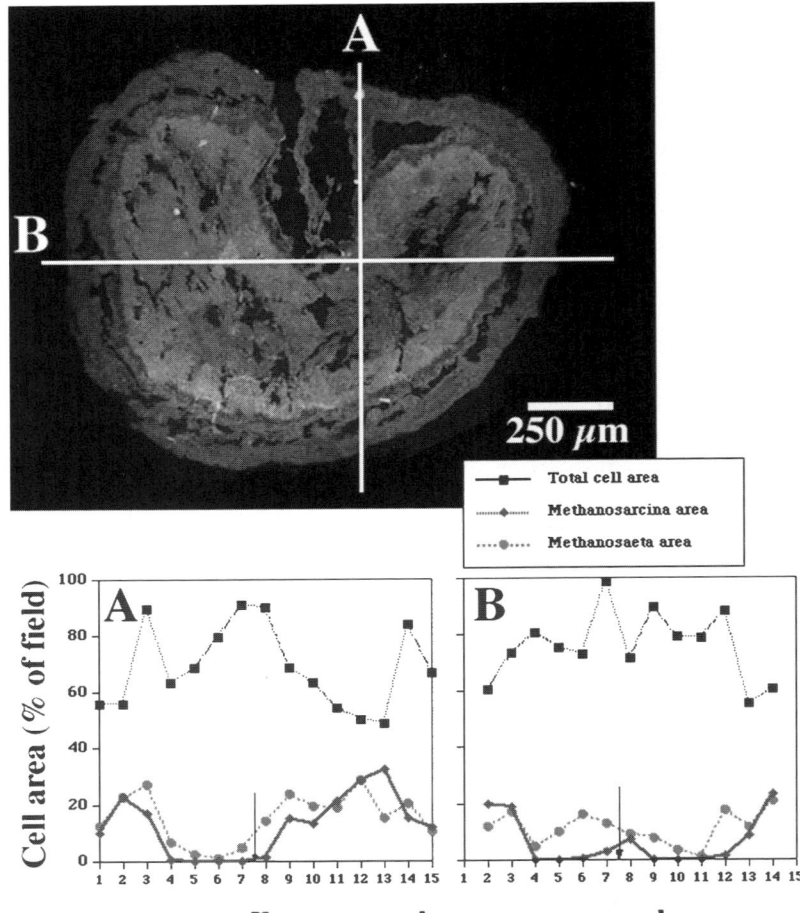

Figure 4.9. 1P-LSM imaging of 8-mm-thick sections of UASB granules that were simultaneously *in situ* hybridized with a MS5 *M. concilii*-specific probe [labelled with TAMRA NHS (red)] and the MB4 *Methanosarcina barkeri*-specific probe [labelled with Cy5 (blue)]. Top: Low-resolution 1P-LSM optical thin section (*XY* plane) showing the layered structure of *Methanosarcina*-enriched granules, in which *M. barkeri* and *M. concilii* are present within the layers. Bottom: Relative quantifications of *M. barkeri* and *M. concilii* cell areas by LSM and digital image analysis. Transects across the centres of the granules were analysed, and the cell area containing both *M. concilii* and *M. barkeri* as well as the total cell area (detected with SYTO 9) was determined. Data are expressed as relative cell area (percentages of the field) versus location within the granule. Frame numbers represent transects of these thin sections through the granule. (See colour plate 13.)

biofilms developed in a special incubator. It may be important to mention that samples which have been physically sectioned and scanned from the side require special software routine for quantification.

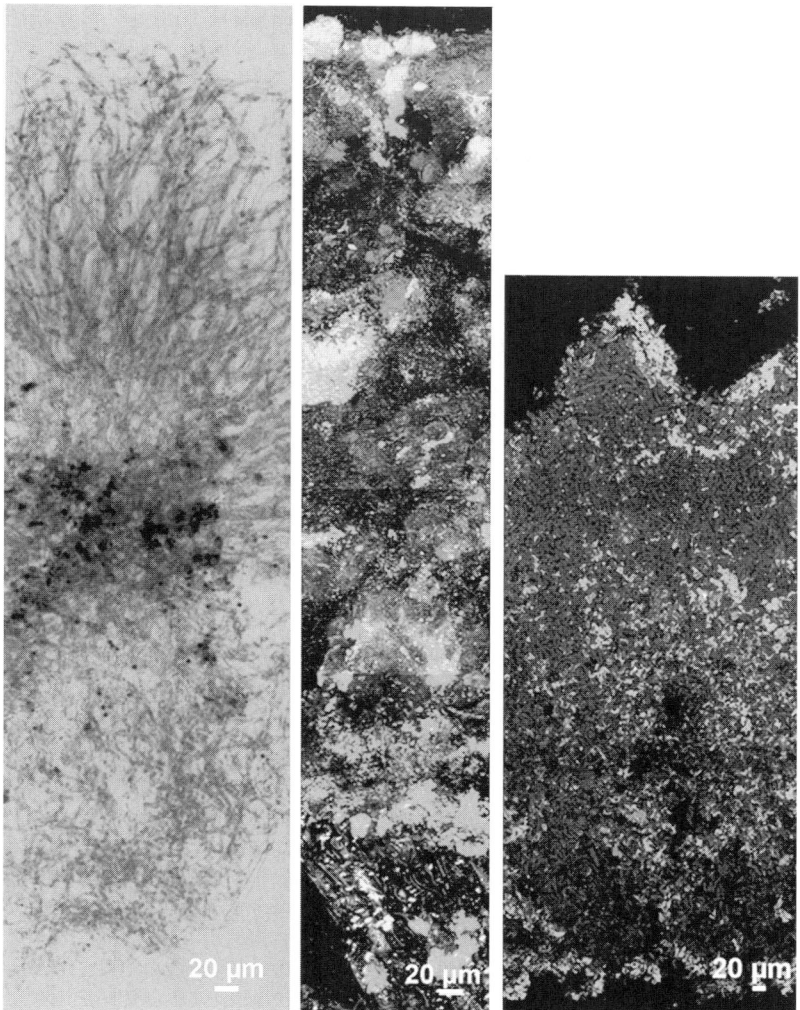

Figure 4.10. Examples of physically sectioned biofilms which were examined by 1P-LSM at several locations across the depth of the structure. For presentation several image series were mounted. (A) Cryosectioning of fungal pellet across the whole aggregate (four image stacks). The pellet was negatively stained using FITC-dextran. The black spots in the central part are spores developing under oxygen limitation. (B) Sectioning of a chemoautotrophic bacterial biofilm grown on pumice in a fluidized bed reactor (six image stacks). Colour allocation: glycoconjugates (= green), nucleic acid stain (= red), co-localised bacteria and glycoconjugates (= yellow), reflection of pumice (= white). (C) Cryosectioning of a 75-day-old freshwater phototrophic biofilm (three image stacks). After sectioning the biofilm was stained with an Alexa-488 labelled lectin. Colour allocation: glycoconjugates (= green), autofluorescence of cyanobacteria (= pink), autofluorescence of algae (= blue). (See colour plate 14.)

◆◆◆◆◆◆ BIOFILM LSM IMAGING

Reflection

Biofilm systems from any habitat show certain intrinsic properties with respect to scattering, general autofluorescence and specific biological autofluorescence. The scattering of a sample can be used as an additional source of information. Scattering may originate from reflective particles associated with a microbiological sample or from the cellular constituents themselves. The cellular reflection can reveal information about inclusions (e.g. sulphur granules) or surface features (e.g. precipitates, exoskeleton of diatoms). The geogenic signal may be collected in order to image the bacteria within the real environmental matrix. Furthermore, the reflection of the substratum may be used for imaging bacteria directly attached to the substratum (e.g. minerals, rocks). Several examples of reflection imaging are given in Figure 4.11. The images show reflection of sulphur granules in Thiomargarita cells, reflection of diatom skeletons, endolithic phototrophic biofilms and reflection of mineral constituents in lotic biofilms.

Autofluorescence

The autofluorescence properties of a microbiological sample will determine the fluorochrome which can be employed for staining of bacterial samples. For example, if the biofilm was exposed to light it usually includes various phototrophic microorganisms which contain chlorophyll A (algae) or both chlorophyll A and phycobilins (cyanobacteria). Chlorophyll A will give a strong emission signal in the far red of the visible light spectrum. Consequently, it is not possible to use a fluorochrome with far red emission. Phycobilins will give a strong emission signal in the red part of the visible spectrum, thus no fluorochrome can be used with a red emission. In the case of phototrophic biofilm constituents one may of course take advantage of the pigments for imaging and even differentiation of different phototrophs (Neu et al., 2004). However, due to the wide variety of newly developed fluorochromes there is usually a staining option with an emission at another appropriate wavelength, e.g. green or blue. As a consequence only one or two stains can be applied in combination with autofluorescence imaging. Nevertheless, this problem may be overcome if the autofluorescence signal and the staining signal can be isolated based on co-localization. Co-localization means that the emission signal of the same object is recorded in two different channels. Again, biofilms with cyanobacteria are a good example for co-localization. The emission signal of phycobilins in the red and chlorophyll A in the far red results in the final image in an overlay of the two channels. Thus, the two colours allocated to the two channels, e.g. red for the red channel and blue for the far red channel result in a pink signal in the overlay (see Figure 4.12). The pink signal can then be identified as the co-localised signal whereas the additional stain in the red would still show up as a red signal in the final overlay. Thereby

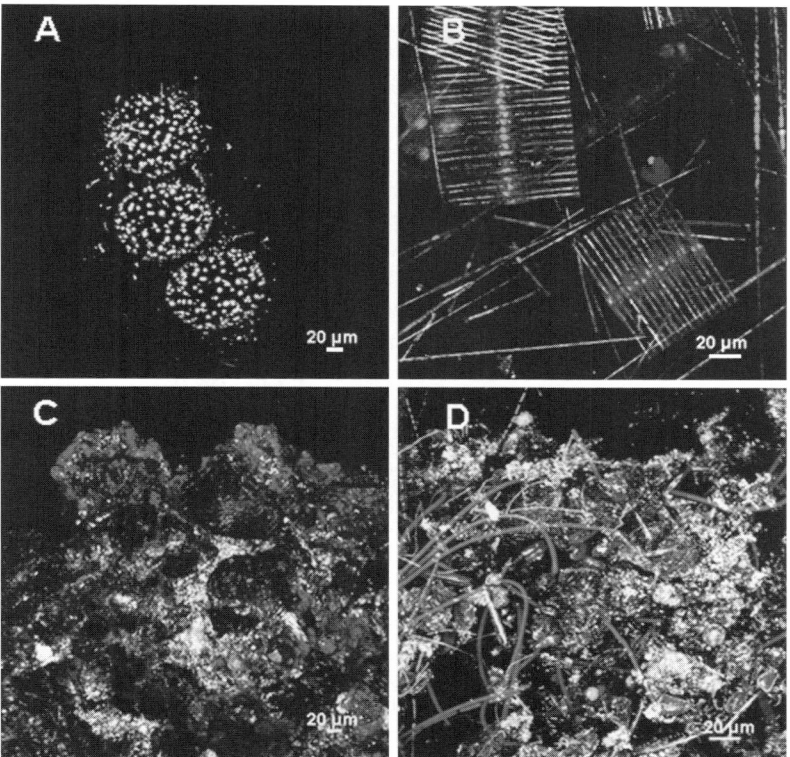

Figure 4.11. Reflection imaging of microbiological samples (1P-LSM). (A) Reflection of sulphur granules in Thiomargarita cells (= white). (B) Reflection of diatoms (*Fragilaria crotonensis*). Colour allocation: reflection (= white), nucleic acid stain (= green), autofluorescence of chlorophyll (= blue). (C) Reflection and autofluorescence of endolithic biofilms with the rock surface at the top of the image. Colour allocation: reflection of rock material (= white), autofluorescence of algae (= blue). (D) Reflection of a river biofilm community including nucleic acid staining. Colour allocation: reflection of mineral compounds (= white), autofluorescence of cyanobacteria (= pink), autofluorescence of algae (= blue), nucleic acid stain (= green). (See colour plate 15.)

the additional stain as well as the co-localised signal can be analysed quantitatively. Figure 4.12 illustrates the imaging and separation of algae and cyanobacteria in river biofilms in combination with digital image processing and analyses.

Cell Staining

In most cases biofilms from environmental habitats are initially stained by using nucleic acid specific stains. By this means the bacteria are visualized and localized and their distribution can be determined qualitatively and quantitatively. Traditionally, AO, 4′,6-diamidino-2-phenylindole (DAPI) and some of the Hoechst dyes have been used for staining nucleic acids of bacteria (Kepner and Pratt, 1994). However, for several reasons they are

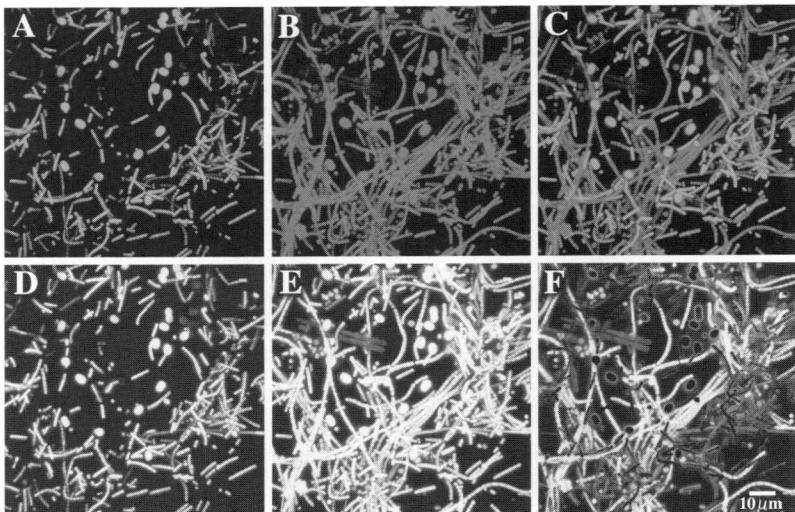

Figure 4.12. Maximum intensity projection of autofluorescence signals in a lotic biofilm examined by confocal laser scanning microscopy (1P-LSM). Images showing single channel signals of (A) red channel = cyanobacteria, (B) far red channel = autofluorescence of all photosynthetic organisms, (C) merged image of red + far red channel showing cyanobacteria (= pink) and algae (= blue), (D) grey scale version of the red channel, (E) grey scale version of far red channel and (F) grey scale image of the result after subtracting the red channel from the far red to image algae alone. (See colour plate 16.)

not necessarily of advantage in LSM. For example, AO stains DNA and RNA and many other biofilm features, DAPI and Hoechst dyes have the necessity of UV lasers for excitation. AO and DAPI also show a very broad emission signal and thus they cannot be employed for multiple staining. Today a range of nucleic acid specific fluorochromes such as the SYTO stains have been developed with excitation and emission in different areas of the visible light spectrum (Haugland, 2002). Several examples of SYTO staining are presented in combination with imaging of reflection, autofluorescence and other staining (see Figure 4.15). They are available with different emissions in the blue, green, orange, red and far red. In our experience SYTO 9 with an emission in the green is an excellent stain for microbiological samples from various locations and environments including freshwater and marine samples. The only restriction we have seen so far was in samples with extremely low pH (pH = 0–1). For staining these types of samples the pH should be adjusted in the range of pH 3–4 (Baffico *et al.*, submitted). Nucleic acid specific stains may also be distinguished in terms of their cell permeation. For example, the SYTO series belongs to the cell permeant stains whereas the POPO, PO-PRO and SYTOX series belong to the cell-impermeant stains which may be used for fixed samples. Apart from these stains other newly developed very sensitive nucleic acid stains have been reported to be useful including SYBRGREEN or SYBRGOLD which have been employed to detect aquatic virus-like particles (Chen *et al.*, 2001; Noble and Fuhrman, 1998). An example of staining virus-like particles using SYBRGREEN is shown in Figure 4.13. Further applications of various nucleic acid stains were compiled elsewhere (Neu and Lawrence, 2002).

Nucleic acid staining protocol using SYTO stains

1. Prepare aliquots of 5 µl in Eppendorf tubes and keep them at −20°C for subsequent use. For staining these aliquots are diluted in 5 ml liquid (tap water, buffer, medium, salt solution, process water).
2. Take the fresh, wet biofilm sample covered with original water, add a few droplets of the SYTO solution and incubate for 5 min at room temperature.
3. The samples may be immediately examined by LSM. There is no background fluorescence and consequently no washing step necessary. If after long observation the fluorescence is fading, re-staining is possible. The SYTO nucleic acid stain can be combined with glycoconjugate staining. Then, the lectin is applied first and the SYTO second.

Figure 4.13. SYBRGREEN staining of a filtered (0.02 µm) river water sample. The image shows stained bacteria (large spots) and virus-like particles (tiny green spots). (See colour plate 17.)

Polymer Staining

Microbial biofilms produce extracellular polymeric substances (EPS) as a major component (Wingender *et al.*, 1999). The chemical and *in situ* analysis of EPS constituents is still a challenge as the different types of polymers cannot be analysed by using a simple and straightforward analytical approach. In fact the analysis is even more difficult if environmental biofilms containing a huge variety of different cell types and thus different EPS compounds have to be analysed (Staudt *et al.*, 2003). One part of the EPS matrix is represented by glycoconjugates. It has been demonstrated that lectins are a very useful probe for the *in situ* characterization of glycoconjugates (Neu and Lawrence, 1997). The technique is already established as a standard method in biofilm research (Neu and Lawrence, 1999b). Furthermore, the approach has been critically evaluated in order to examine the effect of incubation time, lectin concentration, fluors conjugated, carbohydrate inhibition, order of addition and lectin interaction (Neu *et al.*, 2001). It is also possible to combine the technique with fluorescent *in situ* hybridization. This will allow the identification of bacterial groups or species as well as glycoconjugate types (Böckelmann *et al.*, 2002). In the meantime screening of all commercially available lectins has been done on several types of biofilm systems (Staudt *et al.*, 2003). This screening is necessary in order to identify the most suitable lectins for analysing glycoconjugates in a specific type of biofilm. The lectin or lectin panel to be selected is dependent on the question to be answered. There are lectins staining mainly cell surface glycoconjugates of certain bacterial species or other biofilm organisms, e.g. algae, fungi or protozoa (see Figure 4.21A,B). Some lectins may stain the polymer surrounding microcolonies. Other lectins may stain the extracellular matrix within biofilms which is not directly connected to cellular constituents. Recently, the architecture of microbial colonies in lotic biofilms was studied by employing a panel of lectins in combination with other probes for the microhabitat (Lawrence *et al.*, submitted). For lectin-binding analysis single lectins as well as two or three suitable lectins may be combined in order to stain different types of glycoconjugates. Figure 4.14 illustrates the binding of selected lectins to a microcolony and to general biofilm material.

In general, there is no good or bad lectin and the lectins may behave differently depending on the biofilm system investigated. Given a specific biofilm system which has been tested with a selected lectin panel, it may show certain features. However, the same set of lectins will give a different staining pattern in a different type of biofilm. In Figure 4.15 several examples illustrating the different appearance of glycoconjugate signals in biofilms are presented. Due to their significance, data on EPS compounds are now becoming more and more an important part of biofilm modelling (Horn *et al.*, 2001; Kreft and Wimpenny, 2001).

Other types of biofilm polymers may be stained using a variety of fluorescent probes for proteins and lipids, such as SYPRO series or Nile red. Lawrence *et al.* demonstrated differential binding of these probes in

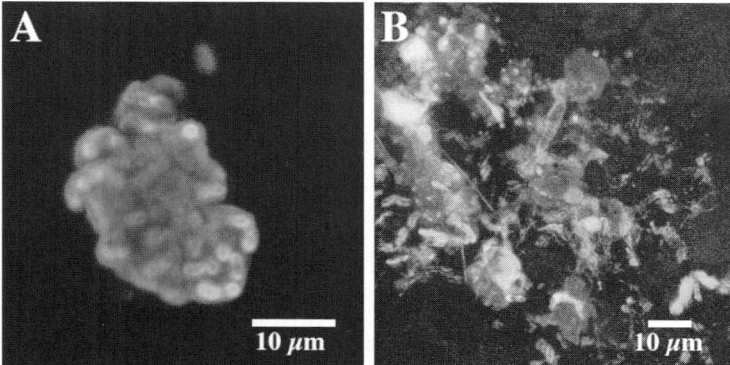

Figure 4.14. 1P-LSM images illustrating the binding of selected lectins. (A) Staining of a bacterial microcolony with two different lectins. (B) Staining of general river biofilm community with three different lectins. Note the discrete binding within the colony showing distinct zones (A) and regions in the overall river biofilm community (B). (See colour plate 18.)

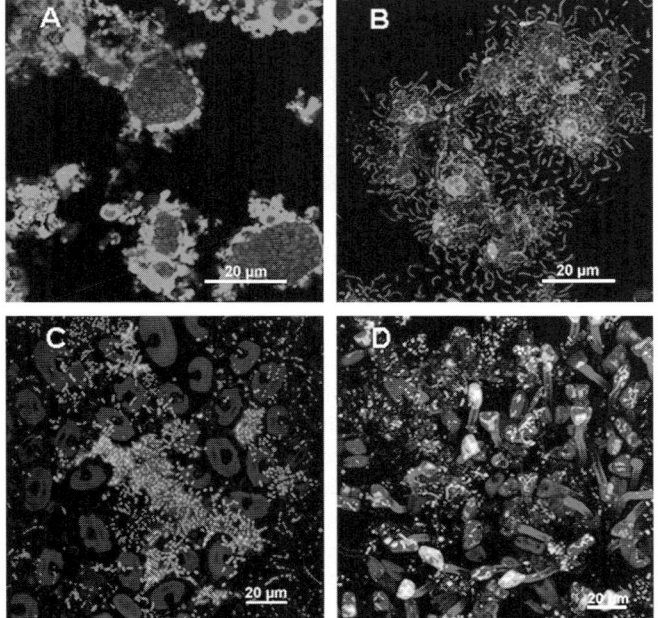

Figure 4.15. Examples illustrating the different appearance of glycoconjugates within biofilms after staining with lectins and 1P-LSM imaging. (A) Microcolonies of a chemoautotrophic biofilm grown in a rotating annular reactor. Colour allocation: lectin stain (= green), nucleic acid stain (= red). (B) Microcolonies of a hetrotrophic biofilm grown in a rotating annular reactor. Colour allocation: lectin stain (= green), nucleic acid stain (= red). (C) Biofilm developed in a creek showing a bacterial/algal microcolony, collected and examined after 30 days. Colour allocation: nucleic acid stain (= green), lectin stain (= red), autofluorescence of algae (= blue). (D) Biofilm developed in a creek showing stalked algae, collected and examined after 9 months. Colour allocation: nucleic acid stain (= green), lectin stain (= red), autofluorescence of algae (= blue). (See colour plate 19.)

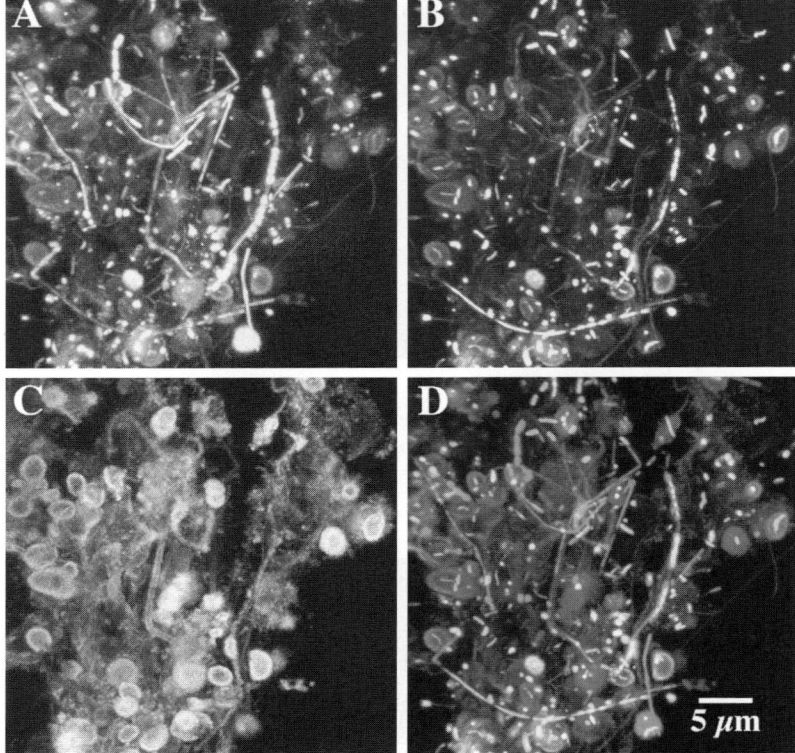

Figure 4.16. A 3-channel 1P-LSM image showing the binding of (A) lipid (Nile red), (B) nucleic acid stain (SYTO 9) and (C) lectin stain in a biofilm community. (D) The combination image showing co-localization of signals from all three probes. (After Lawrence et al., 2003. *Appl. Environ. Microbiol.* 69:5543–5554, with permission.) (See colour plate 20.)

conjunction with lectin staining to characterize the exopolymeric matrix of biofilms (Lawrence et al., 2003). Figure 4.16 is a 3-channel 1P-LSM image showing the binding of lectin, protein and nucleic acid stains and the combination of all three channels showing distribution and co-localization in a river biofilm community.

Glycoconjugate staining protocol using lectins

1. Prepare a stock solution of 1 mg lectin protein per 1 ml buffer by adding filter sterilized (0.2 μm) water with a syringe to the original vial in which the lectin is supplied. Prepare aliquots of 100 μl in Eppendorf tubes and keep them at −20°C for subsequent use. For staining these aliquots are diluted 1:10 with liquid (tap water, buffer, medium, salt solution, process water) in order to have a staining solution at 100 μg/ml.
2. Take the fresh, wet biofilm sample covered with original water, use a filter paper to draw off the excess water from the biofilm surface.

> Then add a few droplets of the lectin working solution to the wet biofilm sample and incubate for 20 min at room temperature.
> 3. In order to remove unbound lectin, the sample has to be carefully washed. For this purpose a filter paper is used, then filter sterilized original water is added again to cover the biofilm. The washing step is repeated four times.
> 4. The samples may then be directly examined by LSM. In most cases a counterstaining using a nucleic acid specific stain, e.g. SYTO with different excitation/emission is applied.

◆◆◆◆◆◆ THREE-DIMENSIONAL DIGITAL IMAGE ANALYSIS

Digital image analysis of microbes has been discussed in an excellent book which the reader is referred to in order to access various topics such as imaging, morphometry, fluorometry, motility and applications (Wilkinson and Shut, 1998). For completion several aspects of image analysis of microbial communities from the original literature have been compiled in Table 4.6.

Digital images are not continuous but they are composed of picture elements (pixels) in two dimensions or of volume elements (voxels) in three dimensions. The information behind a pixel may include a grey value only ranging from 0 to 255 intensity (8 bit). In the case of colour images it may contain information about red, green, blue (RGB) colours (3×8 bit = 24 bit). Thus a single pixel or voxel may contain information

Table 4.6. Digital image analysis of microbial communities excluding biofilm analysis. Some of the routines may be applicable to biofilms

Focus of image analysis	References
Bacterial cell size and shape	Blackburn *et al.* (1998), Liu *et al.* (2001), Vaija *et al.* (1995)
Filaments	Congestri *et al.* (2000), Cox *et al.* (1998), Nedoma *et al.* (2001), Pons *et al.* (1998), Thomas (1992)
Planktonic cells	David and Paul (1989), Koefoed Björnsen (1986), Sieracki *et al.* (1985), Verity *et al.* (1996), Viles and Sieracki (1992)
Adhesion of bacterial cells	An *et al.* (1995), Barthelson *et al.* (1999), Bos *et al.* (1999), Grivet *et al.* (1999)
FISH	Adiga and Chaudhuri (2000), Daims *et al.* (2001), Schönholzer *et al.* (2002), Spear *et al.* (1999)
Soil, thin sections	Fuller *et al.* (2000) and Nunan *et al.* (2001)
General	Adiga, Chaudhuri (2001), Sabri *et al.* (1997), Vecht-Lifshitz and Ison (1992)

Table 4.7. Issues to be considered if LSM data sets are recorded for a specific purpose

Purpose of recording images	LSM settings, comments
2D data	Requires a single scan only, averaging may be longer to reduce noise, zoom may be applied
3D data	Samples size/thickness (μm) and number of scans intended will determine the Z-step, zoom factor is critical due to enhanced bleaching
4D data	Sample has to be stable over a long period of time, fading problems may occur
Perfect image	Resolution in pixels may be set to a higher value (e.g. 1024 × 1024), averaging may be longer to get rid of noise
Routine analysis	To work faster – resolution may be set at an intermediate level (e.g. 512 × 512 pixels), Z-steps maybe set higher, averaging may be set lower
Quantification	Photomultiplier settings have to be kept at the same level throughout the experiment
Statistics	A statistically appropriate number of image series has to be recorded from a single sample, depends on the magnification used
True volumetric measurements	If deconvolution is applied, the point spread function (PSF) may have to be determined, data sets have to be recorded at extremely high resolution in XY and Z, consequently samples are bleached dramatically, as a result huge data files are created, may not be applied routinely

about intensity, colour and 2D or 3D location. Additionally, a single object within a data set may contain information in terms of size, diameter, shape, structure and orientation. Finally, several objects may contain again additional information such as arrangement, connectivity and neighbourhood (Rodenacker *et al.*, 2003). All of these parameters may be measured by digital image analysis.

The major reason for the application of LSM is mostly the creation of 3D data sets. Consequently, the software for digital image analysis has to be able to handle 3D data sets. There are different software options: microscopy software versus separate software, freeware in the internet versus commercial software, visualization versus quantification software, general software versus software developed for biofilms and specific software, e.g. for deconvolution. The problem is that no software does everything and mostly the user has to deal with several software packages. File formats may also be an issue or limitation when moving between the LSM platform and various analyses and display software packages. These issues should be carefully addressed for storage, archiving and analyses purposes.

Digital images of microbiological samples may be collected by LSM for three reasons. First, for 2D, 3D or 4D visualization. Second, by using specific fluorochromes for the analysis of structures and processes. Third, for quantification of the data sets in space and time. Depending on the subsequent use of the image data, the data sets have to be recorded by using appropriate settings at the LSM (see Table 4.7).

Visualization

Each LSM system comes with software for recording images and the software usually includes some basic visualization tools. Most of these tools are also included in other commercial software which is available separately. Experience shows that the microscope software is mainly used for recording images. As the microscope is often occupied for this purpose only, the analysis of digital images is done in many laboratories at a separate computer using separate software packages. For quick visualization of the recorded data sets they may be presented in a variety of ways as transparent view (see Table 4.8). Figures 4.17–4.20 illustrate

Table 4.8. Visualization of transparent LSM data sets

Common term	Comments
Gallery	Series of single optical sections, useful for control and selection of images
Maximum intensity projection (MIP)	Projection of series into one optical plane, an image usually saved in addition to the series as the MIP image is the one which will be remembered
XYZ presentation	XY presentation from the top in combination with XZ and YZ presentation from the side, adjustable to different layers, ideal for inspection of data sets
Ortho slice	Sections through data set, adjustable to different 3D locations
Shadow projection	Data set is presented with an imaginary light source for producing a pseudo 3D effect
Volume view	Data set is presented in 3D at a certain angle, section steps may become visible
Red–green anaglyph	Stereo presentation in one colour, calculated from two images (red and green) by pixel shift, red–green glasses necessary to look at
Stereo pair	Multicolour stereo presentation, calculated from two images by tilting one data set, stereo glasses necessary, some people may see the stereo effect without glasses, some may not see it at all

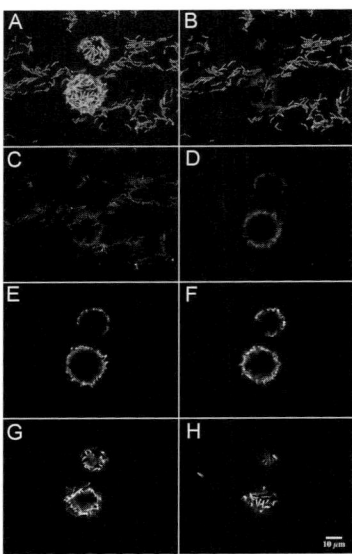

Figure 4.17. A series of 1P-LSM illustrations showing 2D and 3D visualization using the same data set. The images are of a *Rhodococcus* sp. colonizing the surfaces of droplets of diesel fuel during the degradation process. (Whyte *et al.*, 1999. *Appl. Environ. Microbiol.* 65:2961–2968 with permission.) (A) Maximum intensity projection of the Z series. (B)–(H) XY sections at 8 μm intervals selected from the Z series.

several of these 3D visualization techniques using the same data set to produce, Figure 4.17) maximum intensity projection with a series with *XY* images, Figure 4.18) maximum projection with accompanying *XZ* images, Figure 4.19) a maximum projection stereo pair, Figure 4.20) a red–green anaglyph and a 3D rendering of the same data set. The images are of a *Rhodococcus* sp. colonizing the surfaces of droplets of diesel fuel during the degradation process (Whyte *et al.*, 1999).

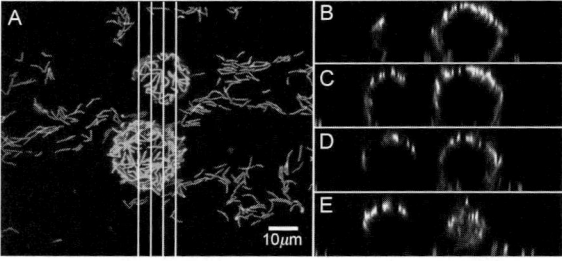

Figure 4.18. A series of 1P-LSM illustrations showing 2D and 3D visualization using the same data set. The images are of a *Rhodococcus* sp. colonizing the surfaces of droplets of diesel fuel during the degradation process. (Whyte *et al.*, 1999. *Appl. Environ. Microbiol.* 65:2961–2968 with permission.) (A) Maximum intensity projection of the diesel droplet with vertical lines indicating the locations of XZ sections shown in the images (B)–(E).

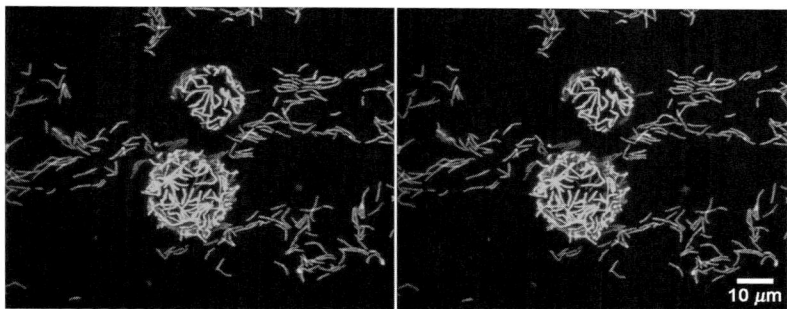

Figure 4.19. A series of 1P-LSM illustrations showing 2D and 3D visualization using the same data set. The images are of a *Rhodococcus* sp. colonizing the surfaces of droplets of diesel fuel during the degradation process. (Whyte *et al.*, 1999. *Appl. Environ. Microbiol.* 65:2961–2968 with permission.) A maximum intensity projection of a stereo pair showing the same data set.

For more advanced visualization of LSM data sets there are several commercial software packages available. All of the programs are very useful and excellent if every feature and option is available. Unfortunately, this often involves not only the purchase of the basic software but also additional software packages from the same or another company (see Table 4.9). Some of them have originally been developed for special graphic computer platforms, e.g. Silicon Graphics. However, in the meantime the situation has changed due to the establishment of advanced computer games and the development of fast graphic cards with lots of memory for the PC market. As a consequence, several companies offer now their programs for PC platforms under WIN or LINUX. Another new development is the current introduction of the 64-bit CPU architecture. This will accelerate

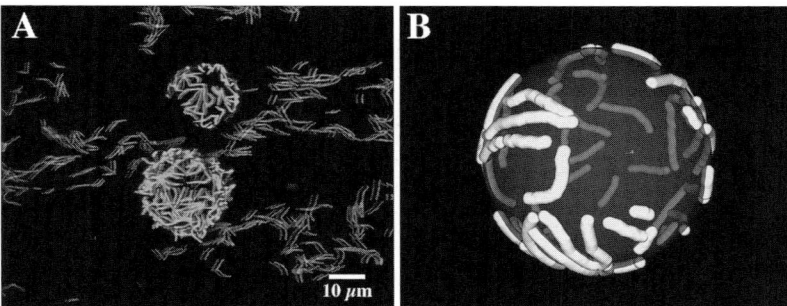

Figure 4.20. A series of 1P-LSM illustrations showing 2D and 3D visualization using the same data set. The images are of a *Rhodococcus* sp. colonizing the surfaces of droplets of diesel fuel during the degradation process. (Whyte *et al.*, 1999. *Appl. Environ. Microbiol.* 65:2961–2968 with permission.) (A) A red–green anaglyph projection of the data set (must be viewed with red–green stereo glasses). (B) A 3D colour rendering of the data stack for the small droplet in the image. (See colour plate 21.)

Table 4.9. Visualization software specifically designed for LSM data sets

Software (company)	Features (in basic package)	Options (to be purchased separately)
AMIRA (TGS)	Highly advanced 3D visualization with many tools	Deconvolution, developer, virtual reality
IMARIS (Bitplane)	Highly advanced 3D visualization	Measurement, time, co-localization, deconvolution (AutoQuant)
VOXBLAST (Vaytek)	Advanced 3D visualization	Deconvolution
VELOCITY (Improvision)	Basic software (free)	Visualization, classification, restoration
METAMORPH (Universal Imaging)	Offline version for normal visualization	Advanced 3D visualization, 3D deconvolution
IMAGE-PRO PLUS (MediaCybernetics)	2D visualization only	3D, rendering, deconvolution

further digital image analysis as the CPU will be able to process the double amount of data and it can handle more memory as well as larger files.

Original LSM data sets are transparent data sets. In opposition to this, in our everyday life we are used to looking at objects which are covered by a non-transparent surface. As a consequence special software tools have been developed to re-calculate the transparent data sets. This treatment is applied during a procedure called surface rendering or isosurface imaging of 3D, originally transparent LSM data sets. The calculation considers the threshold set and the intensity of the signal in relation to the intensity of the background at each neighbouring voxels. As a result the objects are covered by a surface consisting of surface polygons. After this procedure objects are defined in 3D space, they can be counted and the volumes of single objects may be calculated. This feature is usually available in the most advanced visualization packages such as AMIRA, IMARIS and VOLOCITY. However, the same problems present in 2D image analysis are now transposed into 3D space. For example, one of the biggest challenges is still the separation of objects touching each other. This of course becomes critical if objects are supposed to be counted. A variety of solutions to this specific problem have been attempted. In a study of microbial biofilms with co-localised signals an improved procedure for segmentation and object separation has been suggested (Bergner *et al.*, 2002). Further solutions are offered in the book by Wilkinson and Shut (1998). In Figure 4.21 several example for advanced imaging are shown as an isosurface view (see also Figure 4.20B).

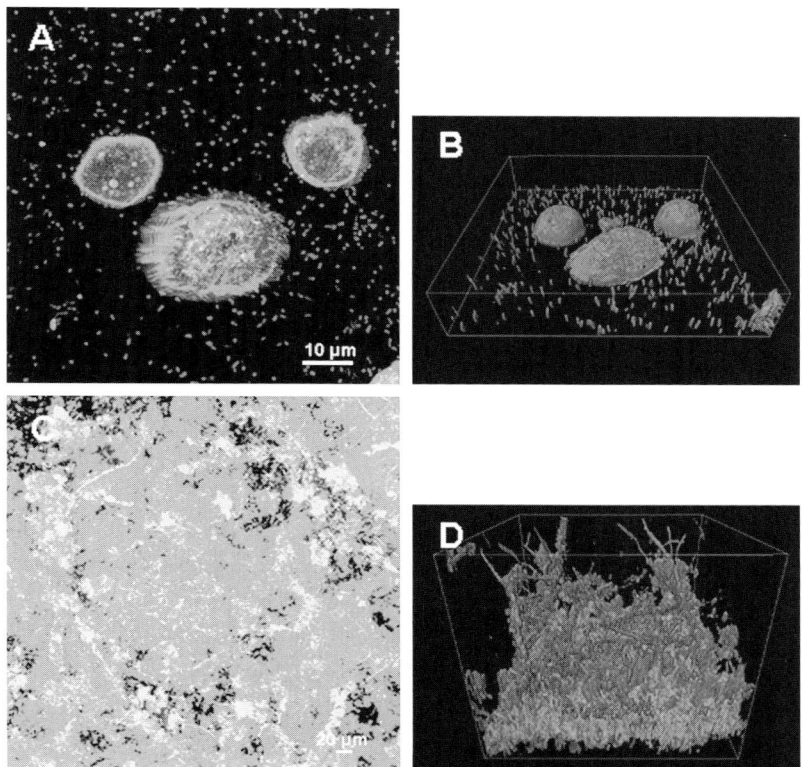

Figure 4.21. Two examples showing maximum intensity projection (A and C) and isosurface projection (B and D) of the same 1P-LSM data set. (A, B) Bacterial cells attached to the substratum which are grazed by protozoan cells. The cell surface of the protozoan cells is stained using an Alexa-488 labelled lectin. (C, D) Heterotrophic biofilm grown in a rotating annular reactor stained with an Alexa-488 lectin and SYTO 60. Colour allocation: lectin staining (= green), nucleic acid stained bacteria (= red). In both examples, the green channel is semi-transparent to visualise bacteria inside the protozoa and inside the biofilm covered by glycoconjugates. (See colour plate 22.)

Quantification

As already indicated laser microscopy images may be quantified in order to calculate volumes and to supply data for modelling. The first step in quantification involves thresholding (segmentation). By this thresholding procedure, the initial image consisting of pixel intensities from 0 to 255 is transferred into a binary image consisting of signal/object and background only. This procedure is extremely critical as the result will depend upon the type of sample, the image information, the way the data set was recorded as well as the operator's identity and day-to-day mood. Therefore, several attempts have been made to automate the thresholding procedure in microbiological image data sets (Adiga and Chaudhuri, 2001; Bergner *et al.*, 2002; Sieracki *et al.*, 1989; Wilkinson, 1998; Wilkinson and Shut, 1998; Xavier *et al.*, 2001; Yang *et al.*, 2001).

Table 4.10. Software for quantitative digital image analysis of biofilm LSM data sets. Some of the software is freely available on Internet

Software	Application	Comments
NIH image (MAC), Scion image (WIN), Image J (Java) = newest version – freely available	General, Neu et al. (2001) (http://rsbinfo.nih.gov/ij/)	Widely distributed, lots of macros or plugins available, automation possible, 2D only, single channel only
Comstat (WIN) MATLAB necessary – freely available	Biofilms, Heydorn et al. (2000)	Quantification, structural parameters, 2D only, single channel only
ISA (WIN) MATLAB necessary – partly available	Biofilms, Yang et al. (2000)	Quantification, structural parameters, 2D only, single channel only
CLSM tool box (web based) now PHLIP MATLAB necessary – later available	Biofilms, Xavier et al. (2003)	Quantification, co-localization, automation, 2D only
MAPPER (platform?) – available?	Biofilms, (www.inel.gov)	Structural parameters
IA (different platforms) IDL necessary – not available	Biofilms, Rodenacker et al. (2003)	Automatic thresholding, quantification, co-localization, specific tools
Microstat (WIN) – not available	Biofilms, Bergner et al. (2002)	Quantification, co-localization, automation
ConAn (WIN) IDL necessary – not available	Biofilms, Staudt et al. (in prep.)	Several thresholding options, quantification, structural parameters, co-localization, multichannel, object orientated volume analysis, automation

In the Internet one may find several useful software packages for image analysis in general or specifically developed for microbial biofilms. The options available for quantification or structural analysis are listed in Table 4.10. Quantification of 3D data sets from 2D single optical sections may be simply done by 2D quantification with subsequent calculation of the volume using the known section step in the axial direction.

Other programs, e.g. the more sophisticated 3D visualization programs (see Table 4.8) first define 3D objects by applying specific algorithms, from these objects they finally calculate the volume.

The quantitative analysis of LSM data sets from microbial biofilms was done with a number of different approaches developed in different research groups (Table 4.10). A convenient option is the application of the free program NIH image or Scion Image or more recently Image J. This program was used to quantify the multichannel data sets from lotic biofilms developed in rotating annular reactors (Lawrence et al., 1998a; Neu et al., 2001). The option of self-written macros or plugins allows automated analysis of numerous data sets. In addition, specific analysis routines may be developed. Other authors employed automated LSM recording in combination with semi-automated image processing. With this approach biofilms of *Pseudomonas fluorescence* and GFP labelled *E. coli* were investigated and quantified (Kuehn et al., 1998). For quantitative image analysis and analysis of structural parameters of pure culture biofilms the program Comstat was developed. This program is written as a script in Matlab and allows the analysis of the following parameters: biovolume, area occupied by bacteria in each layer, thickness distribution, area and volume of microcolonies, fractal dimension, roughness coefficient, diffusion distance and surface to volume ratio (Heydorn et al., 2000). The web-based program, CLSM toolbox, is another quantification tool for biofilms. This program is also based on Matlab and has a number of options: objective threshold selection, microbial colonization profile, biovolume, substratum colonization, average height and biofilm surface area (Xavier et al., 2003). A software routine based on the program IDL has been developed for different biological samples including microbial biofilms and aggregates. The tools allow measurement of various parameters (Rodenacker et al., 2003). The analysis of morphological features of microbial biofilms was a major focus of a further study. For this purpose two categories of features, areal and textural, were analysed using the program image structure analysis (ISA). The features included porosity, fractal dimension, diffusional length, angular second moment, inverse difference moment and textural entropy (Yang et al., 2000). Very recently the quantification of structural features has been critically discussed (Beyenal et al., 2004). One of the major questions in biofilm structural analysis is the meaning and significance of structural parameters in terms of biofilm function and processes.

Restoration

All optical microscopes are hampered by the fact that the resolution in XY is unequal to the resolution in XZ. In other words, the resolution of two objects laterally in modern microscopes is close to perfect whereas the axial resolution is not satisfactory at all. An example of this effect is shown in Figure 4.21. It is obvious that the signals of bacteria on the substratum surface (Figure 4.21B) and the signal of filamentous bacteria on top of a biofilm (Figure 4.21D) are elongated in axial direction. As the axial resolution is very much dependent upon the numerical aperture of the lens, high numerical aperture lenses can significantly improve the axial resolution. However, to balance the remaining differences in XY and XZ resolution, mathematical treatments have to be employed. These calculations are called deconvolution or restoration of images (Shaw and Rawlins, 1991; van der Voort and Strasters, 1995; Van Kempen et al., 1997; Verveer et al., 1999). In the first place this sounds like the solution to the problem, however deconvolution involves several steps which are difficult to achieve. In order to apply deconvolution procedures properly, the point spread function of the microscope/lens has to be determined. Usually this is done by using fluorescent beads. For this purpose the sampling of images must be done according to the Nyquist criteria. This means sampling of data at very high resolution in XY (resolution in pixels) and XZ (resolution in terms of sectioning intervals) direction. For example, the data set has to be recorded at 1024 pixel XY resolution and at 150 nm XZ section intervals. This may be easily done with fluorescent beads, but not necessarily with biological samples stained with fluorochromes. Many biological samples will be extremely bleached and it may be impossible to record data sets from thick samples. In addition, it has been suggested to record the point spread function of fluorescent beads within the biological sample. The reason for this being that the different point spread function will change within different optical layers. This again makes a correct calculation difficult and time consuming. Depending on the size of the data set, the computer available and the number of iterations, the computer may calculate for several hours. As a result, practically speaking, the calculation of deconvolution may be done in specially selected samples and for very few data sets. However, deconvolution may not be employed as a routine procedure for several hundred data sets which are usually recorded in biofilm experiments over time with several parallels and controls. In addition, if deconvolution is not done properly it may introduce more artefacts rather than improving the quality of the data set.

Deconvolution may have to be discussed with respect to quantification. Due to the elongation of the signal in axial direction there is no true volumetric measurement possible on raw LSM data sets. However as already mentioned, high numerical aperture lenses will reduce the elongation of the signal. In addition, if quantification is done on LSM data sets, thresholding (segmentation) will further reduce the axial elongation. As a consequence the investigator has to consider whether the effort of recording images at extremely high resolution is really necessary. If data sets are taken with the same microscopic conditions and subsequently analysed with the same quantification procedure, they are comparable and one can certainly work without deconvolution. However, if true volumetric information is needed, deconvolution is of course necessary.

In any case, in Table 4.11 the software available separately and specifically for deconvolution calculations is listed. In addition to the software in Table 4.11, deconvolution is included in some LSM software packages as well as other general visualization software.

Table 4.11. Software specially designed for deconvolution and image restoration

Software, platform	Algorithms	Source
Huygens (IRIX, LINUX)	Maximum likelihood estimation (MLE), iterative constraint Tikhonov–Miller (TM), quick MLE, quick TM	SVI (http://www.svi.nl)
AMIRA deconv (IRIX, HP-UX, SUN, LINUX, WIN)	Iterative maximum likelihood, blind, non-blind	TGS (http://www.tgs.com)
AutoDeblur (WIN)	Nearest neighbour, inverse filter, non-blind, maximum likelihood, 2D-blind	AutoQuant (http://www.aqi.com)
Microtome (WIN)	Nearest neighbour, constrained iterative, blind, 2D blind	Vaytek (http://www.vaytek.com)

◆◆◆◆◆◆ NEW MICROSCOPIC TECHNIQUES USING LASER EXCITATION

One of the emerging techniques in fluorescence imaging of microbiological samples will very likely be the further establishment of FLIM. Lifetime imaging as compared to intensity imaging has several advantages. These include the independency of the lifetime (τ_f) from the concentration of the fluorochrome and from the output intensity of the laser employed. Another important feature of FLIM is the sensitivity of the fluorochrome lifetime to the micro-environment. Finally, fluorochromes having a similar or identical emission wavelength can be distinguished due to their different lifetimes. FLIM may be done in two different ways, in the frequency domain or in the time domain mode and it may be done using one-photon or two-photon lasers for excitation. Commercially available systems record the lifetime in the time domain mode with a pulsed two-photon laser for excitation and time-correlated single-photon counting (TCSPC) for detection (Clegg et al., 2003; Draaijer et al., 1995; Gadella, 1999). FLIM set-ups may be attached to existing two-photon instruments but stand-alone systems are also offered. There are several applications for FLIM in microbiological samples including: (1) detection of multiple fluorochromes having a similar emission signal (in one channel if intensity imaging is done), (2) measurement of ion concentrations (e.g. pH, calcium), (3) determination of oxygen concentration without using electrodes, (4) probing the microenvironment or binding status of a fluorochrome and (5) by employing fluorescence resonance energy transfer (FRET) performing distance measurements at the nanometer scale thereby enhancing the resolution of the microscope (Centonze *et al.*, 2003).

FLIM has been employed in several cell biological studies and these techniques may also be applied to microbiological samples. For example, the advantage of lifetime imaging over ratiometric/intensity imaging has been shown for pH sensitive fluorochromes. In eukaryotic cells quantitative pH measurements were done with a straightforward calibration (Sanders *et al.*, 1995). The FLIM approach has been described for other fluorochromes with a sensitivity in the low pH range (Lin *et al.*, 2001). Lifetime imaging has also been used to study changes in the cellular autofluorescence of NAD(P)H after exposition of cells to photostress (König *et al.*, 1996). In another application specially prepared oxygen sensor foils were used in order to measure oxygen gradients (Kellner *et al.*, 2002; Liebsch *et al.*, 2000). As fluorochromes are sensitive to their molecular environment they have been employed to distinguish the binding of nucleic specific fluorochromes to DNA and RNA. By this means the dynamic of DNA and RNA in different cell compartments may be investigated (van Zandvoort *et al.*, 2002). FLIM may be used to increase the number of fluorochromes applicable to a biological sample. The separation of fluorochromes with similar emission signals is possible due to their specific lifetime and thus two fluorochromes may be recorded which usually would show up as identical intensity signal in one channel (Carlsson and Liljeborg, 1997). A further application of lifetime imaging is the combination with FRET. By employing two-photon excitation in conjunction with FRET–FLIM high spatial (nanometre) and temporal (nanoseconds) resolution can be achieved. It maybe pointed out again that FRET–FLIM allows quantitative measurements without painful calibration which is usually necessary for intensity imaging (Elangovan *et al.*, 2002).

So far the FLIM technique has been used in a very limited number of microbiological studies. In the very first application of FLIM, AO was used to stain oral biofilms (Sytsma *et al.*, 1998). In a later paper the same group demonstrated pH gradients in oral biofilms using FLIM in combination with the pH sensitive probe carboxyfluorescein (Vroom *et al.*, 1999). The principle of FLIM has been employed in order to detect *Giardia* cysts in samples having a high autofluorescence background. For differentiation of autofluorescence and the cellular signal, a lanthanide chelate having a long lifetime was bound to a *Giardia* specific antibody (Connally *et al.*, 2002). Very recently, lifetime imaging was applied to investigate the activity of bacteria in pure cultures, mixed cultures and in lotic biofilms. In this study, the nucleic acid specific fluorochrome SYTO 13 was used in order to distinguish between cells having a different DNA/RNA ratios (Walczysko *et al.*, submitted). An example of

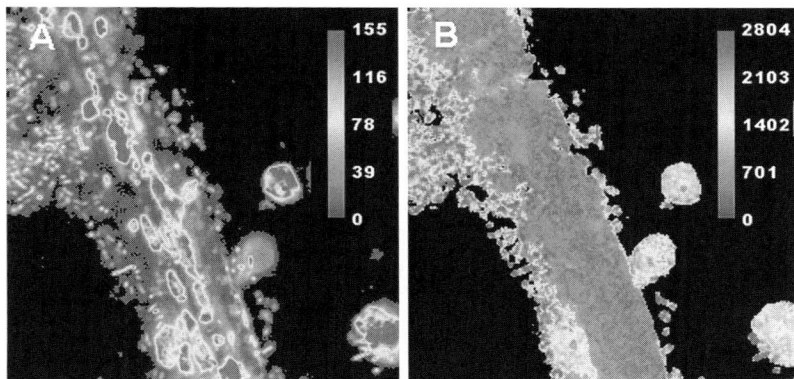

Figure 4.22. Two-photon excitation of lotic bacterial biofilm stained with SYTO 13. The image illustrates a diatom and bacteria colonising inorganic granules. (A) In the intensity image a differentiation of various cells or objects is impossible. The intensity image is displayed as a colour coded signal in arbitrary units from 0 to 155. (B) Lifetime image demonstrating the colour coded lifetime in picoseconds. Due to the longer lifetime of SYTO 13 in stained bacteria (orange–red), they can be distinguished from the autofluorescence of the diatom (blue–green) and the signal of inorganic granules (yellow). (Neu *et al.*, 2004, used with permission.) (See colour plate 23.)

two-photon FLIM measured in a lotic biofilm sample is presented in Figure 4.22.

Fluorescence correlation spectroscopy (FCS) is a technique by which the fluctuations of fluorescence are extracted mathematically. FCS may be combined with fluorescence microscopic techniques and is then called fluorescence correlation microscopy (FCM). By this procedure kinetic processes within a small observation volume can be measured. These processes may include molecular diffusion or chemical reactions. FCS is an important tool for understanding molecular biological processes inside and in between cells (Hink *et al.*, 2003; Müller *et al.*, 2003).

FCS in combination with two-photon excitation was already used to measure the diffusion of latex beads and dextrans in bacterial biofilms. The results showed the significance of the local biofilm structure, the presence of EPS as well as steric and physicochemical interactions between the polymeric fluorescent probes and biofilm matrix polymers (Guiot *et al.*, 2002).

LSM by means of point scanners takes time due to the point scanning mode but also when averaging is necessary due to noise in the image. Traditionally, spinning disk LSM with a Nipkow disk (disk with pinholes) did overcome this limitation. However, the instruments had a poor signal recovery. This problem was solved by introducing spinning disk LSMs with two disks, one with pinholes and one which is equipped with microlenses. The microlens approach was also used in multifocal multiphoton microscopy (MMM). This technique allows high speed imaging at 225 frames per second in combination with two-photon excitation. In addition, it is possible to directly observe the two-photon signal (Bewersdorf *et al.*, 1998; Straub and Hell, 1998).

Two-photon excitation is also used for a new microscopic technique called coherent anti-stokes Raman scattering (CARS) microscopy. This technique takes advantage of the fact that vibrational contrast is an intrinsic property of biological samples. Therefore no fluorescent probes are necessary. Similar to two-photon laser scanning fluorescence microscopy, the CARS signal is generated in the focal region only. The signal accumulation in the CARS process produces a much stronger signal as for example in conventional Raman microscopy. CARS microscopy was demonstrated for the first time by imaging live cells (Zumbusch et al., 1999).

Second harmonic imaging microscopy is another new technique using two-photon excitation. Second harmonic generation (SHG) is a process by which two infrared photons are converted into a single photon of twice the energy. In comparison to two-photon laser scanning fluorescence microscopy, SHG does not involve an excited state and preserves the coherence of the laser light. SHG is restricted to locations lacking a centre of symmetry and is thus ideal for investigating interfaces, e.g. membranes, collagen, myosin and tubulin. SHG signals are also very sensitive to the membrane potential and may be used for physiological studies. Further details on second harmonic imaging microscopy (SHIM) may be found in two recent reviews (Campagnola and Loew, 2003; Millard et al., 2003).

In light microscopy the resolution continuously increased by developing new techniques (Table 4.12). The limitation of axial resolution may be overcome by establishing new approaches in laser microscopy such as 4pi microscopy. The basic principle of 4pi is to increase the numerical aperture of the microscope by using two opposing lenses. As a result the axial resolution of the microscope can be improved dramatically (Lindek et al., 1995). This principle in combination with two-photon excitation and image restoration was further developed and leads to nearly equal resolution in three dimensions (Hell et al., 1997; Hell and Nagorni, 1998). The two high numerical aperture lenses employed for 4pi microscopy are either water (63 × NA 1.2) or oil (100 × NA 1.4) immersion lenses. Consequently, the oil immersion lenses may be used for imaging fixed

Table 4.12. Resolution in light microscopy

Technique/instrument	XY (lateral) resolution (nm)	XZ (axial) resolution (nm)
Light microscope	500	1600
Laser scanning microscope	250	700
Evanescent wave	500	300
Second harmonic microscope	250	–
4pi microscope	230	120
4pi/two photon/deconvolution	100	100
Stimulated emission depletion	100	100
STED/4pi	–	33
Near-field imaging	50	10

biological samples (Schrader *et al.*, 1998), whereas the water immersion lenses are ideal for living biological samples (Bahlmann *et al.*, 2001).

In optical microscopy the resolution is assumed to be in the range of half of the wavelength of the light used for imaging. This means a resolution of about 200 nm laterally and 600 nm axially. Recently, the diffraction resolution barrier has been broken by a technique called stimulated emission depletion (STED). The increased resolution is achieved by "engineering" of the focal spot (PSF) by STED. The set-up involves two synchronised trains of laser pulses, a visible green laser for excitation and an infrared two-photon laser for depletion. The latter STED pulse forces the fluorochrome immediately after excitation into a non-fluorescent state. This depletion is only caused in the outer focal region, resulting in a nearly spherical fluorescent spot of 90–110 nm (Klar *et al.*, 2000). STED may now be combined with 4pi microscopy thereby pushing the resolution in light microscopy towards 33 nm (Dyba and Hell, 2002).

In order to complete the optical techniques developed for high resolution imaging another technique has to be mentioned. Due to the limitation of resolution in optical microscopy with approximately $\lambda/2$ (λ, wavelength) an optical scanning based approach has been designed. Near-field scanning optical microscopy (NSOM) is a technique by which the diffraction limit is overcome by using a sub-wavelength light source. The basic idea is to use an optical fibre with an aperture of 20–50 nm at the tip for illumination of the sample. The tip is then moved across the sample similar to scanning probe microscopy techniques such as atomic force microscopy. With this approach a resolution of approximately 50 nm may be achieved. However so far NSOM has been mainly employed in physics or chemistry and few applications were reported on fixed biological samples (Haydon, 2003; Lewis *et al.*, 2001, 2003).

◆◆◆◆◆◆ INSTEAD OF CONCLUSIONS

In the last century the development of LSM with one-photon and two-photon excitation caused a revolution in optical imaging. Presently optical imaging is pushed even further by increasing the axial resolution and by overcoming the diffraction limit using a combination of different approaches. These highly advanced techniques are not yet commercially available. Nevertheless in the near future they likely will increase our understanding of structures and processes in microbial communities. A second area with potential for future development is digital image analysis. Three-dimensional visualization is already at a reasonable advanced level. However, new procedures for extraction of quantitative and structural information need to be established in order to automatically analyse multichannel data sets with co-localised signals. In addition, correlative fluorescence microscopy with intensity and lifetime imaging will have to be established as a routine method. This combination will allow extraction of the maximum in information from a biological sample. At the end of this chapter reference is made to a series of very recent publications with focus on state of the art in optical imaging (Campagnola and Loew, 2003; Hell, 2003; Jares-Erijman and Jovin, 2003; Lewis *et al.*, 2003; Zipfel *et al.*, 2003).

Acknowledgements

The authors acknowledge the excellent technical support of Ute Kuhlicke (UFZ) and George D. W. Swerhone (NWRI) in the execution of many of the studies referred to herein. Furthermore, we appreciate the collaboration with several colleagues and people collecting samples all over the world (Reiner Bachofen, Matthias Boessmann, Annett Eitner, Andrea Hille, Helmut Roenicke, Heide Schulz, Christian Staudt, Dirk Wagenschein, Petr Walczysko, Markus Weinbauer, Barbara Zippel). The financial support of the NWRI National Water Research Institute (Canada), the UFZ Centre for Environmental Research (Germany) and the Canada–Germany Agreement on research collaboration are acknowledged.

References

Adiga, P. S. U. and Chaudhuri, B. B. (2000). Segmentation and counting of FISH signals in confocal microscopy images. *Micron* **31**, 5–15.

Adiga, P. S. U. and Chaudhuri, B. B. (2001). Some efficient methods to correct confocal images for easy interpretation. *Micron* **32**, 363–370.

Albota, M., Beljonne, D., Bredas, J.-L., Ehrlich, J. E., Fu, J.-Y., Heikal, A. A., Hess, S. E., Kogej, T., Levin, M. D., Marder, S. R., McCord-Maughon, D., Perry, J. W., Röckel, R., Rumi, M., Subramaniam, G., Webb, W. W., Wu, X.-L. and Xu, C. (1998). Design of organic molecules with large two-photon absorption cross sections. *Science* **281**, 1653–1656.

An, Y. H., Friedman, R. J., Draughn, R. A., Smith, E. A., Nicholson, J. H. and John, J. F. (1995). Rapid quantification of staphylococci adhered to titanium surfaces using image analyzed epifluorescence microscopy. *J. Microbiol. Meth.* **24**, 29–40.

Bago, B., Pfeffer, P. E., Zipfel, W., Lammers, P. and Shachar-Hill, Y. (2002). Tracking metabolism and imaging transport in arbuscular mycorrhizal fungi. *Plant Soil* **244**, 189–197.

Bahlmann, K., Jakobs, S. and Hell, S. W. (2001). 4Pi-confocal microscopy of live cells. *Ultramicroscopy* **87**, 155–164.

Barthelson, R., Hopkins, C. and Mobasseri, A. (1999). Quantitation of bacterial adherence by image analysis. *J. Microbiol. Meth.* **38**, 17–23.

Bennett, B. D., Jetton, T. L., Ying, G., Magnuson, M. A. and Piston, D. W. (1996). Quantitative subcellular imaging of glucose metabolism within intact pancreatic islets. *J. Biol. Chem.* **271**, 3647–3651.

Bergner, S., Pohle, R., Al-Zubi, S., Toennies, K., Eitner, A. and Neu, T. R. (2002). In *Segmenting Microorganisms in Multi-modal Volumetric Datasets Using a Modified Watershed Transform* (L. Van Gool, ed.), pp. 429–437. Springer, Heidelberg.

Bestvater, F., Spiess, E., Stobrawa, G., Hacker, M., Feurer, T., Porwol, T., Berchner-Pfannschmidt, U., Wotzlaw, C. and Acker, H. (2002). Two-photon fluorescence absorption and emission spectra of dyes relevant for cell imaging. *J. Microsc.* **208**, 108–115.

Bewersdorf, J., Pick, R. and Hell, S. W. (1998). Multifocal multiphoton microscopy. *Opt. Lett.* **23**, 655–657.

Beyenal, H., Lewandowski, Z. and Harkin, G. (2004). Quantifying biofilm structure: facts and fiction. *Biofouling* **20**, 1–23.

Blackburn, N., Hagström, A., Wikner, J., Cuadros-Hansson, R. and Björnsen, P. K. (1998). Rapid determination of bacterial abundance, biovolume, morphology, and growth by neural network-based image analysis. *Appl. Environ. Microbiol.* **64**, 3246–3255.

Böckelmann, U., Manz, W., Neu, T. R. and Szewzyk, U. (2002). A new combined technique of fluorescent in situ hybridization and lectin-binding-analysis (FISH-LBA) for the investigation of lotic microbial aggregates. *J. Microbiol. Meth.* **49**, 75–87.

Bos, R., van der Mei, H. C. and Busscher, H. J. (1999). Physico-chemistry of initial microbial adhesive interactions – its mechanisms and methods for study. *FEMS Microbiol. Rev.* **23**, 179–230.

Bousso, P., Bhakta, N. R., Lewis, R. S. and Robey, E. (2002). Dynamics of thymocyte–stromal cell interactions visualized by two-photon microscopy. *Science* **296**, 1876–1880.

Bryers, J. D. (2001). Two-photon excitation microscopy for analyses of biofilm processes. *Meth. Enzymol.* **337**, 259–269.

Campagnola, P. J. and Loew, L. M. (2003). Second-harmonic imaging microscopy for visualizing biomolecular arrays in cells, tissues and organisms. *Nat. Biotechnol.* **21**, 1356–1360.

Carlsson, K. and Liljeborg, A. (1997). Confocal fluorescence microscopy using spectral and lifetime information to simultaneously record four fluorophores with high channel separation. *J. Microsc.* **185**, 37–46.

Centonze, V. E. and White, J. G. (1998). Multiphoton excitation provides optical section from deeper within scattering specimens than confocal imaging. *Biophys. J.* **75**, 2015–2024.

Centonze, V. E., Sun, M., Masuda, A., Gerritsen, H. C. and Herman, B. (2003). Fluorescence resonance energy transfer imaging microscopy. *Meth. Enzymol.* **360**, 542–560.

Chen, F., Lu, J. R., Binder, B. J., Liu, Y. C. and Hodson, R. E. (2001). Application of digital image analysis and flow cytometry to enumerate marine viruses stained with SYBR gold. *Appl. Environ. Microbiol.* **67**, 539–545.

Cheng, P., Lin, B., Kao, F., Gu, M., Xu, M., Gan, X., Huang, M. and Wang, Y. (2001). Multi-photon fluorescence microscopy – the response of plant cells to high intensity illumination. *Micron* **32**, 661–669.

Clegg, R. M., Holub, O. and Gohlke, C. (2003). Fluorescence lifetime-resolved imaging: measuring lifetimes in an image. *Meth. Enzymol.* **360**, 509–542.

Congestri, R., Federici, R. and Albertano, P. (2000). Evaluating biomass of Baltic filamentous cyanobacteria by image analysis. *Aquat. Microb. Ecol.* **22**, 283–290.

Conn, P. M. (1999). Confocal microscopy. *Meth. Enzymol.* **307**.

Connally, R., Veal, D. and Piper, J. (2002). High resolution detection of fluorescently labeled microorganisms in environmental samples using time-resolved fluorescence microscopy. *FEMS Microbiol. Ecol.* **41**, 239–245.

Cox, G. and Sheppard, C. (1999). Multiphoton fluorescence microscopy. In *Fluorescent and Luminescent Probes for Biological Activity* (W. T. Mason, ed.), pp. 331–336. Academic Press, San Diego.

Cox, P. W., Paul, G. C. and Thomas, C. R. (1998). Image analysis of the morphology of filamentous micro-organisms. *Microbiology* **144**, 817–827.

Daims, H., Ramsing, N. B., Schleifer, K.-H. and Wagner, M. (2001). Cultivation-independent, semiautomatic determination of absolute bacterial cell numbers in environmental samples by fluorescence in situ hybridization. *Appl. Environ. Microbiol.* **67**, 5810–5818.

David, A. W. and Paul, J. H. (1989). Enumeration and sizing of aquatic bacteria by use of a silicon-intensified target camera linked-image analysis system. *J. Microbiol. Meth.* **9**, 257–266.

Decho, A. W. and Kawaguchi, T. (1999). Confocal imaging of in situ natural microbial communities and their extracellular polymeric secretions using Nanoplast resin. *BioTechniques* **27**, 1246–1252.

Denk, W. (1994). Two-photon scanning photochemical microscopy: mapping ligand-gated ion channel distributions. *Proc. Natl Acad. Sci. USA* **91**, 6629–6633.

Denk, W., Strickler, J. H. and Webb, W. W. (1990). Two-photon laser scanning fluorescence microscopy. *Science* **248**, 73–76.

Denk, W., Piston, D. W. and Webb, W. W. (1995). Two-photon molecular excitation in laser-scanning microscopy. In *Handbook of Biological Confocal Microscopy* (J. B. Pawley, ed.), pp. 445–458. Plenum Press, New York.

Diaspro, A. and Robello, M. (2000). Two-photon excitation of fluorescence for three-dimensional optical imaging of biological structures. *J. Photochem. Photobiol. B: Biol.* **55**, 1–8.

Doyle, R. J. (1999). Biofilms. *Meth. Enzymol.* **310**.

Doyle, R. J. (2001a). Microbial growth in biofilms. Part A. Developmental and molecular biological aspects. *Meth. Enzymol.* **336**.

Doyle, R. J. (2001b). Microbial growth in biofilms. Part B. Special environments and physicochemical aspects. *Meth. Enzymol.* **337**.

Doyle, R. J. and Ofek, I. (1998). *Adhesion of Microbial Pathogens*. Academic Press, San Diego.

Draaijer, A., Sanders, R. and Gerritsen, H. C. (1995). Fluorescence lifetime imaging: a new tool in confocal microscopy. In *Handbook of Biological Confocal Microscopy* (J. B. Pawley, ed.), pp. 491–505. Plenum Press, New York.

Drummond, D. R., Carter, N. and Cross, A. (2002). Multiphoton versus confocal high resolution z-sectioning of enhanced green fluorescent microtubules: increased multiphoton photobleaching within the focal plane can be compensated using a Pockels cell and dual widefield detectors. *J. Microsc.* **206**, 161–169.

Dyba, M. and Hell, S. W. (2002). Focal spots of size $\lambda/23$ open up far-field fluorescence microscopy at 33 nm axial resolution. *Phys. Rev. Lett.* **88**, 163901-1–163901-4.

Elangovan, M., Day, R. N. and Periasamy, A. (2002). Nanosecond fluorescence resonance energy transfer-fluorescence lifetime imaging microscopy to localize the protein interactions in a single living cell. *J. Microsc.* **205**, 3–14.

Fuller, M. E., Streger, S. H., Rothmel, R. K., Mailloux, B. J., Hall, J. A., Constott, T. C., Fredrickson, J. K., Balkwill, D. L. and DeFlaun, M. F. (2000). Development of a vital fluorescent staining method for monitoring bacterial transport in surface environments. *Appl. Environ. Microbiol.* **66**, 4486–4496.

Gadella, Jr., T. W. J. (1999). Fluorescence lifetime imaging microscopy (FLIM): instrumentation and applications. In *Fluorescent and Luminescent Probes for Biological Activity* (W. T. Mason, ed.), pp. 467–479. Academic Press, San Diego.

Gan, X. and Gu, M. (2000a). Spatial distribution of single-photon and two-photon fluorescence light in scattering media: Monte Carlo simulation. *Appl. Opt.* **39**, 1575–1579.

Gan, X. S. and Gu, M. (2000b). Fluorescence microscopic imaging through tissue-like media. *J. Appl. Phys.* **87**, 3214–3221.

Gerritsen, H. C. and de Grauw, C. J. (1999). Imaging of optically thick specimen using two-photon excitation microscopy. *Microsc. Res. Tech.* **47**, 206–209.

Göppert-Mayer, M. (1931). Über Elementarakte mit zwei Quantensprüngen. *Ann. Phys.* **9**, 273–295.

Gray, N. D., Howarth, R., Pickup, R. W., Jones, J. G. and Head, I. M. (2000). Use of combined microautoradiography and fluorescence in situ hybridization to determine carbon metabolism in mixed natural communities of uncultured from the genus Achromatium. *Appl. Environ. Microbiol.* **66**, 4518–4522.

Grivet, M., Morrier, J.-J., Souchier, C. and Barsotti, O. (1999). Automatic enumeration of adherent streptococci or actinomyces on dental alloy by fluorescence image analysis. *J. Microbiol. Meth.* **38**, 33–42.

Gu, M. and Sheppard, C. J. R. (1995). Comparison of three-dimensional imaging properties between two-photon and single-photon fluorescence microscopy. *J. Microsc.* **177**, 128–137.

Gu, M., Gan, X., Kisteman, A. and Xu, M. (2000a). Comparison of penetration depth between single-photon excitation and two-photon excitation in imaging through turbid media. *Appl. Phys. Lett.* **77**, 1551–1553.

Gu, M., Schilders, S. P. and Gan, X. (2000b). Two-photon fluorescence imaging of microspheres embedded in turbid media. *J. Mod. Opt.* **47**, 959–965.

Guiot, E., Georges, P., Brun, A., Fontaine, M. P., Bellon-Fontaine, M.-N. and Briandet, R. (2002). Heterogeneity of diffusion inside microbial biofilms determined by fluorescence correlation spectroscopy under two photon excitation. *Photochem. Photobiol.* **75**, 570–578.

Haugland, R. P. (2002). *Handbook of Fluorescent Probes and Research Chemicals.* Molecular Probes, Eugene.

Haydon, P. G. (2003). Biological near-field microscopy. *Meth. Enzymol.* **360**, 501–508.

Hell, S. W. (2003). Toward fluorescence nanoscopy. *Nat. Biotechnol.* **21**, 1347–1355.

Hell, S. W. and Nagorni, M. (1998). 4pi confocal microscopy with alternate interference. *Opt. Lett.* **23**, 1567–1569.

Hell, S. W., Schrader, M. and van der Voort, H. T. M. (1997). Far-field fluorescence microscopy with three-dimensional resolution in the 100-nm range. *J. Microsc.* **187**, 1–7.

Heydorn, A., Nielsen, A. T., Hentzer, M., Sternberg, C., Givskov, M., Ersbøll, B. K. and Molin, S. (2000). Quantification of biofilm structures by the novel computer program COMSTAT. *Microbiology* **146**, 2395–2407.

Hink, M. A., Borst, J. W. and Visser, A. J. W. G. (2003). Fluorescence correlation spectroscopy of GFP fusion proteins in living plant cells. *Meth. Enzymol.* **361**, 93–112.

Horn, H., Neu, T. R. and Wulkow, M. (2001). Modelling the structure and function of extracellular polymeric substances in biofilms with new numerical techniques. *Water Sci. Technol.* **43**, 121–127.

Huang, C.-T., McFeters, G. A. and Stewart, P. S. (1996). Evaluation of physiological staining, cryoembedding and autofluorescence quenching techniques on fouling biofilms. *Biofouling* **9**, 269–277.

Jameson, D. M., Croney, J. C. and Moens, P. D. J. (2003). Fluorescence: basic concepts, practical aspects and some anecdotes. *Meth. Enzymol.* **360**, 1–43.

Jares-Erijman, E. A. and Jovin, T. M. (2003). FRET imaging. *Nat. Biotechnol.* **21**, 1387–1395.

Kaiser, W. and Garrett, C. G. B. (1961). Two-photon excitation in $CaF_2:Eu_{2+}$. *Phys. Rev. Lett.* **7**, 229–231.

Kawaguchi, T. and Decho, A. (2002). In situ microspatial imaging using two-photon and confocal laser scanning microscopy of bacteria and extracellular polymeric secretions (EPS) within marine stromatolites. *Mar. Biotechnol.* **4**, 127–131.

Kellner, K., Liebsch, G., Klimant, I., Wolfbeis, O. S., Blunk, T., Schulz, M. and Göpferich, A. (2002). Determination of oxygen gradients in engineered tissue using a fluorescent sensor. *Biotechnol. Bioeng.* **80**, 73–83.

Kepner Jr., R. and Pratt, J. R. (1994). Use of fluorochromes for direct enumeration of total bacteria in environmental samples: past and present. *Microbiol. Rev.* **58**, 603–615.

Klar, T. A., Jakobs, S., Dyba, M., Egner, A. and Hell, S. W. (2000). Fluorescence microscopy with diffraction resolution barrier broken by stimulated emission. *Proc. Natl Acad. Sci. USA* **97**, 8206–8210.

Koefoed Björnsen, P. (1986). Automatic determination of bacterioplankton biomass by image analysis. *Appl. Environ. Microbiol.* **51**, 1199–1204.

König, K. (2000). Multiphoton microscopy in life sciences. *J. Microsc.* **200**, 83–104.

König, K., So, P. T. C., Mantulin, W. W., Tromberg, B. J. and Gratton, E. (1996). Two-photon excited lifetime imaging of autofluorescence in cells during UVA and NIR photostress. *J. Microsc.* **183**, 197–204.

Kreft, J.-U. and Wimpenny, J. W. T. (2001). Effect of EPS on biofilm structure and function as revealed by an individual-based model of biofilm growth. *Water Sci. Technol.* **43**, 135–141.

Kuehn, M., Hausner, M., Bungartz, H.-J., Wagner, M., Wilderer, P. A. and Wuertz, S. (1998). Automated confocal laser scanning microscopy and semiautomated image processing for analysis of biofilms. *Appl. Environ. Microbiol.* **64**, 4115–4127.

Lakowicz, J. R. (1999). *Principles of Fluorescence Spectroscopy.* Kluwer Academic/Plenum Publishers, New York.

Lawrence, J. R. and Neu, T. R. (1999). Confocal laser scanning microscopy for analysis of microbial biofilms. *Meth. Enzymol.* **310**, 131–144.

Lawrence, J. R., Korber, D. R., Hoyle, B. D., Costerton, J. W. and Caldwell, D. E. (1991). Optical sectioning of microbial biofilms. *J. Bacteriol.* **173**, 6558–6567.

Lawrence, J. R., Neu, T. R. and Swerhone, G. D. W. (1998a). Application of multiple parameter imaging for the quantification of algal, bacterial and exopolymer components of microbial biofilms. *J. Microbiol. Meth.* **32**, 253–261.

Lawrence, J. R., Wolfaardt, G. and Neu, T. R. (1998b). The study of microbial biofilms by confocal laser scanning microscopy. In *Digital Image Analysis of Microbes* (M. H. F. Wilkinson and F. Shut, eds), pp. 431–465. Wiley, Chichester.

Lawrence, J. R., Korber, D. R., Wolfaardt, G. M., Caldwell, D. E. and Neu, T. R. (2002). Analytical imaging and microscopy techniques. In *Manual of Environmental Microbiology* (C. J. Hurst, R. L. Crawford, G. R. Knudsen, M. J. McInerney and L. D. Stetzenbach, eds), pp. 39–61. ASM, Washington.

Lawrence, J. R., Swerhone, G. D. W., Leppard, G. G., Araki, T., Zhang, X., West, M. M. and Hitchcock, A. P. (2003). Scanning transmission x-ray, laser scanning, and transmission electron microscopy mapping of the exopolymeric matrix of microbial biofilms. *Appl. Environ. Microbiol.* **69**, 5543–5554.

Lee, N., Nielsen, P. H., Andreasen, K. H., Juretschko, S., Nielsen, J. L., Schleifer, K. and Wagner, M. (1999). Combination of fluorescent in situ hybridisation and microautoradiography – a new tool for structure–function analyses in microbial ecology. *Appl. Environ. Microbiol.* **65**, 1289–1297.

Lewis, A., Radko, A., Ben Ami, N., Palanker, D. and Lieberman, K. (2001). Near-field scanning optical microscopy in cell biology. *Trends Cell Biol.* **9**, 70–73.

Lewis, A., Taha, H., Strinkovski, A., Manevitch, A., Khatchatouriants, A., Dekhter, R. and Ammann, E. (2003). Near-field optics: from subwavelength illumination to nanometric shadowing. *Nat. Biotechnol.* **21**, 1378–1386.

Liebsch, G., Klimant, I., Frank, B., Holst, G. and Wolfbeis, O. S. (2000). Luminescence lifetime imaging of oxygen, pH, and carbon dioxide distribution using optical sensors. *Appl. Spectrosc.* **54**, 548–559.

Lin, H.-J., Herman, P., Kang, J. S. and Lakowicz, J. R. (2001). Fluorescence lifetime characterization of novel low-pH probes. *Anal. Biochem.* **294**, 118–125.

Lindek, S., Stelzer, E. H. K. and Hell, S. W. (1995). Two new high-resolution confocal fluorescence microscopies (4pi, theta) with one- and two-photon excitation. In *Handbook of Biological Confocal Microscopy* (J. B. Pawley, ed.), pp. 417–430. Plenum Press, New York.

Liu, J., Dazzo, F. B., Glagoleva, O., Yu, B. and Jain, A. K. (2001). CMEIAS: a computer-aided system for the image analysis of bacterial morphotypes in microbial communities. *Microb. Ecol.* **41**, 173–194.

Marriott, G. and Parker, I. (2003a). Biophotonics. Part A. *Meth. Enzymol.* **360**.

Marriott, G. and Parker, I. (2003b). Biophotonics. Part B. *Meth. Enzymol.* **361**.

Mason, W. T. (1999). *Fluorescent and Luminescent Probes for Biological Activity.* Academic Press, San Diego.

Masters, B. R., So, P. T. C. and Gratton, E. (1999). Multiphoton excitation microscopy and spectroscopy of cells, tissues and human skin in vivo. In *Fluorescent and Luminescent Probes for Biological Activity* (W. T. Mason, ed.), pp. 414–432. Academic Press, San Diego.

Millard, A. C., Campagnola, P. J., Mohler, W., Lewis, A. and Loew, L. M. (2003). Second harmonic imaging microscopy. *Meth. Enzymol.* **361**, 47–69.

Miller, M. J., Wei, S. H., Parker, I. and Cahalan, M. D. (2002). Two-photon imaging of lymphocyte motility and antigen response in intact lymph node. *Science* **296**, 1869–1873.

Minsky, M. (1988). Memoir on inventing the confocal microscope. *Scanning* **10**, 128–138.

Müller, M. (2002). *Introduction to Confocal Fluorescence Microscopy.* Shaker Publishing BV, Maastricht.

Müller, J. D., Chen, Y. and Gratton, E. (2003). Fluorescence correlation spectroscopy. *Meth. Enzymol.* **361**, 69–92.

Nedoma, J., Vrba, J., Hanzl, T. and Nedbalova, L. (2001). Quantification of pelagic filamentous microorganisms in aquatic environments using the line-intercept method. *FEMS Microbiol. Ecol.* **38**, 81–85.

Neu, T. R. and Lawrence, J. R. (1997). Development and structure of microbial biofilms in river water studied by confocal laser scanning microscopy. *FEMS Microbiol. Ecol.* **24**, 11–25.

Neu, T. R. and Lawrence, J. R. (1999a). In situ characterization of extracellular polymeric substances (EPS) in biofilm systems. In *Microbial Extracellular Polymeric Substances* (J. Wingender, T. R. Neu and H.-C. Flemming, eds), pp. 21–47. Springer, Heidelberg.

Neu, T. R. and Lawrence, J. R. (1999b). Lectin-binding-analysis in biofilm systems. *Meth. Enzymol.* **310**, 145–152.

Neu, T. R. and Lawrence, J. R. (2002). Laser scanning microscopy in combination with fluorescence techniques for biofilm study. In *The Encyclopedia of Environmental Microbiology* (G. Bitton, ed.), pp. 1772–1788. Wiley, New York.

Neu, T. R., Swerhone, G. D. W. and Lawrence, J. R. (2001). Assessment of lectin-binding analysis for in situ detection of glycoconjugates in biofilm systems. *Microbiology* **147**, 299–313.

Neu, T. R., Kuhlicke, U. and Lawrence, J. R. (2002). Assessment of fluorochromes for two-photon laser scanning microscopy biofilms. *Appl. Environ. Microbiol.* **68**, 901–909.

Neu, T. R., Walczysko, P. and Lawrence, J. R. (2004). Two-photon imaging for studying the microbial ecology of biofilm systems. *Microbes Environ.* **19**, 1–6.

Neu, T. R., Woelfl, S. and Lawrence, J. R. (2004). Three-dimensional differentiation of photo-autotrophic biofilm constituents by multi-channel laser scanning microscopy (single-photon and two-photon excitation). *J. Microbiol. Meth.* **56**, 161–172.

Nielsen, J. L., Juretschko, S., Wagner, M. and Nielsen, P. H. (2002). Abundance and phylogenetic affiliation of iron reducers in activated sludge as assessed by fluorescence in situ hybridization and microautoradiography. *Appl. Environ. Microbiol.* **68**, 4629–4636.

Nielsen, J. L., Christensen, D., Kloppenborg, M. and Nielsen, P. H. (2003). Quantification of cell-specific substrate uptake by probe-defined bacteria under in situ conditions by microautoradiography and fluorescence in situ hybridization. *Environ. Microbiol.* **5**, 202–211.

Niswender, K. D., Blackman, S. M., Rohde, L., Magnuson, M. A. and Piston, D. W. (1995). Quantitative imaging of green fluorescent protein in cultured cells: comparison of microscopic techniques, use in fusion proteins and detection limits. *J. Microsc.* **180**, 109–116.

Noble, R. T. and Fuhrman, J. A. (1998). Use of SYBR green I for rapid epifluorescence counts of marine viruses and bacteria. *Aquat. Microb. Ecol.* **14**, 113–118.

Nunan, N., Ritz, K., Crabb, D., Harris, K., Wu, K., Crawford, J. W. and Young, I. M. (2001). Quantification of the in situ distribution of soil bacteria by large-scale imaging of thin sections of undisturbed soil. *FEMS Microbiol. Ecol.* **37**, 67–77.

Ouverney, C. C. and Fuhrman, J. A. (1999). Combined microautoradiography–16sRNA probe technique for determination of radioisotope uptake by specific microbial cell types in situ. *Appl. Environ. Microbiol.* **65**, 1746–1752.

Pawley, J. B. (1995). *Handbook of Biological Confocal Microscopy*. Plenum Press, New York.

Periasamy, A., Skoglund, P., Noakes, C. and Keller, R. (1999). An evaluation of two-photon excitation versus confocal and digital deconvolution fluorescence microscopy imaging in Xenopus morphogenesis. *Microsc. Res. Tech.* **47**, 172–181.

Piston, D. W., Masters, B. R. and Webb, W. W. (1994). Three-dimensionally resolved NAD(P)H cellular metabolic redox imaging of the in situ cornea with two-photon excitation laser scanning microscopy. *J. Microsc.* **178**, 20–27.

Piston, D. W., Knobel, S. M., Postic, C., Shelton, K. D. and Magnuson, M. A. (1999). Adenovirus-mediated knockout of a conditional glucokinase gene in isolated pancreatic islets reveals an essential role for proximal metabolic coupling events in glucose-stimulated insulin secretion. *J. Biol. Chem.* **274**, 1000–1004.

Pons, M. N., Drouin, F. J., Louvel, L., Vanhoutte, B., Vivier, H. and Germain, P. (1998). Physiological investigations by image analysis. *J. Biotechnol.* **65**, 3–14.

Ragan, T. M., Huang, H. and So, P. T. C. (2003). In vivo and ex vivo tissue applications of two-photon microscopy. *Meth. Enzymol.* **361**, 481–505.

Rocheleau, S., Greer, C. W., Lawrence, J. R., Cantin, Ch., Laramee, L. and Guiot, S. R. (1999). Differentiation of *Methanosaeta concilii* and *Methanosarcina barkeri* in anaerobic mesophilic granular sludge by fluorescent in situ hybridization and confocal scanning laser microscopy. *Appl. Environ. Microbiol.* **65**, 2222–2229.

Rodenacker, K., Hausner, M. and Gorbushina, A. A. (2003). Quantification and spatial relationship of microorganisms in sub-aquatic and subaerial biofilms. In *Fossil and Recent Biofilms* (W. E. Krumbein, D. M. Paterson and G. A. Zavarzin, eds), pp. 387–399. Kluwer Academic Publishers, Dordrecht.

Russ, J. C. (2002). *The Imaging Processing Handbook*. CRC Press, Boca Raton.

Sabri, S., Richelme, F., Pierres, A., Benoliel, A.-M. and Bongrand, P. (1997). Interest of image processing in cell biology and immunology. *J. Immunol. Meth.* **208**, 1–27.

Sanders, R., Draaijer, A., Gerritsen, H. C., Houpt, P. M. and Levine, Y. K. (1995). Quantitative pH imaging in cells using confocal fluorescence lifetime imaging microscopy. *Anal. Biochem.* **227**, 302–308.

Schilders, S. P. and Gu, M. (2000). Limiting factors on image quality in imaging through turbid media under single-photon and two-photon excitation. *Microsc. Microanal.* **6**, 156–160.

Schönholzer, F., Hahn, D., Zarda, B. and Zeyer, J. (2002). Automated image analysis and in situ hybridization as tools to study bacterial populations in food resources, gut and cast of *Lumbricus terrestris* L. *J. Microbiol. Meth.* **48**, 53–68.

Schrader, M., Bahlmann, K., Giese, G. and Hell, S. W. (1998). 4pi-confocal imaging in fixed biological specimens. *Biophys. J.* **75**, 1659–1668.

Shaw, P. J. and Rawlins, D. J. (1991). The point-spread function of a confocal microscope: its measurement and use in deconvolution of 3-D data. *J. Microsc.* **163**, 151–165.

Sheppard, C. J. R. and Shotton, D. M. (1997). *Confocal Laser Scanning Microscopy*. Bios Scientific Publishers, Oxford.

Sieracki, M. E., Johnson, P. W. and Sieburth, J. M. (1985). Detection, enumeration, and sizing of planktonic bacteria by image-analyzed epifluorescence microscopy. *Appl. Environ. Microbiol.* **49**, 799–810.

Sieracki, M. E., Reichenbach, S. E. and Webb, K. L. (1989). Evaluation of automated threshold selection methods for accurately sizing microscopic fluorescent cells by image analysis. *Appl. Environ. Microbiol.* **55**, 2762–2772.

Spear, R. N., Li, S., Nordheim, E. V. and Andrews, J. H. (1999). Quantitative imaging and statistical analysis of fluorescence in situ hybridization (FISH) of *Aureobasidium pullulans*. *J. Microbiol. Meth.* **35**, 101–110.

Staudt, C., Horn, H., Hempel, D. C. and Neu, T. R. (2003). Screening of lectins for staining lectin-specific glycoconjugates in the EPS of biofilms. In *Biofilms in Medicine, Industry and Environmental Technology* (P. Lens, A. P. Moran, T. Mahony, P. Stoodley and V. O'Flaherty, eds), pp. 308–327. IWA Publishing, UK.

Straub, M. and Hell, S. W. (1998). Multifocal multiphoton microscopy: a fast and efficient tool for 3-D fluorescence imaging. *Bioimaging* **6**, 177–185.

Sytsma, J., Vroom, J. M., de Grauw, C. J. and Gerritsen, H. C. (1998). Time-gated fluorescence lifetime imaging and microvolume spectroscopy using two-photon excitation. *J. Microsc.* **191**, 39–51.

Thomas, C. R. (1992). Image analysis: putting filamentous microorganisms in the picture. *Trends Biotechnol.* **10**, 343–348.

Vaija, J., Lagaude, A. and Ghommidh, C. (1995). Evolution of image analysis and laser granulometry for microbial cell sizing. *Antonie van Leeuwenhoek* **67**, 139–149.

Valeur, B. (2002). *Molecular Fluorescence*. Wiley-VCH, Weinheim.

van der Voort, H. T. M. and Strasters, K. C. (1995). Restoration of confocal images for quantitative image analysis. *J. Microsc.* **178**, 165–181.

Van Kempen, G. M. P., Van Vliet, L. J., Verveer, P. J. and van der Voort, H. T. M. (1997). A quantitative comparison of image restoration methods for confocal microscopy. *J. Microsc.* **185**, 354–365.

van Zandvoort, M. A. M. J., de Grauw, C. J., Gerritsen, H. C., Broers, J. L. V., oude Egbrink, M. G. A., Ramaekers, F. C. S. and Slaaf, D. W. (2002). Discrimination of DNA and RNA in cells by a vital fluorescent probe: lifetime imaging of SYTO13 in healthy and apoptotic cells. *Cytometry* **47**, 226–235.

Vecht-Lifshitz, S. E. and Ison, A. P. (1992). Biotechnological applications of image analysis: present and future prospects. *J. Biotechnol.* **23**, 1–18.

Verity, P. G., Beatty, T. M. and Williams, S. C. (1996). Visualization and quantification of plankton and detritus using digital confocal microscopy. *Aquat. Microb. Ecol.* **10**, 55–67.

Verveer, P. J., Gemkow, M. J. and Jovin, T. M. (1999). A comparison of image restoration approaches applied to three-dimensional confocal and wide-field fluorescence microscopy. *J. Microsc.* **193**, 50–61.

Viles, C. L. and Sieracki, M. E. (1992). Measurement of marine picoplankton cell size by using a cooled, charge-coupled device camera with image-analyzed fluorescence microscopy. *Appl. Environ. Microbiol.* **58**, 584–592.

Vroom, J. M., de Grauw, C. J., Gerritsen, H. C., Bradshaw, A. M., Marsh, P. D., Watson, G. K., Birmingham, J. J. and Allison, C. (1999). Depth penetration and detection of pH gradients in biofilms by two-photon excitation microscopy. *Appl. Environ. Microbiol.* **65**, 3502–3511.

Whyte, L. G., Slagman, S. J., Pietrantonio, F., Bourbonniere, L., Koval, S. F., Lawrence, J. R., Inniss, W. E. and Greer, C. W. (1999). Physiological adaptations involved in alkane assimilation at a low temperature by *Rhodococcus* sp. strain Q15. *Appl. Environ. Microbiol.* **65**, 2961–2968.

Wilkinson, M. H. F. (1998). Automated and manual segmentation techniques in image analysis of microbes. In *Digital Image Analysis of Microbes* (M. H. F. Wilkinson and F. Shut, eds), pp. 135–171. Wiley, Chichester.

Wilkinson, M. H. F. and Shut, F. (1998). *Digital Image Analysis of Microbes*. Wiley, Chichester.

Williams, R. M., Piston, D. W. and Webb, W. W. (1994). Two-photon molecular excitation provides intrinsic 3-dimensional resolution for laser-based microscopy and microphotochemistry. *FASEB J.* **8**, 804–813.

Wingender, J., Neu, T. R. and Flemming, H.-C. (1999). *Microbial Extracellular Polymeric Substances*. Springer, Heidelberg.

Xavier, J. B., Schnell, A., Wuertz, S., Palmer, R., White, D. C. and Almeida, J. S. (2001). Objective threshold selection procedure (OTS) for segmentation of scanning laser confocal microscope images. *J. Microbiol. Meth.* **47**, 169–180.

Xavier, J. B., White, D. C. and Almeida, J. S. (2003). Automated biofilm morphology quantification from confocal laser scanning microscopy imaging. *Water Sci. Technol.* **47**, 31–37.

Xu, C., Williams, R. M., Zipfel, W. and Webb, W. W. (1996). Multiphoton excitation cross-sections of molecular fluorophores. *Bioimaging* **4**, 198–207.

Yang, X., Beyenal, H., Harkin, G. and Lewandowski, Z. (2000). Quantifying biofilm structure using image analysis. *J. Microbiol. Meth.* **39**, 109–119.

Yang, X., Beyenal, H., Harkin, G. and Lewandowski, Z. (2001). Evaluation of biofilm image thresholding methods. *Water Res.* **35**, 1149–1158.

Yu, F. P., Callis, G. M., Stewart, P. S., Griebe, T. and McFeters, G. A. (1994). Cryosectioning of biofilms for microscopic examination. *Biofouling* **8**, 85–91.

Yuste, R. and Denk, W. (1995). Dendritic spines as basic functional units of neuronal integration. *Nature* **375**, 682–684.

Zipfel, W. R., Williams, R. M. and Webb, W. W. (2003). Nonlinear magic: multiphoton microscopy in the biosciences. *Nat. Biotechnol.* **21**, 1369–1377.

Zumbusch, A., Holtom, G. R. and Xie, X. S. (1999). Vibrational microscopy using coherent anti-Stokes Raman scattering. *Phys. Rev. Lett.* **82**, 4142–4145.

5 Applications of Cryo- and Transmission Electron Microscopy in the Study of Microbial Macromolecular Structure and Bacterial–Host Cell Interactions

M I Fernandez[1], M-C Prevost[2], P J Sansonetti[3] and G Griffiths[4]

[1] Department de Biologia Cŀlular i Anatomia Patològica, Facultat de Medicina, Institut d'Investigacions Biomèdiques August Pi i Sunyer (IDIBAPS), University of Barcelona, Casanova, 143, E-08036 Barcelona, Spain; [2] Plate-Forme de Microscopie Electronique, Institute Pasteur, Paris, France; [3] Unité de Pathogénie Microbienne Moléculaire, Institute Pasteur, Paris, France; [4] European Molecular Biology Laboratory, Heidelberg, Germany

◆◆◆

CONTENTS

Introduction
Methodology for transmission electron microscopy
Dehydration, infiltration and embedding
Freeze-substitution
Plastic support films
Contrasting: negative staining
Ultramicrotomy
Cryo-ultramicrotomy
Flat-embedding procedure for cell culture
Freeze-fracture and freeze-etching
Immunocytochemistry: localization of macromolecules
An introduction to cryo-electron microscopy and cryo-electron tomography

◆◆◆◆◆◆ **INTRODUCTION**

Electron microscopy (EM) has played an important role in the study of pathogen cell biology since its early days in the late 1940s, when this approach was first made by biologists. A large number of different techniques fall under the umbrella of EM, such as transmission EM (negative staining, thin sections, freeze-fracture methods) and scanning EM. These have been widely used for analysing microbial pathogens and their host cells (summarized by Griffiths, 2004).

Bacterial pathogens are notoriously difficult to fix using chemical fixatives at ambient temperatures for thin sectioning EM. Among the many artefacts one routinely sees in thin section images of bacteria (free or inside cells), are irregularities in the cell wall and an even, stronger aggregation of the DNA into large clumps. Bacterial visualization is facilitated by the aggregation of ribosomes to the periphery of the cell. Moreover, a regular structure, often quite aesthetic, was given great functional prominence at one time: this was the "mesosome". As it will become clearer below, we now know that mesosomes are artefact by having access to more recent EM technology.

Since the 1970s, based in large part on ideas previously generated by Fernandez Moran (1959, 1960), cryo-based methods were born. It emerged that if one rapidly cooled the specimens with a suitable coolant using the appropriate technology, and then proceeded with the specimen preparation (including chemical fixation) for some days at low temperatures ($<50°C$) before embedding at room temperature in a resin, the thin section often looked quite different in EM compared to conventionally fixed ones. The importance of the need for freeze-substitution to approach the ultrastructure of bacteria was first appreciated by Kellenberger (1987, 1991) who did pioneering work in this area.

But, how can one be sure that the images from freeze-substituted material looked like the living cell? For this, one had to wait for the arrival of a new way in EM specimen preparation, with the concept of vitrification that was driven by Jacques Dubochet. Provided one could solidify the water in the specimen fast enough, one could achieve the vitreous (or amorphous) state of solid water that, by definition, is not crystalline (Dubochet et al., 1988). Two revolutionary new methods for cryo-EM emerged from the work of this group, (i) the bare-grid method for visualizing whole (relatively thin) particles (such as virus or small pathogens) under vitreous conditions. In this method, the electrons penetrate the specimen and provide information about the internal structure. An additional and more recent approach, the tomography takes this approach to a new dimension, as discussed below; (ii) hydrated cryo-section method, where the vitrified specimen is sectioned, and the sections are transferred to the EM under vitreous conditions. Thus, mesosomes could be beautifully admired when the bacteria were previously chemically fixed, but were not likely to be seen under the vitreous preparation conditions (MacDowall et al., 1983). In an excellent agreement with the earlier, ribosomes are always uniformly distributed in the bacterial cytoplasm and the DNA is never clumped when utilizing freeze-substitution studies. The most recent use of hydrated cryo-sectioning methods therefore prevents the appearance of previous artefacts and allows the investigation of new details of bacterial ultrastructure.

For many readers of this chapter the notion of carrying out cryo-EM tomography or the hydrated cryo-sectioning method is probably similar to an experiment in, say rocket technology. Indeed, these include specialized techniques that are not easily undertaken without specialized knowledge. Fortunately, there are also a large number of EM techniques that can be learned by anyone, even without prior specialized knowledge. However, like any technique, it is something best learnt by looking at experienced people actually doing it. The basic method starts with the absolutely trivial (at least in its technology), namely negative staining. This technique can be done within minutes, provided one has a suitable specimen, a clean surface, clean water, a solute of uranyl acetate, forceps and a piece of filter paper. The most commonly applied methods for embedding and sectioning require only the minimal skill for the majority of the steps, which involve only changing solutions. More specialized skills need to be learnt at the cutting step. Even freeze-substitution can be performed using a cheap cryo-container, such as styrofoam box with dry ice. When one has a well-prepared specimen, with either plastic or cryo-sections, one has the option of immunogold labelling. This is a very powerful approach for combining ultrastructural level information with precise localization of antigens.

Few biologists, including microbiologists, would deny that for any system involving cells, and/or pathogens, there are still many important questions to be answered by state-of-the-art ultrastructure methods. Not many would deny, if confronted, that "ultrastructure" means a resolution beyond that obtained by light microscopy. Nevertheless, what is actually happening in practice (perhaps in the majority of papers) is that the use of light microscopy is now being pushed as if it could actually reach the ultrastructural level (Griffiths et al., 1993). In fact, anyone looking at a living cell with the finest confocal- or standard-light microscope would be surprised when given the opportunity to visualize the same preparation of cells in an EM thin section. The contrast is equivalent to the difference between taking a photograph of a forest from a helicopter at 200 m using a telescopic lens, and taking a picture with a simple camera down on the ground. The helicopter view can sometimes see many details on the ground, but not always. Often, the trees will be in the way (in an equivalent fashion when cells are observed by light microscopy, the presence of many organelles will hinder access to the organelles of interest, especially in the peri-nuclear region). However, it is always the camera on the ground that is going to provide most of the details about trees and the surface topography; in our analogy this would be the view provided by EM. It is literally, another world.

The current chapter is a hands-on description of some of the most commonly used methods for analysing pathogens, and pathogen–cell interactions by EM. It is not meant to be a thorough review of all available methods, but rather an account of methods used successfully in our laboratories. The theoretical background is covered superficially, such that only a few of the most important points are mentioned. Techniques such as epoxy resin embedding are highly empirical and there are many viable alternative ways to carry this out. Therefore, we will only summarize standard structural and immunocytochemical methods considered as the most "basic" for any ultrastructural study. Moreover, several excellent books and reviews will be recommended.

◆◆◆◆◆◆ METHODOLOGY FOR TRANSMISSION ELECTRON MICROSCOPY

Fixation for Fine Structure Preservation

Cell or tissue fixation is the first step in the process of sample preparation for EM. The major aim of this step is to stabilize and preserve the structural organization of the cell and tissue, so that the cell components and the ultrastructural interactions remain as close as possible to their original location. Thus, rapid cryo-fixation is the most appropriate method, followed by cryo-sectioning and observation of frozen sections using cryo-EM or freeze-substitution. However, in many studies chemical fixation is required.

Chemical fixation of proteins and lipids is necessary before dehydration in organic solvents and resin embedding is carried out. Fixation permits structural stabilization by cross-linking with cellular constituents and inhibition of enzymatic degradation. However, chemical fixation alters the chemical composition of the sample and this may lead to artefacts, such as extraction, denaturation, steric hindrance and chemical alteration of epitopes, as well as changes in cell volume and shape. Therefore, a major point to be considered when choosing a specific chemical fixative is to reduce or avoid these artefacts and preserve native structure.

When animal and human tissues are analysed by EM, it is critical that the time between the removal of the sample and the initial contact with the fixative should be as brief as possible. The most commonly used methods of fixation include simple immersion and perfusion, which involves a surgical procedure. Although the fixation process is complex, there are some general rules that can help us to preserve the fine structure and antigenicity. Thus, the fixative should be used at low concentration and the time required to fix isolated cells may be shorter than that for tissues. Temperature is also a parameter that needs to be considered. We recommend that the initial fixation should generally be done at the physiological temperature of the system (i.e. at 37°C for mammalian cells). Moreover, the fixative should have a rapid penetration capacity, with an adequate osmolarity and ionic concentration, and a neutral pH. However, several exceptions to these standard rules exist, and each sample should be considered individually, in order to select the best fixation process for the specific biological question of interest.

In the classical epoxide resin embedding protocol, the most commonly used fixatives for the primary fixation steps are the aldehydes (glutaraldehyde and formaldehyde). Fixatives such as acrolein may be used when rapid fixation is essential; imidoesters have also been used in a few studies. In general, including the analysis of numerous bacterial species, three consecutive fixatives are commonly applied: (i) the aldehyde to stabilize cellular proteins, (ii) osmium tetroxide to stabilize lipids and (iii) uranyl acetate to stabilize and contrast membranes. The next part will focus on these fixatives in order to briefly assess their preparation, as well as their effects on fine-structure preservation and influence on antigenicity. Finally, some protocols for fixation will be shown.

Aldehydes

Glutaraldehyde

As mentioned previously, glutaraldehyde permits the stabilization of cytosolic and membrane proteins. Glutaraldehyde is an aliphatic dialdehyde that reacts with many nucleophiles in the cell, generating cross-linked structures. It is soluble in water, ethanol and most organic solvents. Since its introduction by Sabatini *et al.* (1963), this dialdehyde is the most commonly used fixative for routine EM of almost all cells and tissues. Glutaraldehyde in solution is uncharged and is thus able to rapidly cross-link biological membranes. After penetration, it causes three-dimensional (3D) intracellular cross-links that are irreversible. These reactions also release protons and reduce cytoplasmic pH. In addition, a high concentration of this fixative can inhibit the formation of rapid cross-linking. For most procedures, standard fixation of eukaryotic and prokaryotic cells is carried out in a buffered solution with 1–5% glutaraldehyde for 1–2 h at room temperature (RT) (see Griffiths, 1993).

Formaldehyde

Formaldehyde was first used as a fixative by Blum in 1893. It has often been assumed that formaldehyde cross-links proteins less effectively than glutaraldehyde because it has only one functional aldehyde group. This is a misconception because in aqueous solutions, water forms methylene glycol, with two OH groups that can be used for cross-linking. While glutaraldehyde is a better fixative for conventional resin embedding than formaldehyde, it should be noted that complete cross-linking might be unnecessary for some purposes. Thus, the degree of cross-linking required for cryo-sectioning or freeze-substitution may be substantially less. In addition, in many cases formaldehyde gives better immunocytochemical labelling than glutaraldehyde.

Formaldehyde is commercially available in liquid and solid forms. The solid polymer (paraformaldehyde) dissolves slowly in cold water and more rapidly at higher temperatures. The process is increased by the addition of dilute alkalis. Routinely, paraformaldehyde is prepared in water at 60°C, adding a few drops of 1 M NaOH until the solution clears.

It has been suggested that, as for immunofluorescence microscopy (Berod *et al.*, 1981; Eldred *et al.*, 1983), an initial short (5 min approach) fixation with 4% formaldehyde in phosphate buffer at pH 6.5, followed by a subsequent longer (15 min) fixation with 4% formaldehyde in sodium borate buffer (pH 11) might provide a good preservation of cultured cells at the ultrastructural level.

Mixture of glutaraldehyde and formaldehyde

Graham and Karnovsky (1965) was the first to recommend a mixture of glutaraldehyde and formaldehyde for fixation. Although the mechanism of interaction of the two fixatives is not completely understood, mixtures are widely used and have given satisfactory results for many structural and immunocytochemical studies.

Effect of aldehydes on lipid and nucleic acids

Both glutaraldehyde and formaldehyde react and cross-link with amino lipids, although formaldehyde does not fix phospholipids. In fact, when formaldehyde is used for prolonged periods, there may be a considerable loss of phospholipids, or these lipids form artefactual myelin bodies (Baker, 1965). In contrast, glutaraldehyde is able to fix phospholipids to some extent, decreasing its extraction during the dehydration process (Nir and Hall, 1974).

The ability of aldehydes to stabilize nucleic acids by cross-linking of its imino groups, or of proteins associated with them, may be important for *in situ* hybridization methods. It is generally considered that glutaraldehyde and formaldehyde are not very effective cross-linkers of nucleic acids, and are likely to fix these only by cross-linking associated protein molecules.

Buffer Solutions for Aldehydes

Buffer solutions are formed by several ionic components that are able to fix free H^+ ions. These components limit the changes induced by free H^+ ions and thereby maintain a relatively stable pH. The pH range used for EM should be between 6.8 and 7.4. An adequate buffer should (i) be effective within a specific pH range; (ii) be chemically stable; a buffer concentration of 0.1–0.2 M should ideally be used (Griffiths, 1993); (iii) have no or minimal secondary effects, such as toxicity or ionic effects; (v) not react with the fixative.

The most common buffers used for EM studies are cacodylate, carbonate and phosphate buffers, PIPES, HEPES and MOPS (see Hayat, 1981, for general recipes for making buffer for EM).

Osmium Tetroxide

The main effect of osmium tetroxide (OsO_4) is to cross-link lipids, although it also cross-links proteins in a less defined way. Moreover, this fixative increases the contrast of cellular membranes. Since OsO_4 causes a drop in the pH of the sample, appropriate buffering during this step may be applied in order to avoid specimen artefacts caused by acidification. However, all commonly used buffers for EM contain charged molecules that do not readily cross plasma membranes. Thus, this problem needs to be considered. The most common concentration of OsO_4 used for standard TEM is 1%. As osmium tetroxide has a low penetration capability, the time of fixation should be around 1 h at RT, or 1 h at 4°C followed by an additional hour at RT.

Protocol for Standard Fixation

- Fix samples for 1–3 h at RT or overnight at 4°C with 2.5% glutaraldehyde in 0.1 M, pH 7.4 cacodylate buffer; 2.5% paraformaldehyde or 2.5% paraformaldehyde plus 2.5% glutaraldehyde can also be used. To fix bacteria, a fixative pH of 6.8 is preferable. It should be noted that an adequate level of cross-linking might be achieved in a few minutes in an open single cell system.
- Wash samples with 0.1 M, pH 7.4 cacodylate buffer several times, 15 min each at RT
- Post-fix the samples with 1% OsO_4 in the same buffer, 1 h at RT
- Wash intensively with the buffer
- Before fixing with uranyl acetate, wash the samples once with distilled water (dH_2O) and a second time with Michaelis buffer
- Fix the samples with 1% uranyl acetate in Michaelis buffer for 1 h at RT
- Continue with the dehydration and embedding steps (see next section for more details)

Fixation Protocol (Phillips et al., 1992)

This method permits better visualization of membranes

- Fix the samples overnight with 3% glutaraldehyde in 0.1 M, pH 7.2 Sörensen's phosphate buffer
- Wash once, 10 min with the Sörensen buffer
- Wash twice, 10 min each with 0.1 M cacodylate buffer
- Fix samples with 1% OsO_4 in 0.1 M cacodylate buffer, for 1–2 h at RT
- Wash three times, 5 min with 0.1 M cacodylate buffer
- Treat samples for 30 min with 0.2 M cacodylate buffer
- 30% methanol for 30 min
- 2% uranyl acetate in 30% methanol, for 1–24 h at 4°C
- Wash once with 30% methanol
- Dehydration and embedding steps

Tilney's Protocol for Fixation

Recommended for observation of the cytoskeleton

- Wash rapidly with 1 × PBS, and add the fixative 1% glutaraldehyde and 1% OsO_4 in 0.05 M, pH 6.3 phosphate buffer, prepared from the two solutions below;
  Solution A: 0.2 M/l Na_2HPO_4
  $Na_2HPO_4·2H_2O$ – 35.61 g
  $Na_2HPO_4·7H_2O$ – 53.65 g
  $Na_2HPO_4·12H_2O$ – 71.64 g

Solution B: 0.2 M/l $NaHPO_4 \cdot 2H_2O$
   $NaHPO_4 \cdot 2H_2O$ – 27.60 g
   $NaHPO_4 \cdot 2H_2O$ – 31.21 g
Mix 11 ml solution A and 36 ml solution B; adjust the pH to 6.3. Finally, add 25 ml of buffer and 75 ml of dH_2O to have a final 0.05 M buffer.
- Incubate the samples in the fix solution for 40 min on ice
- Wash three times with dH_2O on ice
- Fix overnight with 0.5% uranyl acetate in water
- Dehydration and embedding steps.

◆◆◆◆◆◆ DEHYDRATION, INFILTRATION AND EMBEDDING

Before embedding, removal of free water from the sample is achieved by incubating the sample in a graded series of organic solvents, such as methanol, ethanol or acetone. The dehydration process is then an indispensable step that has to be as rapid as possible in order to preserve structural integrity of the specimen, but it should allow complete removal of water from the sample. For rigid specimens, such as biofilms or even tissues, longer dehydration times are recommended. A standard dehydration method for bacteria involves incubation of fixed samples in a graded acetone or ethanol series for 10–15 min at RT. The 100% acetone or ethanol incubation should be repeated twice (30 min each). Improved staining of the sample is possible when the 70% acetone or ethanol is replaced with 2% uranyl acetate in 70% acetone or ethanol during longer incubation times (3 h). Moreover, to improve the dehydration process, the addition of $CuSO_4$ or $CaCO_3$ (silica gel) as a desiccant to the first 100% acetone steps may be considered (prepare it several hours before use and be careful as the $CuSO_4$ may precipitate). Providing that the sample is no larger than 1 mm (at most!) in any one dimension, a minimum time of 8 h should be allowed for the mixing. The specimen needs to stay in the resin for another 8 h, or more, before polymerizing for another 8 h at 60°C.

After dehydration, an embedding resin infiltrates the sample. These resins solidify when they polymerize. During the embedding process, the resin first infiltrates the sample and secondly high temperatures (i.e. Epon) or UV-induced X-linking (e.g. Lowicryl) polymerizes the resin. It is crucial that sufficient time is allowed for resin infiltration. When resin is used, one important aspect to be considered is the high toxicity. Therefore, handler caution is advised. When immunolabelling is the goal (immunotechniques), methacrylate resins are generally used. The resins Epon, Araldite and Spurr are soluble in acetone, but not in ethanol. However, when ethanol is used as a dehydration agent, a solvent such as propylene oxide has to be used. A detailed protocol for infiltration and embedding with Epon resin is given at the end

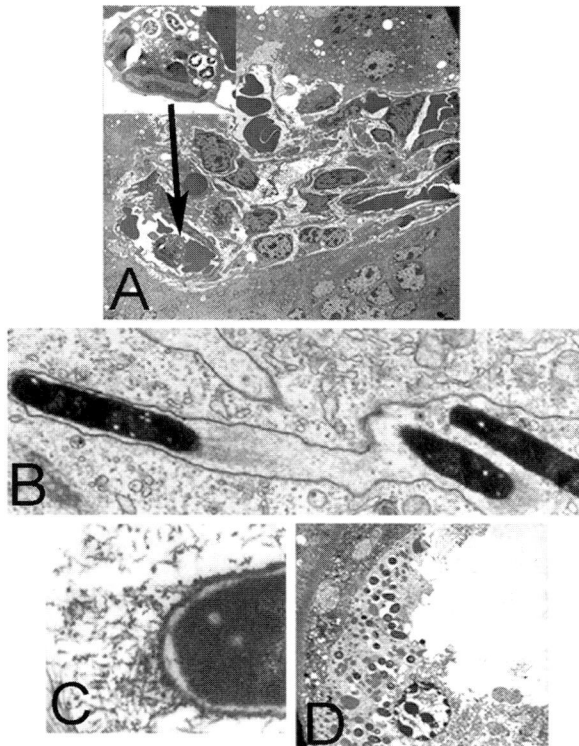

Figure 5.1. Transmission electron microscopy images to show relationship between bacteria and host. (A) Effects of *Shigella* infection on intestinal morphology in a model of shigellosis. Infiltration of immune cells into the lamina propria after bacteria infection. In mice infected with *S. flexneri*, bacteria cross the epithelial barrier. In the lamina propria, some bacteria are observed within PMNs (arrow and insert). (B and C) Actin dependent motility enables the bacterium to move intracellularly and invade adjacent cells. (C) Decoration of actin with the S1 subfragment of myosin at the pole of a bacterium in an infected cell is shown. (D) *Shigella* is observed in the intestinal lumen of newborn-infected mice. Moreover, the intestinal lumen is filled with mucus and cellular debris. (A and D from M.I. Fernandez and P.J. Sansonetti; B and C from P. Gounon and P.J. Sansonetti, Institut Pasteur, France.)

of this section. In Figure 5.1, TEM images of samples embedded in Epon are given.

As mentioned previously, specific embedding for immunocytochemistry should be considered. Lowicryl is the most commonly used resin, as it results in excellent preservation of structure, antigenicity and enzyme activity, due to the low temperatures used during the dehydration and infiltration steps. However, as for all the resins, labelling is absolutely restricted to the surface of the section. Dehydration is possible in both methanol and ethanol, and polymerization can be carried out using UV light. Lowicryl K4M is a polar resin with an embedding temperature around $-35°C$. Lowicryl HM20 resin is a non-polar resin with an embedding temperature as low as $-70°C$.

In addition to Lowicryl, other resins that are suitable for immunolabelling are

(a) *Glycol methacrylate*: water as solvent, UV light polymerization.
(b) *London resin (LR)*: dehydration in methanol or acetone/ethanol, although complete dehydration of the sample is not necessary, UV light polymerization; due to its low viscosity it is useful for infiltration of rigid specimens (insects, plants).
(c) *LR White*: use at RT.
(d) *LR Gold*: use at low temperatures (-4 to $-25°C$).

Standard Protocol for Infiltration with Epon Resin

- 100% ethanol or acetone/propylene oxide (1/1) – 10 min
- Propylene oxide – 10 min
- Propylene oxide/Epon resin 3/1 – 1 h
- Propylene oxide/Epon resin 1/1 – from 1 h to overnight
- Propylene oxide/Epon resin 1/3 – 1 h
- Pure Epon resin – 6–12 h
- Pure Epon resin – 2–6 h

Polymerization should be performed at 60°C for 24 h.

Embedding Procedure in Lowicryl K4M

- Dehydration in methanol or ethanol
- Methanol or ethanol/pre-chilled Lowicryl resin 1/1 – 1 h at $-35°C$
- Lowicryl resin – overnight at $-35°C$
- Lowicryl resin – 2 h at $-35°C$

Polymerize at $-35°C$ for 40 h with UV light, and 3 days at RT with UV light.

◆◆◆◆◆◆ FREEZE-SUBSTITUTION

A suitable method for maintaining structural and antigenic integrity, when sample dehydration has to be avoided, involves a combination of cryo-fixation and freeze-substitution. After this step resin embedding may be performed (Humbel and Müller, 1986). Fixation with 0.3% glutaraldehyde and 0.2% paraformaldehyde may be used. In case of chemical fixation, cryo-protectants, such as 2 M sucrose or dimethylformamide should be used before freezing. Another possibility is to fix samples with uranyl acetate in methanol (Schwarz and Humbel, 1989; Voorhout *et al.*, 1989a,b). Very low osmium tetroxide concentration can also be considered, although for immunolabelling this is antigen dependent. Cacodylate buffer should be avoided as this may reduce the antigenicity of the sample. By freeze-substitution, frozen water has to be replaced by an anhydrous solvent before resin infiltration. Samples can also be processed by a subsequent freeze-drying or critical point drying step (Hippe-Sanwald, 1993; Paul and Beveridge, 1993).

For acceptable freezing, avoiding ice crystal formation in the specimen, small samples can be

(i) immersed into liquid ethane, propane or Freon (McDowall *et al.*, 1983; Nagele *et al.*, 1985; Inoue *et al.*, 1982);
(ii) slammed in a cold metal-block (Murata *et al.*, 1985);
(iii) subjected to high pressure freezing (Moor, 1987). This procedure avoids the volume increase that occurs when water freezes at normal pressures, and the cooling rate required for vitrification is lower. High-pressure freezing machines are now commercially available and this is now the method chosen.

Samples are then transferred to a freeze-substitution system. First, samples are warmed to a temperature that permits resin embedding. The most conventional organic solvent used is methanol or acetone. More information on cryo-fixation and freeze-substitution can be obtained in Robards and Sleytr (1985) and Roos and Morgan (1990).

◆◆◆◆◆◆ PLASTIC SUPPORT FILMS

Usually, the grids used in EM are covered with an electron-translucent support film to provide good stabilization for ultrathin sections of samples embedded in resin. These supports can be (i) plastic films such as collodion, formvar, and butvar; (ii) carbon and carbon-coated plastic films; and (iii) metals such as silicium (we will not describe it in this chapter).

Plastic Support Films

Although plastic support films provide good stabilization for ultrathin sections of resin-embedded samples, they decompose following exposure to the electron beam. To reduce this effect, an evaporated layer of carbon is recommended. Several plastic supports are suitable (collodion, holey carbon, butvar), although formvar and carbon-coated formvar are the most widely used for bacteria and cell aggregates. In this chapter, we will only describe the preparation of formvar films. Formvar is soluble in dichloroethylene, chloroform and dioxane, and is more resistant to electrons than other plastic support films, such as collodion.

Standard protocol for formvar film:

1. Clean a slide with detergent. When drying the slide, a residual layer of the detergent should remain on the surface of the slide.
2. Submerge 3/4 of the slide in a glass funnel that has been filled with a 0.5% solution of formvar in chloroform or ethylene dichloride. Cover with a lid. After 30 s, remove the slide using forceps and drain off.
3. To remove the formvar film from the slide, scratch the edges of the glass slide with a razor blade.

4. Slowly and carefully float off the films in clean double-distilled H$_2$O by lowering the slide into water at an angle of 30–45°.
5. Place the grids on top of the film.
6. Place a piece of parafilm backing paper on the grids and press (very gently). The sandwich is then taken out of the water using forceps, turned upside-down, placed on filter paper and left to dry overnight.

It is absolutely essential to use the cleanest/purest water available (preferably double or even triple glass distilled). It is also important to maintain a clean working environment (e.g. use gloves).

Carbon-coated Plastic Support Films

The most commonly used mineral films are the carbon and carbon-coated plastic films. These are transparent to electrons, have a good resistance and electric conductivity, are thermostable, and are very convenient for working at high resolution. Carbon films are prepared by evaporation and sublimation of carbon on to grids. These girds are previously coated with formvar-carbon. The grids are prepared by removing the formvar support first. This is done by placing the grids in an atmosphere of solvent vapour that dissolves the formvar. Grids are then placed onto a wire mesh in a glass petri dish and the solvent (chloroform or carbon tetrachloride) is placed in the dish below the wire mesh replacing the lid closing the dish. If after a few hours the vapour alone does not remove the film, the process can be accelerated by dipping the coated grids into the solvent prior to closing the dish and placing them onto the mesh to soak. Allow the solvent to completely evaporate before removing the grids, which will now only have carbon on them. Remember that the carbon must be thick enough to be self-supporting. Advantages of this method include high stability to the electron beam and high-resolution examination of the adsorbed specimen. Disadvantages include a hydrophobic surface that is usually not of uniform thickness.

◆◆◆◆◆◆ CONTRASTING: NEGATIVE STAINING

In order to visualize the details of the fine structure, the sample has to be contrasted by the addition of heavy metal ion solutions, such as uranium and lead salts that precipitate onto certain structures. Negative staining in EM was used for the first time to visualize virus particles (for review see Wurtz *et al.* (1992)). This technique is very fast and permits the visualization of macromolecules, subcellular structures and whole organisms such as bacteria. Samples should contain a high concentration of the specimen (bacteria, virus, proteins, lipids or membrane suspensions), and be diluted in buffer or in saline solution (never in sucrose solution since this sugar can interfere with the contrast). Moreover, before contrasting, samples can be fixed in paraformaldehyde (for immuno-EM) or glutaraldehyde. The most commonly used solutions in EM are the

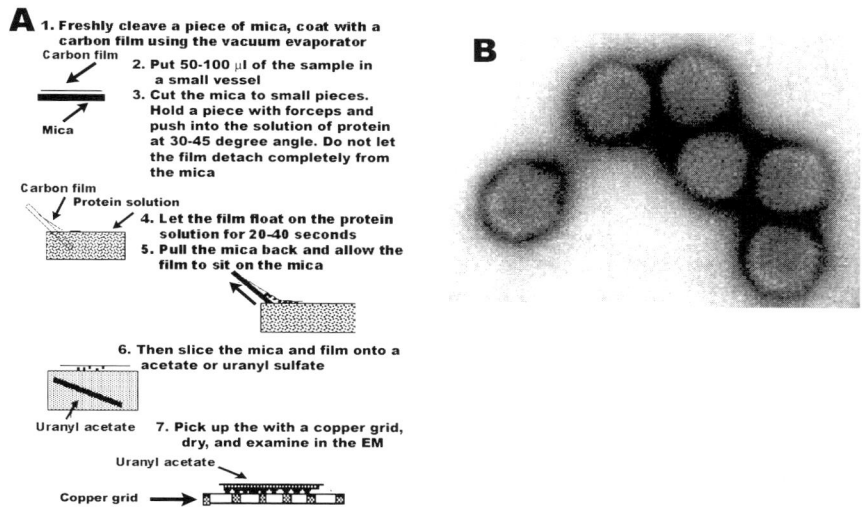

Figure 5.2. (A) Negative staining protocol. Variations of this method can be found in specialized books for EM. (B) An example of negative staining. Adenovirus particles have been adsorbed onto a carbon film that was deposited onto a freshly cleaved mica surface. The film was picked up onto a clean, 200 mesh specimen grid coated with a holey formvar film. The preparation was stained for 1 min with neutral 1% aqueous phosphotungstic acid and photographed in a transmission electron micrograph. (From P. Webster, House Ear Institute, CA, USA.)

uranium salts, phosphotungstic acid and the molybdate salts, generally in (bi)distilled H_2O. An example of the negative staining procedure is shown in Figure 5.2.

Concerning the uranium salts, uranyl acetate (van Bruggen et al., 1960) is prepared as an aqueous solution (0.5–4%), and is used at its natural pH 4.2–5.5. Uranyl acetate is suitable to visualize viruses, appendages of bacteria, bacteria and cells. It reacts quickly, as well as facilitates the stabilization of lipids. Moreover, it may have a fixative effect as a cross-linker for phospholipids when used at a concentration of 0.5% in 50% ethanol after OsO_4 treatment. Uranyl oxalate (Mellema et al., 1997) (0.5% uranyl acetate in 12 mM oxalic acid, pH 5–7 adjustable with ammonia) permits a high contrast and has a good penetration capability. It is frequently used for viruses and small proteins, such as pepsin and proteases. Uranyl formate is prepared in an aqueous solution (0.5–1%) at pH 3.5. This brings a high contrast of fine structures. However, due to its inherent instability it needs to be made up fresh and used rapidly.

Other heavy metal ion solutions commonly used for negative staining include phosphotungstic acid or corresponding Na and K salts (Brenner and Horne, 1959). This solution is suitable to visualize macromolecules, viruses, membrane proteins, and is especially suited to study refractive bacterial inclusion bodies and bacterial cells. It is used in a 0.5–3% aqueous solution (pH 5–8 adjustable with 1 M NaOH or KOH). Phosphotungstic acid permits good contrast and is very stable. However, some structures such as ribosomes, membranes, mitochondria and some viruses (for example, rotavirus) might be affected, and thus a pre-fixation process with aldehydes is required. Moreover, in contrast to uranyl acetate, phosphotungstic acid does not act as a fixative and sample penetration is slower.

Finally, ammonium molybdate (Muscatello and Horne, 1968) is suitable for the same structures as phosphotungstic acid, in addition to cell membranes, membrane-bound systems and protoplasts. An aqueous solution (2–10%) pH 5.2–7.5 is recommended, although it is unstable and can only be used up to 48 h after it is prepared.

Negative staining may be used for standard and more elaborate EM techniques, such as visualizing surface layers on bacteria after freeze-fracture preparation. For a review of negative staining, see Harris and Horne (1994).

◆◆◆◆◆◆ ULTRAMICROTOMY

Resin-embedded Samples

Resin-embedded samples are cut using an ultramicrotome to obtain sections as thick as 50–70 nm. To do this, samples have to be trimmed prior to ultrathin sectioning. The trimming process consists of obtaining a small, flat surface shaped as a pyramid, with a flat top (mesa) as smooth as possible to permit good ultrasectioning. Trimming can be done using a sharp razor blade or with a rotating milling cutter.

Semi-thin and ultrathin sectioning of resin-embedded samples is performed in the ultramicrotome using glass or diamond knives. Glass knives for standard and cryo-ultramicrotomy are made using a knife marker. For more information refer to Griffiths (1993) and Hoppert and Holzenburg (1998). A fluid-filled trough is attached around the knife-edge, and is sealed with warm dental wax. This trough is filled with water during sectioning, and sections are collected from the surface using a grid. Every ultra-microtome has different operating procedures. It is therefore necessary to refer to the manufacturer's manual in each case.

Contrast of the sections is achieved by floating the grids (sections facing downwards) in a drop of staining solution. Uranyl acetate and lead citrate are the most commonly used staining solutions. To avoid lead citrate from precipitating during staining, exclude atmospheric CO_2 during the procedure.

Protocol for contrasting of ultrathin sections

Uranyl acetate

Suspend 4% uranyl acetate in dH_2O several hours before its intended use. Discard non-dissolved uranyl acetate and store at RT in the dark.

Lead citrate (Reynols, 1963)

Dissolve 1.33 g of lead nitrate and 1.76 g of sodium citrate in 80 ml of dH_2O, and shake vigorously for 1 min. Incubate the solution for 30 min, shaking occasionally. Add 80 ml of 1 M NaOH and 12 ml of dH_2O. The working solution should be cleared by centrifugation (10 min at 15 000g). Store at 4°C.

Procedure

- Place a small drop of uranyl acetate solution (clear) on a parafilm strip in a petri dish with KOH or NaOH pellets to maintain an atmosphere free of CO_2. Place the grid with the sections facing downwards. Incubate for 3–15 min.
- Wash the grid in boiled dH_2O, three times, 1 min each.
- Place the grid on a lead citrate drop and incubate for 5–10 min.
- Wash the grid in boiled dH_2O three times, 2 min each.

◆◆◆◆◆◆ CRYO-ULTRAMICROTOMY

Omission of chemical fixation and dehydration in samples prepared by rapid freezing methods permits a good structure preservation of specimens under fully hydrated and uncontaminated conditions. However, cryo-sectioning of hydrated, unfixed samples is technically very difficult. On the other hand, a rapid freezing procedure leads to the formation of water in the vitreous or amorphous form, i.e. without crystalline structure. At very high freezing rates, vitreous water is formed both extracellularly and intracellularly. Therefore, the formation of ice crystals is avoided, preventing shrinkage of the cytoplasm, destruction of cellular structures and intracellular redistribution of ions and macromolecules. For a general review see Roos and Morgan (1990), Griffiths (1993) and Dubochet *et al.* (1988).

In hydrated cryo-sections, cryo-protectors may be used. The most well-known cryo-protectors are polyvinylpyrrolidine (PVP) that does not enter cells, and glycerol and demethylsulfoxide (DMSO), both of which cross membranes freely.

In the thawed frozen section technique or Tokuyasu technique (Tokuyasu, 1973, 1978; Tokuyasu and Singer, 1976) samples are fixed. This technique is considered the most appropriate for immunolabelling and provides excellent molecular and fine structural preservation. The general protocol of the Tokuyasu technique is

- Chemical fixation
- Embedding samples
- Cryo-protection at 4°C and freezing
- Cryo-sectioning
- Transfer of section to coated grids
- Antibody labelling
- Contrasting with heavy metal stains and drying of sections.

The aldehyde fixation process stabilizes the structure and makes the sample more permeable to sucrose (Griffiths et al., 1984). Moreover, fixed samples are protected by the high concentration sucrose infusion (1–2.3 M), which increases the vitrification process using liquid nitrogen. Moreover, a high concentration sucrose solution makes the specimen softer and increases its plasticity. This may be accentuated further when using a mixture of 20% PVP-1.8 M sucrose in 0.1 M phosphate buffer. Increased plasticity facilitates the cutting process. Often, such as for bacterial or cell suspensions, samples have to be embedded to facilitate the processes of cryo-protection, trimming and cutting. Three compounds are extensively used, that include gelatin, fibrinogen and agar compounds. These high molecular weight components are unable to traverse the plasma membrane, and therefore remain extracellular. Briefly, the freezing step initially involves mounting the sucrose-infused fixed sample on a specimen holder, and secondly plunging the mounted sample into liquid nitrogen.

For sectioning of frozen samples, an ultramicrotome equipped with a cryo-chamber is used. This cryo-ultramicrotome, which is available from several companies, maintains the temperature around -140 to $-50°C$. As for resin-embedded samples, sectioning is performed using a glass or diamond knife. Sections are harvested by adsorption on to the surface of a sucrose drop in a small metal loop. An important point to be considered is that during cryo-sectioning, the sample has to remain vitrified. To do this, the temperature of the sample can increase only by $1°C$ during the process, thus the block and knife have to remain frozen ($-100°C$). In addition, the cutting speed should ideally be very slow (<1 mm/s) to avoid fracturing of the section surface. Sections are then transferred onto a coated grid. A commercial anti-static device reduces the effects of the static electricity within the cutting chamber, facilitating section manipulation. For subsequent immunostaining, place the grid onto a 2% ice-cooled gelatin solution. The grid can then be stored on the gelatin surface overnight or be stained immediately. If necessary, grids can also be stored at $4°C$ for 24 h, and even for longer periods in 2.3 M sucrose at $-20°C$. Finally, contrast with heavy metal stains and then dry sections to preserve the structure and prevent the collapse phenomenon upon air-drying of the sections.

Protocol for Contrasting of Cryo-ultrathin Sections

- Dissolve 2% methylcellulose in ice-cold triple-dH$_2$O. Leave the solution for 1–2 days at $4°C$ with periodic stirring, and then centrifuge for 90 min at 100 000g at $4°C$
- Prepare a uranyl acetate solution. Check the correct concentration (from 0.3 to 4%)
- Float the grid on two drops of the staining solution for several seconds and on a third drop for 3–10 min
- Pick up the grid with the wire loop and remove excess staining solution. Dry the grid for 20–30 min.

The thickness of the methylcellulose film determines the section contrast. If the film is too thick, it may be dissolved in cold water for several minutes and the staining can be repeated. In addition to methylcellulose, other staining procedures use polyethyleneglycol-methylcellulose, polyvinyl alcohol and ammonium molybdate (Griffiths, 1993). At all stages during the labelling/staining procedure, the section side of the grid must stay wet whilst the opposite (top) side must remain completely dry.

◆◆◆◆◆◆ FLAT-EMBEDDING PROCEDURE FOR CELL CULTURE

The initial step in the preparation of cell cultures for cryo-sectioning is the detachment of cells from the substrate and pelleting of cells. Consequently, cells loose their *in situ* orientation. Recently, Oorschot *et al.* (2002) developed a novel procedure (the flat-embedding procedure), to sample and cryo-section cultured cells in the same orientation or perpendicular to the plane in which the cells normally grow. Thus, oriented sections of polarized cells can be obtained. In addition, this method also provides the possibility of performing co-related fluorescence and EM studies. Thus, first, the cells are fluorescent labelled using the flat-embedding procedure, and areas of interest are selected. These cells are then washed with PBS and cryo-sections of the selected areas are prepared using the same method. Finally, this method also permits the analyses of dynamic processes in living cells expressing fluorochrome-conjugated proteins (see Chapters 2–4).

◆◆◆◆◆◆ FREEZE-FRACTURE AND FREEZE-ETCHING

These techniques consist of preparing replicas of frozen surfaces and provide 3D information of the fractured specimen. The fracture preferentially occurs between the two leaflets of a lipid bilayer and non-homogeneous areas such as inclusions. See Holt and Beveridge (1982) for a review. Frozen samples are fractured using a commercial freeze-fracture apparatus. Upon fracturing a membrane, four surfaces are identified:

ES   old surfaces that point towards the extracellular surface
PS   old surfaces that point towards the cytoplasm or towards the protoplasm in bacteria
EF   new surfaces that point towards the extracellular side
PF   new surfaces that point towards the intracellular side.

In the case of bacteria, the existence of other structures such as outer membranes and surface layers made this nomenclature more difficult. Thus, for bacterial studies, a more detailed description should be considered.

◆◆◆◆◆◆ IMMUNOCYTOCHEMISTRY: LOCALIZATION OF MACROMOLECULES

The localization of macromolecules in specimens is performed by labelling procedures based on specific binding of electron dense markers, such as colloidal gold. Three types of immunolabelling may be differentiated: (i) pre-embedding labelling, where the immunolabelling reactions are done before plastic embedding of the sample, (ii) post-embedding labelling, when the immunostains are done on plastic sections and (iii) immunolabelling of cryo-sections. In bacteria, post-embedding labelling allows the visualization of antigens at the cell periphery, storage granules, the nucleoid and the cytoplasm.

The antibody should be tested using light microscopy prior to its use for EM. Moreover, this also establishes the appropriate conditions for the processing of the sample for EM, and a decision can be made as to whether the antibody is suitable for use at the EM level. Thus, (i) the effects of the fixative can be assessed by comparing the antigenicity of the antibody in unfixed and fixed samples; (ii) antibody concentrations can be appropriately characterized; (iii) the quality and specificity of the antibody can be tested; (iv) an assessment of whether antiserum purification is required can be made.

On the other hand, as for all immunostaining, in order to reduce the noise it is necessary to first treat the samples with agents that block non-specific labelling, such as gelatin, bovine serum albumin, fetal calf serum or ovalbumin (Tokuyasu and Singer, 1976; Kraehenbühl and Jamieson, 1974; Roth, 1989). Secondly, samples have to be washed carefully by floating the grids on several drops of buffer for a few minutes (5 min each) and even more after incubation with the gold-conjugated antibody or protein A (30 min total) (Slot and Geuze, 1981). In pre-embedding immunostaining more precautions should be taken.

It has to be considered that the amount of antigen present in EM sections may be substantially less than that for conventional immunocytochemistry. Therefore, to visualize as much antigen as possible and to obtain the high signal, with the most "acceptable" noise, antibody concentration is a determinant parameter. Relatively high concentrations of the antibody may likely be used. Griffiths (1993) recommended the followed rough-concentration range for antibodies for labelling cryo-sections: (i) monoclonal-culture supernatants, 1:1–1:10, (ii) antiserum, 1:5–1:200, (iii) IgG fraction, 1:5–1:100. Incubation conditions of 30 min at room temperature are usually sufficient to get an optimal signal.

To exclude false negative or positive signals, several control experiments have to be performed:

- Use only the electron dense marker. This will show non-specific binding of the marker to the specimen.
- Incubate the sample with the first antibody, followed by the secondary antibody or protein A alone, coupled with the electrodense marker.
- Incubate with the pre-immune serum or any serum without the first antibody, followed by the marker.

- Use a sample that does not contain the antigen of interest. In the case where specific components of the cell wall of Gram-negative bacteria are labelled using a polyclonal antibody, other Gram-negative bacteria may be used as control specimens.

Markers for Immunolabelling

Colloidal gold and ferritin

Colloidal gold (5–20 nm diameter) is currently the only widely used particulate marker for immuno-EM (Geuze *et al.*, 1981; Eisman *et al.*, 1995). While the use of its predecessor, the ferritin has declined significantly, it should be noted that ferritin has an advantage over gold: antibodies are covalently conjugated to ferritin but non-covalently adsorbed onto gold particles (Lucocq in Griffiths, 1993). On the other hand, ferritin particles are easily visible on the cell surface as 5 nm diameter particles. However, these particles are difficult to visualize in frozen sections, and ferritin is naturally present in several vertebrate tissues.

By contrast, the use of gold particles not only localizes the antigen, but this method also permits accurate quantification of a specific epitope by estimating the number of particles within a specific structure. These macromolecules are able to form complexes with protein A, antibodies, lectins and polysaccharides, and these complexes are used in the immunolabelling process. Gold particles can be produced by chemical reduction of a gold salt in aqueous solution, producing a saturated solution of elemental gold (for more information on gold particle preparation see Lucocq in Griffiths, 1993). Phosphorus, ascorbic acid, sodium citrate and tannic acid are widely used reducing agents. The quality of the gold preparations has to be carefully checked by EM before particles are coupled and used. Many gold-coupled commercial antibodies are also available. Moreover, using different sizes of colloidal gold particles, simultaneous double labelling using distinct antibodies is feasible. The selection of the particle size should be carefully done. Under normal circumstances, gold particles bigger than 5 nm diameter are used. In cases where very small particles are used, an additional silver enhancement procedure can be useful.

Non-immunological Interactions Used for Labelling

Lectins

Lectins are glycoproteins, mostly derived from plants, that have the capability of binding to specific sugar residues. Commercial lectins, pure or coupled to particulate markers are available. For EM, lectins have to be conjugated to ferritin or colloidal gold. The latter is difficult to use due to the formation of large gold aggregates. It seems better to buy the conjugates from a commercial source that have already been tested than to produce your own conjugates. Examples of lectins include ricin, conA and limulin. As an alternative to direct visualization of conjugated lectins, anti-lectin antibodies may be utilized followed by a conjugated secondary antibody, such as gold-protein A.

Protein A

This highly stable and water-soluble protein is derived from the cell wall of most strains of *Staphylococcus aereus* and is extensively used for immunolabelling procedures. Protein A binds to different antibodies of many species, primarily to the Fc domain of human, rabbit, guinea pig and pig immunoglobulin G (IgG). Protein A coupled to colloidal gold is widely used for TEM. In addition, a three-step method, i.e. antibody followed by a secondary antibody that binds protein A, may be used as a signal amplification system.

Avidin–Biotin

Avidin-bound biotin antibodies can also be used for TEM, although these are more widespread for light microscopy. There are several protocols for this procedure. The first, the LAB technique, utilizes a biotin-labelled molecule, such as an antibody or a lectin, followed by avidin-conjugated to an enzyme such as horseradish peroxidase. The second method is a three-step approach: biotin labelled protein, followed by free avidin, and finally a secondary biotinylated marker protein.

Immunolocalization Using Fluorescent Protein

Selective imaging that allows a correlation of light and EM of molecules would improve the analysis of fundamental processes, such as the assembly, internalization and trafficking of molecules in cells. In *in vitro* dynamic studies, the green fluorescent protein (GFP) of *Aequorea victoria* can be used to localize specific proteins when these are expressed as fusion tag GFPs, and the hybrid proteins are then localized to specific cell compartments in real time using light microscopy (see Chapters 2–4). These proteins can also be localized by TEM using anti-GFP antibodies.

Recently, new technologies have emerged by visualizing GFP using EM. For this, recombinant proteins are labelled by genetically appending or inserting a small motif containing the sequence -Cys-Cys-Xaa-Xaa-Cys-Cys, and then exposing the cells to a non-fluorescent bi-arsenical derivative of a fluorochrome. This compound becomes fluorescent when it binds to the tetracysteine motif. Gaietta *et al.* (2002), synthesized another biarsenical derivative of the red fluorophore resorufin, ReAsH, that becomes highly fluorescent upon binding to tetracysteine motifs. Moreover, ReAsH has the capability to photo-convert DAB to produce an electron dense product visible by TEM. This photo-conversion process consists of an intense illumination of a dye in the presence of oxygen and DAB. This procedure therefore allows to obtain an image of the same cell using both fluorescent and EM. This method also provides the opportunity to perform studies on live cells, as ReAsH can be imaged by confocal or multiphoton microscopy, followed by fixation of cells with glutaraldehyde. The fixed cells are then incubated in a DAB solution, illuminated, post-fixed with OsO_4, and embedded for EM (Gaietta *et al.*, 2002).

◆◆◆◆◆◆ AN INTRODUCTION TO CRYO-ELECTRON MICROSCOPY AND CRYO-ELECTRON TOMOGRAPHY

Three-dimensional imaging techniques in EM have been greatly improved in two areas: (i) electron tomography of cell organelles or cell sections and (ii) reconstruction of macromolecules from single particles. X-ray crystallography has been a highly useful technique to elucidate the structure of large individual proteins. However, X-ray crystallography becomes limited when the structure of multicomponent complexes needs to be analysed. Cryo-EM is a powerful tool to investigate the structure, assembly, conformational changes and dynamic interactions of biological supramolecules, and provides unique information for understanding large, complex biological systems (for review see Frank, 2002; Steven and Aebi, 2003). Electron tomography of frozen-hydrated tissue sections enables the analysis of the 3D structure of the specimen *in situ*, keeping the cells in near-native state. Cryo-tomography has the potential of visualizing the 3D organization of cells and organelles at molecular resolution, and it is a non-invasive method (Baumeister *et al.*, 1999; McEwen and Frank, 2001). By tomography techniques, projections of the specimen from different angles are acquired and then back-projected in a 3D reconstruction, known as the tomogram.

Techniques for specimen preparation and visualization for cryo-EM have been widely developed and improved over the last 30 years. Initially, only highly ordered specimens were studied by cryo-EM, such as helical fibres. In addition, membrane protein structures have been solved using this technique. Applications of cryo-EM techniques (especially in the analysis of molecular complexes in dynamic associations such as ribosomes or transcription complexes) have been reviewed by Nogales and Grigorieff (2001). Single-particle reconstruction, which brings a 3D reconstruction of the specimen, requires that the molecule analysed exists in multiple isolated copies that have the same structure. This method has opened a new field for the application of cryo-EM (Radermacher *et al.*, 1986, 1987a,b; van Heel, 1987), and is still evolving. On the other hand, cryo-electron tomography has also been used to characterize the 3D structure of viruses (Grunewald *et al.*, 2003). For many prokaryotic cells, as they are small enough, sectioning may be unnecessary; and some eukaryotic cells can be grown on a grid flat enough to be examined by cryo-EM (Medalia *et al.*, 2002). In order to study deeper regions of the cell, the hydrated cryo-sections method pioneered by Dubochet is the method of choice (Dubochet *et al.*, 1988).

As mentioned previously, this cryo-EM technique involves rapid freezing of the specimen using liquid ethane in a freeze-plunge apparatus (for more details, see Steward, 1991). The water rapidly turns into vitreous ice and thus, crystals are not formed. After observation by EM, the spatial relationship between all the different orientations of the sample can be defined mathematically. A series of single micrographs of the specimen provides enough information to reconstruct the molecule. Minimizing lens aberrations may be resolved by taking images with different defocus and applying the contrast transfer function correction in the processing of images (Frank, 2002). Moreover, the technique has been improved by using thick sections cut from high-pressure frozen tissues, as well as using improved methods for handling and mounting sections (Frank *et al.*, 2002; Hsieh *et al.*, 2002).

Resolution and quality of the tomogram are dependent on three major factors that have to be carefully considered: (i) orientation of the specimen, (ii) spacing and angular range of the projections and (iii) radiation damage. To characterize the molecular structure, the amount of "noise" in the micrographs should be minimized as this increases the difficulty in elucidating the "real" orientation of the specimen. Several methods have been developed to resolve this orientation problem, such as the random-conical data collection method (Radermacher *et al.*, 1986) and the method of common lines (Crowther, 1971; Penczek *et al.*, 1996; van Heel, 1987). After orientation is established, the angles are defined (Harauz and Ottensmeyer, 1984; Penczek *et al.*, 1994), and the molecule can be mathematically reconstructed (DeRosier and Klug, 1968; Frank, 2002). Images should be collected over a tilt range as wide as possible in order to cover an adequate angular range, and the exposure must be minimized to prevent radiation damage (reviewed in Koster *et al.*, 1997; McEwen and Frank, 2001). To reduce the radiation damage, it is important that each image used for the 3D reconstruction originates from a specimen that has only been observed once. In addition, use the lowest possible electron dose levels (around 10 electrons per A^2). Moreover, automated data-acquisition procedures drastically reduce the exposure. We will not cover theoretical aspects of mathematical reconstruction in this chapter. Rather, refer to a more specialized manuscript (see the whole issue of J. Struct. Biol., April/May; 138(1–2):1–155, 2002).

Regarding characterization of image resolution, data may be divided into two sets and reconstruction is applied to both sets independently. A calculated correlation index is then applied that measures the degree of correlation between the two reconstructions. The Fourier ring correlation is widely used for this purpose (Saxton and Baumeister, 1982; van Heel, 1982). Every 3D analysis should be accompanied by this resolution information, as this is essential in the final data interpretation.

Acknowledgements

We are grateful to F. Lazaro for helpful discussion. P. J. Sansonetti is a Howard Hughes Medical Institute Scholar.

References

Baker, J. R. (1965). The fine structure produced in cells by fixatives. *J. R. Microsc. Soc.* **84**, 115–131.

Baumeister, W., Grimm, R. and Walz, J. (1999). Electron tomography of molecules and cells. *Trends Cell Biol.* **9**, 81–85.

Berod, A., Hartman, B. K. and Pujol, J. F. (1981). Importance of fixation in immunohistochemistry: use of formaldehyde solutions at variable pH for the localization of tyrosine hydroxylase. *J. Histochem. Cytochem.* **29**, 844–850.

Brenner, S. and Horne, R. W. (1959). A negative staining method for high resolution electron microscopy of viruses. *Biochem. Biophys. Acta* **34**, 103–110.

Crowther, R. A. (1971). Procedures for three-dimensional reconstruction of spherical viruses by Fourier synthesis from electron micrographs. *Proc. R. Soc. London Ser. B* **261**, 221–230.

DeRosier, D. and Klug, A. (1968). Reconstruction of 3-dimensional structures from electron micrographs. *Nature* **217**, 130–134.

Dubochet, J., Adrian, M., Chang, J. J., Homo, J. C., Lepault, J., McDowall, A. W. and Schultz, P. (1988). Cryo-electron microscopy of vitrified specimens. *Q. Rev. Biophys.* **21**, 129–228.

Eismann, K., Mlejnek, K., Zipprich, D., Hoppert, M., Gerberding, H. and Mayer, F. (1995). Antigenic determinants of the membrane-bound hydrogenase in *Alcaligenes eutrophus* are exposed toward the periplasm. *J. Bacteriol.* **177**, 6309–6312.

Eldred, W. D., Zucker, C., Karten, H. J. and Yazulla, S. (1983). Comparison of fixation and penetration enhancement techniques for use in ultrastructural immunocytochemistry. *J. Histochem. Cytochem.* **31**, 285–292.

Fernández-Moran, H. (1959). Cryofixation and supplementary low temperature preparation techniques applied to the study of tissue ultrastructure. *J. Appl. Phys.* **30**, 2038.

Fernández-Moran, H. (1960). Low temperature preparation techniques for electron microscopy of biological specimens based on rapid freezing with liquid helium II. *Ann. N.Y. Acad. Sci.* **85**, 689–713.

Frank, J. (2002). Single-particle imaging of macromolecules by cryo-electron microscopy. *Annu. Rev. Biophys. Biomol. Struct.* **31**, 303–319.

Frank, J., Wagenknecht, T., McEwen, B. F., Marko, M., Hsieh, C. E. and Mannella, C. A. (2002). Three-dimensional imaging of biological complexity. *J. Struct. Biol.* **138**, 85–91.

Gaietta, G., Deerinck, T. J., Adams, S. R., Bouwer, J., Tour, O., Laird, D. W., Sosinsky, G. E., Tsien, R. Y. and Ellisman, M. H. (2002). Multicolor and electron microscopic imaging of connexin trafficking. *Science* **296**, 503–507.

Geuze, H. J., Slot, J. W., van der Ley, P., Scheffer, R. T. C. and Griffith, J. M. (1981). Use of colloidal gold particles in double labelling immuno-electron microscopy of ultrathin frozen tissue sections. *J. Cell Biol.* **89**, 653–665.

Graham, R. C. Jr. and Karnovsky, M. J. (1965). The histochemical demonstration of monoamine oxidase activity by coupled peroxidatic oxidation. *J. Histochem. Cytochem.* **13**, 604–605.

Griffiths, G. (1993). *Fine Structure Immuno-cytochemistry*. Springer, Berlin.

Griffiths, G. (2004). *Encyclopedia of Molecular Cell Biology and Molecular Medicine* (R. A. Meyers, ed.), vol. 2, pp. 21–89. Wiley, Weinheim.

Griffiths, G., McDowall, A., Back, R. and Dubochet, J. (1984). On the preparation of cryosections for immunocytochemistry. *J. Ultrastruct. Res.* **89**, 65–78.

Griffiths, G., Parton, R. G., Lucocq, J., van Deurs, B., Brown, D., Slot, J. W. and Geuze, H. J. (1993). The immunofluorescent era of membrane traffic. *Trends Cell Biol.* **3**, 214–219.

Grunewald, K., Desai, P., Winkler, D. C., Heymann, J. B., Belnap, D. M., Baumeister, W. and Steven, A. C. (2003). Three-dimensional structure of herpes simplex virus from cryo-electron tomography. *Science* **302**, 1396–1398.

Harauz, G. and Ottensmeyer, F. P. (1984). Direct three-dimensional reconstruction for macromolecular complexes from electron micrographs. *Ultramicroscopy* **12**, 309–320.

Harris, J. R. and Horne, R. W. (1994). Negative staining: a brief assessment of current technical benefits, and future possibilities. *Micron* **25**, 5–13.

Hayat, M. A. (1981). *Fixation for Electron Microscopy*. Academic Press, London.

Hippe-Sanwald, S. (1993). Impact of freeze substitution on biological electron microscopy. *Microsc. Res. Tech.* **24**, 400–422.

Holt, S. C. and Beveridge, T. J. (1982). Electron microscopy: its development and application to microbiology. *Can. J. Microbiol.* **28**, 1–53.

Hoppert, M. and Holzenburg, A. (1998). *Electron Microscopy in Microbiology*. BIOS Scientific Publishers, Oxford.

Hsieh, C. E., Marko, M., Frank, J. and Mannella, C. A. (2002). Electron tomography analysis of frozen-hydrated tissue sections. *J. Struct. Biol.* **138**, 63–73.

Humbel, B. and Müller, M. (1986). Freeze-substitution and low temperature embedding. In *The Science of Specimen Preparation* (M. Müller, R. P. Becker, A. Boyde and J. J. Wolosewich, eds), pp. 175–183. SEM, AMF O'Hare, Chicago.

Inoue, K., Kurosumi, K. and Deng, Z. P. (1982). An improvement of the device for rapid freezing by use of liquid propane and the application of immunocytochemistry to the resin section of rapid frozen, substitution-fixed anterior pituitary gland. *J. Electron Microsc.* **31**, 93–97.

Koster, A. J., Grimm, R., Typke, D., Hegerl, R., Stoschek, A., Walz, J. and Baumeister, W. (1997). Perspectives of molecular and cellular electron tomography. *J. Struct. Biol.* **120**, 276–308.

Kellenberger, E. (1987). The response of biological macromolecules and supramolecular structures to the physics of specimen cryopreparation. In *Cryotechniques in Biological Electron Microscopy* (R. A. Steinbrecht and K. Zierold, eds), pp. 35–66. Springer, Berlin.

Kellenberger, E. (1991). The potential of cryofixation and freeze substitution: observations and theoretical consideration. *J. Microsc.* **161**, 183–203.

Kraehenbühl, J. P. and Jamieson, J. D. (1974). Localization of intracellular antigens by immunoelectron microscopy. *Exp. Pathol.* **13**, 1–53.

McDowall, A. W., Chang, J. J., Freeman, R., Lepault, J., Walter, C. A. and Dubochet, J. (1983). Electron microscopy of frozen hydrated sections of vitreous ice and vitrified biological samples. *J. Microsc.* **131**, 1–9.

McEwen, B. F. and Frank, J. (2001). Electron tomographic and other approaches for imaging molecular machines. *Curr. Opin. Neurobiol.* **11**, 594–600.

Medalia, O., Weber, I., Frangakis, A. S., Nicastro, D., Gerisch, G. and Baumeister, W. (2002). Macromolecular architecture in eukaryotic cells visualized by cryoelectron tomography. *Science* **298**, 1209–1213.

Mellema, J. E., van Bruggen, E. F. J. and Gruber, M. (1997). Uranyl oxalate as negative stain for electron microscopy proteins. *Biochim. Biophys. Acta* **140**, 180–182.

Moor, H. (1987). Theory and practice of high pressure freezing. In *Cryotechniques in Biological Electron Microscopy* (R. A. Steinbrecht and K. Zierold, eds), pp. 171–191. Springer, Berlin.

Murata, F., Suzuki, S., Tsuyama, S., Suganuma, T., Imada, T. and Furihata, C. (1985). Application of rapid freezing followed by freeze-substitution acrolein fixation for histochemical studies of the rat stomach. *Histochem. J.* **17**, 967–980.

Muscatello, U. and Horne, R. W. (1968). Effect of the tonicity of some negative-staining solutions on elementary structure of membrane-bounded systems. *J. Ultrastruct. Res.* **25**, 73–83.

Nagele, R. G., Kosciuk, M. C., Wang, S. C., Spero, D. A. and Lee, H. (1985). A method for preparing quick-frozen, freeze-substituted cells for transmission electron microscopy and immunocytochemistry. *J. Microsc.* **139**, 291–301.

Nir, I. and Hall, M. O. (1974). The ultrastructure of lipid-depleted rod photoreceptor membranes. *J. Cell Biol.* **63**, 587–598.

Nogales, E. and Grigorieff, N. (2001). Molecular machines: putting the pieces together. *J. Cell Biol.* **152**, F1–F10.

Oorschot, V., de Wit, H., Annaert, W. G. and Klumperman, J. (2002). A novel flat-embedding method to prepare ultrathin cryosections from cultured cells in their in situ orientation. *J. Histochem. Cytochem.* **50**, 1067–1080.

Paul, T. R. and Beveridge, T. J. (1993). Ultrastructure studies of neutral lipid localisation in Streptomyces. *Arch. Microbiol.* **164**, 420–427.

Penczek, P. A., Grassucci, R. A. and Frank, J. (1994). The ribosome at improved resolution: new techniques for merging and orientation refinement in 3D cryo-electron microscopy of biological particles. *Ultramicroscopy* **53**, 251–270.

Penczek, P. A., Zhu, J. and Frank, J. (1996). A common-lines based method for determining orientations for $N > 3$ particle projections simultaneously. *Ultramicroscopy* **63**, 205–218.

Phillips, D. M., Pearce-Pratt, R., Tan, X. and Zacharopoulos, V. R. (1992). Association of mycoplasma with HIV-1 and HTLV-I in human T lymphocytes. *AIDS Res. Hum. Retroviruses* **8**, 1863–1868.

Radermacher, M., Wagenknecht, T., Verschoor, A. and Frank, J. (1986). A new 3-D reconstruction scheme applied to the 50S ribosomal subunit of *E. coli*. *J. Microsc.* **141**, RP1–RP2.

Radermacher, M., Wagenknecht, T., Verschoor, A. and Frank, J. (1987a). Three-dimensional reconstruction from a single-exposure, random conical tilt series applied to the 50S ribosomal subunit of *E. coli*. *J. Microsc.* **146**, 113–136.

Radermacher, M., Wagenknecht, T., Verschoor, A. and Frank, J. (1987b). Three-dimensional structure of the large ribosomal subunit from *E. coli*. *EMBO J.* **6**, 1107–1114.

Reynols, E. S. (1963). The use of lead citrate at high pH as an electron-opaque stain in electron microscopy. *J. Cell Biol.* **17**, 208–213.

Robards, A. W. and Sleytr, U. B. (1985). Practical methods in electron microscopy. In *Low Temperature Methods in Biological Electron Microscopy* (A. M. Glauert, ed.), vol. 10. Elsevier, Amsterdam.

Roos, N. and Morgan, A. J. (1990). *Cryopreparation of thin biological specimens for electron microscopy: methods and applications. Microscopy Handbooks*, vol. 21. Royal Microscopy Society and Oxford University Press, Oxford.

Roth, J. (1989). Postembedding labeling on lowicryl K4M tissue sections: detection and modification of cellular components. *Meth. Cell Biol.* **31**, 513–551.

Sabatini, D. D., Bensch, K. and Barrnett, R. J. (1963). Cytochemistry and electron microscopy: the preservation of cellular structure and enzymatic activity by aldehyde fixation. *J. Cell Biol.* **17**, 19–58.

Saxton, W. O. and Baumeister, W. (1982). The correlation averaging of a regularly arranged bacterial envelope protein. *J. Microsc.* **127**, 127–138.

Schwarz, H. and Humbel, B. M. (1989). Influence of fixatives and embedding media on immunolabeling of freeze-substituted cells. *Scanning Microsc.* (Suppl. 3), 57–64.

Slot, J. W. and Geuze, H. J. (1981). Sizing of protein A-colloidal gold probes for immuno-electron microscopy. *J. Cell Biol.* **90**, 533–536.

Steven, A. C. and Aebi, U. (2003). The next ice age: cryo-electron tomography of intact cells. *Trends Cell Biol.* **13**, 107–110.

Steward, M. (1991). Transmission electron microscopy of vitrified biological macromolecular assemblies. In *Electron Microscopy in Biology – A Practical Approach* (J. R. Harris, ed.), pp. 229–242. Oxford University Press, Oxford.

Tokuyasu, K. T. (1973). A technique for ultracryotomy of cell suspensions and tissues. *J. Cell Biol.* **57**, 551–565.

Tokuyasu, K. T. (1978). A study of positive staining of ultrathin frozen sections. *J. Ultrastruct. Res.* **63**, 287–307.

Tokuyasu, K. T. and Singer, J. S. (1976). Improved procedures for immunoferritin labelling of ultrathin frozen sections. *J. Cell Biol.* **71**, 894–906.

van Bruggen, E. F. J., Wiebinga, E. H. and Gruber, M. (1960). Negative-staining electron microscopy of proteins at pH values below their isoelectric points: its application to hemocyanin. *Biochim. Biophys. Acta* **42**, 171–172.

van Heel, M., Keegstra, W., Schutter, W. and van Bruggen, E. J. F. (1982). Arthropod hemocyanin structures studied by image analysis. In *Life Chemistry Reports, the Structure and Function of Invertebrate Respiratory Proteins* (E. J. Wood, ed.), pp. 69–73. Harwood, London, Suppl. 1.

van Heel, M. (1987). Angular reconstitution: a posteriori assignment of projection directions for 3D reconstruction. *Ultramicroscopy* **21**, 111–124.

Voorhout, W. F., Leunissen-Bijvelt, J. J. M., van der Krift, Th. P. and Verkleij, A. J. (1989a). The application of cryo-ultramicrotomy and freeze-substitution in immuno-gold labelling of hybrid proteins in *Escherichia coli*. A comparison. *Scanning Microsc.* 47–56, Suppl. 3.

Voorhout, W. F., Leunissen-Bijvelt, J. J. M., van der Krift, Th. P. and Verkleij, A. J. (1989b). Immuno-gold labeling of *Escherichia coli* cell envelope components. In *Immuno-gold Labeling in Cell Biology* (A. J. Verkjeij and J. L. M. Leunissen, eds), pp. 292–304. CRC Press, Florida.

Wurtz, M. (1992). Bacteriophage structure. *Electron Microsc. Rev.* **5**, 283–309.

Colour Plates

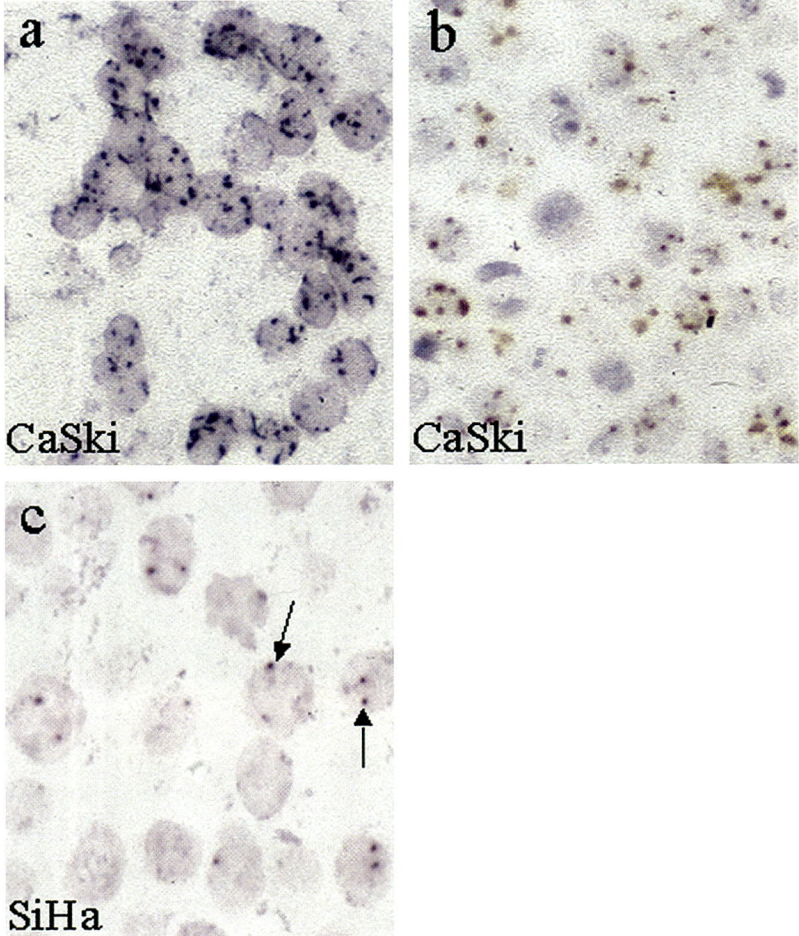

Plate 1. Positive HPV 16 hybridization signal in CaSki cells using biotinylated probe and detection systems (a) no. 5 (mouse anti-biotin/anti-mouse/mouse AP-anti-AP/NBT-BCIP), (b) no. 7 (streptavidin-HRP/biotinylated tyramide/streptavidin-HRP/DAB), and (c) positive HPV 16 hybridization signal (arrows) in SiHa cells using biotinylated probe and detection system no. 5 (mouse anti biotin/anti-mouse/mouse AP-anti-AP/NBT-BCIP). AP, alkaline phosphatase; HRP, horseradish peroxidase; NBT, nitroblue tetrazolium; BCIP, 5-bromo-4 chloro-3-indolylphosphate; DAB, 3',3'-diaminobenzidine. Reprinted with permission from Holm, "A highly sensitive nonisotopic detection method for *in situ* hybridization" in Appl. Immunohistochem. Mol. Morphol. 8(2) 162–165, 2000. Copyright Lippincott Williams & Wilkins. (See Figure 1.2)

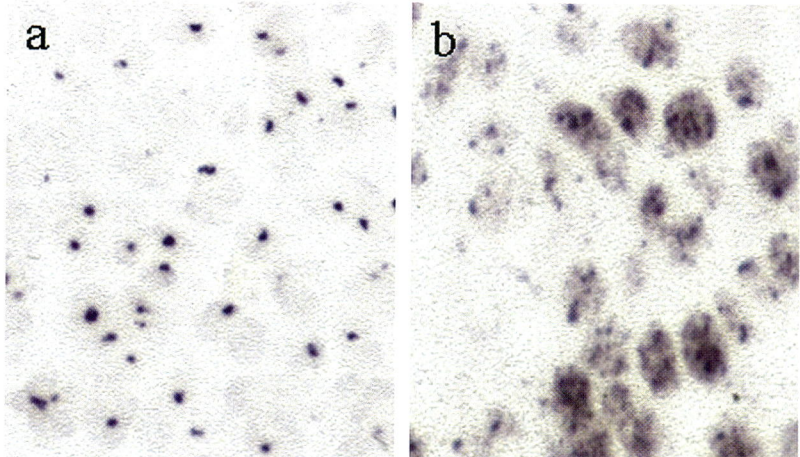

Plate 2. Positive HPV 16 hybridization signal in two cervical carcinomas. A punctuate signal (a) and a punctuate and diffuse signal (b). (See Figure 1.3.)

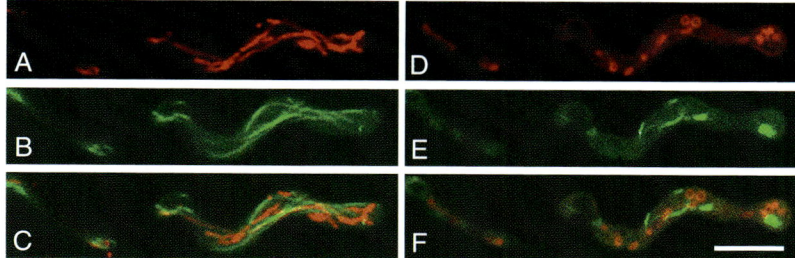

Plate 3. Multi-color imaging combining fluorescent chimeric proteins with a conventional fluorescent dye. Rhodamine B staining (A and D) and expression of a β-tubulin-EYFP chimeric protein (B and E) resulted in the visualization of both microtubules and mitochondria (overlays shown in C and F) in a living *M. grisea* germ tube treated for 0 (A–C) or 16.5 min (D–F) with the anti-microtubule agent benomyl at 20 ppm. Bar, 7 μm. (See Figure 2.5.)

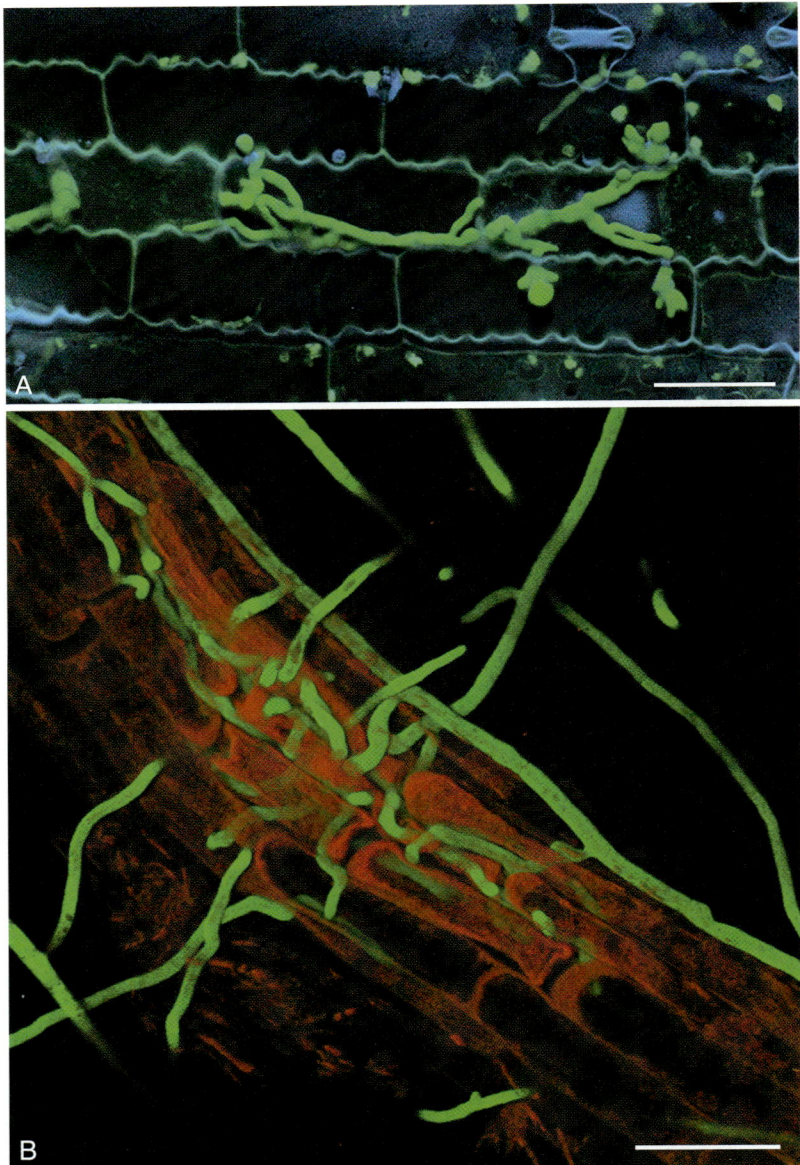

Plate 4. Two-channel multiphoton/confocal imaging of fluorescence-tagged pathogens *in planta*. Cytosolic expression of either AmCyan in *Colletotrichum graminearum* (A) or ZsGreen in *Fusarium oxysporum* (B) facilitated imaging of host–pathogen interactions within a maize leaf (A) or *Arabidopsis* root (B). Single photon excitation with either a 458 or 488 nm laser was sufficient to image AmCyan and ZsGreen, respectively. Significant autofluorescence of plant cell walls was generated in a separate channel by multiphoton excitation at 730 nm. Images A and B were based on a series of optical sections rendered as shadow projections using Zeiss LSM Image Visart 3D rendering software. Bar for A, 70 μm; B, 37 μm. (See Figure 2.6.)

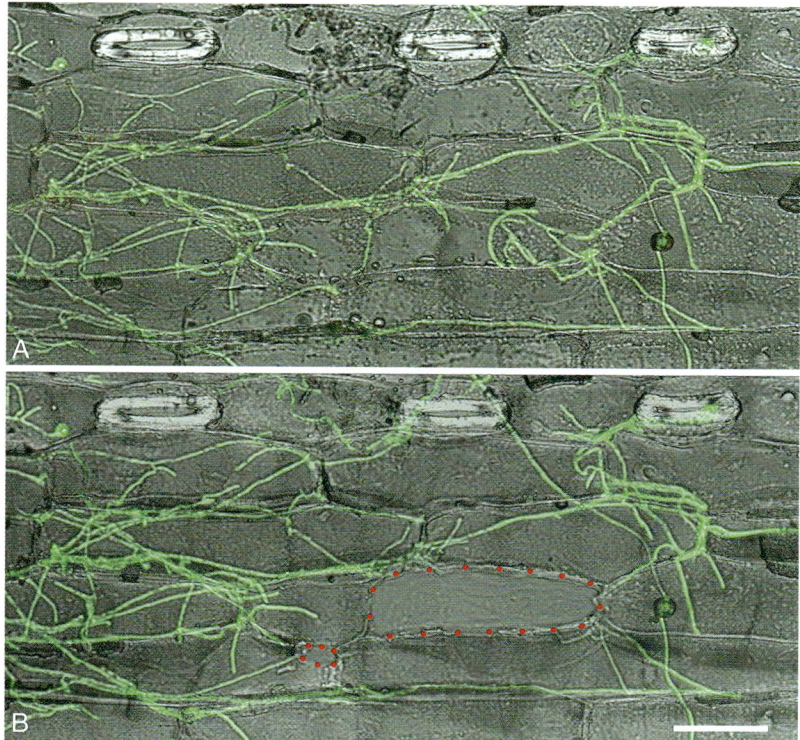

Plate 5. Fluorescent protein-assisted laser microdissection. Expression of ZsGreen in the cytoplasm of *M. grisea* facilitated the identification of infection hyphae within barley leaf epidermal cells (A). One capture event recovered the contents of a single infected epidermal cell in a two-step procedure combining laser ablation and separation (B, outlined by red dots). A second ablation, lower and to the left (also outlined by red dots), showed the location where a tip of a single infection hypha was precisely excised and captured. Bar, 45 μm. (See Figure 2.7.)

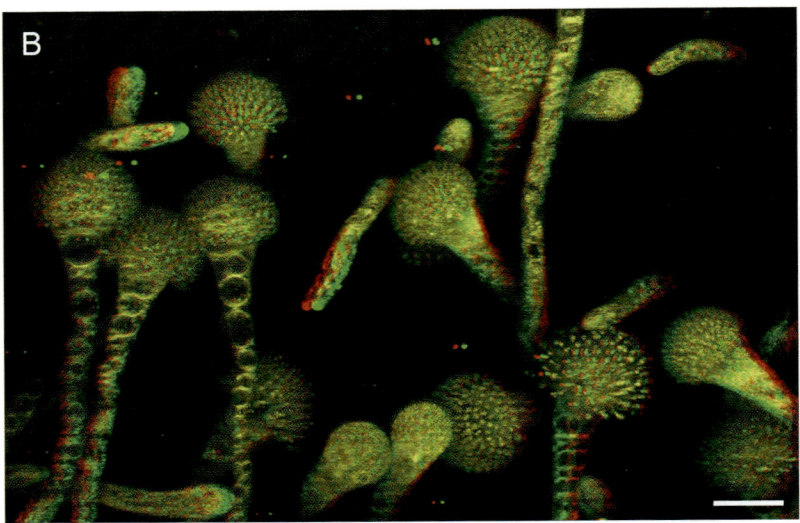

Plate 6. (B) 3D projection of 50 × 0.5 μm optical sections (requires viewing with green/red stereo glasses). Bar = 20 μm (See Figure 3.2B).

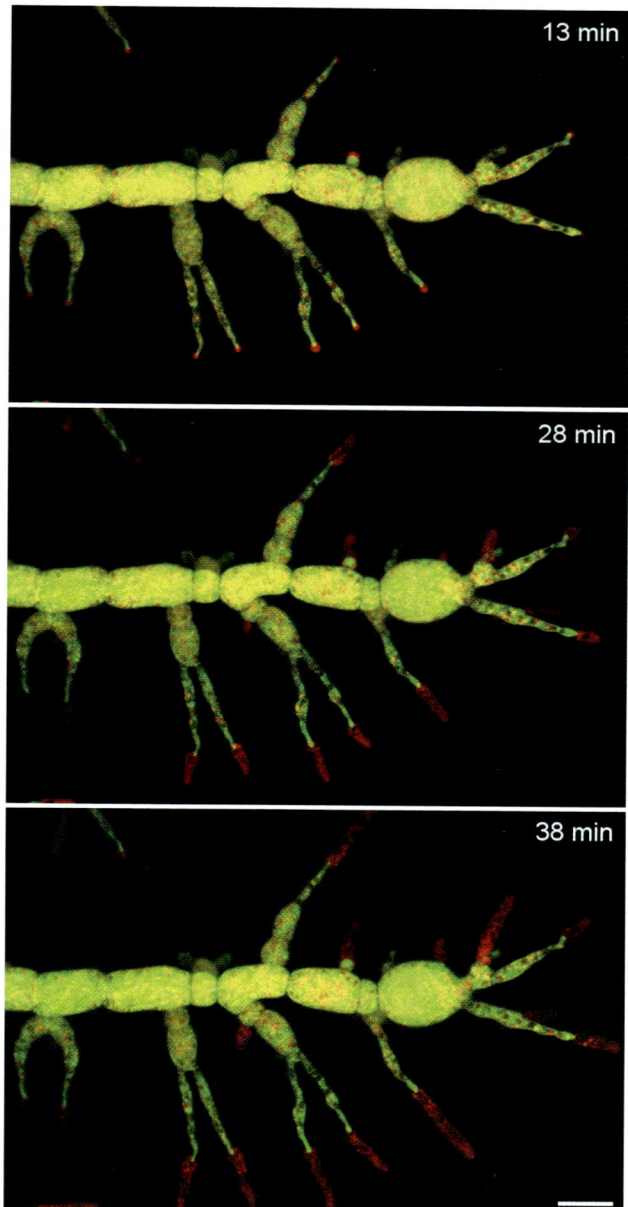

Plate 7. Hyperbranching *cot-1* mutant of *Neurospora crassa* stained with FM4-64 grown at 37°C for 3 h and shifted back to 25°C. Each image represents a projection of 30 × 1.0 µm² optical sections through the hypha. The times represent the periods after which the hyphae were shifted to 25°C. Note the bulbous compartments of the main hypha and the finely tapered branches (yellow-green) which become wider during recovery of the wild type hyphal phenotype (red). Bar = 20 µm. (See Figure 3.11.)

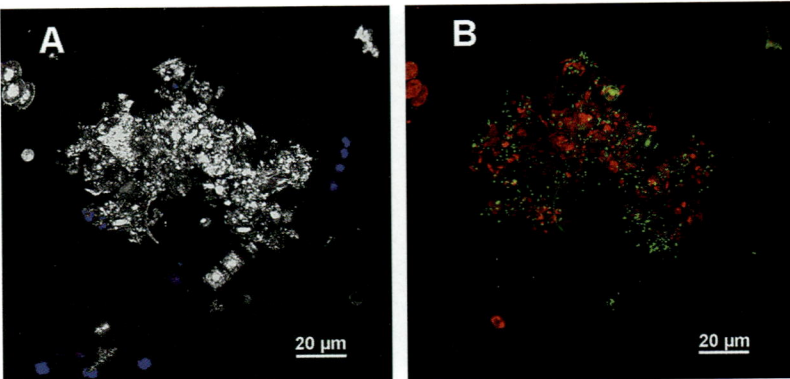

Plate 8. 1P-LSM maximum intensity projection of dual channels from the same lotic aggregate. (A) shows the reflection (= white) and autofluorescence (= blue). (B) shows the nucleic acid stained bacteria (= green) and the lectin stained glycoconjugates (= red). (See Figure 4.2.)

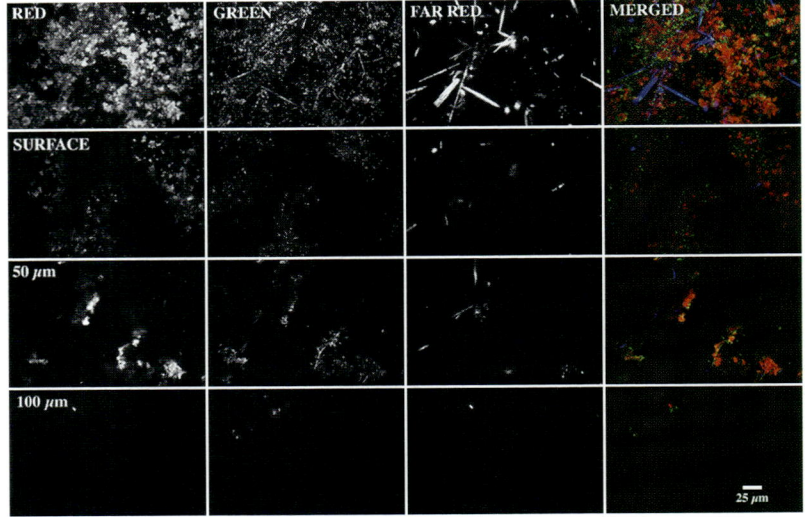

Plate 9. 1P-LSM images showing the distribution of bacteria, photosynthetic organisms (algae and cyanobacteria) and exopolymer in a river biofilm. The series illustrates the use of a combination of fluorescent probes (nucleic acid and lectin) in combination with autofluorescence to determine the vertical distribution of bacteria, total photosynthetic organisms and exopolymer in a river biofilm community. (See Figure 4.3.)

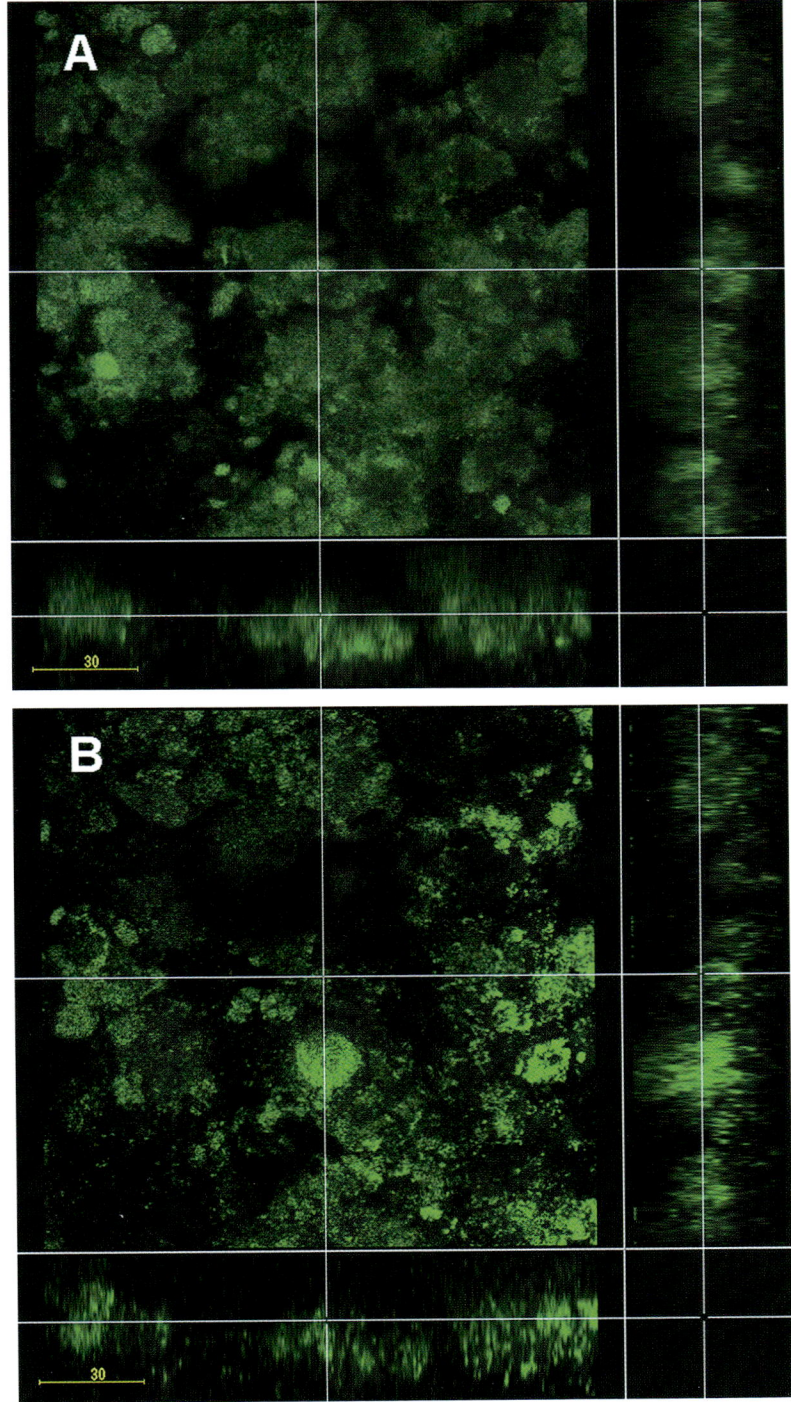

Plate 10. 1P-LSM (A) and 2P-LSM (B) of the same location in *XY* and *XZ* of laboratory biofilms grown on pumice with EDTA medium. Biofilms were stained with the nucleic acid specific stain SYTO 40. (See Figure 4.6.)

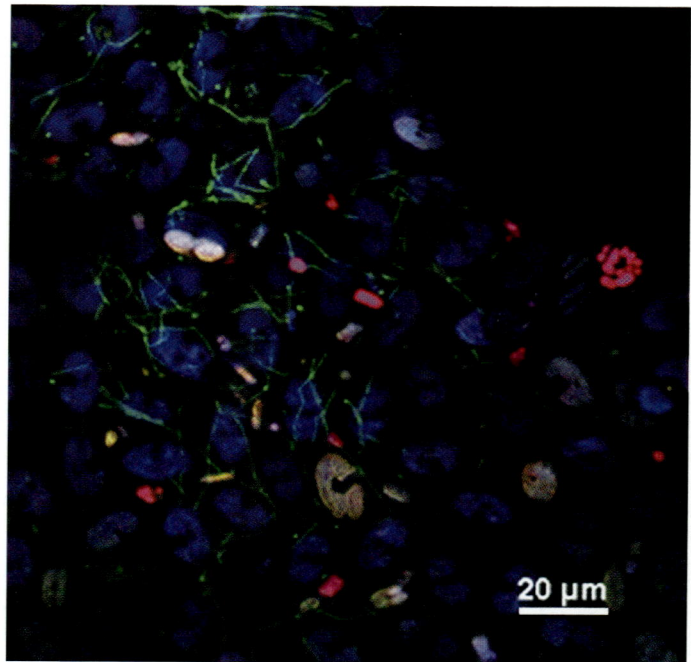

Plate 11. 2P-LSM of a SYTO 9 stained lotic biofilm community with excitation at 800 nm. Colour allocation: nucleic acid stain (= green), autofluorescence of cyanobacteria (= pink), autofluorescence of algae (= blue). (See Figure 4.7.)

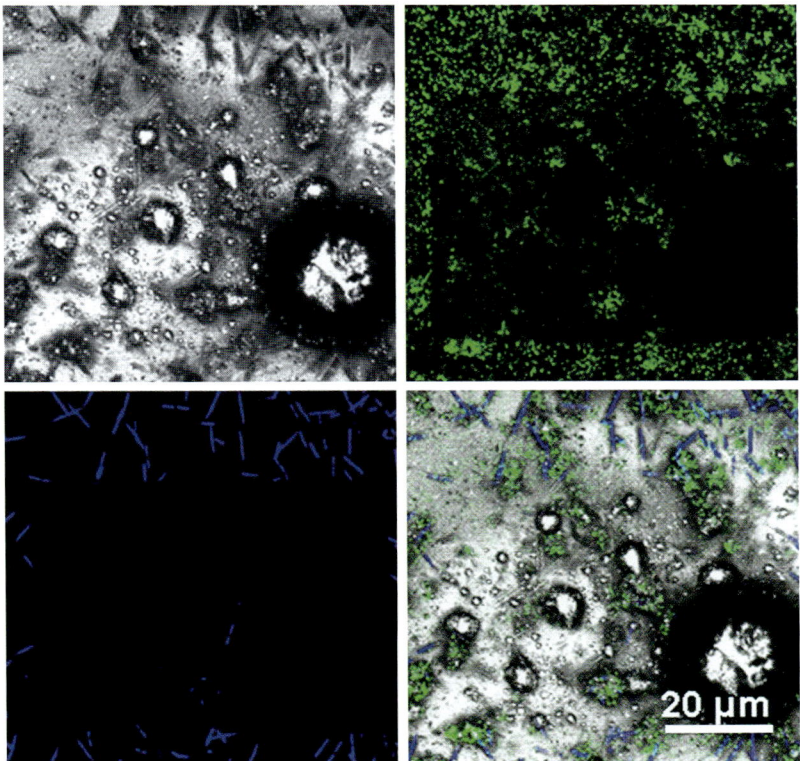

Plate 12. Effect of wrong settings in two-photon excitation on biofilm and substratum surface. The images show the bleaching and damaging effect in different channels and as an overlay. Take notice of the bubbles created by melting of the polycarbonate substratum. Colour allocation: reflection (= white), nucleic acid stain (= green), autofluorescence of algae (= blue). See also text for further details. (See Figure 4.8.)

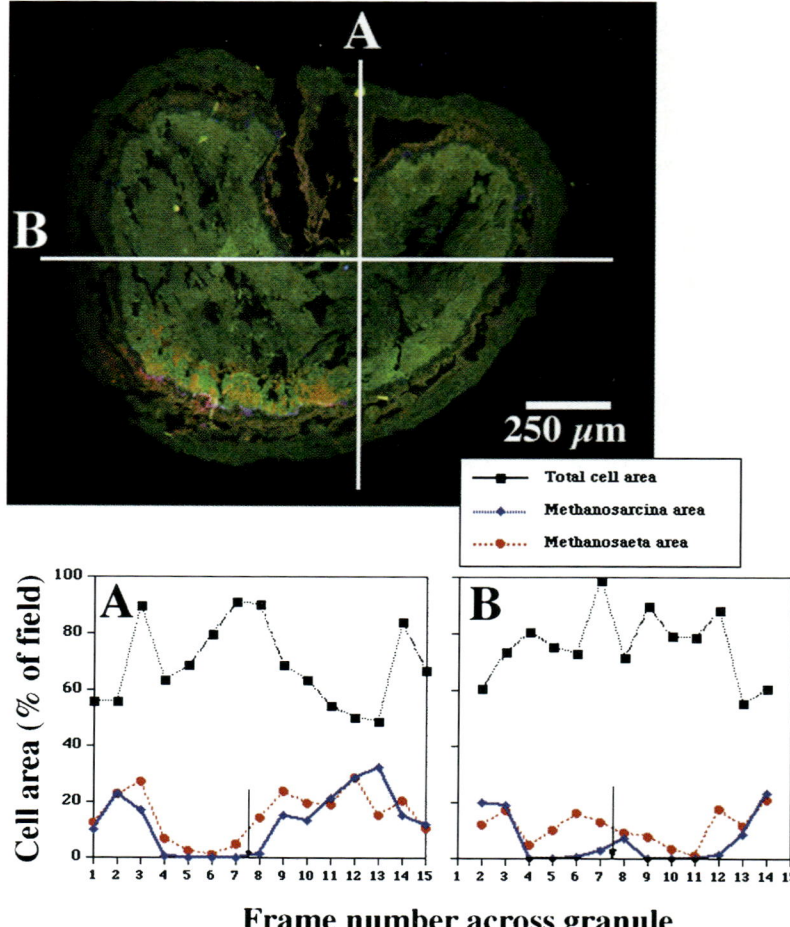

Plate 13. 1P-LSM imaging of 8-mm-thick sections of UASB granules that were simultaneously *in situ* hybridized with a MS5 *M. concilii*-specific probe [labelled with TAMRA NHS (red)] and the MB4 *Methanosarcina barkeri*-specific probe [labelled with Cy5 (blue)]. Top: Low-resolution 1P-LSM optical thin section (*XY* plane) showing the layered structure of *Methanosarcina*-enriched granules, in which *M. barkeri* and *M. concilii* are present within the layers. Bottom: Relative quantifications of *M. barkeri* and *M. concilii* cell areas by LSM and digital image analysis. Transects across the centres of the granules were analysed, and the cell area containing both *M. concilii* and *M. barkeri* as well as the total cell area (detected with SYTO 9) was determined. Data are expressed as relative cell area (percentages of the field) versus location within the granule. Frame numbers represent transects of these thin sections through the granule. (See Figure 4.9.)

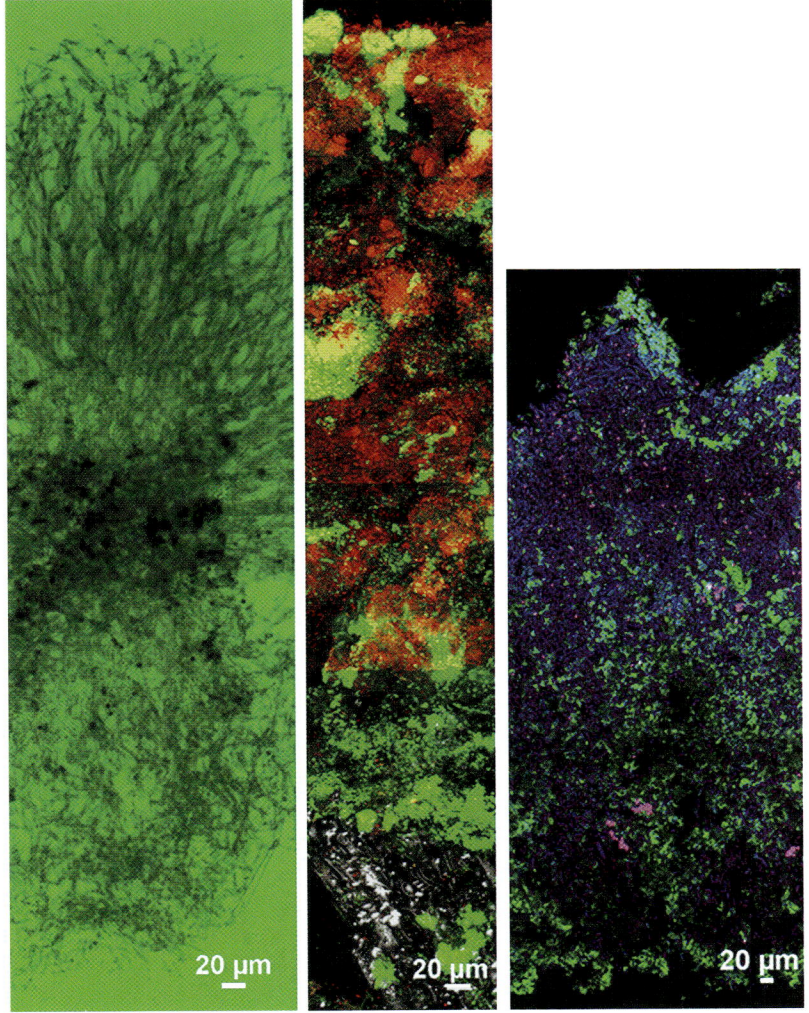

Plate 14. Examples of physically sectioned biofilms which were examined by 1P-LSM at several locations across the depth of the structure. For presentation several image series were mounted. (A) Cryosectioning of fungal pellet across the whole aggregate (four image stacks). The pellet was negatively stained using FITC-dextran. The black spots in the central part are spores developing under oxygen limitation. (B) Sectioning of a chemoautotrophic bacterial biofilm grown on pumice in a fluidized bed reactor (six image stacks). Colour allocation: glycoconjugates (= green), nucleic acid stain (= red), co-localised bacteria and glycoconjugates (= yellow), reflection of pumice (= white). (C) Cryosectioning of a 75-day-old freshwater phototrophic biofilm (three image stacks). After sectioning the biofilm was stained with an Alexa-488 labelled lectin. Colour allocation: glycoconjugates (= green), autofluorescence of cyanobacteria (= pink), autofluorescence of algae (= blue). (See Figure 4.10.)

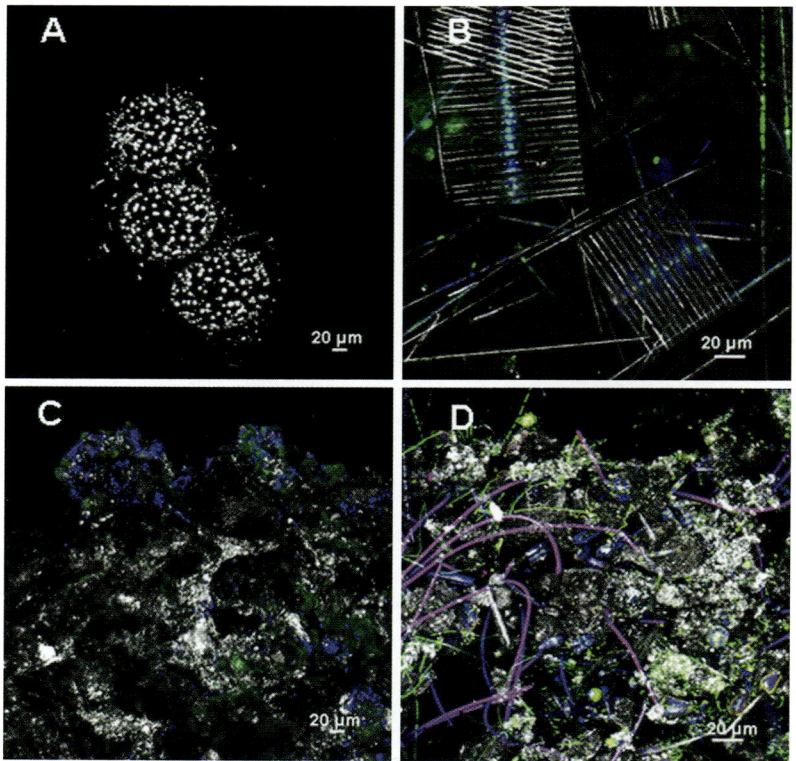

Plate 15. Reflection imaging of microbiological samples (1P-LSM). (A) Reflection of sulphur granules in Thiomargarita cells (= white). (B) Reflection of diatoms (*Fragilaria crotonensis*). Colour allocation: reflection (= white), nucleic acid stain (= green), autofluorescence of chlorophyll (= blue). (C) Reflection and autofluorescence of endolithic biofilms with the rock surface at the top of the image. Colour allocation: reflection of rock material (= white), autofluorescence of algae (= blue). (D) Reflection of a river biofilm community including nucleic acid staining. Colour allocation: reflection of mineral compounds (= white), autofluorescence of cyanobacteria (= pink), autofluorescence of algae (= blue), nucleic acid stain (= green). (See Figure 4.11.)

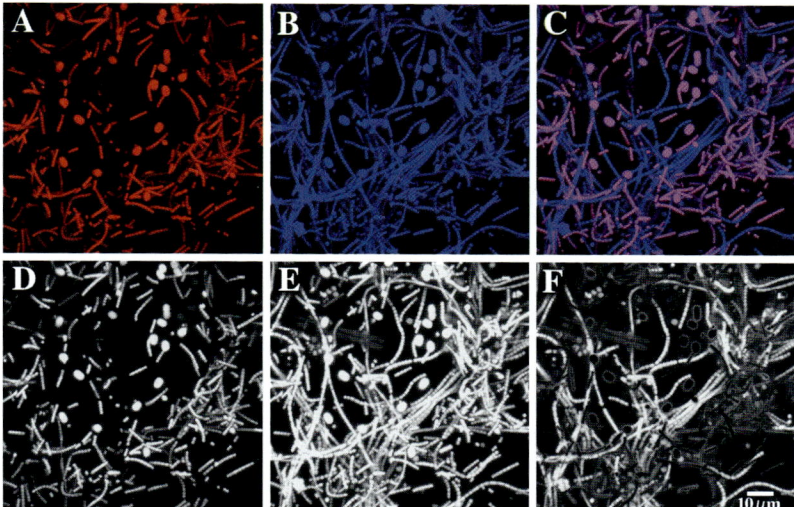

Plate 16. Maximum intensity projection of autofluorescence signals in a lotic biofilm examined by confocal laser scanning microscopy (1P-LSM). Images showing single channel signals of (A) red channel = cyanobacteria, (B) far red channel = autofluorescence of all photosynthetic organisms, (C) merged image of red + far red channel showing cyanobacteria (= pink) and algae (= blue), (D) grey scale version of the red channel, (E) grey scale version of far red channel and (F) grey scale image of the result after subtracting the red channel from the far red to image algae alone. (See Figure 4.12.)

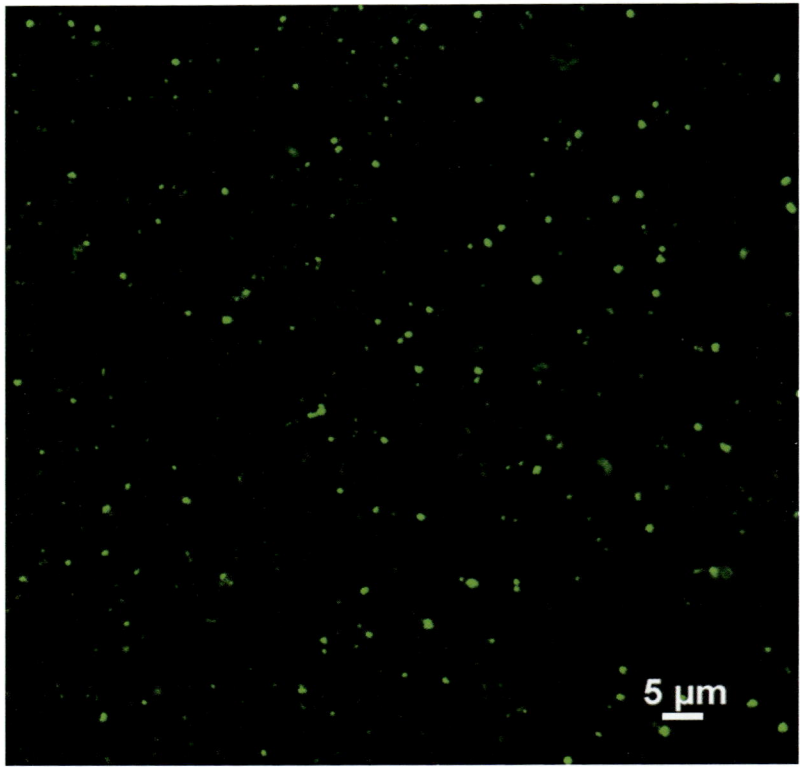

Plate 17. SYBRGREEN staining of a filtered (0.02 μm) river water sample. The image shows stained bacteria (large spots) and virus-like particles (tiny green spots). (See Figure 4.13.)

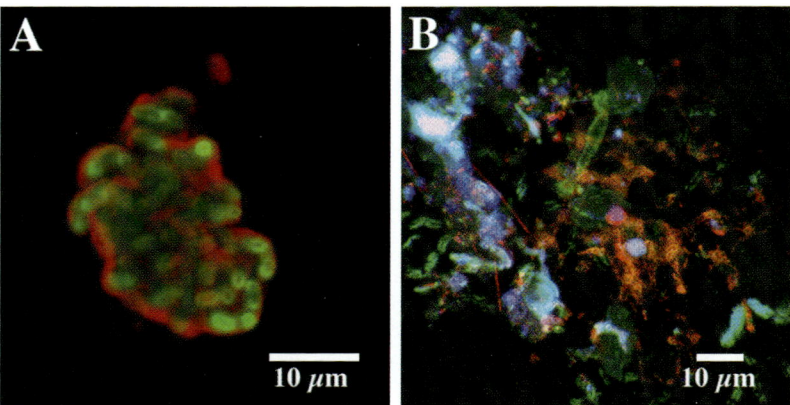

Plate 18. 1P-LSM images illustrating the binding of selected lectins. (A) Staining of a bacterial microcolony with two different lectins. (B) Staining of general river biofilm community with three different lectins. Note the discrete binding within the colony showing distinct zones (A) and regions in the overall river biofilm community (B). (See Figure 4.14.)

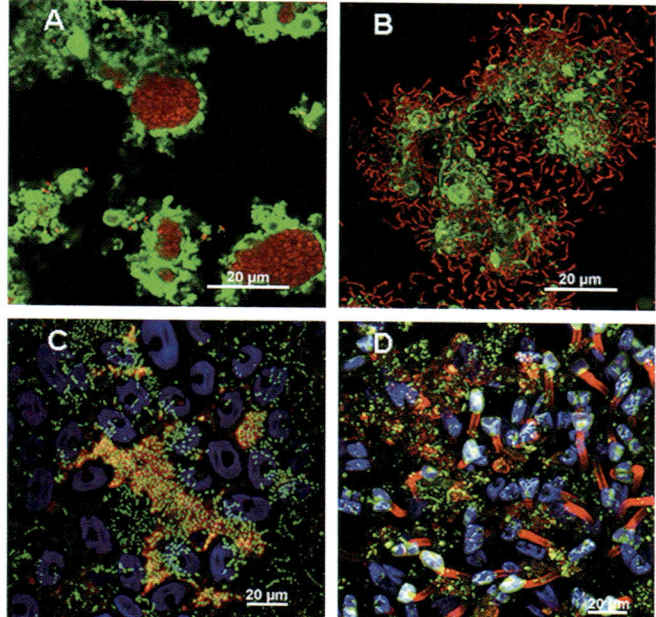

Plate 19. Examples illustrating the different appearance of glycoconjugates within biofilms after staining with lectins and 1P-LSM imaging. (A) Microcolonies of a chemoautotrophic biofilm grown in a rotating annular reactor. Colour allocation: lectin stain (= green), nucleic acid stain (= red). (B) Microcolonies of a hetrotrophic biofilm grown in a rotating annular reactor. Colour allocation: lectin stain (= green), nucleic acid stain (= red). (C) Biofilm developed in a creek showing a bacterial/algal microcolony, collected and examined after 30 days. Colour allocation: nucleic acid stain (= green), lectin stain (= red), autofluorescence of algae (= blue). (D) Biofilm developed in a creek showing stalked algae, collected and examined after 9 months. Colour allocation: nucleic acid stain (= green), lectin stain (= red), autofluorescence of algae (= blue). (See Figure 4.15.)

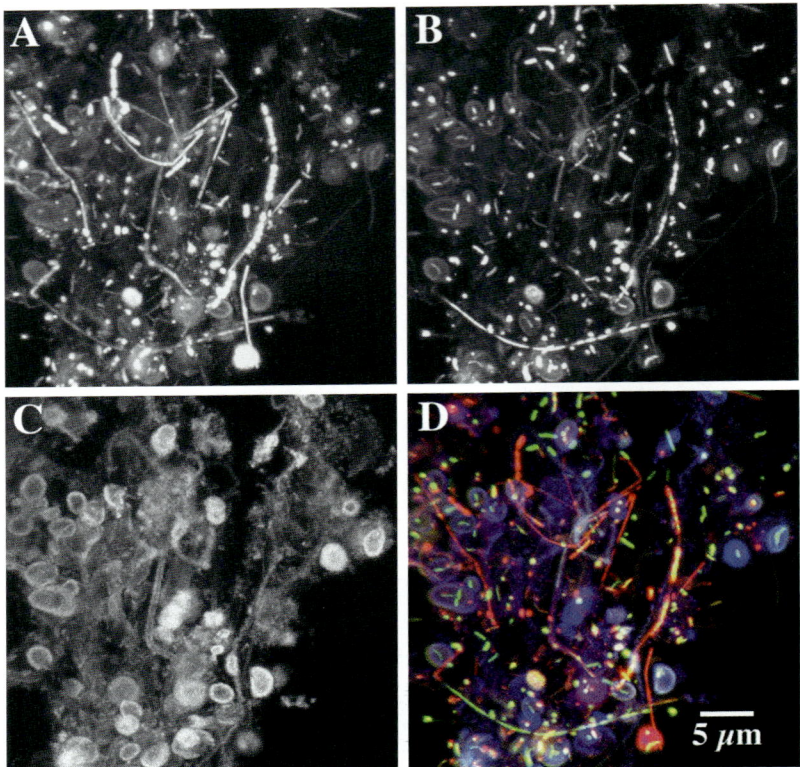

Plate 20. A 3-channel 1P-LSM image showing the binding of (A) lipid (Nile red), (B) nucleic acid stain (SYTO 9) and (C) lectin stain in a biofilm community. (D) The combination image showing co-localization of signals from all three probes. (After Lawrence *et al.*, 2003. *Appl. Environ. Microbiol.* 69:5543–5554, with permission). (See Figure 4.16.)

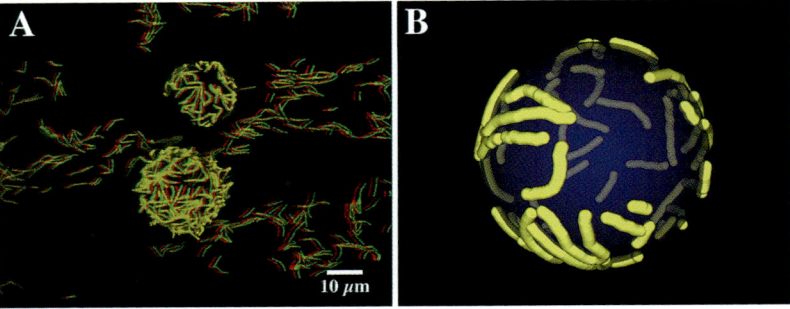

Plate 21. A series of 1P-LSM illustrations showing 2D and 3D visualization using the same data set. The images are of a *Rhodococcus* sp. colonizing the surfaces of droplets of diesel fuel during the degradation process. (Whyte *et al.*, 1999. *Appl. Environ. Microbiol.* 65:2961–2968 with permission.) (A) A red–green anaglyph projection of the data set (must be viewed with red–green stereo glasses). (B) A 3D colour rendering of the data stack for the small droplet in the image. (See Figure 4.20.)

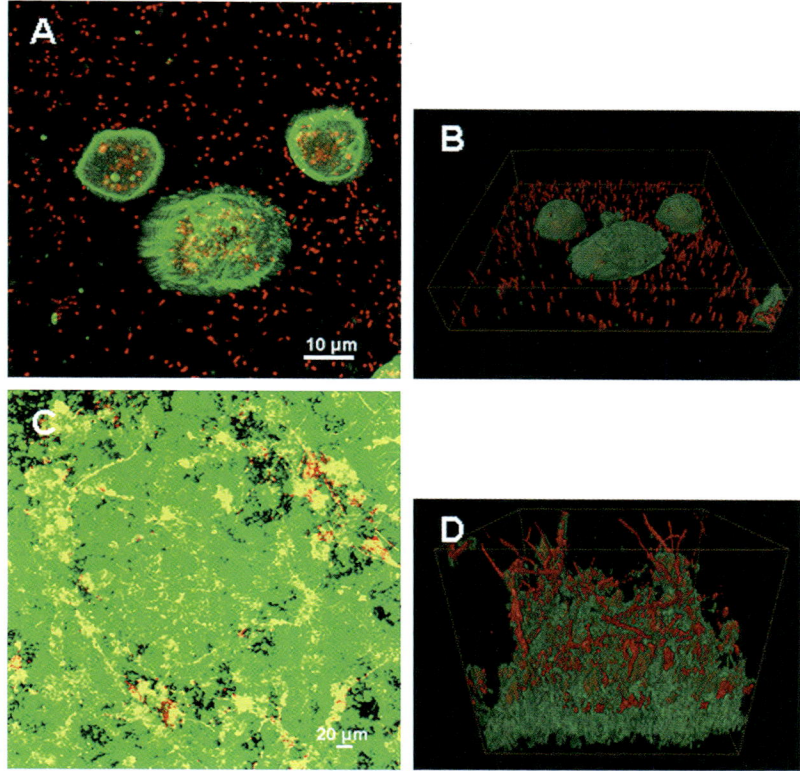

Plate 22. Two examples showing maximum intensity projection (A and C) and isosurface projection (B and D) of the same 1P-LSM data set. (A, B) Bacterial cells attached to the substratum which are grazed by protozoan cells. The cell surface of the protozoan cells is stained using an Alexa-488 labelled lectin. (C, D) Heterotrophic biofilm grown in a rotating annular reactor stained with an Alexa-488 lectin and SYTO 60. Colour allocation: lectin staining (= green), nucleic acid stained bacteria (= red). In both examples, the green channel is semi-transparent to visualise bacteria inside the protozoa and inside the biofilm covered by glycoconjugates. (See Figure 4.21.)

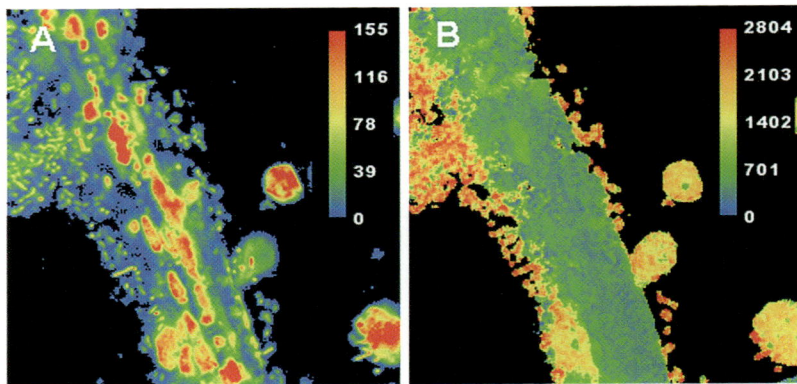

Plate 23. Two-photon excitation of lotic bacterial biofilm stained with SYTO 13. The image illustrates a diatom and bacteria colonising inorganic granules. (A) In the intensity image a differentiation of various cells or objects is impossible. The intensity image is displayed as a colour coded signal in arbitrary units from 0 to 155. (B) Lifetime image demonstrating the colour coded lifetime in picoseconds. Due to the longer lifetime of SYTO 13 in stained bacteria (orange–red), they can be distinguished from the autofluorescence of the diatom (blue–green) and the signal of inorganic granules (yellow). (See Figure 4.22.)

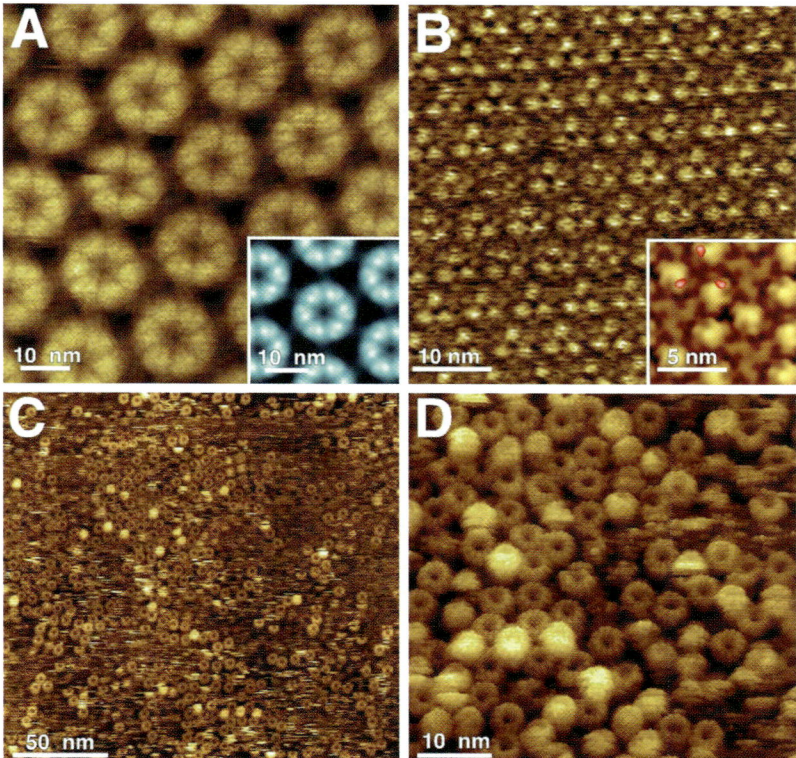

Plate 24. High-resolution topographs of crystalline and non-crystalline bacterial membrane proteins. (A) The outer surface of the HPI-layer of *D. radiodurans*. The inset shows the symmetrized average of (A). (B) Extracellular surface of purple membrane from *H. salinarum*. The inset shows the symmetrized average of (B). (C) Densely packed region of transmembrane rotors of ATP synthases from *I. tartaricus*. (D) same as (C), but at higher magnification. All topographs were recorded using contact mode AFM in buffer solution at room temperature. Their full gray levels correspond to vertical heights of about 3 nm (A), 1.2 nm (B), 2 nm (C and D). (See Figure 6.3.)

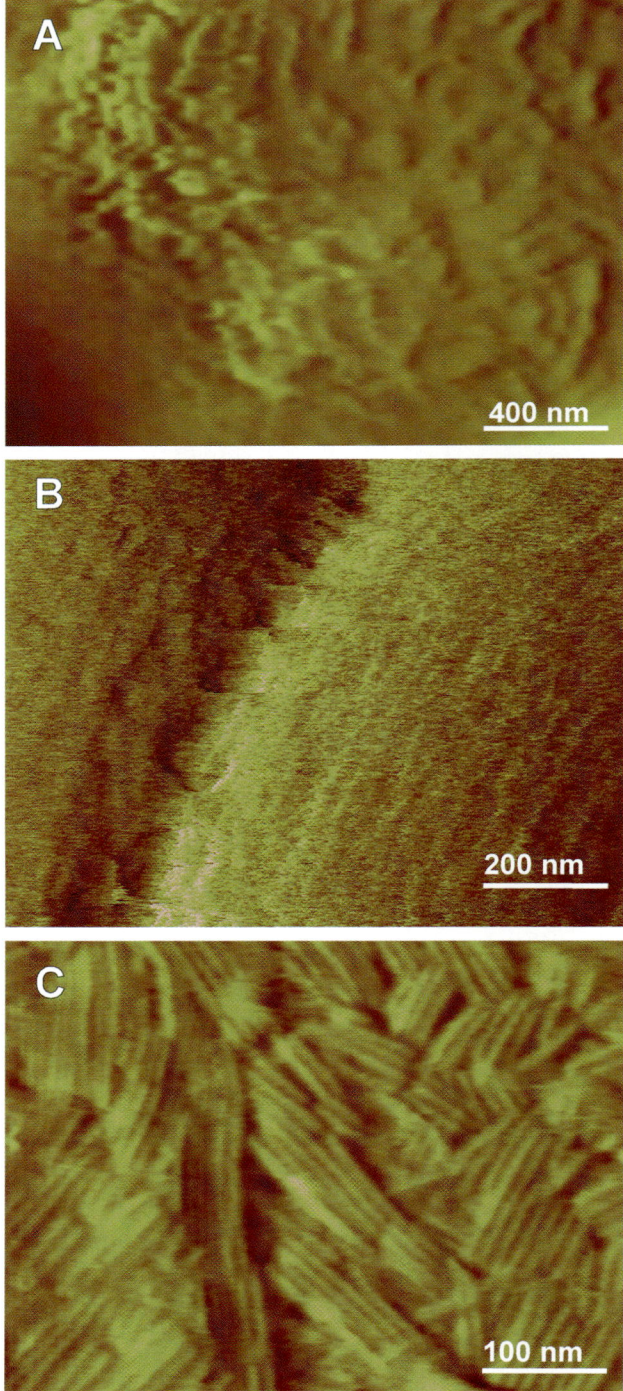

Plate 25. Observing cell surfaces at high resolution. AFM deflection images of bacterial and fungal cell surfaces acquired in aqueous solution: (A) *D. radiodurans*, (B) *L. lactis* and (C) *A. oryzae*. (See Figure 6.5.)

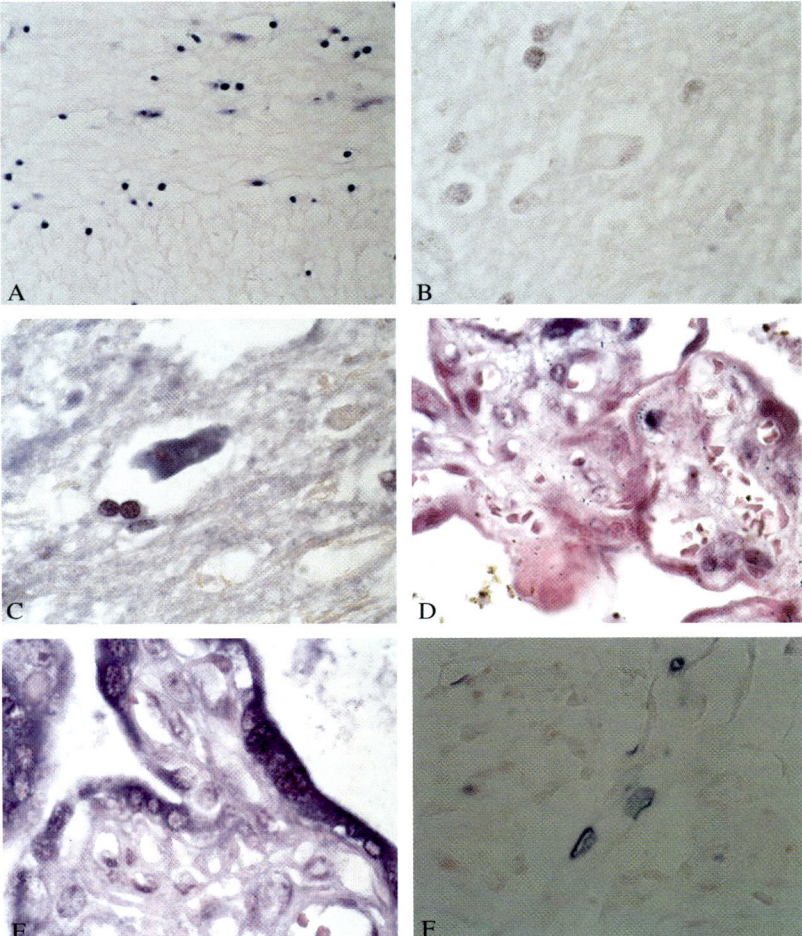

Plate 26. The importance of the exact localization of the signal with RT *in situ* PCR. Panels A and B show the expected results for the no DNase control and DNase, no RT (or irrelevant primers) control in a case of rabies encephalitis. Note the strong nuclear based signal for A and the lack of any signal for B. Panel C shows that rabies RNA localizes to the same brain tissue after optimal protease digestion to the cytoplasm of large neurons. Panels D, E and F show that the signal for RT *in situ* PCR for different targets after optimal protease and sufficient DNase digestion will show specific localization patterns. Panel D shows the detection of Klebsiella in the villi of a case of severe in utero bacterial infection; note the rod shaped forms. Panel E shows the cytoplasmic localization to trophoblasts of the placenta in a case of fatal in utero infection by coxsackie virus while panel F shows the cytoplasmic localization to endothelial cells in a case of fatal rotaviral sepsis. (Magnification: Panel A. 400 × . Panel B. C. D. E. F. 1000 × .). (See Figure 9.1.)

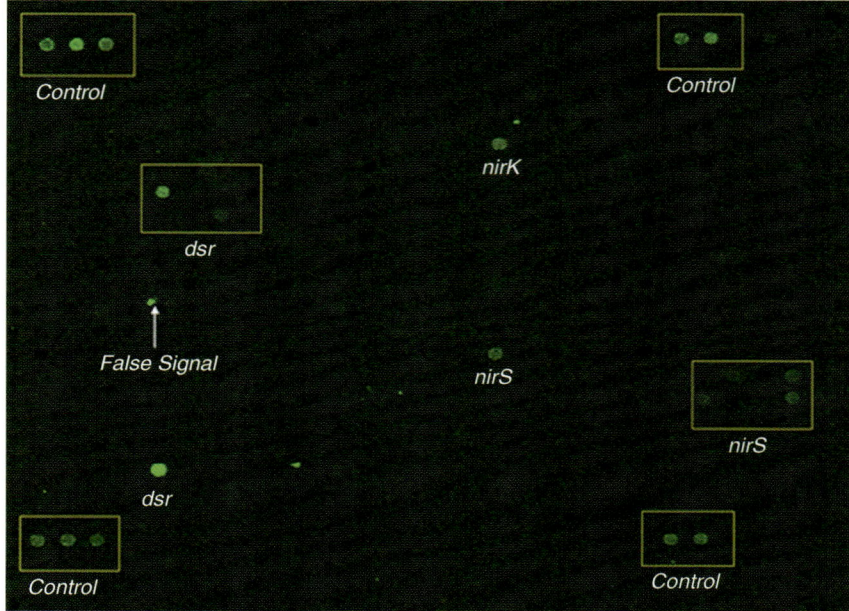

Plate 27. 50mer oligonucleotide microarray hybridization with a marine sediment genomic DNA sample. 3 ug of extracted marine sediment DNA was labeled Cy5 using methods outlined in the protocols outlined in this chapter. A small portion of an array image is shown to illustrate typical FGA images and true versus false signals. (See Figure 11.2.)

6 Microbial Surfaces Investigated Using Atomic Force Microscopy

Yves F Dufrêne[1] and Daniel J Müller[2]

[1] Unité de chimie des interfaces, Université Catholique de Louvain, Croix du Sud 2/18, B-1348 Louvain-la-Neuve, Belgium; [2] BioTechnological Center, University of Technology Dresden, c/o Max-Planck-Institute of Molecular Cell Biology and Genetics, Pfotenhauerstr. 108, D-01307 Dresden, Germany

◆◆

CONTENTS

Introduction
Methodology
Imaging applications
Force spectroscopy applications
Conclusion and perspectives

◆◆◆◆◆◆ INTRODUCTION

Most microorganisms possess a thick, elastic cell wall outside the plasma membrane. During evolution, microbial cells have developed a large diversity in their cell wall supramolecular architecture, reflecting specific adaptations to environmental and ecological pressures (Beveridge and Graham, 1991; Beveridge, 1999; Sleytr and Beveridge, 1999). Because microbial surfaces are in direct contact with the external environment, they are vital to the organisms and play key roles in determining cellular shape and growth, enabling the organisms to resist turgor pressure, acting as molecular sieves, and mediating molecular recognition and cellular interactions. Cellular interaction processes are ubiquitous phenomena, encountered not only in the natural environment but also in medicine (biofilms on implants, microbial–host interactions) and biotechnology (cellular interactions in fermentation technology, removal of heavy metals).

Understanding the structure, properties and functions of microbial surfaces has been hampered by the lack of sensitive and high-resolution surface analysis techniques. Probing cell surface structures at high resolution has traditionally involved electron microscopy. Of particular interest are cryo-methods, which allow examination of cell surface structures trapped in conditions close to the native state (Beveridge and Graham, 1991; Beveridge, 1999). However, these sophisticated methods are very demanding in terms of sample preparation and analysis. Characterization of the cell surface chemical composition and physical properties can be achieved by means of cell wall biochemical analysis, electron microscopy techniques and surface analysis methods (Beveridge and Graham, 1991; Mozes *et al.*, 1991). Yet, a major weakness of these techniques is that they involve cell manipulation (extraction, drying, labeling) prior to examination and, most often, provide averaged information obtained from many molecules or cells. These limitations emphasize the need to develop new, non-destructive cell surface analysis techniques that can probe the surface *in situ* and at the molecular level (Pembrey *et al.*, 1999).

In this context, the atomic force microscope (AFM) (Binnig *et al.*, 1986) provides new avenues to microscopists to study microbial surfaces. Besides its capability to image biological specimens in buffer solution, the AFM has an outstanding signal-to-noise ratio being superior to any electron or light microscope. Thus, within the past few years, remarkable advances have been made applying AFM to visualize native bacterial surface structures of two-dimensional (2D) protein crystals, densely packed membrane proteins and individual cells. Accordingly, cell surfaces were observed across dimensions resolving macroscopic features of a native cell surface down to single membrane proteins. Even individual substructures of proteins could be observed at a spatial resolution < 1 nm. Since biological systems can be observed under physiologically relevant conditions, it was possible to observe their dynamics by time-lapse AFM. Examples of biological functions observed by AFM are cell motility (Kuznetsov *et al.*, 1997; Rotsch *et al.*, 1999), the gating of nuclear pore complexes (Oberleithner *et al.*, 2000; Stoffler *et al.*, 1999; Wang and Clapham, 1999), of bacterial surface layers (Engel and Müller, 2000), and of bacterial membrane channels (Müller and Engel, 1999; Müller *et al.*, 2002a).

Besides being applied as a microscope, the AFM can be used to measure molecular forces with piconewton (10^{-12} N) accuracy. Here, the force spectroscopy mode has proved to be a powerful tool to measure physical properties and molecular interactions of microbial systems, providing fundamental insights into the structure–function relationships of cell surfaces.

In this chapter, we discuss the principle and methodology of AFM and highlight various applications offered by the technique in microbial cell surface research, emphasizing both imaging and force spectroscopy measurements. To this end, we discuss a selection of recent data obtained from our and other laboratories which we hope will help the readers to identify the potential of the technique in their respective field.

◆◆◆◆◆◆ METHODOLOGY

Principle

How does AFM work? In contrast to optical microscopes, AFM uses a very sharp tip to contour the 3D surface of a specimen (Figure 6.1A).

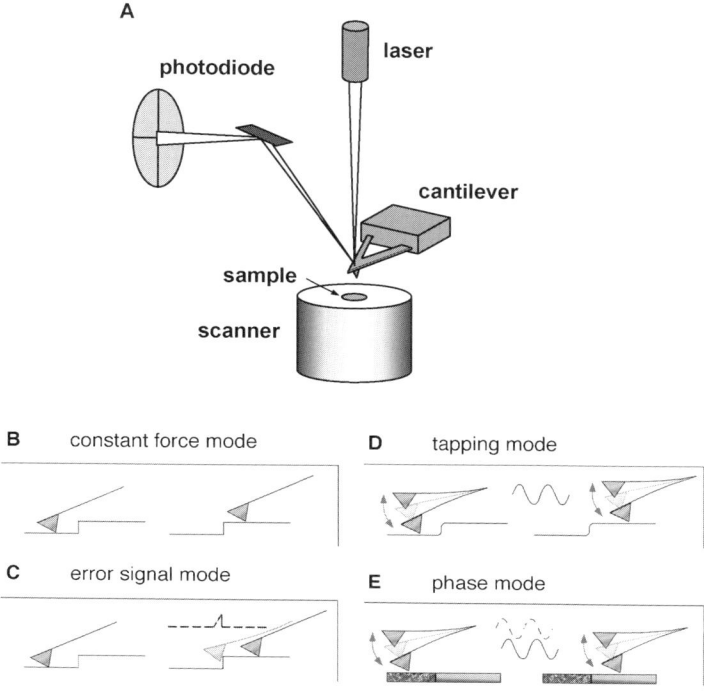

Figure 6.1. Basic elements (A) and imaging modes (B–E) of the AFM. AFM imaging is performed by measuring the force between a very sharp tip and the sample surface (A). The sample is mounted on a piezoelectric scanner which ensures 3D positioning with high resolution and the force is monitored by attaching the tip to a flexible cantilever and measuring the cantilever deflection using an optical method (laser, photodiode). Several imaging modes are available. In the constant force mode (B), the AFM tip is raster scanned over the sample surface while a feedback system holds the cantilever deflection constant. The accuracy at which the force is held constant depends on differences detected in the deflection signal, the parameters of the feedback loop and the reaction speed of the system. If the tip scans too fast over sharp edges the deflection shows bigger deviations than normal (C). This error signal provides an image that reveals the edges of the surface topograph. Scanning the sample with an oscillating tip is the basic principle of tapping mode imaging (D). Small reductions of the oscillation amplitude are used from the feedback loop to control the vertical cantilever movement. Because the tip–sample contact is disrupted periodically, the friction forces are eliminated. The phase difference between the measured oscillation (solid wave) and the oscillation driving the cantilever (dotted wave) depends on the mechanical, chemical and structural properties of the sample (E). Thus, the phase signal is sensitive to changes in surface properties.

To this end, a piezoelectric ceramic raster scans the AFM tip (or sample) with sub-Ångstrom accuracy while the force interacting between tip and specimen is monitored with piconewton sensitivity. This force is measured by the deflection of a soft cantilever detected by a laser beam usually focused on the free end of the cantilever and reflected into a photodiode. Both AFM cantilever and its attached tip are typically made of silicon or silicon nitride.

Several imaging modes have been developed to operate the AFM (Figure 6.1B–E). They differ mainly in the mechanism moving the tip over the sample. In the constant force mode (Figure 6.1B), the AFM tip is raster scanned over the sample while the force applied to the tip is kept constant. However, abrupt changes of the tip–sample interaction caused by steep height variations, chemical or electrostatic differences of the sample cause fluctuations of the cantilever deflection. The feedback loop of the AFM tries to compensate these fluctuations by piezo movements. The faster the system of feedback loop, cantilever and piezo reacts, the more accurate an AFM can adjust and control the forces applied. Parameters of the feedback loop can be optimized, by minimizing the deviation of the cantilever deflection (error signal mode (Putman et al., 1992); Figure 6.1C). It is particularly important to optimize the feedback loop, which enables accurate contouring of biological surfaces at minimal forces. Otherwise the scanning tip easily loses contact with the surface (large fluctuations away from sample) or deforms the biological surface in an uncontrolled manner (large fluctuations towards the sample).

In tapping mode (also named dynamic or intermittent mode) AFM (Figure 6.1D), an oscillating tip is scanned over the surface. If the parameters have been adjusted correctly, the tip touches the sample surface only at the very end of its downward movement. Interactions between the tip and sample change the cantilever oscillation, which is used by the feedback loop to control the amplitude damping. The difference between the oscillation activated and that performed by the cantilever is visualized by the phase signal (Figure 6.1E). It has been shown that various interactions, being of chemical, electrostatic, viscous, elastic, and topographic origin can cause significant changes in the phase signal. Thus, using the phase signal to differentiate these interactions provides a great potential to learn about molecular interactions. However, the complexity of the interactions acting at the same time often makes them difficult to be identified unambiguously. Due to its intermittent sample contact, tapping mode AFM (TMAFM) can significantly reduce lateral forces that may interact between tip and sample.

In force spectroscopy, the cantilever deflection is recorded as a function of the vertical displacement of the piezoelectric scanner, i.e. as the sample is pushed towards the tip and retracted from it. The result is a plot showing the cantilever deflection (d) vs the scanner displacement (z). A force–distance curve is then obtained by converting the cantilever deflection into a force (F) using Hooke's law ($F = -kd$, where k is the cantilever spring constant) and subtracting the deflection from the scanner displacement to obtain the distance ($z - d$). The point of contact (zero separation distance) is determined as the position of the vertical linear parts of the curve in the contact region. Because AFM does not provide an independent measurement of the tip–sample separation distance, determining the point of contact can sometimes be a delicate task, especially when long-range surface forces or sample deformation effects dominate. Force–distance curves can be recorded either at single, well-defined locations of the (x, y) plane or at multiple locations to yield a so-called "force-volume image" (for more details see Butt et al., 1995;

Heinz and Hoh, 1999a). In doing so, spatially resolved maps of sample properties and molecular interactions forces can be produced.

Sample Preparation

One critical issue in biological AFM is sample preparation. The sample must be well attached to a solid support in order to resist the lateral forces that may be exerted by the scanning tip. Most importantly, the biological object should not change its position while being scanned. Many biological processes occur within much shorter time ranges than the imaging time of commercially available AFMs (ranging from about 30 s to several minutes); this fact is often underestimated by the AFM operator.

Various immobilization strategies have been established for microbial specimens. For reconstituted membrane proteins and membrane patches from cells, the most frequently used procedure is based on physical adsorption in aqueous solution (Müller et al., 1997). For adsorption, the electrostatic double layer (EDL) interaction (Israelachvili, 1991) acting between the supporting surface and the biological object should be considered. In cases where both surfaces are negatively charged they naturally repel each other. This electrostatic repulsion is compensated by counter-ions accumulated in close proximity (nm range) to the surface. The compensation of surface charges increases with the electrolyte valency and concentration. As a result, the repulsion between both the surfaces decreases. At sufficiently high electrolyte concentrations, the object can approach the support close enough to allow both surfaces to snap into van der Waals attraction and the sample adsorbs onto the support. For large and flat samples, the adsorption is sufficiently high to immobilize the membrane patches such that they can be imaged at subnanometer resolution using AFM. Immobilization by adsorption may, however, not always be sufficient to allow subnanometer resolution of single, isolated molecules. For example, the Hansma's group showed that DNA adsorbed to mica in different ways depending on the electrolyte type and concentration, and under certain electrolyte compositions, the DNA molecules retained their ability to diffuse laterally (Hansma, 2001). To tightly tether biological molecules to supporting surfaces, several procedures have been introduced to establish covalent bonds between biological objects and their support (Amrein and Müller, 1999; Shlyakhtenko et al., 2003; Wagner, 1998). Besides showing the disadvantage of requiring more preparation steps, covalent immobilization has to be carefully controlled and one should always be concerned about the possible influence of particular chemical linkage on the functionality of the biological sample.

Giving an approximate guide of how to adsorb bacterial protein layers to mica, we will discuss some examples in detail. Purple membrane of *Halobacterium salinarum* is negatively charged (Alexiev et al., 1994). Therefore, it shows an EDL repulsion towards negatively charged mica surfaces. To adsorb the membrane to mica, an aqueous solution buffered at pH 7.8 (20 mM Tris–HCl) containing a protein concentration of ≈ 10 μg/ml was prepared (Müller et al., 1997). As monovalent electrolyte, KCl was added at a given concentration (it should be noted that NaCl and LiCl work as well). A small drop of this solution (≈ 20 μl) was deposited onto a mica support having a diameter of ≈ 1 cm. After allowing the membranes to adsorb for a distinct time range of 10–15 min, the sample was rinsed by the same buffer solution but containing no biological specimen. Then the sample was imaged by AFM. According to their negatively charged surfaces, no purple membranes adsorbed to the mica at monovalent electrolyte concentrations $\ll 50$ mM. However, purple membranes adsorbed to mica after enhancing the monovalent electrolyte concentrations above 50 mM which was sufficient to compensate the electrostatic repulsion between both the surfaces. Similarly, membrane patches of reconstituted OmpF porin from *Escherichia coli* adsorbed to mica at monovalent electrolyte concentrations above 50 mM. This suggested that purple membrane and OmpF porin membranes have negative charges of similar magnitude (Müller and Engel, 1997). In contrast to purple membrane and reconstituted OmpF porin, the hexagonally packed intermediate (HPI) layer of *Deinococcus radiodurans* exposes a negative surface charge density about one order of magnitude lower (Müller and Engel, 1997). Therefore, HPI layers adsorbed to the mica surface at much lower monovalent salt concentration, ranging between 5 and 10 mM.

For microbial cells, sample preparation is generally more delicate. As microbial cells are fairly rigid spherical or rod-like particles, the cell–support contact area is very small, leading most of the time to cell detachment by the scanning tip. Several immobilization methods have been developed to solve this issue. Air drying and chemical fixation may be used to promote cell attachment, but these treatments are not recommended since they are likely to cause significant denaturation of the specimen. More gentle immobilization procedures have been proposed. In the agar gel method (Gad and Ikai, 1995), a few drops of a highly concentrated cell suspension is deposited on a clean cover glass and then molten agar is dropped on the cell layer. Another piece of cover glass is quickly put onto the agar. After solidification, the sample is turned upside down and the lower cover glass is pulled from the agar surface. Most cells are localized on that side of the agar disc and weakly captured cells can be easily removed by washing with water. The cell-free side of the disc is carefully dried on a sheet of tissue and then placed on the sample holder.

In the porous membrane method, a concentrated cell suspension is gently sucked through an isopore polycarbonate membrane with pore size slightly smaller than that of the cell (Kasas and Ikai, 1995; Dufrêne et al., 1999). After cutting the filter, the lower part is carefully dried on a sheet of tissue and the specimen is then attached to the sample holder using a small piece of adhesive tape. This approach is fairly simple and straightforward and can be used to image single bacterial, yeast and fungal cells under aqueous conditions while minimizing denaturation of the surface molecules.

High-resolution Imaging

The effects of the force applied between the AFM tip and the biological object have been studied on various examples ranging from living cells to native membrane proteins. All investigations have in common that forces ranging significantly above 100 pN can lead to the reversible or irreversible distortion of the relatively soft biological system. Thus, to reveal AFM topographs at molecular resolution (≈ 1 nm), the applied force has to be minimized to values below 100 pN and kept constant within small deviations. Approaches to achieve high-resolution topographs differ mainly on the AFM imaging mode used. The two major imaging modes currently applied to observe substructures of individual proteins are tapping and constant force modes.

In TMAFM, the essential key to achieve high-resolution topographs was the activation of the AFM cantilever at its resonance frequency in buffer solution (Schäffer et al., 1996), adjusting a sufficiently small cantilever amplitude of ≈ 1 nm and an amplitude change corresponding to an applied force ≤ 100 pN (Möller et al., 1999). These conditions allowed recording topographs in buffer solution exhibiting a lateral resolution of ≈ 1.5 nm and vertical resolution of ≈ 0.1 nm. TMAFM topographs of native membrane proteins and bacterial S-layers correlated well to their 3D structures obtained from X-ray or electron crystallography.

Even higher spatial resolution is currently achieved using constant force imaging of membrane proteins and S-layers. To reproducibly reveal a lateral resolution of ≈ 0.5 nm and a vertical resolution of ≈ 0.1 nm, the forces interacting between the AFM tip and the molecules must be precisely controlled for each experiment (Müller et al., 1999b). It was shown that the force applied externally to the scanning AFM tip can be damped by adjusting the EDL repulsion with the biological sample. This repulsion interacts in the range of several nm and thus distributes over a relatively large sample area. The damping force can be adjusted by the electrolyte and pH of the buffer solution until it compensates almost completely the force externally applied to the AFM tip (≈ 100 pN). As a result, the net force interacting locally between the fragile substructures of the protein and the AFM tip can be minimized substantially down to $\ll 100$ pN. These localized interactions (<0.5 nm) can be used to contour the biological object at subnanometer resolution. High-resolution topographs of native membrane proteins recorded following the previous procedure showed an excellent agreement to their atomic structure (Figure 6.2) (Müller and Engel, 1999).

Tip Artifacts in AFM Imaging

The size and shape of the AFM tip play an important role in determining the resolution and reliability of the topograph. Subnanometer resolution can be routinely acquired on flat surface layers using commercial probes with a 10–50 nm radius of curvature. The unexpected

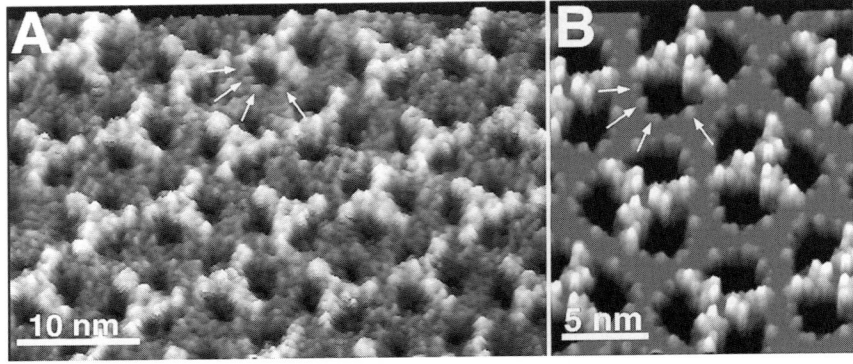

Figure 6.2. Comparing high-resolution AFM topograph of a membrane protein with its atomic model. AFM topograph of the outer membrane protein OmpF porin from *E. coli* (A) recorded in buffer solution at room temperature. The topograph recorded in contact mode exhibits a lateral resolution of ≈ 0.5 nm and a vertical resolution of ≈ 0.1 nm. Individual protrusions of single OmpF porin trimers were clearly resolved at subnanometer resolution and demonstrate the exceptionally high signal-to-noise ratio of the AFM, which is superior to any other optical microscopic technique. For comparison, the atomic model derived from X-ray crystallographic data is displayed (B).

high resolution is thought to result from nanoscale protrusions or asperities extending from the tip (Müller *et al.*, 1999b). A common source of artifact in imaging is the broadening of the surface features due to the finite size of the AFM tip. This effect is important when imaged heights vary in the range of the tip radius, i.e. 10–50 nm. In this case, the sample interacts with the sides of the tip and the resulting topograph will be a combination of the sample structure and the tip geometry (Schwarz *et al.*, 1994). Similarly, asymmetric tip asperities can result in an apparent asymmetric distortion of features imaged. Another common problem is the shadowing or multiplication of small structures produced by multiple tip effects. Here, every single asperity of a tip containing multiple asperities can potentially produce an individual topograph. The final topograph reveals multiple images of the same region, each one being laterally and vertically slightly shifted relative to the other.

◆◆◆◆◆◆ IMAGING APPLICATIONS

Observing Two-dimensional Protein Crystals at Submolecular Resolution

S-layers are 2D crystalline monomolecular assemblies of proteins or glycoproteins with molecular weight ranging from 40 to 200 kDa, and represent one of the most common cell surface structures in bacteria (Sleytr, 1997; Sleytr *et al.*, 1997; Sleytr and Beveridge, 1999). They usually exhibit oblique, square or hexagonal lattice symmetry with the morphological unit cell being repeated between 3 and 30 nm. S-layers are often extremely stable and easy to prepare, and their 2D regular arrays have proved to be particularly well-suited for high-resolution AFM imaging (Karrasch *et al.*, 1994; Müller *et al.*, 1996; Xu *et al.*, 1996; Scheuring *et al.*, 2002b). Such topographs can provide insights into the structural assembly of individual proteins into the final S-layer under physiological conditions (Gyorvary *et al.*, 2003). This makes AFM a complementary tool for X-ray and electron crystallography.

The first S-layer imaged by AFM at submolecular resolution was the HPI layer from *D. radiodurans* (Karrasch *et al.*, 1994). This early work of the Engel's group, showed for the first time that high-resolution AFM topographs of native proteins can exhibit an excellent correlation to their 3D models revealed by established structural methods. However, the prerequisites to reveal high-resolution topographs have to be controlled precisely (see above). Figure 6.3A shows the outer surface of the HPI layer. From biochemical experiments it is known that the HPI layer is

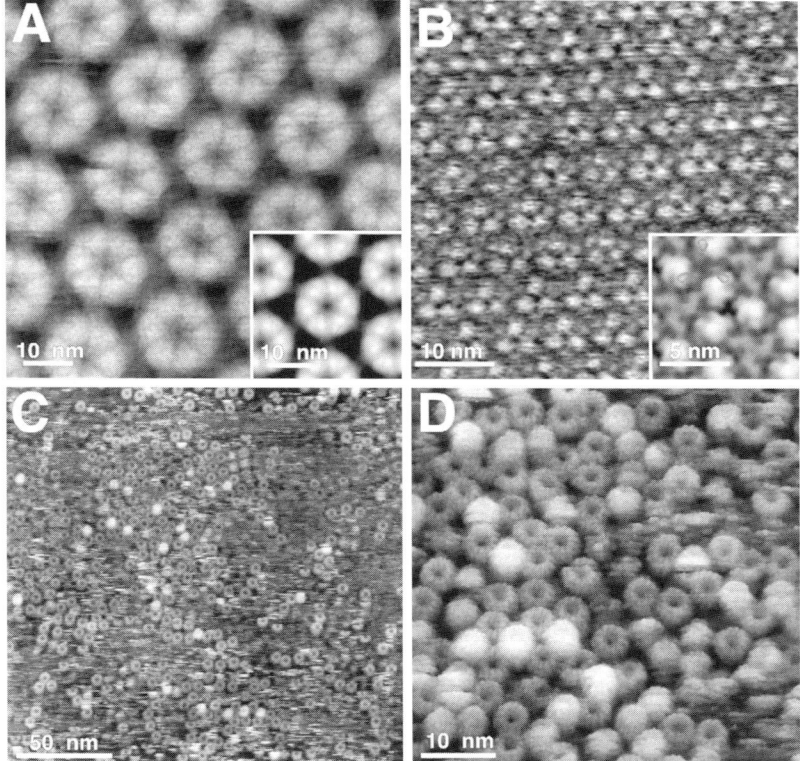

Figure 6.3. High-resolution topographs of crystalline and non-crystalline bacterial membrane proteins. (A) The outer surface of the HPI-layer of *D. radiodurans*. The inset shows the symmetrized average of (A). (B) Extracellular surface of purple membrane from *H. salinarum*. The inset shows the symmetrized average of (B). (C) Densely packed region of transmembrane rotors of ATP synthases from *I. tartaricus*. (D) same as (C), but at higher magnification. All topographs were recorded using contact mode AFM in buffer solution at room temperature. Their full gray levels correspond to vertical heights of about 3 nm (A), 1.2 nm (B), 2 nm (C and D). (See Colour Plate 24.)

composed only from one type of protomer (Peters et al., 1987). The AFM topograph showed six protomers forming one donut-shaped ring. These donuts were arranged in a hexagonal arrangement. A single protomer established connections with directly neighbored protomers of the same donut. Additionally, each protomer exhibited an emanating arm, which was connected with that of one protomer of a neighbored donut-shaped ring.

In the meantime, many other AFM studies have been performed on the HPI layer, which discovered conformational changes of HPI layer pores and detected intermolecular forces at which subunit protomers interact with each other in the S-layer (see below for details). Since the HPI layer has been well characterized by AFM, it is expected that newly developed and future scanning probe microscopy techniques will be applied to this system adding novel insights on how this S-layer may work. Recent observations of other S-layers by AFM (Xu et al., 1996; Moll et al., 2002; Scheuring et al., 2002b; Gyorvary et al., 2003) showed that this method boroughs the potential to explore the multiple properties of the extremely large S-layer family (Sleytr, 1997; Sleytr et al., 1997).

In the past, AFM has been extensively used to observe bacterial membrane proteins that have been 2D crystallized. In most cases, the membrane proteins had to be solubilized, purified and reconstituted back into a lipid membrane where they assembled into 2D crystals under certain conditions (Kühlbrandt, 1992; Stahlberg et al., 2001a). Examples of crystallized membrane proteins that were imaged using AFM are Aquaporin Z (Scheuring et al., 1999) and OmpF porin (Figure 6.2; Müller and Engel, 1999) from *E. coli*, human Aquaporin-1 (Fotiadis et al., 2002), major intrinsic protein (MPI) (Fotiadis et al., 2000), reaction center (RC) light-harvesting complex 1 (LH1) from *Rhodospirillum rubrum* (Fotiadis et al., 2004), light-harvesting complex LH2 from *Rubrivivax gelatinosus* (Scheuring et al., 2001) and from *Rhodobacter sphaeroides* (Scheuring et al., 2003), photosystem 1 complex (Fotiadis et al., 1998), sodium driven rotors of the ATP synthase from *Ilyobacter tartaricus* (Stahlberg et al., 2001b), and halorhodopsin from *H. salinarum* (Persike et al., 2001). In rare cases, as observed for bacteriorhodopsin, the membrane proteins naturally occur as 2D crystals in the bacterial cell membrane. Here, the membrane patches can be isolated using well-known procedures (Oesterhelt and Stoeckenius, 1974) previous to imaging (Figure 6.3B).

Visualizing Non-crystallized Membrane Proteins at Submolecular Resolution

Thus far, we have discussed examples of 2D protein crystals imaged by AFM. In this section, we will show that it is also possible to image non-crystalline membrane proteins at high resolution. The AFM preparation steps of these samples are similar to those developed for crystals. The samples simply have to be adsorbed onto a supporting surface in buffer solution at ambient temperature. In contrast, the biochemical preparation steps are much simpler, since they do not require crystallization of the protein. Figure 6.3C shows an AFM topograph of sodium driven rotors of the ATP synthase from *I. tartaricus* after their reconstitution into a lipid membrane. The topograph shows the loose assembly of single transmembrane rotors. About 60–70% of the bilayer was covered with membrane proteins comparing closely to the protein density of native cell membranes. After recording the overview topograph, it was possible to scan a selected area at higher magnification (Figure 6.3D). Details of the individual rotors were clearly resolved; each one having an outer diameter of ≈ 5.4 nm. This allowed the rotor subunits to be counted and their stoichiometry to be directly accessed (Stahlberg et al., 2001b). The lateral resolution of these topographs lies between 0.6 and 0.7 nm.

Various other densely packed membrane proteins and membrane-associated proteins have been observed using high-resolution AFM in buffer solution. Examples of transmembrane protein studies are bacterio-opsin (Möller et al., 2000), bovine rhodopsin (Fotiadis et al., 2003) and proton-driven rotors of spinach chloroplast ATP synthase (Seelert et al., 2000). Membrane-associated proteins observed were mainly toxins secreted from bacteria such as cholera toxin (Mou et al., 1995), staphylococcal α-hemolysin (Czajkowsky et al., 1998b), pertussis toxin (Yang et al., 1994), and vacuolating toxin from *Helicobacter pylori* (Czajkowsky et al., 1999).

Monitoring Conformational Changes

AFM allows biological systems to be observed in physiologically relevant environments (Drake et al., 1989). Thus, the biologically important question arises whether it is also possible to directly observe their conformational changes. The first conformational changes observed at subnanometer resolution were induced by the external force applied by the AFM tip while scanning the sample (Müller et al., 1995). Here, the AFM tip could be used as a molecular tweezer, enabling to reversibly push away single subunits of the protein, thereby accessing subunits otherwise hidden. The force between 100 and 200 pN, which was required for pushing away distinct molecular structures, allowed an estimation of local flexibilities of the protein (Müller et al., 1998). In a complementary approach, different conformations of individual proteins imaged by AFM were compared and classified (Scheuring et al., 2002a). While some protein domains always appeared in the same conformation, others showed an increased variability. From these variabilities energy landscapes of individual protein domains could be calculated.

Function-related conformational changes were imaged on several membrane proteins at high resolution. The first such study was focused on the HPI layer of *D. radiodurans*. AFM topographs showed that individual pores of this S-layer can co-exist in plugged and unplugged conformations (Müller et al., 1996). This heterogeneous mixture of conformations within one system was not observed before, since conventional structural methods such as electron microscopy and X-ray crystallography integrate over many proteins to obtain a sufficient signal-to-noise ratio for image reconstitution. Time-lapse AFM images of the S-layer showed that the pores reversibly switched from their open to closed conformation. It is assumed that S-layers, besides having various functions, may also work as molecular sieves (Baumeister et al., 1988; Sleytr and Sára, 1986). However, since additional experimental data are lacking, the function of the switching HPI layer pores remains to be understood.

OmpF porin, the outer membrane protein of E. coli, establishes a diffusion-driven channel for the transport of hydrophilic solutes across the bacterial membrane (Schirmer, 1998). It was shown that these channels are reversibly gated by applying a critical transmembrane potential (Schindler and Rosenbusch, 1978). For a long time, it was assumed that OmpF may be gated by an internal polypeptide loop constricting the transmembrane channel (Klebba and Newton, 1998). Experiments inhibiting the movement of this loop showed that it does not move upon channel gating. Thus, other mechanisms had to be considered for gating. Remarkably, AFM topographs of reconstituted OmpF porin showed conformational changes upon applying a critical transmembrane potential or by decreasing the pH to extremely low values of $\approx$pH 2. Here, large extracellular loops normally protruding into the aqueous solution reversibly collapsed onto the channel entrance (Müller and Engel, 1999). Interestingly, drastic pH changes induced the same conformational change of the protein.

These experiments, and many others, demonstrate that the AFM can be applied to observe conformational changes of proteins. Because of the slow scanning time required to record a topograph ($\approx$30 s), the direct visualization of faster conformational changes is currently limited. Therefore, most functions observed by AFM were of static (on/off) or relatively slow nature. Recent developments of fast scanning AFMs (Ando et al., 2001; Humphris et al., 2003) may be used in the future to resolve protein structure and function at much higher temporal resolution.

Looking at the Surface of Native Cells

With AFM, the surface morphology of individual cells and biofilms can be visualized in the native, hydrated state, thus providing data that are complementary to those obtained with conventional microscopy techniques.

The association of microorganisms with solid surfaces leads to the formation of biofilms, a process which can have either beneficial effects, such as in biotechnology (wastewater treatment, bioremediation, immobilized cells in reactors), or detrimental effects, such as in industrial systems (fouling, contamination) and in medicine (accumulation on teeth, implants and prosthetic devices). AFM has proved useful in bacterial biofilm studies for measuring critical dimensions, surface roughness and the organization of hydrated extracellular polymeric substances (EPS). For instance, AFM was used to image biofilms of *Pseudomonas* species and of a marine isolate (Beech, 1996), to reveal the structure of adhesive trails of gliding diatoms (Higgins *et al.*, 2000) and to visualize adsorbed EPS produced by the bacterium *Azospirillum brasilense* (van der Aa and Dufrêne, 2002). In geomicrobiology, AFM has proved useful to resolve pits at the surface of clay minerals in relation with microbially mediated dissolution (Maurice *et al.*, 2001).

Using appropriate immobilization procedures, it is possible to examine single cells in the native state. Figure 6.4A presents a 3D height image showing three *D. radiodurans* bacterial cells immobilized in a porous polymer membrane. A fairly rough morphology was observed, consistent with the fact that the outermost surface of the bacterium is coated with a soft, 40 nm thick carbohydrate layer (Baumeister *et al.*, 1986). In contrast, for the yeast *Saccharomyces cerevisiae*, the most thoroughly investigated eukaryotic microorganism in the world, the surface was smooth (Figure 6.4B), in agreement with the presence of mannoproteins in the outermost part of the wall and with electron microscopy observations in the dried state (Koch and Rademacher, 1980).

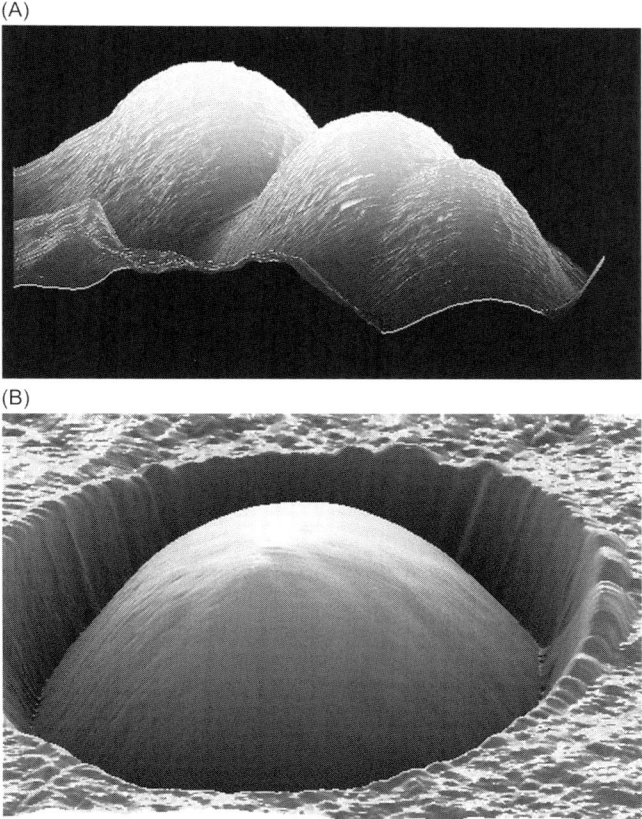

Figure 6.4. Imaging living cells. AFM topographs of three *D. radiodurans* cells (A) and of a *S. cerevisiae* cell (B) immobilized in a porous membrane. Topographs were recorded in aqueous solutions and exhibited a scan range of 5 μm × 5 μm and vertical ranges of 750 nm (A) and 500 nm (B).

At high resolution, very different morphologies can be observed depending on the organism (for reviews, see Dufrêne, 2002; 2003). For bacteria, the presence of soft, flexible macromolecules may strongly limit the image resolution. An example of this is shown in Figure 6.5A for *D. radiodurans*. Here, the image contrast is essentially dominated by the interaction between the scanning tip and the carbohydrate coat. In an attempt to resolve the inner HPI layer, several enzymatic treatments were applied to the cells to digest the outer carbohydrate layer. None of these treatments yielded a clean, crystalline surface indicating either that the carbohydrate material was strongly anchored into the HPI layer or that the latter had been removed by the treatment. This example illustrates one of the key limitations of AFM imaging in microbiology: conventional imaging modes are not suited for imaging soft macromolecular architectures such as polysaccharide layers or appendages.

Figure 6.5B shows a high-resolution image obtained for the bacterium *Lactococcus lactis* during the course of the septation process. As opposed to the previous organism, the cell surface was quite smooth and apparently not altered by the scanning tip, a finding that is consistent with the presence of a peptidoglycan layer known to be a major determinant of cell shape and rigidity. However, the morphology of the septum region that separates two dividing cells was more irregular, with a number of holes clearly visible. These holes may be related to the digestion of the peptidoglycan cell wall by the action of autolysins, a process that eventually leads to cell separation. Hence, because it provides 3D resolution on native samples, AFM might become an important complementary tool to electron microscopy to elucidate the mechanisms of cell growth and cell division.

The outermost surface of many fungal spores and hyphae is known to be covered by a thin layer of regularly arranged proteinaceous rodlets, which are 5–10 nm thick and up to 250 nm in length. This was confirmed for several organisms using high-resolution AFM imaging. An example of this is shown in Figure 6.5C where it can be seen that the surface of *Aspergillus oryzae* spores was covered by a layer of regularly arranged rodlets having a periodicity of 10 nm (van der Aa *et al.*, 2001). These crystalline-like nanostructures may play an important role in determining the biological functions of the spore.

Interestingly, critical dimensions such as the thickness of cell wall layers can be estimated from AFM height images. Figure 6.6 shows that the outer rodlet layer on *A. oryzae* was actually heterogeneous (van der Aa *et al.*, 2002). Therefore, vertical cross-sections could be used to provide a direct measure of the thickness of the outer cell wall layer (≈ 35 nm).

Using different immobilization procedures, a number of other microorganisms have been imaged in buffer solution, and include the freshwater diatom *Pinnularia viridis* (Crawford *et al.*, 2001), lactic acid bacteria (Schaer-Zammaretti and Ubbink, 2003), *Staphylococcus aureus* and *E. coli* (Doktycz *et al.*, 2003), and *Enterococcus hirae* and *Pseudomonas aeruginosa* (Yao *et al.*, 2002). AFM has also emerged as a valuable tool to

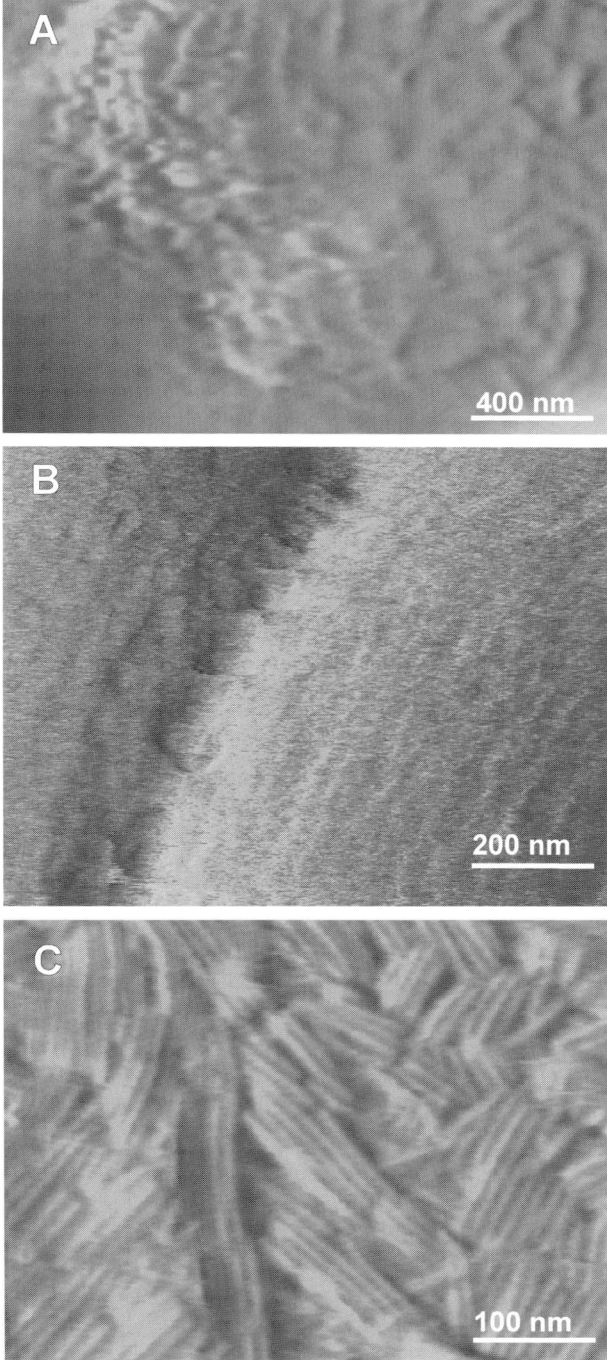

Figure 6.5. Observing cell surfaces at high resolution. AFM deflection images of bacterial and fungal cell surfaces acquired in aqueous solution: (A) *D. radiodurans*, (B) *L. lactis* and (C) *A. oryzae*. (See Colour Plate 25.)

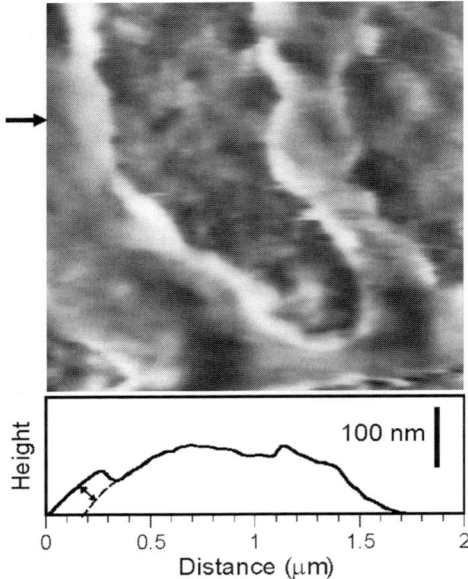

Figure 6.6. Measuring critical dimensions such as cell wall thicknesses. AFM topograph (vertical range: 200 nm) of a dormant spore of *A. oryzae* recorded in buffer solution. The vertical cross-section taken along the dashed line revealed the presence of an outer cell wall layer exhibiting a thickness of ≈35 nm. Reprinted with permission from van der Aa et al., 2002. Copyright 2002 Elsevier Science Ltd.

investigate the crystallization and structural biology of viruses (Malkin et al., 2002; Malkin et al., 2003).

Tracking Cell Surface Dynamics in Real Time

One additional feature of particular interest to cellular microbiologists is the possibility to monitor cell surface structural changes in real time. For instance, the changes resulting from growth and germination can be followed. For *A. oryzae* spores, high-resolution images revealed that dramatic changes of the surface morphology occurred upon germination, the rodlet layer changing into a layer of soft material (van der Aa et al., 2001). This soft material was attributed to cell wall polysaccharides, in agreement with previous structural and chemical studies, and could be manipulated at the single molecule level using force spectroscopy (see below).

The time-dependent effect of external agents such as solvents, ions, chemicals, enzymes and antibiotics on the cell surface can also be followed. In this context, the influence of enzymes on the surface morphology of *S. cerevisiae* was recently investigated using real-time AFM imaging (Ahimou et al., 2003). Images collected at fixed time intervals following addition of protease revealed progressive changes of the cell-surface topography. With time, the protein-rich cell wall became eroded and showed large depressions, about 500 nm in diameter, surrounded by protruding edges, about 50 nm in height. By contrast, no modification of the cell surface was noted upon addition of amyloglucosidase, which was consistent with the cell wall biochemical composition. We anticipate that these real-time experiments should be useful in cell wall enzyme digestion studies as well as in medicine to assess the effect of antibiotics on microbial cell surface morphology.

◆◆◆◆◆◆ FORCE SPECTROSCOPY APPLICATIONS

We have seen that AFM imaging offers a range of new possibilities for exploring the cell surface structure. In addition, AFM provides a means to quantitatively probe biomolecular interactions and physical properties at high lateral resolution. These properties have traditionally been difficult to explore at the subcellular level because of the small size of microorganisms. In the following sections, we show that AFM force spectroscopy can be applied to microbial systems to get physical information with a sensitivity and a resolution that were not previously accessible.

Measuring Cell Wall Elasticity

While the shape of animal cells is determined by the cytoskeleton, that of microbial cells is determined by mechanically strong cell walls. Cell wall stiffness is also important for cell growth and division processes that involve the incorporation of new polymers in localized regions of the wall (Beveridge and Graham, 1991).

AFM force measurements enable obtaining direct, quantitative information on cell wall elastic properties. One approach consists of pressing isolated cell wall material into a groove in a solid surface with the AFM tip and then deducing the elastic modulus from the applied force vs depression distance curve. This method was used to measure the elastic modulus of the sheath of the archeon *Methanospirillum hungatei* GP1 (Xu *et al.*, 1996). The large modulus values (20–40 GPa) indicated that this single-layered structure of unusual strength could withstand an internal pressure of 400 atm. In another study, elastic moduli of 25 MPa were measured for murein sacculi of Gram-negative bacteria in the hydrated state, in excellent agreement with theoretical calculation of the elasticity of the peptidoglycan network (Yao *et al.*, 1999).

Alternatively, sample elasticity of living cells can be obtained by recording force–distance curves on whole cells, converting the force vs distance curves into force vs indentation curves and analyzing them with theoretical models (Radmacher, 2002). In yeast, significant variation of cell wall mechanical properties is expected during cell division. While chitin is usually not found in the cell walls, it accumulates in the very localized region of the wall involved in

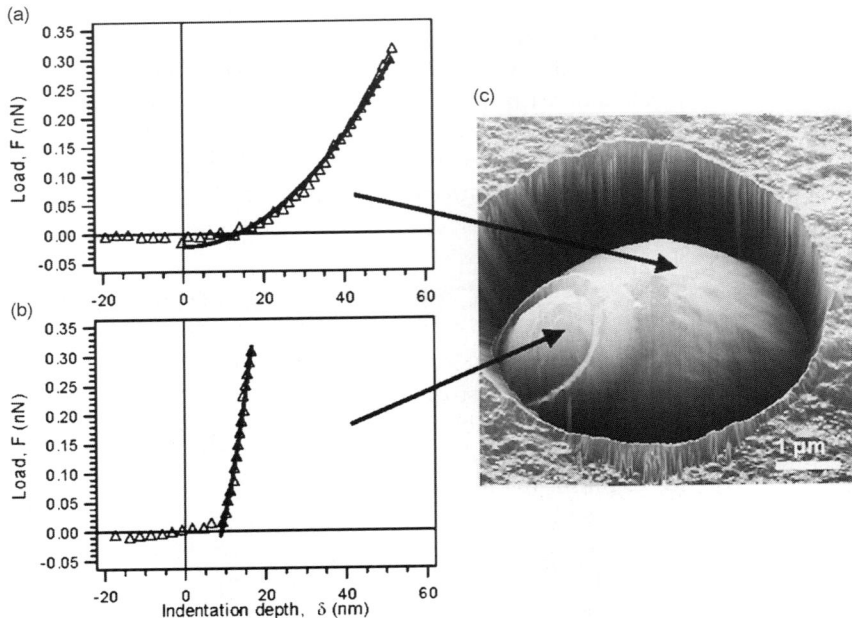

Figure 6.7. Measuring cell wall elasticity at the sub-cellular level. Typical load–indentation curves obtained for *S. cerevisiae* after cell division on the mother cell wall (a) and bud scar (b) regions. The curves were well fitted with the Hertz model (solid lines), providing an estimation of the Young's moduli. The 3D height image (vertical range: 1 μm) of the cell clearly showed a circular bud scar left after detachment of the daughter cell (c). Adapted with permission from Touhami *et al.*, 2003a. Copyright 2003 Am. Chem. Soc.

budding where it is believed to stiffen the cell wall (Sloat and Pringle, 1978). This was recently confirmed by recording spatially resolved maps of elasticity in parallel with topographic images for the surface of *S. cerevisiae* cells (Touhami *et al.*, 2003a). Figure 6.7 reveals that the force vs indentation curves obtained for the bud scar region and for the surrounding cell wall were quite different, stiffer behavior being observed for the bud scar. Fitting the curves with the Hertz model yielded Young's modulus values of 6.1 ± 2.4 and 0.6 ± 0.4 MPa for the bud scar and surrounding cell surface, respectively. This provided the first demonstration that in yeast, the bud scar is 10 times stiffer than the surrounding cell wall, a finding which is consistent with the accumulation of chitin in this area.

Probing Surface Forces and Surface Properties

The AFM detects molecular forces interacting between the tip and the sample. There is a whole set of different types of interactions which can contribute to these forces (Israelachvili, 1991). In principle, it is possible to localize and characterize the nature of theses forces on the biological sample if the particular interaction can be modulated. For example, the electrostatic force interacting between the charged AFM tip (negatively in case of silicon or silicon nitride) and locally charged areas of the sample can be compensated by the electrolyte of the buffer solution. While the electrostatic repulsion between the AFM tip and sample is maximal at low electrolyte concentration the repulsion can be minimized by increasing the electrolyte concentration. Thus, recording the same biological sample at different electrolyte concentrations allows substracting topographs to locate the areas of electrostatic repulsion. This effect, which was first investigated by using the AFM in the force spectroscopy mode (Butt, 1991; Butt et al., 1995), is used quite often to detect local surface charges across dimensions. Biological systems imaged and electrostatically characterized by AFM range from membranes (Heinz and Hoh, 1999b; Müller and Engel, 1997) or DNA (Czajkowsky et al., 1998a), to transmembrane channels of OmpF porin (Philippsen et al., 2002). Simulations of the environments, however, showed that the AFM tip locally influences the dielectric environment, which should be considered when calculating electric potentials from the experimental data.

Cell surface hydrophobicity and surface charge are involved in a variety of interfacial interactions including cell–support, cell–cell, cell–virus, antigen–antibody, cell–drug and cell–ions interactions. Therefore, experimental methods have been developed to assess these properties and include microelectrophoresis, colloid titration, isoelectric focusing of cells, ion-exchange chromatography and surface conductivity, water contact angle measurements on cell lawns, adhesion to hydrocarbons, partitioning in aqueous two-phase systems and hydrophobic interaction chromatography (Doyle and Rosenberg, 1990; Mozes et al., 1991). Because these assays only provide averaged information obtained on a large ensemble of cells, developing complementary tools for probing properties at the subcellular level is an important issue in current cell surface research.

For the first time, chemical functionalization of AFM tips makes it possible to map the local surface hydrophobicity and charges of individual cells. In one such study, AFM tips modified with self-assembled monolayers terminated with hydroxyl and methyl groups were used to investigate the surface of *Phanerochaete chrysosporium* spores (Dufrêne, 2000). Remarkably, the functionalized tips enabled direct visualization of surface features to a lateral resolution of 10 nm. Multiple force–distance curves recorded over the spore surface with both hydrophilic and hydrophobic tips always showed no adhesion. Comparison with data obtained on functionalized model substrate led to the conclusion that the spore surface was uniformly hydrophilic, a finding which was related to the biological functions of spores (protection, dispersion).

In other works, tips functionalized with ionizable groups were used to probe cell surface charges. The surfaces of most living cells, including microorganisms, carry a net negative charge at physiological pH due to the ionization of amino, carboxyl and phosphate groups. Up to now, charge properties have been essentially characterized by the zeta potential deduced from electrophoretic mobility measurements on large populations of cells. To assess the complementarity of AFM in such studies, Ahimou and co-workers (2002) recorded force–distance curves between AFM tips functionalized with carboxyl groups and *S. cerevisiae*. The results were strongly dependent on pH: while no adhesion was measured in the retraction curves at neutral/alkaline pH, significant adhesion forces were recorded at acidic pH. Adhesion maps always showed fairly homogeneous contrast, indicating that the cell surface properties were homogeneous. The change of adhesion forces as a function of pH was interpreted as resulting from a change of cell surface electrostatic properties, a claim which was supported by the correlation obtained between the AFM titration curve constructed by plotting the adhesion force vs pH and the electrophoretic mobility vs pH curve

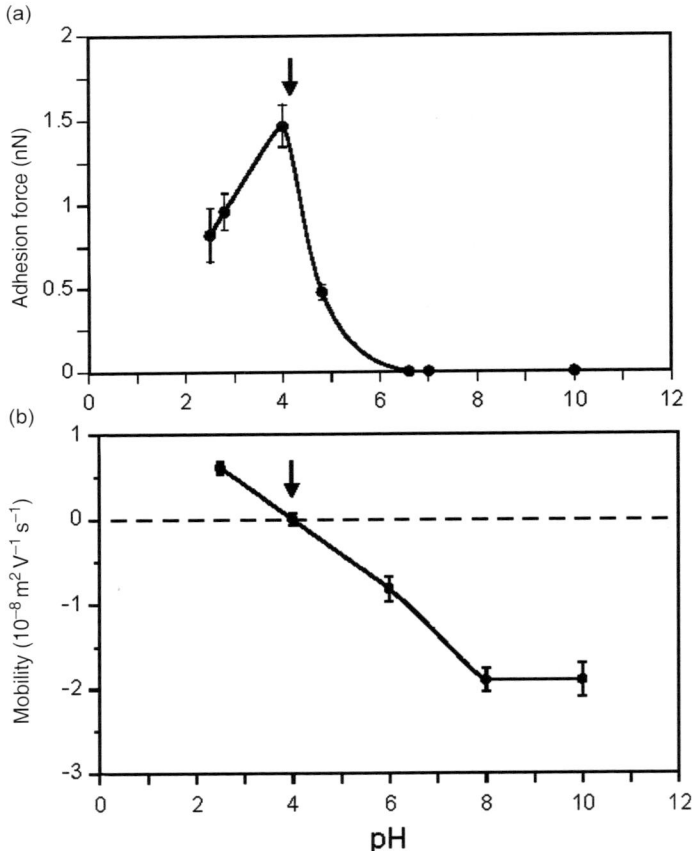

Figure 6.8. Use of chemically functionalized tips for probing cell surface charges. (a) Adhesion force titration curve obtained by plotting the adhesion force measured between the surface of *S. cerevisiae* and a carboxyl-terminated tip as a function of pH. (b) Electrophoretic mobility vs pH curve obtained from large populations of *S. cerevisiae* cells. A good correlation was found between the two techniques, the largest adhesion force being observed at the cell isoelectric point. Adapted with permission from Ahimou *et al.*, 2002. Copyright 2002 Am. Chem. Soc.

obtained by classical microelectrophoresis. In Figure 6.8, it can be seen that the microelectrophoresis curve showed an isoelectric point (p*I*) of 4 consistent with the presence of an amino-carboxyl surface (Dengis and Rouxhet, 1997). At pH < p*I*, the surface is positively charged due to the presence of NH_3^+ groups while above pH 4, an increase in the negative value of the mobility is noted which is essentially due to the dissociation of weak COOH groups. A good agreement was observed between the position of the maximum of the AFM titration curve and the cell p*I* (Figure 6.8), suggesting that the modified tips are sensitive to local isoelectric points (Ahimou *et al.*, 2002).

Manipulation of Single Molecules

The remarkable force sensitivity of AFM, which falls within the piconewton range, permits the measurement of intra- and intermolecular forces at the single molecule level. These experiments, in which single biomolecules are manipulated, provide new, quantitative information on the mechanical properties of the molecule, on conformational transitions along the chains, on protein folding mechanisms, and on the forces involved in supramolecular assembly (Clausen-Schaumann *et al.*, 2000; Janshoff *et al.*, 2000; Zlatanova *et al.*, 2000).

Applied to bacterial S-layers, the combination of AFM imaging and force spectroscopy (Müller *et al.*, 1999a) allowed the manipulation of single subunits of the S-layer, detecting their intermolecular forces. Using this approach, the S-layer was first imaged at high resolution. Then the AFM tip was pushed onto the S-layer surface allowing peptide ends of individual proteins to adsorb. Once a molecular bridge was formed, the peptide end could be used to extract the protein from its assembly. During separation of the tip from the surface, a force spectrum was recorded giving insight into the binding force of the protein. Re-imaging the S-layer after manipulation allowed an unambiguous correlation of the force curve to the protein (Figure 6.9). When applied to the S-layers of *D. radiodurans* (Müller *et al.*, 1999a) and *Corynebacterium glutamicum* (Scheuring *et al.*, 2002b), the anchoring forces of their single protomers were shown to be between 70 and 310 pN. These values are particularly important in order to understand the stability of the S-layers.

Combined single molecule AFM imaging and force spectroscopy can also be applied to image single membrane proteins and to unfold their secondary structure elements (Oesterhelt *et al.*, 2000). Similar to the previous approach, single bacteriorhodopsin molecules from *H. salinarum* were imaged at subnanometer resolution allowing individual polypeptide loops to be resolved (Figure 6.10). The tip was then pushed slightly onto the membrane allowing the C-terminal end of one bacteriorhodopsin molecule to adsorb. Separating the tip from the surface enabled the secondary structure elements to unfold one after the other, thereby permitting measurement of their stability against mechanical unfolding (Figure 6.10B). Finally, re-imaging of the same

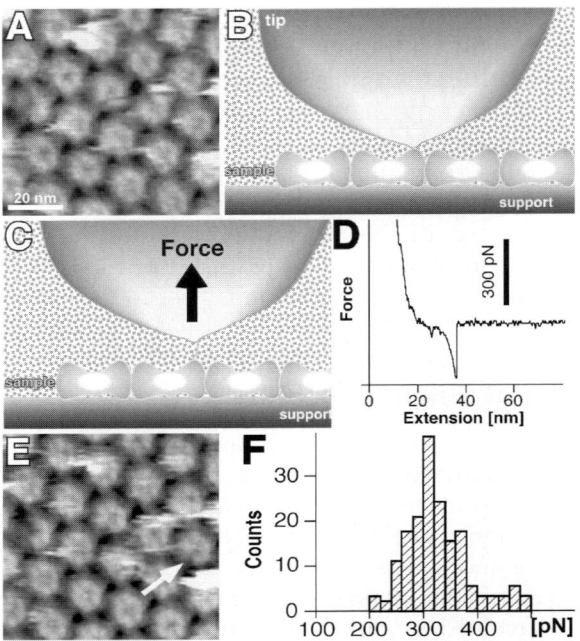

Figure 6.9. Combined single-molecule imaging and force spectroscopy of a protein from a bacterial S-layer. AFM topograph (A) of the inner surface of the regular bacterial S-layer of *D. radiodurans*. Identical proteins were assembled into hexameric pores coexisting in plugged or unplugged conformations (Müller *et al.*, 1996). After recording the topograph, the AFM tip was pushed onto the S-layer (B) for about 1 s. In ≈25% of all approaches the N-terminal end of a protomer adsorbed onto the tip. This molecular bridge allowed the application of a mechanical force on the protein by retracting the AFM tip (C). The force applied was recorded by the cantilever deflection. Occasionally, the force-extension curve recorded (D) showed a single adhesion peak of ≈310 pN after pulling the polypeptide for about 18 nm. Re-imaging of the same surface (E) showed that one single protomer has been removed (arrow). Histogram of adhesion forces detected to remove single protomers (F). The Gaussian distribution of 198 measurements is centered at 312 ± 43 pN. Contact mode topographs were recorded in buffer solution and exhibited a vertical range of 3 nm.

surface allowed a correlation of the force curve to the individual molecule. Besides observing that individual transmembrane helices and single polypeptide can exhibit a sufficient mechanical stability (Figure 6.10D), these measurements demonstrate that a single bacteriorhodopsin may choose various pathways to unfold (Müller *et al.*, 2002b). This probability of a protein choosing distinct unfolding pathways depends on small changes in the temperature (Janovjak *et al.*, 2003). It is expected that other environmental conditions may also exhibit an influence on the unfolding pathways and that this change occurs upon shifting the energy landscape of the protein.

Although single molecule force spectroscopy studies on living cells remain challenging, efforts are being made in this direction. For instance, the elastic properties of fungal cell surface macromolecules

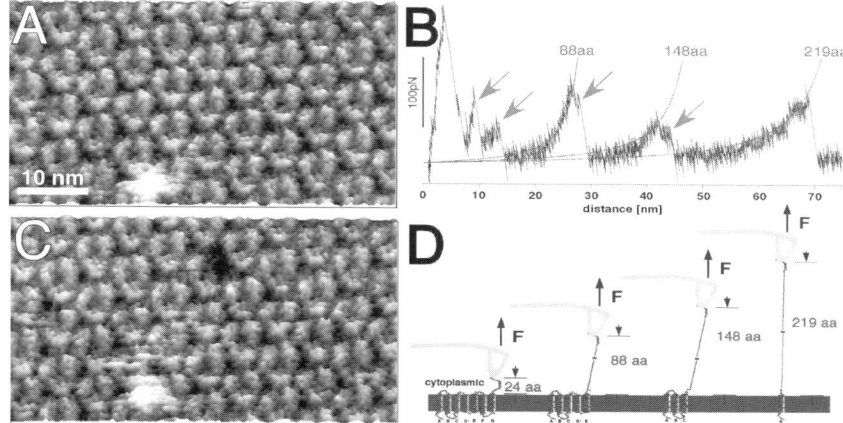

Figure 6.10. Unfolding individual bacteriorhodopsins. (A) Topograph of the cytoplasmic purple membrane surface. Similar to the experiments on the *D. radiodurans* S-layer (Figure 6.9), a single bacteriorhodopsin molecule was selected and the AFM tip was pushed onto its surface allowing a molecular bridge to be established. (B) By separating probe and membrane surface, the force spectrum exhibited discontinuous changes. The length of the force spectra extended far out to distances up to 75 nm, corresponding to the length of one entirely unfolded protein. For analysis, the force spectra can be divided into main peaks, which were fitted using the worm-like chain (WLC) model (solid curves) and into side peaks (arrows). (C) After recording the force spectra, a topograph of the manipulated surface area was captured. The image clearly shows that one single bacteriorhodopsin monomer is missing. (D) The main peaks of the force spectra were correlated to present the predominant pair-wise unfolding of transmembrane α-helices. Statistical analysis of force curves recorded under identical conditions showed that the unfolding spectra of bacteriorhodopsin exhibit individual, but highly reproducible, unfolding patterns. Side peaks of the force spectra (B, arrows), occurring at distinct probabilities, were interpreted to represent the unfolding process of individual polypeptide loops and helices (Müller *et al.*, 2002b). Topographs were recorded in buffer solution (300 mM KCl, 10 mM Tris–HCl, pH 7.8) using contact mode AFM. They exhibited a vertical full gray level range of 1.5 nm and were displayed as relief tilted by 5°.

were measured in relation to cellular interactions (van der Aa *et al.*, 2001). During germination in liquid medium, spores of *A. oryzae* aggregate. Although aggregation is expected to have a marked influence on the performance of *A. oryzae* in bioreactors, little is known about the molecular mechanisms of this process. In this context, force spectroscopy was applied to probe the change of stickiness and elasticity of cell surface macromolecules of *A. oryzae* during germination (van der Aa *et al.*, 2001). Force–distance curves showed no/weak adhesion forces on dormant spores, which was consistent with the presence of a hydrophilic, crystalline rodlet layer at the surface. By contrast, it can be seen in Figure 6.11 that force curves obtained on germinating spores showed attractive forces of 400 ± 100 pN magnitude, along with characteristic elongation forces. Because the cell surface is known to enrich in polysaccharides during germination, the measured elongation

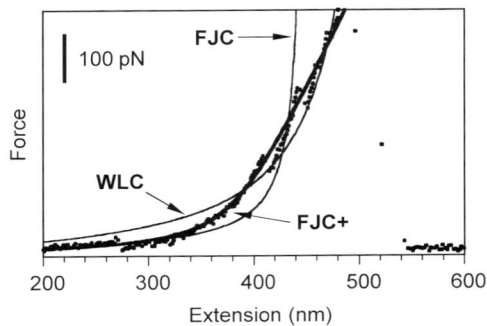

Figure 6.11. Pulling on cell surface macromolecules. Force–distance curve recorded between a silicon nitride tip and the surface of a germinating *A. oryzae* spore. The elongation force was well described by an extended freely jointed chain model (FJC + ; thick line) with parameters consistent with the stretching of a single polysaccharide chain. By contrast, FJC and WLC models (thin lines) failed to adequately describe the data. Reproduced with permission from van der Aa *et al.*, 2001. Copyright 2001 Am. Chem. Soc.

forces were attributed to the molecular stretching of flexible polysaccharide chains (van der Aa *et al.*, 2001).

Quantitative information on the molecule elasticity can be obtained by fitting AFM elongation forces with various models from statistical mechanics (for a review, see Janshoff *et al.*, 2000). In the worm-like chain (WLC) model, the polymer is described as an irregular curved filament, which is linear on the scale of the persistence length, a parameter which represents the stiffness of the molecule. Hence, molecules with low persistence length have a tendency to form coils. Entropic (reduction in the number of conformations) and enthalpic (bond deformation, bond rupture) contributions are combined. The WLC model has been successfully applied to describe the stretching by AFM of single DNA and protein molecules (Janshoff *et al.*, 2000). However, this model did not adequately describe the *A. oryzae* force data (Figure 6.11).

In the freely jointed chain (FJC) model, the polymer has a purely entropic elastic response and is considered as consisting of a series of rigid, orientationally independent statistical (Kuhn) segments, connected through flexible joints. The segment length, referred to as the Kuhn length, is a direct measure of the chain stiffness. An extended FJC model has been developed in which Kuhn segments can stretch and align under force. The polymer consists of a series of elastic springs characterized by a given segment elasticity. The extended FJC model has been successfully applied to describe the elastic behavior of polysaccharides, amylose and dextran, as probed by single molecule force spectroscopy (Rief *et al.*, 1997; Marszalek *et al.*, 1998). Figure 6.11 shows that elongation forces obtained on *A. oryzae* fitted well with an extended FJC model, with parameters (Kuhn length, segment elasticity) that were consistent with the stretching of individual polysaccharides molecules (Rief *et al.*, 1997). Presumably, the stickiness and flexibility of the cell surface polysaccharide chains play a role in promoting cell–cell interactions via macromolecular bridging.

In the past 2 years, such single molecule pulling experiments have been applied to other microbes, including bacteria (Lower *et al.*, 2001; Abu-Lail and Camesano, 2002) and yeast (Touhami *et al.*, 2003b). Most of these experiments have been recently reviewed (Abu-Lail and Camesano, 2003).

Measuring Receptor–Ligand Interactions at the Single Molecule Level

Molecular recognition, which involves multiple non-covalent bonds, plays a central role in cellular behavior and immunology. Attempts to investigate specific molecular recognition forces have been limited by the lack of suitable, sensitive techniques. For the first time, researchers can actually measure these forces at the single molecule level. Basically, this implies attaching specific biomolecules on the AFM tip and on solid supports, and recording force–distance curves between the modified surfaces (Zlatanova *et al.*, 2000; Hinterdorfer, 2002). When anchoring the biomolecules on the tip and support surfaces, several important issues are to be considered: the binding of the biomolecules should be much stronger than the intermolecular force being studied; the attached biomolecules should keep sufficient mobility so that they can freely interact with complementary molecules; unspecific adsorption on the modified surfaces should be inhibited to minimize the contribution of unspecific adhesion to the measured forces; for oriented systems, site-directed coupling in which the molecule has a defined orientation may be desired (Hinterdorfer, 2002).

In the microbial world, specific interactions between receptors and ligands mediate numerous important cellular interaction processes (Sharon and Lis, 1989). For instance, the flocculation (i.e. aggregation) of the yeast *S. carlsbergensis* in fermentation technology involves site-specific interactions between cell surface lectins and mannose residues (Figure 6.12A). To measure these interactions at the single molecule level, gold-coated AFM tips were functionalized with thiol-terminated hexasaccharide molecules (Touhami *et al.*, 2003b; Figure 6.12B). The advantage of this strategy lies in that it provides enhanced accessibility of the carbohydrate moieties to the lectin-binding sites. As seen in Figure 6.12C, the force curves recorded between *S. carlsbergensis* in flocculating conditions and the carbohydrate tips showed single or multiple unbinding forces. The histogram of the largest unbinding forces (Figure 6.12D) displayed an asymmetric distribution centered at 121 ± 53 pN ($n = 100$). Blocking experiments were performed to demonstrate the specificity of the measured adhesion forces; in the presence of mannose, no/little adhesion was detected, indicating that mannose had blocked the lectin receptor sites. Further controls were carried out using non-flocculating cells; here again, the curves showed poor adhesion that lead us to conclude that the measured 121 pN force reflected the specific interactions involved in yeast flocculation. In future, this kind of single

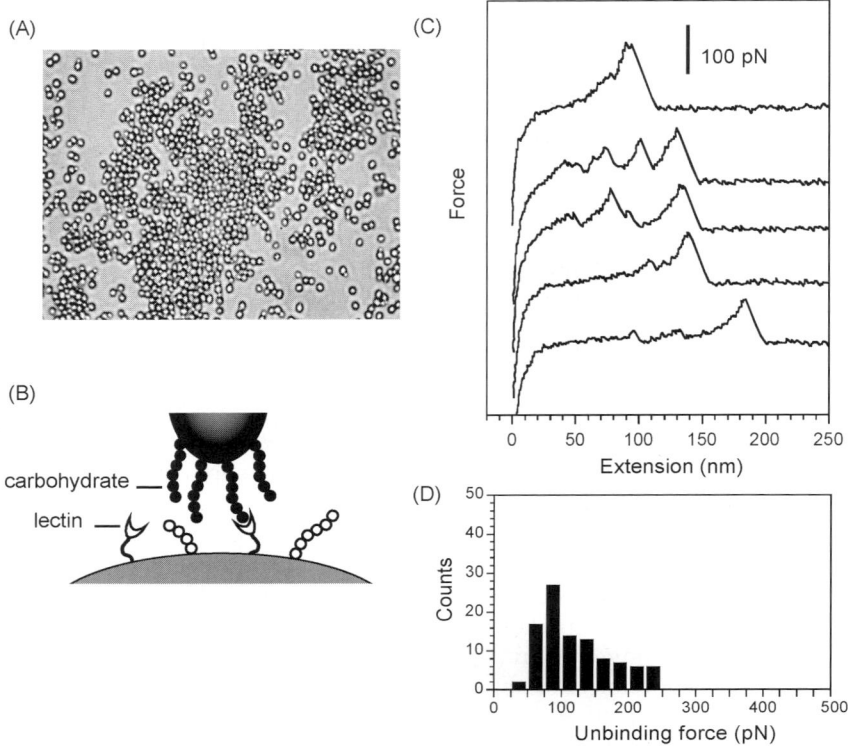

Figure 6.12. Measuring receptor–ligand interactions: application to fermentation technology. Flocculation of the yeast *S. carlsbergensis* (A) is mediated through specific lectin–carbohydrate interactions. To probe these interactions, force measurements were performed between single cells and AFM tips functionalized with carbohydrate molecules (B). Set of force–distance curves recorded at different locations (C) and histogram of the largest unbinding forces (D) revealed a mean adhesion force of 121 ± 53 pN attributed to the rupture of single lectin–carbohydrate complexes. Adapted with permission from Touhami *et al.*, 2003b. Copyright 2003 Soc. Gen. Microbiol.

molecule experiments may help in controlling cellular interactions in biotechnological processes.

Molecular recognition is also a key event in pathogenesis, where the infectious process is generally initiated by the interaction between microbial adhesins and specific receptors on the host cell surface. An example of this is found in tuberculosis that is caused by the infectious agent *Mycobacterium tuberculosis*. Despite the importance of the disease, the molecular bases leading to the pathogenicity of *M. tuberculosis* remain poorly understood. It has been shown that *M. tuberculosis* bacilli exhibit on their outermost surface a heparin-binding hemagglutinin adhesin (referred to as HBHA) that is involved in the adhesion to epithelial cells. In particular, the lysine-rich repeated motifs in the carboxyl-terminal end of the adhesin are responsible for the binding to heparan sulfate chains present on the surface of human pneumocytes (Pethe *et al.*, 2000). We recently applied force spectroscopy to better understand the molecular basis of the HBHA–heparin interaction. The surface functionalization procedure involved in attaching HBHA adhesins onto an AFM tip and heparin onto a solid support is shown in Figure 6.13A. Because HBHA proteins interact through their carboxyl-terminal ends with heparin, we used the site-directed NTA (nitrilotriacetate)–His (histidine) system to attach them through their amino-terminal ends. Biotinylated heparin was attached on gold supports via a sandwich layer of streptavidin and biotinylated bovine serum albumin (BBSA). As illustrated in Figure 6.13B and C, most force curves recorded between HBHA-terminated tip and the heparin-terminated supports showed adhesion forces of ~80 pN magnitude, attributed to the interaction between individual HBHA–heparin complexes. Current studies are focusing on determining whether information on affinity and rate constants can be obtained from these results and whether these measurements can be applied to living systems. The above example indicates that force spectroscopy provides new opportunities to investigate the molecular interactions between microbial pathogens and host

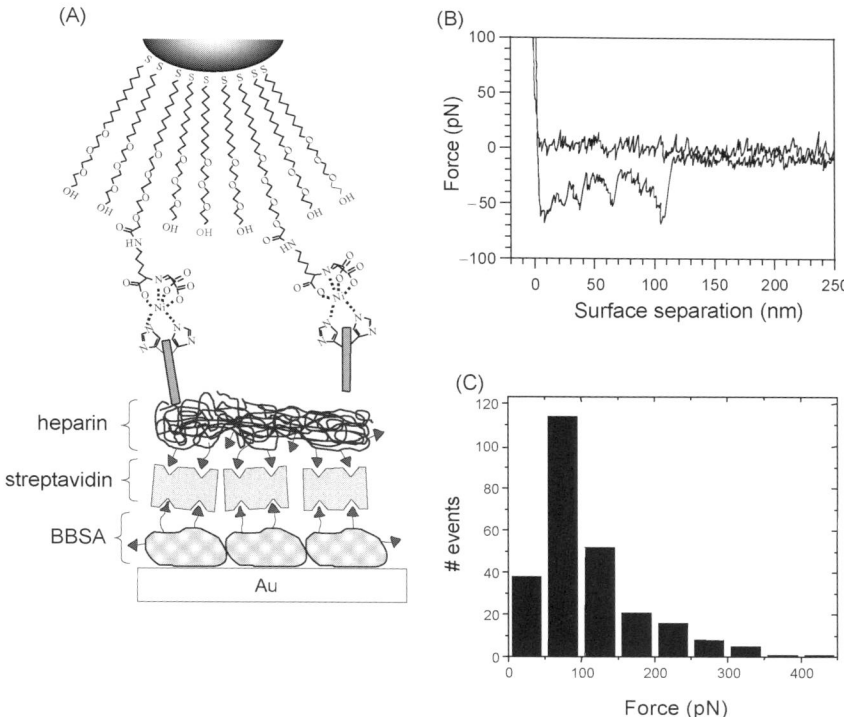

Figure 6.13. Measuring receptor–ligand interactions: application to pathogenesis. *M. tuberculosis* produces a surface adhesin HBHA which is involved in the adhesion to epithelial cells. To probe HBHA–heparin interactions on a single molecule basis, force–distance curves were recorded between a HBHA-terminated AFM tip and a heparin-coated solid support (A). The typical force–distance curve (B) and adhesion histogram (C) showed a mean adhesion force of ~80 pN magnitude reflecting the rupture of the HBHA–heparin bond.

cells, which may ultimately help in developing new therapeutic approaches.

Finally, it is important to mention that the "cell probe" method is an elegant alternative approach for probing receptor–ligand interactions and other cell surface interactions. Here, native cells are linked to the AFM probe and force curves are then recorded between the modified probe and solid supports or immobilized cells (Benoit et al., 2000; Bowen et al., 2001; Lower et al., 2001).

◆◆◆◆◆◆ CONCLUSION AND PERSPECTIVES

The data discussed in this chapter demonstrate that AFM has become a powerful addition to the range of instruments available to study microbial cell surfaces. AFM imaging is particularly useful for visualizing, under physiological conditions, the surface structure of isolated cell layers and of whole cells, with unprecedented signal-to-noise ratio and resolution. Conformational changes of protein layers can be detected at subnanometer resolution, which in turn may be correlated to function. Time-lapse AFM imaging offers a means to follow dynamic events occurring at cell surfaces. Besides topographical data, AFM can also provide a wealth of information on cell surface physical properties and biomolecular interactions. Force spectroscopy can be used to probe quantitatively local cell wall elasticity, surface forces, surface charge and hydrophobicity. The piconewton force sensitivity of the instrument enables it to manipulate single cell surface molecules and to measure their inter- and intramolecular interactions, providing new insights into the molecular bases of protein folding, of protein–protein assembly and of receptor–ligand interactions. Remarkably, these force spectroscopy experiments can be combined with high-resolution imaging, thereby making it possible to observe the manipulated specimens.

In view of these unique capabilities, we anticipate that in future AFM should engage the interest of microbiologists from all horizons, whether fundamental, environmental, clinical or industrial. However, it must be emphasized that, today, accurate data collection and interpretation often remains delicate and that a number of technological issues still need to be addressed. For S-layers, the relatively slow imaging speed of commercially available AFMs (>30 s required to image proteins at high resolution) represents a crucial bottleneck for exploring an increasing number of conformational changes. The situation is even worse for living cells, where the time currently required to record a topograph is in the order of minutes due to the highly heterogeneous character of the cell surface. In the past few years, important technological developments towards solving this problem have been made. Fast speed AFMs have been introduced, which enable recording up to 200 frames per second (Ando et al., 2001; Humphris et al., 2003; Viani et al., 2000). Establishing these techniques for imaging biological samples at (sub)molecular resolution will open a novel way of exploring cell surface processes *in vitro* and *in vivo*.

Another key issue when dealing with living cells is that the image resolution is often limited because the extremely soft cell surface is deformed by the scanning tip. The use of novel dynamic imaging modes which minimize the applied force (Humphris and Miles, 2002), combined with improved sample preparation methods, may help to solve this problem, thereby providing reliable topographs at much higher resolution than is currently available. With such further technological developments (faster imaging, lower interaction forces), it should become possible to observe, for example, the binding of individual antibodies or drugs to distinct locations of the cell surface, function-related conformational changes of surface proteins, or local alterations of the cell surface topography associated with cellular processes such as cell growth and division.

With regards to force spectroscopy, it is hoped that in the near future standard protocols for attaching biomolecules and cells to AFM cantilevers, together with (semi-)automatic force spectroscopy experiments and analysis, will enable the measurement of molecular interactions as a daily routine. Such approaches will assist in the decoding of molecular mechanisms of cell surface processes, and more importantly, may allow screening for parameters to manipulate or to control them. These measurements would find applications in a variety of fields, ranging from basic cell surface research, to environmental, clinical and industrial microbiology.

Acknowledgements

Y.F.D. is a Research Associate of the Belgian National Foundation for Scientific Research (FNRS). The support of the FNRS, of the Foundation for Training in Industrial and Agricultural Research (FRIA), of the Federal Office for Scientific, Technical and Cultural Affairs (Interuniversity Poles of Attraction Programme), of the European Union, of the State of Saxiona and of the Research Department of Communauté Française de Belgique (Concerted Research Action) is gratefully acknowledged. Y.F.D. thanks C. Boonaert, B. van der Aa, F. Ahimou, A. Touhami, F. Denis, I. Burton, V. Dupres, S. Derclaye, P. Rouxhet, F. Menozzi, D. Raze, P. Hols, B. Nysten, M. Asther, A. Vasella and N. Abbott, who were involved in the experiments discussed here. D.J.M. thanks Pierre-Henry Puech for critical discussion of the manuscript.

References

Abu-Lail, N. I. and Camesano, T. A. (2002). Elasticity of *Pseudomonas putida* KT2442 surface polymers probed with single-molecule force microscopy. *Langmuir* **18**, 4071–4081.

Abu-Lail, N. I. and Camesano, T. A. (2003). Polysaccharide properties probed with atomic force microscopy. *J. Microsc.-Oxf.* **212**, 217–238.

Ahimou, F., Denis, F. A., Touhami, A. and Dufrêne, Y. F. (2002). Probing microbial cell surface charges by atomic force microscopy. *Langmuir* **18**, 9937–9941.

Ahimou, F., Touhami, A. and Dufrêne, Y. F. (2003). Real-time imaging of the surface topography of living yeast cells by atomic force microscopy. *Yeast* **20**, 25–30.

Alexiev, U., Marti, T., Heyn, M. P., Khorana, H. G. and Scherrer, P. (1994). Surface charge of bacteriorhodopsin detected with covalently bound pH indicators at selected extracellular and cytoplasmic sites. *Biochemistry* **33**, 298–306.

Amrein, M. and Müller, D. J. (1999). Sample preparation techniques in scanning probe microscopy. *Nanobiology* **4**, 229–256.

Ando, T., Kodera, N., Takai, E., Maruyama, D., Saito, K. and Toda, A. (2001). A high-speed atomic force microscope for studying biological macromolecules. *Proc. Natl Acad. Sci. USA* **98**, 12468–12472.

Baumeister, W., Barth, M., Hegerl, R., Guckenberger, R., Hahn, M. and Saxton, W. O. (1986). Three-dimensional structure of the regular surface layer (HPI layer) of *Deinococcus radiodurans*. *J. Mol. Biol.* **187**, 241–253.

Baumeister, W., Wildhaber, I. and Engelhardt, H. (1988). Bacterial surface proteins: some structural, functional and evolutionary aspects. *Biophys. Chem.* **29**, 39–49.

Beech, I. B. (1996). The potential use of atomic force microscopy for studying corrosion of metals in the presence of bacterial biofilms – an overview. *Int. Biodet. Biodeg.* **37**, 141–149.

Benoit, M., Gabriel, D., Gerisch, G. and Gaub, H. E. (2000). Discrete interactions in cell adhesion measured by single-molecule force spectroscopy. *Nat. Cell. Biol.* **2**, 313–317.

Beveridge, T. J. (1999). Structures of gram-negative cell walls and their derived membrane vesicles. *J. Bacteriol.* **181**, 4725–4733.

Beveridge, T. J. and Graham, L. L. (1991). Surface layers of bacteria. *Microbiol. Rev.* **55**, 684–705.

Binnig, G., Quate, C. F. and Gerber, C. (1986). Atomic force microscope. *Phys. Rev. Lett.* **56**, 930–933.

Bowen, W. R., Lovitt, R. W. and Wright, C. J. (2001). Atomic force microscopy study of the adhesion of *Saccharomyces cerevisiae*. *J. Coll. Interf. Sci.* **237**, 54–61.

Butt, H.-J. (1991). Electrostatic interaction in atomic force microscopy. *Biophys. J.* **60**, 777–785.

Butt, H.-J., Jaschke, M. and Ducker, W. (1995). Measuring surface forces in aqueous electrolyte solution with the atomic force microscope. *Bioelectrochem. Bioenerg.* **38**, 191–201.

Clausen-Schaumann, H., Seitz, M., Krautbauer, R. and Gaub, H. E. (2000). Force spectroscopy with single bio-molecules. *Curr. Opin. Chem. Biol.* **4**, 524–530.

Crawford, S. A., Higgins, M. J., Mulvaney, P. and Wetherbee, R. (2001). Nanostructure of the diatom frustule as revealed by atomic force and scanning electron microscopy. *J. Phycol.* **37**, 543–554.

Czajkowsky, D., Allen, M., Elings, V. and Shao, Z. (1998a). Direct visualization of surface charge in aqueous solution. *Ultramicroscopy* **74**, 1–5.

Czajkowsky, D. M., Sheng, S. and Shao, Z. (1998b). Staphylococcal alpha-hemolysin can form hexamers in phospholipid bilayers. *J. Mol. Biol.* **276**, 325–330.

Czajkowsky, D. M., Iwamoto, H., Cover, T. L. and Shao, Z. (1999). The vacuolating toxin from *Helicobacter pylori* forms hexameric pores in lipid bilayers at low pH. *Proc. Natl Acad. Sci. USA* **96**, 2001–2006.

Dengis, P. B. and Rouxhet, P. G. (1997). Surface properties of top- and bottom-fermenting yeast. *Yeast* **13**, 931–943.

Doktycz, M. J., Sullivan, C. J., Hoyt, P. R., Pelletier, D. A., Wu, S. and Allison, D. P. (2003). AFM imaging of bacteria in liquid media immobilized on gelatin coated mica surfaces. *Ultramicroscopy* **97**, 209–216.

Doyle, R. J. and Rosenberg, M. (1990). *Microbial Cell Surface Hydrophobicity.* American Society for Microbiology, Washington, DC.

Drake, B., Prater, C. B., Weisenhorn, A. L., Gould, S. A. C., Albrecht, T. R., Quate, C. F., Cannell, D. S., Hansma, H. G. and Hansma, P. K. (1989). Imaging crystals, polymers, and processes in water with the atomic force microscope. *Science* **243**, 1586–1588.

Dufrêne, Y. F. (2000). Direct characterization of the physicochemical properties of fungal spores using functionalized AFM probes. *Biophys. J.* **78**, 3286–3291.

Dufrêne, Y. F. (2002). Atomic force microscopy, a powerful tool in microbiology. *J. Bacteriol.* **184**, 5205–5213.

Dufrêne, Y. F. (2003). Recent progress in the application of atomic force microscopy imaging and force spectroscopy to microbiology. *Curr. Opin. Microbiol.* **6**, 317–323.

Dufrêne, Y. F., Boonaert, C. J. P., Gerin, P. A., Asther, M. and Rouxhet, P. G. (1999). Direct probing of the surface ultrastructure and molecular interactions of dormant and germinating spores of *Phanerochaete chrysosporium*. *J. Bacteriol.* **181**, 5350–5354.

Engel, A. and Müller, D. J. (2000). Observing single biomolecules at work with the atomic force microscope. *Nat. Struct. Biol.* **7**, 715–718.

Fotiadis, D., Müller, D. J., Tsiotis, G., Hasler, L., Tittmann, P., Mini, T., Jenö, P., Gross, H. and Engel, A. (1998). Surface analysis of the photosystem I complex by electron and atomic force microscopy. *J. Mol. Biol.* **283**, 83–94.

Fotiadis, D., Hasler, L., Müller, D. J., Stahlberg, H., Kistler, J. and Engel, A. (2000). Surface tongue-and-groove contours on lens MIP facilitate cell-to-cell adherence. *J. Mol. Biol.* **300**, 779–789.

Fotiadis, D., Suda, K., Tittmann, P., Jeno, P., Philippsen, A., Muller, D. J., Gross, H. and Engel, A. (2002). Identification and structure of a putative Ca^{2+}-binding domain at the C terminus of AQP1. *J. Mol. Biol.* **318**, 1381–1394.

Fotiadis, D., Liang, Y., Filipek, S., Saperstein, D. A., Engel, A. and Palczewski, K. (2003). Atomic-force microscopy: rhodopsin dimers in native disc membranes. *Nature* **421**, 127–128.

Fotiadis, D., Qian, P., Philippsen, A., Bullough, P. A., Engel, A. and Hunter, C. N. (2004). Structural analysis of the reaction center light-harvesting complex I photosynthetic core complex of *Rhodospirillum rubrum* using atomic force microscopy. *J. Biol. Chem.* **279**, 2063–2068.

Gad, M. and Ikai, A. (1995). Method for immobilizing microbial cells on gel surface for dynamic AFM studies. *Biophys. J.* **69**, 2226–2233.

Gyorvary, E. S., Stein, O., Pum, D. and Sleytr, U. B. (2003). Self-assembly and recrystallization of bacterial S-layer proteins at silicon supports imaged in real time by atomic force microscopy. *J. Microsc.* **212**, 300–306.

Hansma, H. G. (2001). Surface biology of DNA by atomic force microscopy. *Annu. Rev. Phys. Chem.* **52**, 71–92.

Heinz, W. F. and Hoh, J. H. (1999a). Spatially resolved force spectroscopy of biological surfaces using the atomic force microscope. *Tibtech.* **17**, 143–150.

Heinz, W. F. and Hoh, J. H. (1999b). Relative surface charge density mapping with the atomic force microscope. *Biophys. J.* **76**, 528–538.

Higgins, M. J., Crawford, S. A., Mulvaney, P. and Wetherbee, R. (2000). The topography of soft, adhesive diatom "trails" as observed by atomic force microscopy. *Biofouling* **16**, 133–139.

Hinterdorfer, P. (2002). Molecular recognition studies using the atomic force microscope. *Meth. Cell Biol.* **68**, 115–139.

Humphris, A. D. L. and Miles, M. J. (2002). Developments in dynamic force microscopy and spectroscopy. *Meth. Cell Biol.* **68**, 337–355.

Humphris, A. D. L., Hobbs, J. K. and Miles, M. J. (2003). Ultrahigh-speed scanning near-field optical microscopy capable of over 100 frames per second. *Appl. Phys. Lett.* **83**, 6–8.

Israelachvili, J. (1991). *Intermolecular & Surface Forces*, 2nd edn. Academic Press, London.

Janovjak, H., Kessler, M., Gaub, H., Oesterhelt, D. and Müller, D. J. (2003). Unfolding pathways of native bacteriorhodopsin depend on temperature. *EMBO J.* **22**, 5220–5229.

Janshoff, A., Neitzert, M., Oberdörfer, Y. and Fuchs, H. (2000). Force spectroscopy of molecular systems-single molecule spectroscopy of polymers and biomolecules. *Angew. Chem. Int. Ed.* **39**, 3213–3237.

Karrasch, S., Hegerl, R., Hoh, J., Baumeister, W. and Engel, A. (1994). Atomic force microscopy produces faithful high-resolution images of protein surfaces in an aqueous environment. *Proc. Natl Acad. Sci. USA* **91**, 836–838.

Kasas, S. and Ikai, A. (1995). A method for anchoring round shaped cells for atomic force microscope imaging. *Biophys. J.* **68**, 1678–1680.

Klebba, P. E. and Newton, S. M. (1998). Mechanisms of solute transport through outer membrane porins: burning down the house. *Curr. Opin. Microbiol.* **1**, 238–247.

Koch, Y. and Rademacher, K. H. (1980). Chemical and enzymatic changes in the cell walls of *Candida albicans* and *Saccharomyces cerevisiae* by scanning electron microscopy. *Can. J. Microbiol.* **26**, 965–970.

Kühlbrandt, W. (1992). Two-dimensional crystallization of membrane proteins. *Q. Rev. Biophys.* **25**, 1–49.

Kuznetsov, Y. G., Malkin, A. J. and McPherson, A. (1997). Atomic force microscopy studies of living cells: visualization of motility, division, aggregation, transformation, and apoptosis. *J. Struct. Biol.* **120**, 180–191.

Lower, S. K., Hochella, M. F. and Beveridge, T. J. (2001). Bacterial recognition of mineral surfaces: nanoscale interactions between *Shewanella* and α-FeOOH. *Science* **292**, 1360–1363.

Malkin, A. J., Plomp, M. and McPherson, A. (2002). Application of atomic force microscopy to studies of surface processes in virus crystallization and structural biology. *Acta Crystallogr. Sect. D – Biol. Crystallogr.* **58**, 1617–1621.

Malkin, A. J., McPherson, A. and Gershon, P. D. (2003). Structure of intracellular mature vaccinia virus visualized by *in situ* atomic force microscopy. *J. Virol.* **77**, 6332–6340.

Marszalek, P. E., Oberhauser, A. F., Pang, Y. P. and Fernandez, J. M. (1998). Polysaccharide elasticity governed by chair-boat transitions of the glucopyranose ring. *Nature* **396**, 661–664.

Maurice, P. A., Vierkorn, M. A., Hersman, L. E., Fulghum, J. E. and Ferryman, A. (2001). Enhancement of kaolinite dissolution by an aerobic *Pseudomonas mendocina* bacterium. *Geomicrobiol. J.* **18**, 21–35.

Moll, D., Huber, C., Schlegel, B., Pum, D., Sleytr, U. B. and Sara, M. (2002). S-layer-streptavidin fusion proteins as template for nanopatterned molecular arrays. *Proc. Natl Acad. Sci. USA* **99**, 14646–14651.

Möller, C., Allen, M., Elings, V., Engel, A. and Müller, D. J. (1999). Tapping mode atomic force microscopy produces faithful high-resolution images of protein surfaces. *Biophys. J.* **77**, 1050–1058.

Möller, C., Büldt, G., Dencher, N., Engel, A. and Müller, D. J. (2000). Reversible loss of crystallinity on photobleaching purple membrane in presence of hydroxylamine. *J. Mol. Biol.* **301**, 869–879.

Mou, J. X., Yang, J. and Shao, Z. F. (1995). Atomic force microscopy of cholera toxin B-oligomers bound to bilayers of biologically relevant lipids. *J. Mol. Biol.* **248**, 507–512.

Mozes, N., Handley, P. S., Busscher, H. J. and Rouxhet, P. G. (1991). *Microbial Cell Surface Analysis: Structural and Physicochemical Methods*. VCH Publishers, New York.

Müller, D. J. and Engel, A. (1997). The height of biomolecules measured with the atomic force microscope depends on electrostatic interactions. *Biophys. J.* **73**, 1633–1644.

Müller, D. J. and Engel, A. (1999). Voltage and pH-induced channel closure of porin OmpF visualized by atomic force microscopy. *J. Mol. Biol.* **285**, 1347–1351.

Müller, D. J., Büldt, G. and Engel, A. (1995). Force-induced conformational change of bacteriorhodopsin. *J. Mol. Biol.* **249**, 239–243.

Müller, D. J., Baumeister, W. and Engel, A. (1996). Conformational change of the hexagonally packed intermediate layer of *Deinococcus radiodurans* imaged by atomic force microscopy. *J. Bacteriol.* **178**, 3025–3030.

Müller, D. J., Amrein, M. and Engel, A. (1997). Adsorption of biological molecules to a solid support for scanning probe microscopy. *J. Struct. Biol.* **119**, 172–188.

Müller, D. J., Fotiadis, D. and Engel, A. (1998). Mapping flexible protein domains at subnanometer resolution with the AFM. *FEBS Lett.* **430**, 105–111.

Müller, D. J., Baumeister, W. and Engel, A. (1999a). Controlled unzipping of a bacterial surface layer with atomic force microscopy. *Proc. Natl Acad. Sci. USA* **96**, 13170–13174.

Müller, D. J., Fotiadis, D., Scheuring, S., Müller, S. A. and Engel, A. (1999b). Electrostatically balanced subnanometer imaging of biological specimens by atomic force microscopy. *Biophys. J.* **76**, 1101–1111.

Müller, D. J., Hand, G. M., Engel, A. and Sosinsky, G. (2002a). Conformational changes in surface structures of isolated Connexin26 gap junctions. *EMBO J.* **21**, 3598–3607.

Müller, D. J., Kessler, M., Oesterhelt, F., Moeller, C., Oesterhelt, D. and Gaub, H. (2002b). Stability of bacteriorhodopsin alpha-helices and loops analyzed by single-molecule force spectroscopy. *Biophys. J.* **83**, 3578–3588.

Oberleithner, H., Schillers, H., Wilhelmi, M., Butzke, D. and Danker, T. (2000). Nuclear pores collapse in response to CO_2 imaged with atomic force microscopy. *Pflugers Arch.* **439**, 251–255.

Oesterhelt, D. and Stoeckenius, W. (1974). Isolation of the cell membrane of *Halobacterium halobium* and its fraction into red and purple Membrane. *Meth. Enzymol.* **31**, 667–678.

Oesterhelt, F., Oesterhelt, D., Pfeiffer, M., Engel, A., Gaub, H. E. and Müller, D. J. (2000). Unfolding pathways of individual bacteriorhodopsins. *Science* **288**, 143–146.

Pembrey, R. S., Marshall, K. C. and Schneider, R. P. (1999). Cell surface analysis techniques: what do cell preparation protocols do to cell surface properties? *Appl. Environ. Microbiol.* **65**, 2877–2894.

Persike, N., Pfeiffer, M., Guckenberger, R., Radmacher, M. and Fritz, M. (2001). Direct observation of different surface structures on high-resolution images of native halorhodopsin. *J. Mol. Biol.* **310**, 773–780.

Peters, J., Peters, M., Lottspeich, F., Schäfer, W. and Baumeister, W. (1987). Nucleotide sequence of the gene encoding the *Deinococcus radiodurans* surface protein, derived amino acid sequence, and complementary protein chemical studies. *J. Bacteriol.* **169**, 5216–5223.

Pethe, K., Aumercier, M., Fort, E., Gatot, C., Locht, C. and Menozzi, F. D. (2000). Characterization of the heparin-binding site of the mycobacterial heparin-binding hemagglutinin adhesin. *J. Biol. Chem.* **275**, 14273–14280.

Philippsen, A., Im, W., Engel, A., Schirmer, T., Roux, B. and Müller, D. J. (2002). Imaging the electrostatic potential of transmembrane channels: atomic probe microscopy on OmpF porin. *Biophys. J.* **82**, 1667–1676.

Putman, C. A. J., van der Werft, K., de Grooth, B. G., van Hulst, N. F., Greve, J. and Hansma, P. K. (1992). A new imaging mode in the atomic force microscope based on the error signal. *SPIE* **1639**, 198–204.

Radmacher, M. (2002). Measuring the elastic properties of living cells by the atomic force microscope. *Meth. Cell Biol.* **68**, 67–90.

Rief, M., Oesterhelt, F., Heymann, B. and Gaub, H. E. (1997). Single molecule force spectroscopy on polysaccharides by atomic force microscopy. *Science* **275**, 1295–1297.

Rotsch, C., Jacobson, K. and Radmacher, M. (1999). Dimensional and mechanical dynamics of active and stable edges in motile fibroblasts investigated by using atomic force microscopy. *Proc. Natl Acad. Sci. USA* **96**, 921–926.

Schaer-Zammaretti, P. and Ubbink, J. (2003). Imaging of lactic acid bacteria with AFM-elasticity and adhesion maps and their relationship to biological and structural data. *Ultramicroscopy* **97**, 199–208.

Schäffer, T. E., Cleveland, J. P., Ohnesorge, F., Walters, D. A. and Hansma, P. K. (1996). Studies of vibrating atomic force microscope cantilevers in liquid. *J. Appl. Phys.* **80**, 3622–3627.

Scheuring, S., Ringler, P., Borgina, M., Stahlberg, H., Müller, D. J., Agre, P. and Engel, A. (1999). High resolution topographs of the *Escherichia coli* waterchannel aquaporin Z. *EMBO J.* **18**, 4981–4987.

Scheuring, S., Reiss-Husson, F., Engel, A., Rigaud, J. L. and Ranck, J. L. (2001). High-resolution AFM topographs of *Rubrivivax gelatinosus* light-harvesting complex LH2. *EMBO J.* **20**, 3029–3035.

Scheuring, S., Müller, D. J., Stahlberg, H., Engel, H. A. and Engel, A. (2002a). Sampling the conformational space of membrane protein surfaces with the AFM. *Eur. Biophys. J.* **31**, 172–178.

Scheuring, S., Stahlberg, H., Chami, M., Houssin, C., Rigaud, J. L. and Engel, A. (2002b). Charting and unzipping the surface layer of *Corynebacterium glutamicum* with the atomic force microscope. *Mol. Microbiol.* **44**, 675–684.

Scheuring, S., Seguin, J., Marco, S., Levy, D., Breyton, C., Robert, B. and Rigaud, J. L. (2003). AFM characterization of tilt and intrinsic flexibility of *Rhodobacter sphaeroides* light harvesting complex 2 (LH2). *J. Mol. Biol.* **325**, 569–580.

Schindler, H. and Rosenbusch, J. P. (1978). Matrix protein from *Escherichia coli* outer membranes forms voltage-controlled channels in lipid bilayers. *Proc. Natl Acad. Sci. USA* **75**, 3751–3755.

Schirmer, T. (1998). General and specific porins from bacterial outer membranes. *J. Struct. Biol.* **121**, 101–109.

Schwarz, U. D., Haefke, H., Reimann, P. and Guntherodt, H. J. (1994). Tip artefacts in scanning force microscopy. *J. Microsc.* **173**, 183–197.

Seelert, H., Poetsch, A., Dencher, N. A., Engel, A., Stahlberg, H. and Müller, D. J. (2000). Proton powered turbine of a plant motor. *Nature* **405**, 418–419.

Sharon, N. and Lis, H. (1989). Lectins as cell recognition molecules. *Science* **246**, 227–234.

Shlyakhtenko, L. S., Gall, A. A., Filonov, A., Cerovac, Z., Lushnikov, A. and Lyubchenko, Y. L. (2003). Silatrane-based surface chemistry for immobilization of DNA, protein-DNA complexes and other biological materials. *Ultramicroscopy* **97**, 279–287.

Sleytr, U. B. (1997). Basic and applied S-layer research: an overview. *FEMS Microbiol. Rev.* **20**, 5–12.

Sleytr, U. B. and Beveridge, T. J. (1999). Bacterial S-layers. *Trends Microbiol.* **7**, 253–260.

Sleytr, U. B. and Sára, M. (1986). Ultrafiltration membranes with uniform pores from crystalline bacterial cell envelope layers. *Appl. Microbiol. Biotechnol.* **25**, 83–90.

Sleytr, U. B., Bayley, H., Sara, M., Breitwieser, A., Kupcu, S., Mader, C., Weigert, S., Unger, F. M., Messner, P., Jahn-Schmid, B., Schuster, B., Pum, D., Douglas, K., Clark, N. A., Moore, J. T., Winningham, T. A., Levy, S., Frithsen, I., Pankovc, J., Beale, P., Gillis, H. P., Choutov, D. A. and Martin, K. P. (1997). Applications of S-layers. *FEMS Microbiol. Rev.* **20**, 151–175.

Sloat, B. F. and Pringle, J. R. (1978). A mutant of yeast defective in cellular morphogenesis. *Science* **200**, 1171–1173.

Stahlberg, H., Fotiadis, D., Scheuring, S., Remigy, H., Braun, T., Mitsuoka, K., Fujiyoshi, Y. and Engel, A. (2001a). Two-dimensional crystals: a powerful approach to assess structure, function and dynamics of membrane proteins. *FEBS Lett.* **504**, 166–172.

Stahlberg, H., Müller, D. J., Suda, K., Fotiadis, D., Engel, A., Matthey, U., Meier, T. and Dimroth, P. (2001b). Bacterial ATP synthase has an undecameric rotor. *EMBO Rep.* **21**, 1–5.

Stoffler, D., Goldie, K. N., Feja, B. and Aebi, U. (1999). Calcium-mediated structural changes of native nuclear pore complexes monitored by time-lapse atomic force microscopy. *J. Mol. Biol.* **287**, 741–752.

Touhami, A., Nysten, B. and Dufrêne, Y. F. (2003a). Nanoscale mapping of the elasticity of microbial cells by atomic force microscopy. *Langmuir* **19**, 4539–4543.

Touhami, A., Hoffmann, B., Vasella, A., Denis, F. A. and Dufrêne, Y. F. (2003b). Aggregation of yeast cells: direct measurement of discrete lectin-carbohydrate interactions. *Microbiol. SGM* **149**, 2873–2878.

van der Aa, B. C. and Dufrêne, Y. F. (2002). *In situ* characterization of bacterial extracellular polymeric substances by AFM. *Colloids Surf. B: Biointerf.* **23**, 173–182.

van der Aa, B. C., Michel, R. M., Asther, M., Zamora, M. T., Rouxhet, P. G. and Dufrêne, Y. F. (2001). Stretching cell surface macromolecules by atomic force microscopy. *Langmuir* **17**, 3116–3119.

van der Aa, B. C., Asther, M. and Dufrêne, Y. F. (2002). Surface properties of *Aspergillus oryzae* spores investigated by atomic force microscopy. *Colloids Surf. B: Biointerf.* **24**, 277–284.

Viani, M. B., Pietrasanta, L. I., Thompson, J. B., Chand, A., Gebeshuber, I. C., Kindt, J. H., Richter, M., Hansma, H. G. and Hansma, P. K. (2000). Probing protein–protein interactions in real time. *Nat. Struct. Biol.* **7**, 644–647.

Wagner, P. (1998). Immobilization strategies for biological scanning probe microscopy. *FEBS Lett.* **430**, 112–115.

Wang, H. and Clapham, D. E. (1999). Conformational changes of the *in situ* nuclear pore complex. *Biophys. J.* **77**, 241–247.

Xu, W., Mulhern, P. J., Blackford, B. L., Jericho, M. H., Firtel, M. and Beveridge, T. J. (1996). Modeling and measuring the elastic properties of an archaeal surface, the sheath of *Methanospirillum hungatei*, and the implication for methane production. *J. Bacteriol.* **178**, 3106–3112.

Yang, J., Mou, J. X. and Shao, Z. F. (1994). Structure and stability of pertussis toxin studied by *in situ* atomic force microscopy. *FEBS Lett.* **338**, 89–92.

Yao, X., Jericho, M., Pink, D. and Beveridge, T. (1999). Thickness and elasticity of Gram-negative murein sacculi measured by atomic force microscopy. *J. Bacteriol.* **181**, 6865–6875.

Yao, X., Walter, J., Burke, S., Stewart, S., Jericho, M. H., Pink, D., Hunter, R. and Beveridge, T. J. (2002). Atomic force microscopy and theoretical considerations of surface properties and turgor pressures of bacteria. *Colloids Surf. B: Biointerf.* **23**, 213–230.

Zlatanova, J., Lindsay, S. M. and Leuba, S. H. (2000). Single molecule force spectroscopy in biology using the atomic force microscope. *Prog. Biophys. Mol. Biol.* **74**, 37–61.

7 Positron Emission Tomography Imaging of Clinical Infectious Diseases

Christophe Van de Wiele, Olivier De Winter, Hamphrey Ham and Rudi Dierckx

Department of Nuclear Medicine, University Hospital Ghent, De Pintelaan 185, 9000 Ghent, Belgium

◆◆

CONTENTS

Introduction
Acute osteomyelitis and spondylodiscitis
Chronic osteomyelitis
Prosthetic joint infections
Fever of unknown origin
AIDS
General conclusions

◆◆◆◆◆◆ INTRODUCTION

Positron emission tomography (PET) is a molecular imaging technique based on the use of compounds labelled with positron emitter isotopes that decay, producing a coincident gamma ray pair that can be externally detected by a PET scanner. The most widely used PET tracer in oncology is 2-^{18}Fluoro-2-deoxy-glucose (FDG) (Kritz, 1999). The rationale behind its use is the finding of an increased rate of glucose consumption in malignant tissues, due to an increase of *glycolysis*, and of the number of glucose transporters expressed on malignant cells (Ak *et al.*, 2000; Bar-Shalom *et al.*, 2000; Coleman, 1998). After injection, FDG is transported by facilitated diffusion into neoplastic cells where it is phosphorylated by hexokinase and subsequently is trapped as it is not a substrate for the subsequent enzymatic driven pathways for glucose metabolism. As the neoplastic cells accrue larger amounts of FDG due to their increased metabolism, increased activity is detected that delineates the hypermetabolic tumour from the surrounding normal tissues.

Since its first application in the detection of primary brain tumours, PET has been increasingly used for its ability to detect primary malignant tumours, but also for its ability to detect both regional and distant metastases, distinguish benign from malignant tissue or recurrent cancer from treatment-related scarring, and document response to therapy (Delbeke, 1999). Although FDG-PET is reported to be a sensitive and specific technique in ontological imaging, it is well known that inflammatory and infectious lesions can cause false-positive results (Bakheet and Powe, 1998). Various types of inflammatory cells such as macrophages, lymphocytes and neutrophil granulocytes as well as fibroblasts have been shown to avidly take up FDG, especially under conditions of activation. It even appears that on autoradiography, the FDG distribution in certain tumours is highest in the reactive inflammatory tissue, i.e. the activated macrophages and leukocytes surrounding the neoplastic cells (Brown et al., 1995; Kubota et al., 1992).

Over the past few years, the finding of increased FDG accumulation by inflammatory and infectious cells has resulted in a number of promising reports on the potential of FDG-PET imaging in different types of infection and inflammation (Kubota et al., 1992; Guhlmann et al., 1998; Skehan et al., 1999; Stumpe et al., 2000). Results deriving from these studies are described and discussed in this chapter.

◆◆◆◆◆◆ ACUTE OSTEOMYELITIS AND SPONDYLODISCITIS

Both experimental and clinical studies suggest that FDG-PET may be useful for diagnosing acute osteomyelitis or spondylodiscitis. In the absence of complicating factors, the added value relative to the combination of physical examination, biochemical alterations in combination with three-phase bone scanning or especially MRI, is expected to be limited, as these techniques bear a high sensitivity ($>90\%$) for this indication (Palestro and Torres, 1997; Kaiser and Holland, 1998; Erdman et al., 1991). However, PET may have a role in rare doubtful cases, such as the differentiation between spondylodiscitis and erosive degenerative disk disease, where both MRI and bone scan may be falsely positive (Kaiser and Holland, 1998; Stabler et al., 1998; Champsaur et al., 2000; Even-Sapir and Martin, 1994). In these cases, a negative PET will exclude infection (Guhlmann et al., 1998).

◆◆◆◆◆◆ CHRONIC OSTEOMYELITIS

The diagnosis of chronic osteomyelitis is complex. Although many techniques have been proposed for the noninvasive evaluation of chronic osteomyelitis, clinicians are often confronted with an indeterminate diagnosis and the clinical strategy adopted is often limited to a "wait and see" policy or empirical antibiotic treatment (Segreti et al., 1998; Spangehl et al., 1998; Zimmerli, 1995). Inflammatory parameters (C-reactive protein, erythrocyte sedimentation rate and leucocytes) lack sensitivity (especially in low-grade infections) and specificity (Zimmerli, 1995; Perry, 1996; Sanzen and Sundberg, 1997; Shih et al., 1987). CT and MRI provide excellent anatomic detail. Although MRI is extremely helpful in unoperated cases, it is of limited value in the presence of metallic implants as well as for discriminating between oedema and active infection after surgery (Erdman et al., 1991; Crim and Seeger, 1994; Seabold and Nepola, 1999; Ledermann et al., 2000).

Combined three-phase bone scintigraphy and leucocyte scan has a good clinical accuracy (79–100%) when considering the peripheral skeleton (Palestro and Torres, 1997; Becker, 1995; Datz, 1994; Kaim et al., 1997; Krznaric et al., 1996; Peters, 1998), however, its accuracy decreases (i) in low-grade chronic infections (Seabold and Nepola, 1999; Becker, 1995) (lower sensitivity), (ii) in the presence of periskeletal soft tissue infection due to the limited resolution of conventional nuclear imaging (lower sensitivity and specificity), (iii) in the central skeleton due to the presence of normal bone marrow and the possibility of so-called "cold lesions" (lower sensitivity and specificity) (Guhlmann et al., 1998; Even-Sapir and Martin, 1994; Becker, 1995; Datz, 1994; Krznaric et al., 1996; Palestro et al., 1991) and (iv) after trauma or surgery due to the presence of ectopic haematopoietic bone marrow (lower specificity) (Knockaert et al., 1994). To avoid false-positive studies due to ectopic bone marrow, the combination of leucocyte scanning with bone marrow scanning (99m-technetium sulphur colloid) has been proposed. Congruency between leucocyte and bone marrow scanning indicates the presence of bone marrow, while discongruency in the presence of a positive leucocyte scan suggests the presence of infection. In the vertebral column, a combination of bone and gallium scan has been proposed to improve both sensitivity and specificity (Palestro, 1994). However, the need for two or even three (bone scan/leucocyte scan/bone marrow scan or bone scan/gallium scan) techniques is not practical, adds to the cost and patient radiation dose and is time consuming. Thus, an equally specific and sensitive all-in-one technique would be most welcome.

Guhlmann et al. (1998) studied 51 patients suspected of having chronic osteomyelitis in the peripheral ($n = 36$) or central ($n = 15$) skeleton prospectively with static FDG-PET imaging and combined ^{99m}Tc-AGAb/^{99m}Tc-methylene diphosphonate bone scanning within 5 days. The images obtained were evaluated in a blinded and independent manner by visual interpretation, which was graded on a five-point scale of two observer's confident diagnosis of osteomyelitis. Receiver operating characteristic (ROC) curve analysis was performed for both imaging modalities. The final diagnosis was established by means of bacteriologic culture of surgical specimens and histopathologic analysis ($n = 31$) or by biopsy and clinical follow-up over 2 years ($n = 20$). Of 51 patients, 28 had osteomyelitis and 23 did not. According to the unanimous evaluation of both readers, FDG-PET correctly identified 27 of the 28 positives and 22 of the 23 negatives (IS identified 15 of 28 positives and 17 of 23 negatives, respectively). On the basis of ROC analysis, the overall accuracies of FDG-PET and IS in the detection of chronic osteomyelitis were 96/96% and 82/88%, respectively.

Kälicke et al. (2000) evaluated the clinical usefulness of fluorine-18 fluorodeoxyglucose positron emission tomography (FDG-PET) in acute and chronic osteomyelitis and inflammatory spondylitis. The study population comprised 21 patients suspected of having acute or chronic osteomyelitis or inflammatory spondylitis. Fifteen of these patients subsequently underwent surgery. FDG-PET results were correlated with histopathological findings. The remaining six patients, who underwent conservative therapy, were excluded from any further evaluation due to the lack of histopathological data. The histopathological findings revealed osteomyelitis or inflammatory spondylitis in all 15 patients: seven patients had acute osteomyelitis and eight patients had chronic osteomyelitis or inflammatory spondylitis. FDG-PET yielded 15 true-positive results. However, the absence of negative findings in this series may raise questions concerning selection criteria.

Our group reported on 60 patients suffering from a variety of suspected chronic orthopaedic infections (De Winter *et al.*, 2001). In this prospective study, the presence or absence of infection was determined by surgical exploration in 15 patients and long-term clinical follow-up in 28 patients. As opposed to the study by Guhlmann *et al.*, patients with recent surgery were not excluded. Considering only those patients with suspected chronic osteomyelitis, FDG-PET was correct in 40 of 43 patients. There were three false-positive findings, 17 true-negative findings and no false-negative findings. This resulted in a sensitivity of 100%, a specificity of 85% and an accuracy of 93%. Two of three false-positive findings occurred in patients who had been operated recently (6 weeks and 4 months, respectively).

Zhuang *et al.* (2000) studied the accuracy of FDG-PET for the diagnosis of chronic osteomyelitis. Twenty-two patients with possible osteomyelitis (five in the tibia, five in the spine, four in the proximal femur, four in the pelvis, two in the maxilla and two in the feet) that underwent FDG-PET imaging and in whom operative or clinical follow-up data were available were included for analysis. The final diagnosis was made by surgical exploration or clinical follow-up during a 1-year period. FDG-PET correctly diagnosed the presence or absence of chronic osteomyelitis in 20 of 22 patients but produced two false-positive results, respectively, two cases of recent osteotomies, resulting in a sensitivity of 100%, a specificity of 87.5% and an accuracy of 90.9%. It is, however, unclear from their report in how many patients histopathologic or microbiologic studies were available.

Finally, Chacko *et al.* (2002) retrospectively analysed the accuracy of FDG-PET for diagnosing infection in a large population of patients and in a variety of clinical circumstances, including suspicion of chronic osteomyelitis in 56 patients. Final diagnosis was made on the basis of surgical pathology and clinical follow-up for a minimum of 6 months. Among the patients suspected of having chronic osteomyelitis the accuracy was 91.2%. Thus, although limited, available data deriving from these four studies appear promising. The fact that FDG-PET is not disturbed by the presence of metallic implants and is able to differentiate between scar tissue and active inflammation constitutes a major advantage when compared to CT and MRI. As opposed to radiolabelled leucocytes or radiolabelled antibodies, FDG is likely to penetrate easier and faster in lesions than cellular tracers or antibodies (Chianelli et al., 1997). Aside from the potential for higher sensitivity, taking into account available data, a negative PET scan virtually rules out osteomyelitis (De Winter et al., 2001; Chacko et al., 2002). Specificity seems to be especially high if patients presenting with recently (less than 3–4 months) traumatized or operated bone are excluded (De Winter et al., 2001; Zhuang et al., 2000; Meyer et al., 1994). In spite of its excellent accuracy for diagnosing chronic osteomyelitis, FDG-PET may currently only be advocated for assessing the central skeleton. This especially as for the peripheral skeleton, more available techniques like leucocyte scanning are as adequate (Guhlmann et al., 1998; Kaim et al., 1997) and especially cheaper. In the above cited study by Guhlmann et al. (1998), comparing the combination of bone scan and leucocyte scan with FDG-PET, the latter proved significantly more accurate in the central skeleton. For the peripheral skeleton, available data suggested that FDG-PET was better, but statistical significance could not be reached.

◆◆◆◆◆◆ PROSTHETIC JOINT INFECTIONS

Modern preventive measures have lowered the rate of prosthetic infection due to preoperative contamination to 0.5–2% (Lew and Waldvogel, 1997). However, the reported incidence of infected hip prosthesis is as high as 30% when considering revision arthroplasty (Maderazo et al., 1988; Hunter et al., 1979). In addition to the clinical evaluation, many diagnostic modalities have been used, such as blood chemistry (sedimentation, C-reactive protein), plain radiography of the pelvis, radiographic contrast examination of the prosthetic joint and joint aspiration followed by bacterial cultures. However, their success has been limited because of the relatively high frequency of false-negative results. In most cases, additional tests are required to diagnose infection before revision surgery.

Radiographic methods and three-phase bone scanning are not able to differentiate between septic and aseptic loosening. The imaging gold standard (accuracy >90%) is considered to be the combination of leucocyte scan and sulphur colloid bone marrow scan, in which a dissociation in uptake pattern is diagnostic for infection. However, leucocyte scanning is laborious, and thus expensive, and the labelling technique requires the manipulation of human blood. Scanning is ideally postponed until 24 h after injection. Moreover, in the case of a positive scan, additional bone marrow scanning is warranted to maintain specificity, adding to the complexity of the technique. Theoretically, FDG-PET overcomes most of these problems, as the whole procedure can be performed safely in less than 2 h.

Zhuang et al. (2001) studied 74 prostheses in 62 patients in whom infection was suspected after artificial hip or knee placement was studied with this technique. Obtained FDG-PET images were interpreted as positive for infection if tracer uptake was increased at the bone–prosthesis interface. Final diagnosis was made by surgical exploration or clinical follow-up for 1 year. The sensitivity, specificity and accuracy of PET for detecting infection associated with knee prostheses were 90.9, 72.0 and 77.8%, respectively. The sensitivity, specificity and accuracy of PET for detecting infection associated with hip prostheses were 90, 89.3 and 89.5%, respectively. Overall, the sensitivity was 90.5% and the specificity was 81.1% for detection of lower limb infections.

Manthey *et al.* (2002) studied 23 patients with 28 prostheses, 14 hip and 14 knee prostheses, who had a complete operative or clinical follow-up. High glucose uptake in the bone prostheses interface was considered as positive for infection, an intermediate uptake as suspect for loosening, and uptake only in the synovia was considered as synovitis. The imaging results were compared with operative findings or clinical outcome. PET correctly identified three hip and one knee prostheses as infected, two hip and two knee prostheses as loosening, four hip and nine knee prostheses as synovitis and two hip and one knee prostheses as unsuspected for loosening or infection. In three patients covered with an expander after explantation of an infected prosthesis PET revealed no further evidence of infection in concordance with the clinical follow-up. PET was false negative for loosening in one case.

Vanquickenborne *et al.* (2003) compared the accuracy of fluorine-18 labelled 2-fluoro-2-deoxy-D-glucose positron emission tomography ((18)FDG-PET) with that of technetium-99m hexamethylpropylene amine oxime leucocyte scintigraphy (LS) for the detection of infected hip prosthesis in 17 patients. Patients were prospectively included and underwent (99m)Tc-methylene diphosphonate bone scintigraphy (BS), LS and an (18)FDG-PET scan within a 2-week period. Seven volunteers with 10 asymptomatic hip prostheses were used as a control group and underwent BS and an (18)FDG-PET scan. Bacteriology of samples obtained by surgery or by needle aspiration and/or clinical follow-up for up to 6 months were used as the gold standard. The combined analysis of the planar BS and LS resulted in 75% sensitivity and 78% specificity. The SPET LS images showed a better lesion contrast, resulting in a 88% sensitivity and a 100% specificity, while 24-h planar images were of no additional value. The analysis of PET images alone resulted in 88% sensitivity and 78% specificity. The combination of (18)FDG-PET and BS images resulted in 88% sensitivity and 67% specificity. Given the presence of small errors near the edge of the metal, which can induce significant artefacts in the corrected emission image, the authors decided to use the data without attenuation correction. In this preliminary study, (18)FDG-PET scans alone showed the same sensitivity as combined BS and LS, although the specificity was slightly lower.

Finally, Van Acker *et al.* (2001) prospectively studied 21 patients who underwent a three-phase BS, LS and FDG-PET examination for exclusion of infection of total knee prosthesis using operative findings, culture results and clinical outcome as gold standard. LS alone had specificity for infection of 53% [positive predictive value (PPV) 42%, sensitivity 100%], compared with 73% for PET scan (PPV 60%, sensitivity 100%). Considering only lesions at the bone–prosthesis interface that were also present on the third phase of the bone scan, a specificity of 93% (PPV 83%) for LS was found. Using these criteria, a specificity of 80% (PPV 67%) was obtained for PET scan. Two out of three false-positive PET scans were due to loosening of the total knee prostheses. It was concluded that LS in combination with bone scintigraphy had a high specificity in the detection of infected total knee prostheses. FDG-PET seemed to offer no additional benefit.

◆◆◆◆◆◆ FEVER OF UNKNOWN ORIGIN

Fever of unknown origin (FUO) has been defined as recurrent fever of 38.3°C or higher lasting more than 3 weeks and remaining undiagnosed after appropriate in- or out-patient evaluation for a minimum of 3 days or three outpatient visits (Durack and Street, 1991). With modern diagnostic techniques available, the face of FUO has changed, with a relatively lower incidence of infections (20–25%) and neoplasms (around 10%) and proportionally more noninfectious inflammatory diseases (20–25%) observed compared to earlier studies (Gelfand, 2001; Knockaert et al., 1994; De Kleijn et al., 1997).

From the perspective of medical imaging, the potential to perform whole-body screening is of major relevance. The goal of radionuclide imaging in patients presenting with FUO is to localize a potential focus causing fever, which can subsequently be investigated by other diagnostic modalities. Whole-body gallium scintigraphy is currently considered as the radionuclide investigation of first choice because it images acute, chronic and granulomatous inflammation as well as various malignant diseases. In a large series of 145 patients investigated for FUO by whole-body gallium scintigraphy, gallium scanning contributed to the final diagnosis in 29% of the cases, a much higher yield than ultrasound (6%) or CT scanning (14%). On the basis of their findings, the use of gallium scintigraphy was suggested as a second-step (as opposed to a last resort) procedure in the evaluation of FUO (Gelfand, 2001).

As compared to 67Ga-scanning, whole-body FDG-PET has several advantages: rapid reporting, superior image quality and resolution, higher lesion-to-background ratios at early time points and low liver-, abdominal and bone marrow-uptake, resulting in optimal imaging conditions. Moreover, the radiation dose to the patient is lower.

While a number of reports have dealt with heterogeneous populations of patients with suspected infection, few papers have addressed the added value of FDG-PET in the diagnosis of FUO.

Blockmans et al. (2001) described the diagnostic contribution of [(18)F]fluoro-deoxyglucose (FDG) positron emission tomography (PET) scan in 58 consecutive cases of FUO as compared to gallium scintigraphy. A final diagnosis was established for 38 patients (64%). Forty-six FDG-PET scans (79%) were abnormal; 24 of these abnormal scans (41% of the total number of scans) were considered helpful in diagnosis, and 22 (38% of the total number) were considered noncontributory to the diagnosis. In a subgroup of 40 patients (69%), both FDG-PET and gallium scintigraphy were performed. FDG-PET scan and gallium scintigraphy were normal in 23 and 33% of these cases, respectively, helpful in diagnosis in 35 and 25%, respectively, and noncontributory in 42% each. All foci of abnormal gallium accumulation were also detected by use of an FDG-PET scan. The authors concluded that FDG-PET is a valuable second-step technique in patients with FUO because it yields diagnostic information in up to 40% of the patients in whom the probability of a definite diagnosis was only 64%. FDG-PET scan compared favourably with gallium scintigraphy for this indication.

Lorenzen *et al.* (2001) performed FDG-PET in 16 patients with FUO in whom conventional diagnostics had not been conclusive. In 12 patients, (75%) nonphysiological accumulations of FDG were found which led to the final diagnosis in 11 patients (69%). FDG-PET was negative in four patients (25%). Two of these patients had rheumatic fever, while in the other two patients the origin of fever could not be detected within 3 months after PET by any other laboratory or imaging means. These findings point to the high sensitivity of FDG whole-body PET for the detection of morphologically assessable foci as an origin of FUO. Moreover, they suggest a high negative predictive value of FDG-PET in the setting of FUO, since in no patient with a negative FDG-PET could a morphological origin of the fever be determined.

More recently, Bleeker-Rovers *et al.* (2004) assessed the value of FDG-PET in patients with FUO and patients with suspected focal infection or inflammation. All FDG-PET scans ordered because of FUO or suspected focal infection or inflammation in the last 4 years were reviewed. These results were compared with the final diagnosis. Thirty-five FDG-PET scans were performed in 35 patients with FUO. A final diagnosis was established in 19 patients (54%). Of the total number of scans, 37% were clinically helpful. The PPV of FDG-PET in these patients was 87% and the negative predictive value was 95%.

Finally, Kjaer *et al.* (2004) compared prospectively, on a head-to-head basis, the diagnostic value of FDG-PET and indium-111 granulocyte scintigraphy in patients with FUO. Nineteen patients with FUO underwent both FDG-PET and (111)In-granulocyte scintigraphy within 1 week. The diagnostic values of FDG-PET and granulocyte scintigraphy were evaluated with regard to identification of a focal infectious/inflammatory or malignant cause of FUO. The sensitivity of granulocyte scintigraphy and FDG-PET were 71 and 50%, respectively. The specificity of granulocyte scintigraphy was 92% (71–100%), which was significantly higher than that of FDG-PET, at 46% (34–62%). Positive and negative predictive values for granulocyte scintigraphy were both 85%. Positive and negative predictive values for FDG-PET were 30 and 67%, respectively. In this series, the poorer performance of FDG-PET was in particular attributable to a high percentage of false positive scans.

On the basis of these limited studies, it can be assumed that FDG-PET may have an added value to conventional screening in up to 40–70% of the patients with true FUO. Though a prospective comparison in a large series to date has not been published, this is higher than the 29% found by

Knockaert *et al.* (1994) for gallium scanning. As for the series by Kjaer *et al.* this was a small study designed for head-to-head comparison of two methods. Accordingly, the performance of each method had wide confidence intervals and hence, requires further evaluation in larger series.

In conclusion, FDG-PET may prove to be preferable to gallium scintigraphy in the diagnostic work-up of patients with FUO. A cost-effectiveness study comparing an FDG-PET-based strategy (second-step) versus a conventional strategy, however, eventually followed by third-line FDG-PET, is warranted.

◆◆◆◆◆◆ AIDS

The presentation of FUO in human immunodeficiency virus (HIV) infection may be considered a separate clinical entity, as the frequency distribution of etiologic disease is quite different. HIV-infected patients may at a certain time develop symptoms of fever, weight loss or deterioration of mental function, which classifies them as having the acquired immunodeficiency syndrome (AIDS) (Ancell-Park, 1993). Many pathologies, mostly neoplastic and/or infectious, may be underlying. Often, clinical examination, chest radiograph, haematological and biochemical analyses and blood and urine cultures reveal the causative pathology. However, additional imaging remains necessary in a number of cases to guide further diagnostic work-up. Especially when thoracic pathology is suspected, gallium scanning has proven useful (Palestro, 1994). The use of gallium scanning, however, poses several problems on top of the disadvantages that have been discussed above. Evaluation of brain and abdomen is difficult, and the appearance of persistent generalized lymphadenopathy and lymphoma may be similar. Moreover, compared to FDG imaging, the time to complete a study (24–96 h) is much longer (Seabold *et al.*, 1997).

Hoffman *et al.* (1993) studied 11 individuals with AIDS and central nervous system (CNS) lesions with ^{18}F-fluoro-2-deoxyglucose (FDG) and positron emission tomography (PET). FDG-PET was able to accurately differentiate between a malignant (lymphoma) and nonmalignant aetiology for the CNS lesions with significantly higher uptake occurring in the lymphomatous lesions than in the nonmalignant lesions. Both qualitative visual inspection of the images as well as semiquantitative analysis using count ratios were performed and revealed similar results. These results were more recently reproduced by other authors (Villringer *et al.*, 1995).

O'Doherty *et al.* (1997) examined 80 HIV patients using PET. Fifty-seven patients had half-body scans with [18F]fluorodeoxyglucose (FDG), and 23 patients had brain studies performed with FDG. Fourteen patients also had [11C]methionine studies (2 chest, 1 abdomen and 11 brain) performed. Thirteen patients with lymphoma had the extent of the disease clearly identified in both nodal and extranodal sites. Patients with a variety of infections (*Cryptococcus neoformans*, *Pseudomonas aeruginosa*, *Mycobacterium tuberculosis* and *Mycobacterium avium intracellulare*) had disease localized for appropriate biopsy or sampling procedures. A half-body FDG-PET scan had a sensitivity of 92% and a specificity of 94% for localization of focal pathology that needed treatment. High uptake of FDG (greater than liver) had a PPV for pathology needing treatment of 95%. FDG brain studies showed that 16 patients with CD4 T-lymphocyte counts less than 200 cells/ml had reduced cortical uptake compared with that in basal ganglia. FDG scans were abnormal in all 19 patients with focal space occupying lesions identified by magnetic resonance scans. The standardized uptake values (SUVs) over cerebral lesions due to toxoplasma were in the range of 0.14–3.7 (13 patients) and due to lymphoma were in the range of 3.9–8.7 (six patients). Three more patients with progressive multi-focal leukoencephalopathy had SUVs in the range of 1.0–1.5 over the lesions. Another patient had a low-grade oligodendroglioma (SUV = 2.9). Carbon-11-methionine uptake also was high in patients with cerebral lymphoma but did not add to the discrimination between toxoplasmosis and lymphoma in these patients obtained with the FDG scan.

More recently, Scharko *et al.* used FDG-PET to study anatomical correlates of HIV-1 infection in man (O'Doherty *et al.*, 1997). Whole-body FDG-PET images from 15 patients with HIV-1 showed distinct lymphoid tissue activation in the head and neck during acute disease, a generalized pattern of peripheral lymph node activation at midstages, and involvement of abdominal lymph nodes during late disease. Unexpectedly, HIV-1 progression was evident by distinct anatomical correlates, suggesting that lymphoid tissues are engaged in a predictable sequence.

In conclusion, FDG-PET seems to be a promising tool in the work-up of HIV complications. Also, understanding the sequence of lymphoid tissue involvement effectuated by HIV-1 infection, as evidenced by FDG-PET, could encourage use of surgical and/or radiological intervention to supplement chemotherapy. However, cost-effectiveness studies are needed to justify its use as a second-line strategy.

◆◆◆◆◆◆ GENERAL CONCLUSIONS

Promising results have been obtained with FDG-PET in the field of clinical infectious diseases. Available studies suggest that the specificity of FDG-PET will be limited by the fact that FDG accumulates in sterile inflammatory lesions and tumours. Thus, depending on the clinical setting, the problem of specificity may theoretically limit the use of FDG-PET in infectious diseases. However, in most patients, a thorough medical history makes the presence of tumour unlikely, and sterile inflammations such as chronic polyarthritis, vasculitis and tumours often appear at sites or show distribution patterns that are suggestive of these diseases.

Based on available data, the following indications for FDG-PET may prove of clinical relevance in the near future:

- Diagnosis of chronic osteomyelitis, especially in the central, bone-marrow-containing skeleton. The presence of metallic implants poses no problems for diagnosis. An interval of 3–6 months postsurgery should be allowed to prevent false-positive findings. For the moment, there are not enough arguments to recommend the technique in the assessment of infection in joint prostheses.
- FUO, where it may be preferable over gallium scanning as an adjunct to other imaging techniques.
- AIDS, especially for the differential diagnosis of central nervous system lesions and for the early detection of complications.

However, cost–benefit studies and larger series will be required prior to the implementation of FDG-PET in the current diagnostic strategies.

References

Ak, I., Stokkel, M. P. M. and Pauwels, E. K. J (2000). Positron-emission tomography with 18F-fluorodeoxyglucose. Part 2. The clinical value in detecting and staging primary tumors. *J. Cancer Res. Clin. Oncol.* **126**, 560–574.

Ancell-Park, R. (1993). Expanded European AIDS case definition. *Lancet* **341**, 441.

Bakheet, S. M. and Powe, J. (1998). Benign causes of 18-FDG uptake on whole body imaging. *Semin. Nucl. Med.* **28**, 352–358.

Bar-Shalom, R., Valdivia, A. Y. and Blaufox, M. D. (2000). PET imaging in oncology. *Semin. Nucl. Med.* **30**, 150–185.

Becker, W. (1995). The contribution of nuclear medicine to the patient with infection. *Eur. J. Nucl. Med.* **22**, 1195–1211.

Bleeker-Rovers, C. P., De Kleijn, E. M., Corstens, F. H., Van Der Meer, J. W. and Oyen, W. J. (2004). Clinical value of FDG PET in patients with fever of unknown origin and patients suspected of focal infection or inflammation. *Eur. J. Nucl. Med. Mol. Imaging* **31**, 29–37.

Blockmans, D., Knockaert, D., Maes, A., De Caestecker, J., Stroobants, S., Bobbaers, H. and Mortelmans, L. (2001). Clinical value of [18F]fluoro-deoxyglucose positron emission tomography for patients with fever of unknown origin. *Clin. Infect. Dis.* **32**, 191–196.

Brown, R. S., Leung, J. Y., Fisher, S. J., Frey, K. A., Ethier, S. P. and Wahl, R. L. (1995). Intratumoral distribution of tritiated fluorodeoxyglucose in breast carcinoma: are inflammatory cells important? *J. Nucl. Med.* **36**, 1854–1861.

Chacko, T. K., Zhuang, H., Stevenson, K., Moussavian, B. and Alavi, A. (2002). The importance of the location of fluorodeoxyglucose uptake in periprosthetic infection in painful hip prostheses. *Nucl. Med. Commun.* **23**, 851–855.

Champsaur, P., Parlier-Cuau, C., Juhan, V., Daumen-Legre, V., Chagnaud, C., Lafforgue, P., Laredo, J. D. and Kasbarian, M. (2000). Differential diagnosis in infective spondylodiscitis and erosive degenerative disk disease. *J. Radiol.* **81**, 516–522.

Chianelli, M., Mather, S. J., Martin-Comin, J. and Signore, A. (1997). Radiopharmaceuticals for the study of inflammatory processes: a review. *Nucl. Med. Commun.* **18**, 437–455.

Coleman, R. E. (1998). Clinical PET in oncology. *Clin. Positron Imaging* **1**, 15–30.

Crim, J. T. and Seeger, L. L. (1994). Imaging evaluation of osteomyelitis. *Crit. Rev. Diagn. Imaging* **35**, 201–256.

Datz, F. L. (1994). Indium-111-labeled leukocytes for the detection of infection: current status. *Semin. Nucl. Med.* **24**, 92–109.

De Kleijn, E. M., Vandenbroucke, J. P. and Van Der Meer, J. W. (1997). Fever of unknown origin. I. A prospective multi-center study of 167 patients with FUO, using fixed epidemiologic entry criteria. *Medicine* **76**, 392–400.

Delbeke, D. (1999). Oncological applications of FDG PET imaging. *J. Nucl. Med.* **40**, 1706–1715.

De Winter, F., van de Wiele, C., Vogelaers, D., de Smet, K., Verdonk, R. and Dierckx, R. A. (2001). F-18 fluorodeoxyglucose positron emission tomography: a highly accurate imaging modality for the diagnosis of chronic musculoskeletal infections. *Am. J. Bone Joint Surg.* **83A**, 651–660.

Durack, D. T. and Street, A. C. (1991). Fever of unknown origin – reexamined and redefined. *Curr. Clin. Top. Infect. Dis.* **11**, 35–51.

Erdman, W. A., Tamburro, F., Jayson, H. T., Weatherall, P. T., Ferry, K. B. and Peshock, R. M. (1991). Osteomyelitis: characteristics and pitfalls of diagnosis with MR imaging. *Radiology* **180**, 533–539.

Even-Sapir, E. and Martin, R. H. (1994). Degenerative disc disease: a cause for diagnostic dilemma on In-111 WBC studies in suspected osteomyelitis. *Clin. Nucl. Med.* **19**, 388–392.

Gelfand, J. A. (2001). Fever of unknown origin. In *Harrison's Principles of Internal Medicine* (E. Braunwald, A. S. Fauci, D. L. Kaspen, S. L. Hausen, D. L. Longo and J. L. Jameson, eds), pp. 804–809. McGraw-Hill, New York.

Guhlmann, A., Brecht-Krauss, D., Suger, G., Glatting, G., Kotzerke, J., Kinzl, L. and Reske, S. N. (1998). Fluorine-18-FDG PET and technetium-99m antigranulocyte antibody scintigraphy in chronic osteomyelitis. *J. Nucl. Med.* **39**, 2145–2152.

Hoffman, J. M., Waskin, H. A., Schifter, T., Hanson, M. W., Gray, L., Rosenfeld, S. and Coleman, R. E. (1993). FDG-PET in differentiating lymphoma from nonmalignant central nervous system lesions in patients with AIDS. *J. Nucl. Med.* **34**, 567–575.

Hunter, G. A., Welsh, R. P., Cameron, H. U. and Bailey, W. H. (1979). The results of revision of total hip arthroplasty. *Br. J. Bone Joint Surg.* **61B**, 419–421.

Kaim, A., Maurer, T., Ochsner, P., Jundt, G., Kirsch, E. and Mueller-Brand, J. (1997). Chronic complicated osteomyelitis of the appendicular skeleton diagnosed with technetium-99m labelled monoclonal antigranulocyte antibody-immunoscintigraphy. *Eur. J. Nucl. Med.* **24**, 732–778.

Kaiser, J. A. and Holland, B. A. (1998). Imaging of the cervical spine. *Spine* **23**, 2701–2712.

Kälicke, T., Schmitz, A., Risse, J. H., Arens, S., Keller, E., Hansis, M., Schmitt, O., Biersack, H. J. and Grunwald, F. (2000). Fluorine-18 fluorodeoxyglucose PET in infectious bone diseases: results of histologically confirmed cases. *Eur. J. Nucl. Med.* **27**, 524–528.

Kjaer, A., Lebech, A. M., Eigtved, A. and Hojgaard, L. (2004). Fever of unknown origin: prospective comparison of diagnostic value of (18)F-FDG PET and (111)In-granulocyte scintigraphy. *Eur. J. Nucl. Med. Mol. Imaging* Jan 17 (Epub ahead of print).

Knockaert, D. C., Mortelmans, L. A., De Roo, M. C. and Bobbaers, H. J. (1994). Clinical value of gallium-67 scintigraphy in evaluation of fever of unknown origin. *Clin. Infect. Dis.* **18**, 601–605.

Kritz, F. L. (1999). PET scanning moves into community hospitals. *J. Nucl. Med.* **40**, 11N–12N.

Krznaric, E., De Roo, M. D., Verbruggen, A., Stuyck, J. and Mortelmans, L. (1996). Chronic osteomyelitis: diagnosis with technetium-99m-D,L-hexamethylpropylene amine oxime labeled leukocytes. *Eur. J. Nucl. Med.* **23**, 792–797.

Kubota, R., Yamada, S., Kubota, K., Ishiwata, K., Tamahashi, N. and Ido, T. (1992). Intratumoral distribution of fluorine-18-fluorodeoxyglucose in vivo: high accumulation in macrophages and granulation tissues studied by microautoradiography. *J. Nucl. Med.* **33**, 1972–1980.

Ledermann, H. P., Kaim, A., Bongartz, G. and Steinbrich, W. (2000). Pitfalls and limitations of magnetic resonance imaging in chronic posttraumatic osteomyelitis. *Eur. Radiol.* **10**, 1815–1823.

Lew, D. P. and Waldvogel, F. A. (1997). Osteomyelitis. *N. Engl. J. Med.* **336**, 999–1007.

Lorenzen, J., Buchert, R. and Bohuslavizki, K. H. (2001). Value of FDG PET in patients with fever of unknown origin. *Nucl. Med. Commun.* **22**, 779–783.

Maderazo, E. G., Judson, S. and Pasternak, H. (1988). Later infections of total joint prostheses: a review and recommendation for prevention. *Clin. Orthop.* **229**, 131–142.

Manthey, N., Reinhard, P., Moog, F., Knesewitsch, P., Hahn, K. and Tatsch, K. (2002). The use of [18F]fluorodeoxyglucose positron emission tomography to differentiate between synovitis, loosening and infection of hip and knee prostheses. *Nucl. Med. Commun.* **23**, 645–653.

Meyer, M., Gast, T., Raja, S. and Hubner, K. (1994). Increased F-18 FDG accumulation in an old fracture. *Clin. Nucl. Med.* **19**, 13–14.

O'Doherty, M. J., Barrington, S. F., Campbell, M., Lowe, J. and Bradbeer, C. S. (1997). PET scanning and the human immunodeficiency virus-positive patient. *J. Nucl. Med.* **38**, 1575–1583.

Palestro, C. J. (1994). The current role of gallium imaging in infection. *Semin. Nucl. Med.* **24**, 128–141.

Palestro, C. J. and Torres, M. A. (1997). Radionuclide imaging in orthopaedic infections. *Semin. Nucl. Med.* **27**, 33433–33435.

Palestro, C. J., Kim, C. K., Swyer, A. J., Vallabhajosula, S. and Goldsmith, S. J. (1991). Radionuclide diagnosis of vertebral osteomyelitis: indium-111-leukocyte and technetium-99m-methylene diphosphonate bone scintigraphy. *J. Nucl. Med.* **32**, 1861–1865.

Perry, M. (1996). Erythrocyte sedimentation rate and C reactive protein in the assessment of suspected bone infection: are they reliable indices? *J. R. Coll. Surg. Edinb.* **41**, 116–118.

Peters, A. M. (1998). The use of nuclear medicine in infections. *Br. J. Radiol.* **71**, 252–261.

Sanzen, L. and Sundberg, M. (1997). Periprosthetic low-grade hip infections. Erythrocyte sedimentation rate and C-reactive protein in 23 cases. *Acta Orthop. Scand.* **68**, 461–465.

Seabold, J. E. and Nepola, J. V. (1999). Imaging techniques for the evaluation of postoperative orthopedic infections. *Q. J. Nucl. Med.* **43**, 21–28.

Seabold, J. E., Palestro, C. J., Brown, M. L., Datz, F. L., Forstrom, L. A., Greenspan, B. S., McAfee, J. G., Schauwecker, D. S. and Royal, H. D. (1997). Procedure guidelines for gallium scintigraphy in inflammation. Society of Nuclear Medicine. *J. Nucl. Med.* **38**, 994–997.

Segreti, J., Nelson, J. A. and Trenholme, G. M. (1998). Prolonged suppressive antibiotic therapy for infected orthopaedic prostheses. *Clin. Infect. Dis.* **27**, 711–713.

Shih, L. Y., Wu, J. J. and Yang, D. J. (1987). Erythrocyte sedimentation rate and C-reactive protein values in patients with total hip arthroplasty. *Clin. Orthop.* **225**, 238–246.

Skehan, S. J., Issenman, R., Mernagh, J., Nahmias, C. and Jacobson, K. (1999). 18F-fluorodeoxyglucose positron emission tomography in diagnosis of paediatric inflammatory bowel disease. *Lancet* **354**, 836–837.

Spangehl, M. J., Younger, A. S. E., Masri, B. A. and Duncan, C. P. (1998). Diagnosis of infection following total hip arthroplasty. *Am. J. Bone Joint Surg.* **79A**, 1578–1588.

Stabler, A., Baur, A., Kruger, A., Weiss, M., Helmberger, T. and Reiser, M. (1998). Differential diagnosis of erosive osteochondrosis and bacterial spondylitis: magnetic resonance tomography. *Rofo Fortschr. Geb. Rontgenstr. Neuen Bildgeb. Verfahr.* **168**, 421–428.

Stumpe, K. D., Dazzi, H., Schaffner, A. and von Schulthess, G. K. (2000). Infection imaging using whole-body FDG-PET. *Eur. J. Nucl. Med.* **27**, 822–832.

Van Acker, F., Nuyts, J., Maes, A., Vanquickenborne, B., Stuyck, J., Bellemans, J., Vleugels, S., Bormans, G. and Mortelmans, L. (2001). FDG-PET, 99mTc-HMPAO white blood cell SPET and bone scintigraphy in the evaluation of painful total knee arthroplasties. *Eur. J. Nucl. Med.* **28**, 1496–1504.

Vanquickenborne, B., Maes, A., Nuyts, J., Van Acker, F., Stuyck, J., Mulier, M., Verbruggen, A. and Mortelmans, L. (2003). The value of (18)FDG-PET for the detection of infected hip prosthesis. *Eur. J. Nucl. Med. Mol. Imaging* **30**, 705–715.

Villringer, K., Jager, H., Dichgans, M., Ziegler, S., Poppinger, J., Herz, M., Kruschke, C., Minoshima, S., Pfister, H. W. and Schwaiger, M. (1995). Differential diagnosis of CNS lesions in AIDS patients by FDG-PET. *J. Comput. Assist. Tomogr.* **19**, 532–536.

Zhuang, H., Duarte, P. S., Pourdehand, M., Shnier, D. and Alavi, A. (2000). Exclusion of chronic osteomyelitis with F-18 fluorodeoxyglucose positron emission tomography. *Clin. Nucl. Med.* **25**, 281–284.

Zhuang, H., Duarte, P. S., Pourdehnad, M., Maes, A., Van Acker, F., Shnier, D., Garino, J. P., Fitzgerald, R. H. and Alavi, A. (2001). The promising role of ^{18}F-FDG PET in detecting infected lower limb prosthesis implants. *J. Nucl. Med.* **42**, 44–48.

Zimmerli, W. (1995). Role of antibiotics in the treatment of infected joint prosthesis. *Orthopade* **24**, 308–313.

8 Biosensor Characterization of Structure–Function Relationships in Viral Proteins

L Choulier, D Altschuh, G Zeder-Lutz and M H V Van Regenmortel

UMR7100, CNRS, Ecole Supérieure de Biotechnologie de Strasbourg, Boulevard Sébastien Brandt, BP 10413, 67412 Illkirch Cedex, France

◆◆

CONTENTS

Introduction
Description of BIACORE instrument
Mapping of viral epitopes
Use of biosensors for selecting probes suitable for viral diagnosis
Establishing correlations between the structure and binding activity of viral antigens
Prediction of antigen–antibody interaction kinetics
Engineering of antibody fragments

◆◆◆◆◆◆ INTRODUCTION

In order to understand how biosensors can be used to analyse structure–function relationships in viral proteins, it is helpful to first examine what is meant by a structure–function relationship.

What is the Connection Between the Structure and Function of a Protein?

Which aspect of the structure of a protein is relevant when analysing a structure–function relationship is rarely made clear. Since, the secondary and tertiary structures of proteins are always better conserved than their sequence, it is frequently found that members of a protein family that possess not more than 20% sequence identity have three-dimensional structures that are practically superimposable. Only a small number of residues tend to be highly conserved, either because they are critical for maintaining the protein-fold or are essential for functional activity. During evolution, proteins may lose some functions and acquire new ones and the residues implicated in a function will thus not necessarily be conserved even if the protein-fold remains the same. Conservation of protein-fold is thus not necessarily correlated with retention of function, and a link between structure and function may only be apparent if attention is restricted to the functional binding site region instead of the whole protein molecule (Van Regenmortel, 2002a). Since proteins are able, on average, to interact specifically with as many as five different partners through a variety of binding sites, it is also necessary to identify which site is relevant for the function that is being analysed.

Another difficulty in analysing correlations between protein structure and function lies in the fact that structure should be thought of as a dynamic rather than a static property. The tertiary structure of a protein is not determined solely by its sequence, since it is also influenced by the physico-chemical environment. Even the quaternary structure of a virus particle should not be viewed as a static configuration, since small changes in pH and temperature can lead to a rearrangement of parts of the viral subunits forming the viral capsid. As a result, regions of the subunits that appear in the X-ray crystallographic structure to be buried may become transiently exposed to antibodies, cellular receptors and proteases (Roivanen et al., 1993; Bothner et al., 1998). When these dynamic features of virus particles are not taken into account, it is impossible to account for some of the observed interactions between virions and various biomolecules.

A further difficulty lies in the ambiguous meaning of the term function itself. This term is used in various ways and a possible correlation with structure will depend on which aspects of function and of biological organization are being considered (Van Regenmortel, 2001; Brun et al., 2004). Biochemists tend to regard the function of a protein simply as what the molecule does and how it acts, i.e. its functioning or activity at the molecular level (Murzin and Patthy, 1999). In this restrictive sense of function, the only activity that is considered is binding activity, and function is then taken to be synonymous with binding. However, functions can also be defined at the level of the cell or the organism (Bork et al., 1998) in which case they acquire a meaning only with respect to the biological system as a whole, for instance by contributing to its health, survival or reproduction. In a similar way, the activity of an enzyme such as trypsin can be analysed purely at the chemical level, i.e. in terms of which peptide bonds in a protein are cleaved. However, the biological function of trypsin emerges only at the cellular and physiological levels through its participation in protein degradation and digestion processes. In the present chapter, the only protein function that will be considered is binding activity and our analysis of the relationship between structure and function will be restricted to the relations that exist between the atomic structure of binding sites and their binding activity. In spite of the prevailing paradigm *structure determines function*, it is important to realize that there is no unique causal link between structure and function. As discussed elsewhere (Van Regenmortel, 2001, 2002b), the structure of the binding surfaces of two interacting molecules should not be interpreted as the cause of the binding reaction, since causal relations are relations between successive events and not between two material objects or between a structure and a binding event. Instead of searching for a non-existing causal link between structure and function, investigators should analyse the multiplicity of factors that influence binding activity.

Biosensors are Useful for Analysing Correlations Between Structure and Binding Activity

In recent years, the development of biosensor instruments based on the optical phenomenon of surface plasmon resonance (SPR) has made it much easier to measure the binding activity of proteins. These instruments, which measure the binding between a ligand immobilized on a sensor chip and an analyte introduced in a flow passing over the sensor surface, make it possible to visualize the binding process on a computer screen as a function of time. The binding of the analyte is followed by the increase in refractive index caused by the mass of bound species. None of the reactants are labelled which avoids the artefactual changes in binding properties that usually occur when molecules are labelled. Since the interaction is measured in real time, it is possible to determine both kinetic rate constants and equilibrium affinity constants (Granzow, 1994; Morton and Myszka, 1998). Biosensors have become the method of choice for measuring the binding characteristics of viral proteins (Van Regenmortel *et al.*, 1994; Rich and Myszka, 2003a) and for establishing correlations between the structure and binding activity of proteins (Van Regenmortel, 2001).

◆◆◆◆◆◆ DESCRIPTION OF BIACORE INSTRUMENT

The most widely used biosensor instrument is the BIACORE developed by Biacore AB (Uppsala, Sweden) which exists in several versions and which has been used in about 90% of the studies published so far (Myszka, 1999a; Rich and Myszka, 2000, 2001, 2002, 2003b). The instrument consists of an optical detector system, exchangeable sensor chips, a processing unit and a personal computer for control and evaluation. The processing unit contains the SPR monitor, an integrated microfluidic cartridge and an autosampler for dispensing samples automatically.

The sensor chip is a glass slide coated on one side with a gold film which is covered with a 100 nm dextran layer. The dextran is usually carboxymethylated which allows the immobilization of molecules containing primary amines. In addition, other immobilization strategies like aldehyde, hydrazide group and sulfhydryl group coupling and chelate linkage of oligohistidine tags can also be used (Johnsson *et al.*, 1995).

The hydrophilic dextran matrix provides a flexible anchor for ligand immobilization and allows interactions to occur very much like in solution. Although it has been suggested that kinetic rate constants could be affected by the slow diffusion of analyte into the dextran matrix, there is experimental evidence that identical kinetics are observed when ligands are immobilized to surfaces in the presence or absence of a dextran layer (Karlsson and Falt, 1997). It has been shown that the use of an immobilized ligand in the BIACORE, under appropriate experimental conditions, leads to equilibrium binding constants that do not differ significantly from those measured in solution by titration calorimetry or analytical ultracentrifugation (Kortt *et al.*, 1999; Rich and Myszka, 2000; Day *et al.*, 2002). The range of kinetic parameters that can be measured reliably is 10^3–10^7 M^{-1} s^{-1} for the on-rate and 10^{-5}–10^{-1} s^{-1} for the off-rate (Karlsson, 1999). Since the temperature can be finely controlled in the range 4–40°C, it is possible to derive thermodynamic parameters for binding interactions from kinetic and affinity measurements (Roos *et al.*, 1998).

Four independent flow cells are present on each sensor chip. Normally one of the cells is used as a reference surface, which must be submitted to the same activation and deactivation steps as the reaction surface. This reference surface which is used to check for bulk refractive index changes, non-specific binding, detector drift and injection noise is essential for collecting high quality kinetic data (Myszka, 1999b).

Changes in SPR signal corresponding to the mass of bound analyte are monitored continuously and are visualized on the computer screen as a plot of resonance units (RU) versus time, known as a sensorgram (Figure 8.1). For proteins, a signal of 1 RU corresponds to a surface concentration change of 1 pg mm^{-2}. After each analysis, the sensor chip can be regenerated by introducing a small volume of a suitable dissociating agent which removes the analyte from the immobilized ligand. Chips can withstand high salt concentrations, extremes of pH and

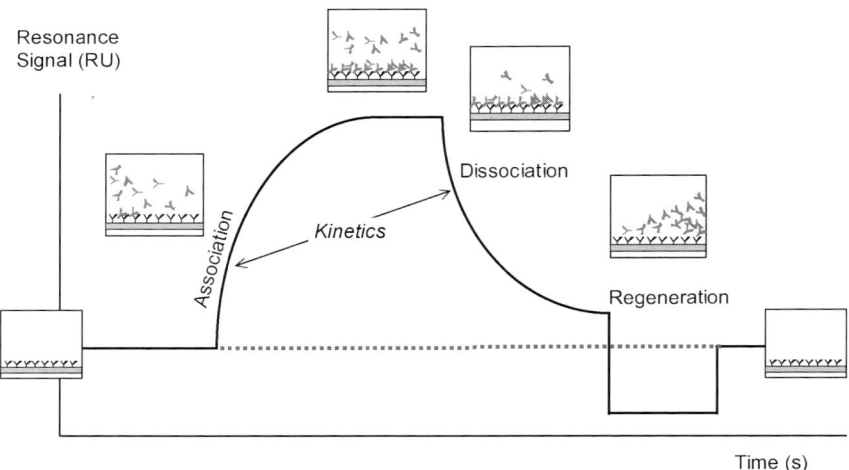

Figure 8.1. Sensorgram presenting a biomolecular interaction as a function of time. The baseline corresponds to the flow of buffer over the immobilized ligand. When the analyte is injected, the signal increases during the association phase as analyte molecules bind to the ligand. The dissociation phase starts when buffer is injected instead of analyte.

the use of organic solvents. Usually about 100 analytical cycles can be carried out on the same ligand surface.

In order to obtain reliable kinetic data, low concentrations of immobilized ligand must be used and a number of possible system-dependent artefacts must be avoided. Artefacts may arise from mass transport effects (Myszka et al., 1997), aggregation, heterogeneity induced by surface immobilization of the ligand, avidity and matrix effects as well as non-specific binding (Myszka, 1999b). Binding data should be analysed using global curve fitting and numerical integration and a theoretical best-fit curve to the primary data should be presented to show that an appropriate interaction model was used to interpret the data (Karlsson and Roos, 1997; Morton and Myszka, 1998).

◆◆◆◆◆◆ MAPPING OF VIRAL EPITOPES

The antigenic reactivity of viruses resides in restricted parts of the virion surface, known as antigenic determinants or epitopes, which are recognized by the binding sites or paratopes of antibody molecules. The particles of small viruses consist of nucleocapsids built up of a layer of protein subunits enclosing the nucleic acid, whereas larger viruses have an additional envelope which contains glycoprotein spikes. The regions of the viral protein recognized by the antibody are usually described in terms of amino acid residues although it is at the level of atomic interactions that the recognition takes place. When it is said that a residue of the epitope makes contact with a residue of the paratope, it is of course only a small number of the atoms of each residue that participate in the interaction.

Continuous and Discontinuous Epitopes

Since, antigenic reactivity is usually ascribed to particular residues in a protein, this has led to the classification of epitopes as either continuous or discontinuous, depending on whether the residues involved in the epitopes are contiguous in the polypeptide chain or not. Continuous epitopes, also referred to as linear or sequential epitopes, correspond to short peptide fragments of the protein that can bind to antibodies raised against the intact protein. Discontinuous epitopes, on the other hand, are made up of residues that are brought into spatial proximity by the folding of the peptide chain. The distinction between continuous and discontinuous epitopes is a fuzzy one, since a short peptide recognized by an antibody may be called a continuous epitope even if it corresponds in reality to a continuous region within a discontinuous epitope (Van Regenmortel, 1998). In general, about 90% of the monoclonal antibodies raised against intact proteins are considered to be directed against discontinuous epitopes because they do not react with any peptide fragment of the parent protein. When a protein is fragmented into peptides, the residues that make up the different discontinuous epitopes are mostly scattered on individual peptides. In rare instances will a linear peptide fragment contain a sufficient number of residues of the original discontinuous epitope to enable it to bind to antibodies raised against the intact protein.

Nowadays, continuous epitopes of proteins are frequently identified by analysing the antigenic reactivity of sets of overlapping synthetic peptides encompassing the entire sequence of the protein antigen (Rodda and Tribbick, 1996). Another approach consists in testing the antigenic reactivity of sets of peptides obtained from a combinatorial library. If a peptide in such a library is found to bind to an antiprotein monoclonal antibody and the same or very similar peptide sequence is present in the protein immunogen used to raise the antibody, the peptide will be considered as a continuous epitope of the protein. On the other hand, if the sequence of the antigenically active peptide is not present in the immunogen used to raise the antibody, the peptide is said to correspond to a mimotope, i.e. a peptide that is believed to mimic either a discontinuous epitope of the protein or a continuous epitope showing little or no sequence similarity with the active peptide. The term mimotope was coined by Geysen *et al.* (1986) and was originally defined as a peptide capable of binding to the paratope of an antibody but unrelated in sequence to the epitope used to induce the antibody. Subsequently, the term mimotope has been used more broadly and is now applied to any mimic of an epitope. The reason why two peptides showing little or no sequence similarity are sometimes able to react with the same antibody is due to the fact that only a minority of the atoms of any residue actually participate in the interaction. In addition, if atoms of the main chain rather than of the side chain of a residue interact with the antibody, there need be little sequence similarity in two cross-reacting peptides.

Cryptotopes, Neotopes and Metatopes

It has been known for many years that the epitopes present on intact virus particles differ from those present on the dissociated viral coat protein subunits. Since, the state of aggregation of protein subunits influences their antigenic properties, a number of terms were introduced to distinguish between the epitopes found at the surface of virions and dissociated subunits (Van Regenmortel, 1966). When subunits polymerize, a portion of the protein surface is buried and this leads to the disappearance of certain epitopes called cryptotopes (Jerne, 1960). Such epitopes become accessible to antibodies only after virus particles have dissociated into monomeric coat protein subunits. On the other hand, new epitopes known as neotopes arise in the polymerized viral subunits as a result of the quaternary structure. Neotopes may arise as a result of conformational changes in the monomers induced by intersubunit bonds or by the juxtaposition of residues from neighbouring subunits that are recognized as a single entity by the antibody. In practice, the label neotope is given to any epitope of the virus that is absent in the monomeric

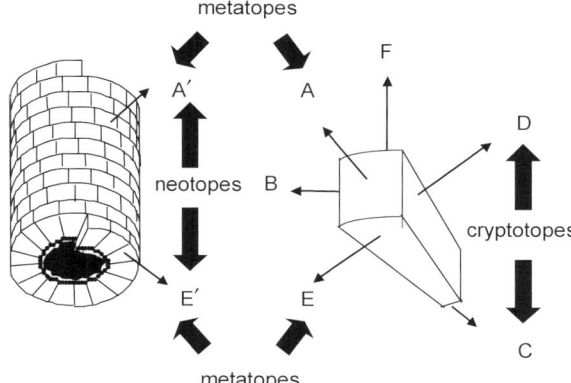

Figure 8.2. Schematic model of the protein subunits of TMV in monomeric form and in the virus particle. An intact TMV particle has a length of 300 nm and comprises 2130 protein subunits. Neotopes are found on surfaces A' and E'. Metatopes are found on surfaces A, A', E and E'. The surface E' represents about 1.5% of the total accessible surface of a virus peptide. Cryptotopes are found on surfaces B, C and D. The type of epitope present on surface F has not been defined. The E' extremity of the particle contains the 5' end of the RNA (From Saunal and Van Regenmortel, 1995a, with permission.).

subunits forming the viral capsid. A third term, metatope, was introduced for epitopes found in both dissociated and polymerized forms of the viral protein.

Initially, the existence of antibodies specific for neotopes, cryptotopes and metatopes was established by antiserum cross-absorption experiments (Van Regenmortel, 1966) but nowadays these different types of epitopes are more easily identified by means of monoclonal antibodies (Mabs). The location of neotopes, metatopes and cryptotopes on the surface of monomeric protein subunits and assembled particles of tobacco mosaic virus (TMV) is shown in Figure 8.2. Information on the location of a limited number of neotopes and metatopes on TMV particles was initially obtained by immunoelectron microscopy (Dore *et al.*, 1988) but subsequent biosensor mapping experiments showed that neotopes and metatopes were present on parts of the viral surface (surface E' and A' respectively), where they had not been found by electron microscopy (Saunal and Van Regenmortel, 1995a).

Mapping of Viral Epitopes with Biosensors

The mapping of epitopes on the surface of viral proteins is commonly done by double-sandwich enzyme-linked immunosorbent assays (ELISA) using all possible pairwise combinations of a panel of Mabs. In such assays, one Mab is first adsorbed to a well of a microtitre plate, the antigen is then allowed to bind to the Mab and a second labelled Mab is tested for its ability to bind the antigen. If the second Mab binds, it is concluded that the two Mabs recognize distinct, non-overlapping epitopes. Epitope mapping with a biosensor follows essentially the same steps except that no labelling of the second Mab is required. The binding of each successive unlabelled reactant is visualized in real time, whereas in ELISA it is only the binding of the second, labelled Mab that is detected. In the absence of a signal in ELISA it is impossible to infer which binding step failed to occur and each component of the system must be analysed separately to discover the one that is at fault. Another advantage of using the BIACORE for epitope mapping is that each antigen–antibody binding event can be quantified and that the stoichiometry can be calculated as a molar ratio (MR) of the reactants, using the following relationship:

$$MR = (RU_{Mab}/RU_{Ag})(MW_{Ag}/MW_{Mab}).$$

Expressing the data as MR has the advantage that SPR signals can be normalized (Daiss and Scalice, 1994). From the binding stoichiometry observed with different Mabs specific for TMV, it was possible, for instance, to establish whether Mabs recognized the E' extremity of the particle (1.5% of the total surface; Figure 8.2) or the entire length of virions (surface A' constituting 97% of the surface). Such an experiment is illustrated in Figure 8.3 which shows the binding of two anti-metatope Mabs to TMV particles captured by anti-neotope Mabs. The MR of bound Mab 25/TMV was found to be about 10, indicating that this antibody recognized only one end of the virion (surface E', Figure 8.2). The MR of Mab 5V/TMV was about 300, indicating that this antibody recognized the surface A' of the particle (Saunal and Van Regenmortel, 1995a). When anti-neotope Mabs were analysed in the same way, some of the antibodies gave high MRs of bound Mab/TMV indicating that they were binding to surface A'. Other anti-neotope Mabs gave MRs of 5–15 even when a considerable excess of Mab was used, indicating that they were binding to surface E' (Saunal and Van Regenmortel, 1995a).

The first successful biosensor mapping of viral epitopes was carried out with the core protein p24 of human immunodeficiency virus (HIV) using a panel of 30 Mabs (Fägerstam et al., 1990). By testing all possible pairwise combinations of Mabs an epitope map was constructed corresponding to a two-dimensional map of circles representing the footsteps of antibody molecules. A total of 17 groups of Mabs showing different reaction patterns were identified. By sequential injection of different Mabs that recognize non-overlapping sites, it was possible to concurrently bind five Mabs to one p24 molecule (Fägerstam et al., 1990; Malmqvist, 1996). An approximate location of the epitopes on the surface of the p24 molecule was achieved by inhibiting the binding of different Mabs with synthetic peptides corresponding to discrete regions of the p24 molecule.

Different results are sometimes observed when the order in which pairs of Mabs are added is changed. It is therefore necessary in the case of a negative result, to test the pair of Mabs also in the reverse order (Fägerstam et al., 1990). Non-reciprocal reaction patterns may arise because of large differences in the kinetic constants of the two Mabs or because the binding of the first Mab induces a conformational change in

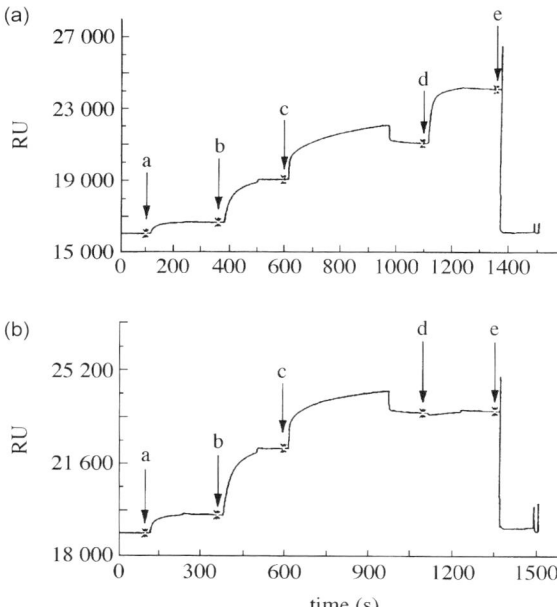

Figure 8.3. Sensorgrams used for determining the location of neotopes and metatopes on TMV particles. Phase (a) corresponds to the injection of capturing antineotope Mab 253P. About 700 RU of Mab were bound. Phase (b) corresponds to the injection of TMV (about 2500 RU). Phase (c) corresponds to the saturation of rabbit anti-mouse globulins by the addition of non-specific ascitic fluid (1500–2500 RU bound). In the top panel phase (d) corresponds to the binding of Mab 5V. The molar ratio of reactants showed that about 300 antibody molecules were bound per virus particle. In the bottom panel B, phase (d) corresponds to the binding of Mab 25P to one extremity of the viral rods. About 10 antibody molecules were bound per virus particle. Phase (e) corresponds to the regeneration of the sensor surface with a pulse of 100 mM HCL (From Saunal and Van Regenmortel, 1995a, with permission.).

the antigen which prevents it from being recognized by the second Mab (Daiss and Scalice, 1994; Saunal and Van Regenmortel, 1995b). A necessary control in pairwise mapping experiments is to use the same Mab both as first and second antibody. If the same antibody is able to bind twice to the antigen molecule, it indicates the presence of oligomers, a frequent occurrence with viral subunits that are prone to self-aggregation. If this aggregation cannot be controlled, epitope mapping is not feasible.

Mapping of the Epitopes of TMV Protein

A number of cryptotopes of TMV protein (TMVP) could be located on the surface of the protein because many anti-cryptotope antibodies are able to react with peptide fragments of TMVP. This kind of epitope mapping was not feasible with neotopes and metatopes because none of the corresponding Mabs reacted with linear peptide fragments of TMVP (Van Regenmortel, 1999). However, using immunoelectron microscopy with gold-labelled antibodies, several metatopes could be located on the extremity of the virus particle that contains the 5' end of the RNA (surface E' in Figure 8.2) while neotopes were found along the entire length (surface A') of the particle (Dore et al., 1988). Since the anti-metatope antibodies recognized the extremity of the virus particles that is known to disassemble first during the infection process (Wilson, 1984), experiments were done to assess whether these antibodies were able to block the disassembly of virions and the translation of the viral RNA. It was found that about half of the anti-metatope antibodies that were analysed strongly inhibited disassembly and RNA translation while the others inhibited only weakly or not at all (Saunal et al., 1993). This difference in functional activity of various anti-metatope antibodies was explained by subsequent BIACORE mapping experiments (Saunal and Van Regenmortel, 1995b).

Earlier ELISA experiments had given an erroneous localization of some of the epitopes because the double-sandwich ELISA format that was used could not exclude the possibility that the conformation of the viral protein was altered by the initial binding of the capturing antibody (Dekker et al., 1989; Van Regenmortel, 1999). Such a conformational change may allow the second Mab to recognize captured TMVP, although it would have been unable to recognize free TMVP in solution. Such erroneous interpretations arise because it is not possible with ELISA to visualize each of the successive binding steps in the assay, as is feasible in biosensor experiments. Using BIACORE, it was indeed possible to show that TMVP captured by a first anti-metatope Mab underwent a conformational change which then allowed it to be recognized by an anti-neotope Mab (Dubs et al., 1992; Saunal and Van Regenmortel, 1995b). The induction of the neotope conformation in monomeric viral subunits following binding of the subunit to a first antibody apparently mimics at least partly the conformational change that occurs in the subunits when they polymerize into virions. However, not all anti-neotope Mabs become capable of recognizing dissociated TMVP after it has reacted with a first anti-metatope antibody (Saunal and Van Regenmortel, 1995b). In such cases, the anti-neotope antibodies probably recognize an epitope formed by the juxtaposition of residues from two neighbouring subunits.

Two-site binding assays with BIACORE produced the epitope map of TMVP shown in Figure 8.4. The combining sites of the antibodies were assumed to cover a surface of 600 Å^2 in a circular footprint (Braden and Poljak, 1995). The presence of metatopes also on surface A of the viral subunit increased the range of two-site binding assays that could be performed with TMVP monomers, since it was possible to present the antigen in two different orientations using capturing anti-metatope Mabs specific for either surface A or E. A number of Mabs recognizing the larger surface E were able to bind simultaneously to the same viral subunit whereas surface A was unable to bind more than one Mab molecule (Figure 8.4).

The TMVP epitope map achieved with BIACORE was useful for understanding the mechanism by which certain antibodies are able to inhibit the co-translational disassembly of TMV by ribosomes. No

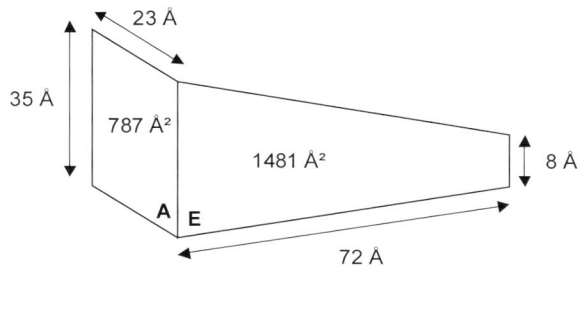
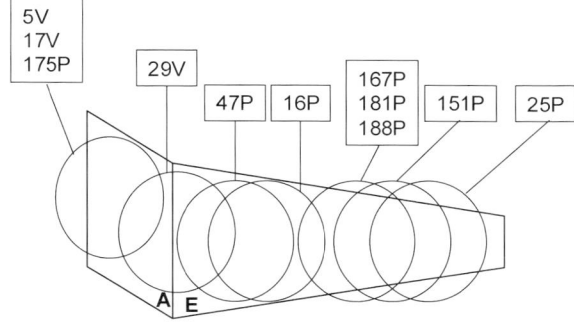

Figure 8.4. Schematic representation of the coat protein subunit of TMV, showing surfaces A and E. The epitope map was constructed from pairwise interaction data obtained with BIACORE. Three clusters of epitopes can be distinguished which are recognized, respectively, by Mabs 5V, 17V and 175P, Mabs 16P and 47P and Mabs 25P, 151P, 167P, 181P and 188P. Mab 29V overlaps two of the epitope clusters (From Saunal and Van Regenmortel, 1995b, with permission.).

correlation was observed between the affinity of the antibodies measured with BIACORE and their inhibitory capacity. On the other hand, the antibodies (Mabs 25P, 151P, 167P, 181P and 188P, Figure 8.4), which possessed the strongest inhibitory capacity were found to bind to the region of surface E closest to the central axis of polymerized TMVP, which interacts with the viral RNA. It seems likely, therefore, that the inhibitory Mabs act by sterically preventing the interaction between RNA and ribosomes (Saunal and Van Regenmortel, 1995b).

Kinetic Mapping of Viral Epitopes

During the course of two-site binding assays used for epitope mapping, there is continuous dissociation and reassociation of the binding partners. When the kinetic dissociation rate constant (k_{off}) of the first capturing Mab1 is about $10^{-3}\,s^{-1}$, approximately half of the trapped antigen molecules will dissociate during the 10 min injection time of Mab2. As a result, the region on the dissociated antigen molecule previously occupied by Mab1 becomes available to Mab2 and if the two respective epitopes are sufficiently close on the antigen surface, this will prevent reassociation of the Mab2–antigen complex to Mab1, leading to a lower RU response. When such a decrease in RU value is observed independently of the binding sequence of Mabs, it is concluded that the two Mabs recognize overlapping epitopes.

In view of the dynamic competition between different Mabs, this type of epitope mapping has been called kinetic mapping (Saunal and Van Regenmortel, 1995b). If the k_{off} value of Mab1 is faster than $10^{-2}\,s^{-1}$, no meaningful data can be obtained from the experiment, since practically all antigen molecules will have dissociated during the time frame of the experiment.

From the epitope map obtained in these studies it seemed that three Mabs molecules should be able to be accommodated simultaneously on one viral subunit. This was verified by showing, for instance, that TMVP captured by Mab 47P, was still able to accommodate both Mabs 17V and 151P (Figure 8.4).

It is interesting to note that the Mabs specific for surfaces E and E' originated from mice immunized with viral protein (Mab numbers with suffix P) whereas most Mabs specific for surfaces A and A' were obtained by immunization with intact virus (Mabs numbers with suffix V). This finding supports the suggestion of Dore *et al.* (1990) that at the high concentration of TMVP used for immunizing mice, the viral protein was aggregated in the form of bipolar double disks that present mostly surface E' to the immune system.

◆◆◆◆◆◆ USE OF BIOSENSORS FOR SELECTING PROBES SUITABLE FOR VIRAL DIAGNOSIS

Biosensors are useful for selecting diagnostic reagents suitable for viral diagnosis by high throughput immunoassays (Richalet-Sécordel *et al.*, 1996). Mabs intended for use in ELISA should have a sufficiently low k_{off} (lower than $10^{-3}\,s^{-1}$) to prevent them from dissociating during the washing step of the assay. This can be assessed by using a few microlitres of culture supernatant during the early stages of hybridoma cell culture. Diagnostic assays based on double-antibody sandwich ELISA require pairs of Mabs able to bind concurrently to the viral antigen, and biosensors are convenient instruments for selecting such Mabs (Fägerstam *et al.*, 1990; Tosser *et al.*, 1994). Any deleterious effect resulting from the labelling of Mabs can also be easily ascertained.

Although it is possible to immobilize whole virus particles on a sensor chip (Dubs *et al.*, 1991; Saunal and Van Regenmortel, 1995a; Lea *et al.*, 1998; Abad *et al.*, 2002), it is more common to immobilize viral proteins or synthetic peptides corresponding to viral epitopes (Van Cott *et al.*, 1992; Altschuh *et al.*, 1992; Gomes *et al.*, 2002). Conformational changes that occur in viral proteins as a result of oligomerization or following interaction with antibodies or other ligands can be detected by the change in binding properties revealed in a biosensor experiment (Dubs et al., 1992; Glaser and Hausdorf, 1996; Zeder-Lutz et al., 2001; Myszka et al., 2000).

Biosensors are also particularly useful for measuring the active concentration of antigen or antibody molecules when they are present in their native conformation in crude unpurified samples or untreated immune sera (Benito and Van Regenmortel, 1998). The method developed by Karlsson et al. (1994) requires a calibration curve and binding partners with a k_{on} faster than $10^5 \, M^{-1} \, s^{-1}$, whereas the method of Richalet-Sécordel et al. (1997) based on BIACORE measurements made at different flow rates, does not need a calibration curve and is applicable to systems with a k_{on} as low as $10^3 \, M^{-1} \, s^{-1}$. The active concentration of biomolecules measured with a biosensor is nearly always lower than the concentration determined by conventional methods such as spectrophotometry which do not differentiate between native and denatured molecules. In the case of recombinant proteins, it is not rare that the active concentration is less than 10% of the nominal concentration (Zeder-Lutz et al., 1999).

Instead of immobilizing viral antigens on the sensor chip, it is also possible to insert viral epitopes in permissive sites of a carrier protein such as the maltose binding protein of *Escherichia coli* (Benito and Van Regenmortel, 1998; Coëffier et al., 2000). The BIACORE can be used to determine which insertion sites are optimal for allowing the viral epitopes to adopt a conformation recognized by neutralizing Mabs. The corresponding carrier molecule presenting the inserted viral epitopes can then be tested as a vaccine candidate by assessing its ability to elicit neutralizing antibodies (Coëffier et al., 2000).

In order to study the interaction between viral proteins and membrane receptors, an elegant approach consists of incorporating membrane proteins in the lipid envelope of a retrovirus such as murine leukaemia virus (Hoffman et al., 2000). Many cellular membrane proteins can be incorporated into retroviral envelopes during the process of budding from the cell surface and the resulting viral pseudotypes have been shown to present the receptors in their native conformation. This allows receptor–ligand interactions to be analysed with BIACORE in a format that presents the membrane proteins in their native, lipid environment.

Synthetic peptides have been used extensively as diagnostic probes for detecting viral antibodies produced during infection and for raising antipeptide antibodies capable of detecting viral antigens in infected tissues (Leinikki et al., 1993). Biosensors are useful for optimizing the binding activity of synthetic peptides with respect to peptide length, sequence and conformation (Richalet-Secordel et al., 1994, 1996). Various linear and cyclic peptides corresponding to immunodominant epitopes of the gp120 and gp41 proteins of HIV have been optimized for use in different immunoassay formats, for instance as free inhibitor peptides or conjugated peptides (Mani et al., 1994; Richalet-Sécordel et al., 1996). Since the binding parameters observed with these peptides were similar to those observed with recombinant gp120 (Ferrer et al., 1999), the screening of virus isolates is usually done with more easily obtained peptides than with recombinant proteins. Biosensor measurements have shown that the neutralizing capacity of HIV antibodies is usually correlated with their affinity and dissociation rate constants (Van Cott et al., 1994; Richalet-Sécordel et al., 1996).

◆◆◆◆◆◆ ESTABLISHING CORRELATIONS BETWEEN THE STRUCTURE AND BINDING ACTIVITY OF VIRAL ANTIGENS

When searching for correlations between the structure and binding activity of viral proteins, it is customary to modify the proteins by mutagenesis and to assess the effect of mutations on the binding affinity. Biosensors have replaced most other biochemical techniques in this type of study.

When residues of the antigen present at the binding interface are mutated, it is commonly found that only a few of the substitutions lead to an affinity that is reduced by more than 100-fold. This is usually interpreted to mean that only a small fraction of the residues assigned to the epitope on the basis that they make contact with the antibody are actually contributing to the binding free energy. However, perturbations are not energy determinations (DeLano, 2002). The approach may also identify key residues for maintaining the overall geometry of the contact interface (Altschuh, 2002). On the other hand, substitutions of residues that are not in contact at the interface are also frequently found to affect binding affinity, presumably because these substitutions produce structural changes that propagate far beyond the mutated region (Ben Khalifa et al., 2000). Furthermore, multiple substitutions frequently have non-additive or cooperative effects on binding affinity (Mateu et al., 1992; Rauffer-Bruyère et al., 1997).

Studies Using Monoclonal Antibodies

Many studies of viral epitopes utilize panels of synthetic peptides in which amino acid replacements are easily introduced (Gomes et al., 2002). A detailed protocol for the direct kinetic analysis of small peptides interacting with immobilized Mabs has been published (Gomes and Andreu, 2002).

Biosensors are commonly used to study the influence of different types of residue substitutions on the interaction between Mabs and viral peptides. In the case of four anti-peptide Mabs recognizing the region 128–135 of TMVP, kinetic measurements showed that minor side chain replacements (for instance, the removal of a methyl group in mutants I129L and E131D of the peptide) could have a major effect on the dissociation rate of the antibody (Zeder-Lutz et al., 1993). A similar major effect amounting to a three log difference in k_{off} value was observed when a single Tyr residue was replaced by Phe in the light chain of a scFv specific for influenza neuraminidase (Dougan et al., 1998).

Studies Using Recombinant Antibodies

The effect on binding kinetics of the same five residue substitutions, introduced either in peptide 134–151 of TMVP or in the parent protein was also analysed with BIACORE (Choulier et al., 1999). The equilibrium binding affinities ($K_a = k_{on}/k_{off}$) of the wild type peptide–Fab and protein–Fab interactions were found to be similar (around $10^7 \, M^{-1}$). In four out of five cases, the effect on K_a of the substitutions was identical when the epitope was presented in the form of a peptide or as the parent protein, indicating that binding specificity was not drastically affected by epitope presentation (Figure 8.5). However, kinetic rate parameters were 5–10 times larger for the Fab–peptide than for the Fab–protein interaction, suggesting a different binding mechanism. These results demonstrated that the peptides were able to mimic correctly some but not all properties of the protein/Fab-57P interaction.

Recombinant antibody fragments are powerful tools for the detection, identification, purification and neutralization of viruses. In order to use antibody fragments successfully in these various applications, it is necessary to have a good understanding of the parameters that influence their binding properties. Some recombinant antibodies directed to HIV and respiratory syncytial virus have been used as therapeutic agents

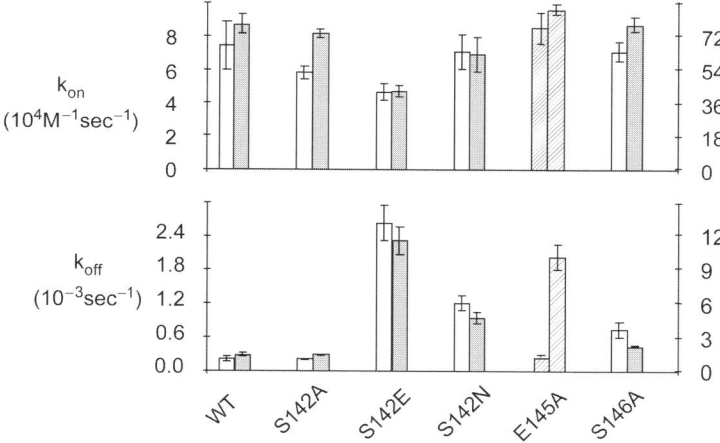

Figure 8.5. Comparison of the effect on k_{on} and k_{off} of identical substitutions in TMV protein (white bars) and peptide 134–151 (grey bars) when the different antigens interact with Fab 57P. The Y-axis on the left and right refer, respectively, to kinetic parameters for protein and peptide antigens. Data obtained with substitution E145A, which has a different effect in the protein and the peptide, are shown by hatched bars (Adapted from Choulier et al., 1999, with permission.).

(Hudson and Souriau, 2003; Presta, 2003). Two recombinant antibody fragments recognizing viral antigens have been studied extensively in our laboratory in an attempt to assess the effects of antigen, antibody and buffer modifications on the binding kinetics. These two antibodies are the Fab 57P, directed against TMVP which interacts with a peptide corresponding to residues 134–146 of the viral coat protein, and the scFv-1F4 which recognizes both the protein E6 of the human papillomavirus 16 and a peptide corresponding to its N-terminal sequence (Giovane et al., 1999).

The phage display (Smith and Scott, 1993) and the multiple peptide synthesis (Frank, 1992) techniques were used to identify peptide sequences recognized by the two recombinant antibodies. Kinetic rates were measured for the interaction between a number of peptide variants and the antibody fragments. The two systems were found to respond differently when various parameters that can influence binding were modified (Choulier et al., 2002a). For the Fab 57P system, a functional epitope comprising four key positions (N_{140}, R_{141}, S_{143} and F_{144}) with well-defined physico-chemical properties was clearly delineated. Adjacent positions only marginally influenced Fab binding (Choulier et al., 2001). In contrast, such a clear-cut epitope delineation of the E6 peptide was not possible with scFv-1F4, since many different epitope sequences possessed significant binding activity (Choulier et al., 2002a). The binding characteristics of the scFv-1F4 epitope may be more appropriate than those of the Fab 57P epitope for designing and engineering new peptidic antigens possessing a higher affinity useful in the efficient purification of recombinant antibodies.

◆◆◆◆◆◆ PREDICTION OF ANTIGEN–ANTIBODY INTERACTION KINETICS

Attempts to predict the kinetics of peptide–antibody interactions have been made using a multivariate quantitative structure–activity relationship (QSAR) approach that involves modifications in peptide sequence and buffer composition (Hansch and Fujita, 1964). The QSAR approach establishes mathematical models that relate variations in physico-chemical properties of residues (i.e. hydrophobicity, size and electronic properties) with variations in binding activity and kinetics. The models are used first to identify which properties of the peptides influence binding and then to predict what the binding will be when various residues are introduced in the peptides.

The method was tested with a series of 16- and 19-residue long peptides (corresponding to residues 134–151 of TMVP) that are recognized by the recombinant anti-TMVP Fab fragment 57P (Chatellier et al., 1996). For multivariate peptide design, three positions (S_{142}, E_{145} and S_{146}) were selected because their modification was known to moderately affect binding, without abolishing it entirely. Experiments in 19 buffers

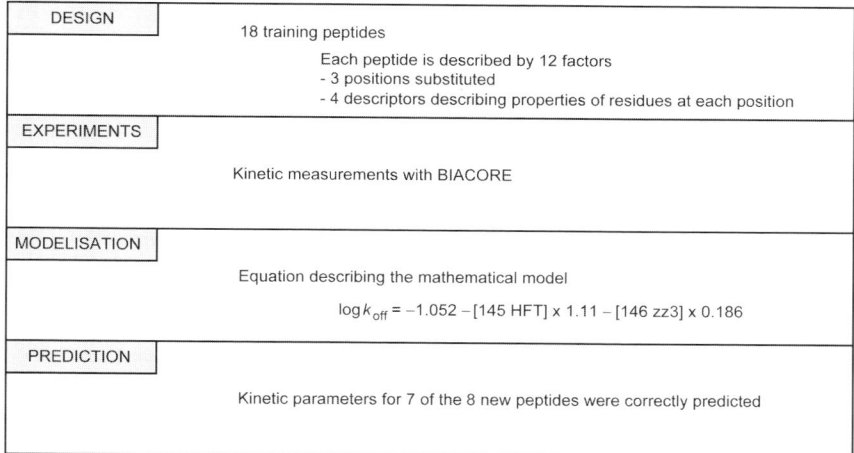

Figure 8.6. QSAR analysis relating peptide sequence to kinetics of dissociation. The experiments involved the interaction of Fab 57P with 18 designed peptides. Substituted positions 142, 145 and 146 of the peptide were described by the four descriptors zz1, zz2, zz3 and HFT (see text for details). The equation describing the mathematical model of interaction was used to predict the k_{off} of eight new peptides from data obtained with 18 training peptides.

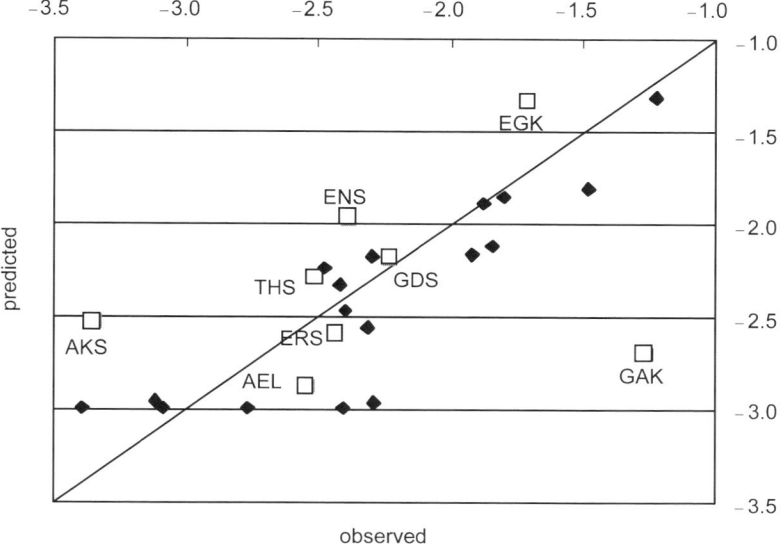

Figure 8.7. Predicted versus observed log k_{off} for training peptides (diamonds) and the eight new peptides identified by the residues at position 142, 145 and 146 (squares), using the model derived from the data of the training peptides. The observed k_{off} range varied 147-fold for all mutants (Adapted from Choulier et al., 2002b, with permission.).

were also performed to assess the sensitivity of the interactions to change in buffer composition.

Figure 8.6 describes the QSAR analysis approach used for relating peptide sequence and kinetics of dissociation. From a limited set of experiments, it was possible to determine the physico-chemical nature of the residues and the influence of the chemical environment on the kinetics (Andersson et al., 2001). Amongst the four factors used to describe the peptide sequence [zz1, zz2, zz3, interpreted, respectively, as describing hydrophobicity, size and electronic properties of the amino acids (Sandberg et al., 1998), and a descriptor for helix forming tendency (HFT, Deleage and Roux, 1987)], zz3 and HFT were found to influence on- and off-rates most strongly. The presence of urea, DMSO and NaCl in the buffer influenced binding properties, while changes in pH and the presence of EDTA and KSCN had no effect. The predictive capacity of these mathematical models was demonstrated by comparing predicted and observed kinetic parameters for new peptides or peptides in new buffers (Figure 8.7; Choulier et al., 2002b). Although the predictions are so far valid only for the particular protein under study and cannot be generalized for other antigen–antibody systems, these results open new perspectives for the predictive optimization of interaction kinetics.

◆◆◆◆◆◆ ENGINEERING OF ANTIBODY FRAGMENTS

The modification of antibody framework (non-CDR) positions may show similar effects in different antibodies if these positions play similar roles, in which case predictive antibody engineering may be feasible. We have analysed the effects of modifying conserved or co-variant (Choulier et al., 2000) positions (Figure 8.8) located either at the VL/VH interface of antibody fragments (Chatellier et al., 1996; Ben Khalifa et al., 2000; Hugo et al., 2003) or at the surface of their variable region (Weidenhaupt et al., 2002; Hugo et al., 2002). In general, the substitutions only marginally affected the kinetic parameters of the antigen–antibody interactions and the effects differed in different antibodies. However, these studies led to the identification of a VL/VH position (L34, Kabat numbering) whose mutation produces similar effects in scFv-1F4 and Fab 57P: the expression levels of active antibody fragments were enhanced four to nine-fold, and the dissociation rate parameters of the interactions were increased (Table 8.1), suggesting the possibility of predictive antibody engineering. Furthermore, modifications of surface charges in scFv-1F4 influenced the half-life of the fragments in periplasmic extracts (Hugo et al., 2002). It was concluded that surface charges may drastically affect the level of active scFvs in complex protein mixtures, a finding of considerable importance when engineering scFvs for various biotechnological applications.

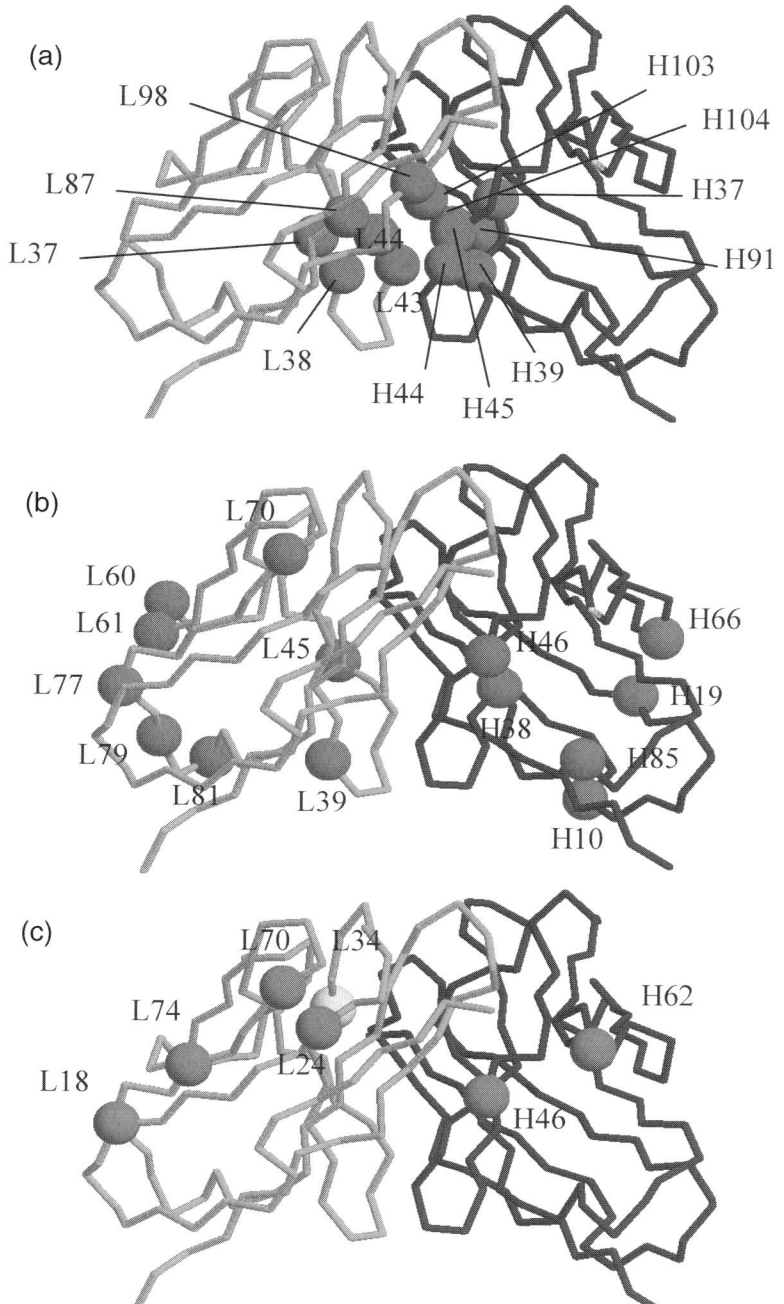

Figure 8.8. Location of positions analysed in the mutational studies, shown as balls on the backbone tracing of an Fv fragment. The light (VL) and heavy (VH) chain variable regions are coloured cyan and blue, respectively. Mutational analysis was carried out on positions indicated by spheres. (a) VL/VH interface residues (Chatellier et al., 1996; Ben Khalifa et al., 2000). (b) Conserved surface charged (Weidenhaupt et al., 2002). (c) Co-variant surface residues (Hugo et al., 2002, 2003, with permission.).

Table 8.1. Kinetic parameters for the interaction between the peptide antigens and the antibody fragments (adapted from Hugo et al., 2003, with permission)

Molecule	k_{on} (10^6 M^{-1} s^{-1})	Ratio k_{on} (mut)/k_{on} (WT)	k_{off} (10^{-4} s^{-1})	Ratio k_{off} (mut)/k_{off} (WT)
ScFv 1F4-WT	1.7 ± 0.3		7.4 ± 1.5	
ScFv 1F4-Q^{L34}S	1.3 ± 0.2	0.8	17.3 ± 0.2	2.3
BCCP-scFv1F4-WT [a] purified	1.2 ± 0.02		7.4 ± 0.5	
BCCP-scFv1F4-Q^{L34}S purified	0.8	0.7	17.7	2.4

Molecule	k_{on} (10^5 M^{-1} s^{-1})	Ratio k_{on} (mut)/k_{on} (WT)	k_{off} (10^{-3} s^{-1})	Ratio k_{off} (mut)/k_{off} (WT)
Fab 57P-WT	11.3 ± 0.4		1.2 ± 0.3	
Fab 57P-N^{L34}S	7.8	0.7	100	83.3
Fab 57P-WT purified	8.9 ± 0.3		1.7 ± 0.1	
Fab 57P-N^{L34}S purified	5.1 ± 0.3	0.6	97 ± 10	57

[a] Antibody fragments fused to *E. coli* biotin carboxy carrier protein (BCCP, Sibler et al., 1999).

References

Abad, L. W., Neumann, M., Tobias, L., Obenauer-Kutner, L., Jacobs, S. and Cullen, C. (2002). Development of a biosensor-based method for detection and isotyping of antibody responses to adenoviral-based gene therapy vectors. *Anal. Biochem.* **310**, 107–113.

Altschuh, D. (2002). Cyclosporin A as a model antigen: immunochemical and structural studies. *J. Mol. Recognit.* **15**, 277–285.

Altschuh, D., Dubs, M. C., Weiss, E., Zeder-Lutz, G. and Van Regenmortel, M. H. V. (1992). Determination of kinetic constants for the interaction between a monoclonal antibody and peptides using surface plasmon resonance. *Biochemistry* **31**, 6298–6304.

Andersson, K., Choulier, L., Hämäläinen, M. D., Van Regenmortel, M. H. V., Altschuh, D. and Malmqvist, M. (2001). Predicting the kinetics of peptide–antibody interactions using a multivariate experimental design of sequence and chemical space. *J. Mol. Recognit.* **14**, 62–71.

Benito, A. and Van Regenmortel, M. H. V. (1998). Biosensor characterization of antigenic site A of foot-and-mouth disease virus presented in different vector systems. *FEMS Immunol. Med. Microbiol.* **21**, 101–115.

Ben Khalifa, M., Weidenhaupt, M., Choulier, L., Chatellier, J., Rauffer-Bruyère, N., Altschuh, D. and Vernet, T. (2000). Effects on interaction kinetics of mutations at the VH–VL interface of Fabs depend on the structural context. *J. Mol. Recognit.* **13**, 127–139.

Bork, P., Dandekar, T., Diaz-Lazcoz, Y., Eisenhaber, F., Huynen, M. and Yuan, Y. (1998). Predicting function: from genes to genomes and back. *J. Mol. Biol.* **283**, 707–725.

Bothner, B., Dong, X. F., Bibbs, L., Johnson, J. E. and Suizdak, G. (1998). Evidence of viral capsid dynamics using limited proteolysis and mass spectrometry. *J. Biol. Chem.* **273**, 673–676.

Braden, B. C. and Poljak, R. J. (1995). Structural features of the reactions between antibodies and protein antigens. *FASEB J.* **9**, 9–16.

Brun, C., Baudot, A., Guénoche, A. and Jacq, B. (2004). The use of protein–protein interaction networks for genome wide protein function comparisons and predictions. In *Methods in Proteome and Protein Analysis* (R. M. Kemp, J. J. Calvete and T. Choli-Papadopoulou, eds), pp. 103A–112A. Springer, Berlin.

Chatellier, J., Rauffer-Bruyère, N., Van Regenmortel, M. H. V., Weiss, E. and Altschuh, D. (1996). Comparative interaction kinetics of two recombinant Fab fragments and of the corresponding antibodies directed to the coat protein of tobacco mosaic virus. *J. Mol. Recognit.* **9**, 39–51.

Choulier, L., Rauffer-Bruyere, N., Ben Khalifa, M., Martin, F., Vernet, T. and Altschuh, D. (1999). Kinetic analysis of the effect on Fab binding of identical substitutions in a peptide and its parent protein. *Biochemistry* **38**, 3530–3537.

Choulier, L., Lafont, V., Hugo, N. and Altschuh, D. (2000). Covariance analysis of protein families: the case of the variable domains of antibodies. *Proteins* **41**, 475–484.

Choulier, L., Laune, D., Orfanoudakis, G., Wlad, H., Janson, J.-C., Granier, C. and Altschuh, D. (2001). Delineation of a linear epitope by multiple peptide synthesis and phage display. *J. Immunol. Meth.* **249**, 253–264.

Choulier, L., Orfanoudakis, G., Robinson, P., Laune, D., Ben Khalifa, M., Granier, C., Weiss, E. and Altschuh, D. (2002a). Comparative properties of two peptide–antibody interactions as deduced from epitope delineation. *J. Immunol. Meth.* **259**, 77–86.

Choulier, L., Andersson, K., Hämäläinen, M. D., Van Regenmortel, M. H. V., Malmqvist, M. and Altschuh, D. (2002b). QSAR studies applied to the prediction of antigen–antibody interaction kinetics as measured by BIACORE. *Protein Eng.* **15**, 373–382.

Coëffier, E., Clement, J. M., Cussac, V., Khodaei-Boorane, N., Jehanno, M., Rojas, M., Dridi, A., Latour, M., El Habib, R., Barre-Sinoussi, F., Hofnung, M. and Leclerc, C. (2000). Antigenicity and immunogenicity of the HIV-1 gp41 epitope ELDKWA inserted into permissive sites of the MalE protein. *Vaccine* **19**, 684–693.

Daiss, J. L. and Scalice, E. R. (1994). Epitope mapping on BIAcore: theoretical and practical considerations. *Meth. Companion Meth. Enzymol.* **6**, 143–156.

Day, Y. S., Baird, C. L., Rich, R. L. and Myszka, D. G. (2002). Direct comparison of binding equilibrium, thermodynamic, and rate constants determined by surface- and solution-based biophysical methods. *Protein Sci.* **11**, 1017–1025.

Dekker, E. L., Porta, C. and Van Regenmortel, M. H. (1989). Limitations of different ELISA procedures for localizing epitopes in viral coat protein subunits. *Arch. Virol.* **105**, 269–286.

DeLano, W. L. (2002). Unraveling hot spots in binding interfaces: progress and challenges. *Curr. Opin. Struct. Biol.* **12**, 14–20.

Deleage, G. and Roux, B. (1987). An algorithm for protein secondary structure prediction based on class prediction. *Protein Eng.* **1**, 289–294.

Dore, I., Weiss, E., Altschuh, D. and Van Regenmortel, M. H. V. (1988). Visualization by electron microscopy of the location of tobacco mosaic virus epitopes reacting with monoclonal antibodies in enzyme immunoassay. *Virology* **162**, 279–289.

Dore, I., Ruhlmann, C., Oudet, P., Cahoon, M., Caspar, D. L. and Van Regenmortel, M. H. V. (1990). Polarity of binding of monoclonal antibodies to tobacco mosaic virus rods and stacked disks. *Virology* **176**, 25–29.

Dougan, D. A., Malby, R. L., Gruen, L. C., Kortt, A. A. and Hudson, P. J. (1998). Effects of substitutions in the binding surface of an antibody on antigen affinity. *Protein Eng.* **11**, 65–74.

Dubs, M. C., Altschuh, D. and Van Regenmortel, M. H. V. (1991). Interaction between viruses and monoclonal antibodies studied by surface plasmon resonance. *Immunol. Lett.* **31**, 59–64.

Dubs, M. C., Altschuh, D. and Van Regenmortel, M. H. V. (1992). Mapping of viral epitopes with conformationally specific monoclonal antibodies using biosensor technology. *J. Chromatogr.* **597**, 391–396.

Fägerstam, L. G., Frostell, A., Karlsson, R., Kullman, M., Larsson, A., Malmqvist, M. and Butt, H. (1990). Detection of antigen–antibody interactions by surface plasmon resonance. Application to epitope mapping. *J. Mol. Recognit.* **3**, 208–214.

Ferrer, M., Sullivan, B. J., Godbout, K. L., Burke, E., Stump, H. S., Godoy, J., Golden, A., Profy, A. T. and van Schravendijk, M. R. (1999). Structural and functional characterization of an epitope in the conserved C-terminal region of HIV-1 gp120. *J. Pept. Res.* **54**, 32–42.

Frank, R. (1992). Spot-synthesis: an easy technique for the positionally addressable, parallel chemical synthesis on a membrane support. *Tetrahedron* **48**, 9217–9232.

Geysen, M. H., Rodda, S. J. and Mason, T. J. (1986). A priori delineation of a peptide which mimics a discontinuous antigenic determinant. *Mol. Immunol.* **23**, 709–715.

Giovane, C., Travé, G., Briones, A., Lutz, Y., Wasylyk, B. and Weiss, E. (1999). Targetting of the N-terminal domain of the human papillomavirus type 16 E6 oncoprotein with monomeric ScFvs blocks the E6-mediated degradation of cellular p53. *J. Mol. Recognit.* **12**, 141–152.

Glaser, R. W. and Hausdorf, G. (1996). Binding kinetics of an antibody against HIV p24 core protein measured with real-time biomolecular interaction analysis suggest a slow conformational change in antigen p24. *J. Immunol. Meth.* **189**, 1–14.

Gomes, P. and Andreu, D. (2002). Direct kinetic assay of interactions between small peptides and immobilized antibodies using a surface plasmon resonance biosensor. *J. Immunol. Meth.* **259**, 217–230.

Gomes, P., Giralt, E., Ochoa, W., Verdaguer, N. and Andreu, D. (2002). Probing degeneracy in antigen–antibody recognition at the immunodominant site of foot-and-mouth disease virus. *J. Pept. Res.* **59**, 221–231.

Granzow, R. (1994). Biomolecular interaction analysis. In *Methods: A Companion to Methods in Enzymology* (R. Granzow, ed.), pp. 95–205. Academic Press, London.

Hansch, C. and Fujita, T. (1964). p-s-p analysis. A method for the correlation of biological activity and chemical structure. *J. Am. Chem. Soc.* **86**, 1616–1626.

Hoffman, T. L., Canziani, G., Jia, L., Rucker, J. and Doms, R. W. (2000). A biosensor assay for studying ligand-membrane receptor interactions: binding of antibodies and HIV-1 Env to chemokine receptors. *Proc. Natl Acad. Sci. USA* **97**, 11215–11220.

Hudson, P. J. and Souriau, C. (2003). Engineered antibodies. *Nat. Med.* **9**, 129–134.

Hugo, N., Lafont, V., Beukes, M. and Altschuh, D. (2002). Functional aspects of co-variant surface charges in an antibody fragment. *Protein Sci.* **11**, 2697–2705.

Hugo, N., Weidenhaupt, M., Beukes, M., Xu, B., Janson, J. C., Vernet, T. and Altschuh, D. (2003). VL position 34 is a key determinant for the engineering of stable antibodies with fast dissociation rates. *Protein Eng.* **16**, 381–386.

Jerne, N. K. (1960). Immunological speculations. *Annu. Rev. Microbiol.* **14**, 341–358.

Johnsson, B., Lofas, S., Lindquist, G., Edstrom, A., Muller Hillgren, R. M. and Hansson, A. (1995). Comparison of methods for immobilization to carboxymethyl dextran sensor surfaces by analysis of the specific activity of monoclonal antibodies. *J. Mol. Recognit.* **8**, 125–131.

Karlsson, R. (1999). Affinity analysis of non-steady-state data obtained under mass transport limited conditions using BIAcore technology. *J. Mol. Recognit.* **12**, 285–292.

Karlsson, R. and Falt, A. (1997). Experimental design for kinetic analysis of protein–protein interactions with surface plasmon resonance biosensors. *J. Immunol. Meth.* **200**, 121–133.

Karlsson, R. and Roos, H. (1997). Reaction kinetics. In *Principles and Practice of Immunoassay* (C. P. Price and D. J. Newman, eds), 2nd edn., pp. 101–122. Macmillan, London.

Karlsson, R., Roos, H., Fägerstam, L. and Persson, B. (1994). Kinetic and concentration analysis using BIA technology. *Meth. Companion Meth. Enzymol.* **6**, 99–110.

Kortt, A. A., Nice, E. and Gruen, L. C. (1999). Analysis of the binding of the Fab fragment of monoclonal antibody NC10 to influenza virus N9 neuraminidase from tern and whale using the BIAcore biosensor: effect of immobilization level and flow rate on kinetic analysis. *Anal. Biochem.* **273**, 133–141.

Lea, S. M., Powell, R. M., McKee, T., Evans, D. J., Brown, D., Stuart, D. I. and van der Merwe, P. A. (1998). Determination of the affinity and kinetic constants for the interaction between the human virus echovirus 11 and its cellular receptor, CD55. *J. Biol. Chem.* **273**, 30443–30447.

Leinikki, P., Lehtinen, M., Hyoty, H., Parkkonen, P., Kantanen, M. L. and Hakulinen, J. (1993). Synthetic peptides as diagnostic tools in virology. *Adv. Virus Res.* **42**, 149–186.

Malmqvist, M. (1996). Epitope mapping by label-free biomolecular interaction analysis. *Methods* **9**, 525–532.

Mani, J. C., Marchi, V. and Cucurou, C. (1994). Effect of HIV-1 peptide presentation on the affinity constants of two monoclonal antibodies determined by BIAcore technology. *Mol. Immunol.* **31**, 439–444.

Mateu, M. G., Andreu, D., Carreno, C., Roig, X., Cairo, J. J., Camarero, J. A., Giralt, E. and Domingo, E. (1992). Non-additive effects of multiple amino acid substitutions on antigen–antibody recognition. *Eur. J. Immunol.* **22**, 1385–1389.

Morton, T. A. and Myszka, D. G. (1998). Kinetic analysis of macromolecular interactions using surface plasmon resonance biosensors. *Meth. Enzymol.* **295**, 268–294.

Murzin, A. G. and Patthy, L. (1999). Sequences and topology. From sequence to structure to function. *Curr. Opin. Struct. Biol.* **9**, 359–362.

Myszka, D. G. (1999a). Survey of the 1998 optical biosensor literature. *J. Mol. Recognit.* **12**, 390–408.

Myszka, D. G. (1999b). Improving biosensor analysis. *J. Mol. Recognit.* **12**, 279–284.

Myszka, D. G., Morton, T. A., Doyle, M. L. and Chaiken, I. M. (1997). Kinetic analysis of a protein antigen–antibody interaction limited by mass transport on an optical biosensor. *Biophys. Chem.* **64**, 127–137.

Myszka, D. G., Sweet, R. W., Hensley, P., Brigham-Burke, M., Kwong, P. D., Hendrickson, W. A., Wyatt, R., Sodroski, J. and Doyle, M. L. (2000). Energetics of the HIV gp120–CD4 binding reaction. *Proc. Natl Acad. Sci. USA* **97**, 9026–9031.

Presta, L. (2003). Antibody engineering for therapeutics. *Curr. Opin. Struct. Biol.* **13**, 519–525.

Rauffer-Bruyère, N., Chatellier, J., Weiss, E., Van Regenmortel, M. H. V. and Altschuh, D. (1997). Cooperative effects of mutations in a recombinant Fab on the kinetics of antigen binding. *Mol. Immunol.* **34**, 165–173.

Rich, R. L. and Myszka, D. G. (2000). Survey of the 1999 surface plasmon resonance biosensor literature. *J. Mol. Recognit.* **13**, 388–407.

Rich, R. L. and Myszka, D. G. (2001). Survey of the year 2000 commercial optical biosensor literature. *J. Mol. Recognit.* **14**, 273–294.

Rich, R. L. and Myszka, D. G. (2002). Survey of the year 2001 commercial optical biosensor literature. *J. Mol. Recognit.* **15**, 352–376.

Rich, R. L. and Myszka, D. G. (2003a). Spying on HIV with SPR. *Trends Microbiol.* **11**, 124–133.

Rich, R. L. and Myszka, D. G. (2003b). A survey of the year 2002 commercial optical biosensor literature. *J. Mol. Recognit.* **16**, 351–382.

Richalet-Sécordel, P. M., Zeder-Lutz, G., Plaue, S., Sommermeyer-Leroux, G. and Van Regenmortel, M. H. V. (1994). Cross-reactivity of monoclonal antibodies to a chimeric V3 peptide of HIV-1 with peptide analogues studied by biosensor technology and ELISA. *J. Immunol. Meth.* **176**, 221–234.

Richalet-Sécordel, P. M., Poisson, F. and Van Regenmortel, M. H. V. (1996). Uses of biosensor technology in the development of probes for viral diagnosis. *Clin. Diagn. Virol.* **5**, 111–119.

Richalet-Sécordel, P. M., Rauffer-Bruyère, N., Christensen, L. L., Ofenloch-Haehnle, B., Seidel, C. and Van Regenmortel, M. H. V. (1997). Concentration measurement of unpurified proteins using biosensor technology under conditions of partial mass transport limitation. *Anal. Biochem.* **249**, 165–173.

Rodda, S. J. and Tribbick, G. (1996). Antibody-defined epitope mapping using the multipin method of peptide synthesis. *Meth. Companion Meth. Enzymol.* **9**, 473–481.

Roivanen, M., Piirainen, L., Rysä, T., Närvänen, A. and Hovi, T. (1993). An immunodominant N-terminal region of VP1 protein of poliovirus that is buried in crystal structure can be exposed in solution. *Virology* **195**, 762–765.

Roos, H., Karlsson, R., Nilshans, H. and Persson, A. (1998). Thermodynamic analysis of protein interactions with biosensor technology. *J. Mol. Recognit.* **11**, 204–210.

Sandberg, M., Eriksson, L., Jonsson, J., Sjostrom, M. and Wold, S. (1998). New chemical descriptors relevant for the design of biologically active peptides. A multivariate characterization of 87 amino acids. *J. Med. Chem.* **41**(14), 2481–2491.

Saunal, H. and Van Regenmortel, M. H. V. (1995a). Mapping of viral conformational epitopes using biosensor measurements. *J. Immunol. Meth.* **183**, 33–41.

Saunal, H. and Van Regenmortel, M. H. V. (1995b). Kinetic and functional mapping of viral epitopes using biosensor technology. *Virology* **213**, 462–471.

Saunal, H., Witz, J. and Van Regenmortel, M. H. V. (1993). Inhibition of in vitro cotranslational disassembly of tobacco mosaic virus by monoclonal antibodies to the viral coat protein. *J. Gen. Virol.* **74**, 897–900.

Sibler, A. P., Kempf, E., Glacet, A., Orfanoudakis, G., Bourel, D. and Weiss, E. (1999). In vivo biotinylated recombinant antibodies: high efficiency of labelling and application to the cloning of active anti-human IgG1 Fab fragments. *J. Immunol. Meth.* **224**, 129–140.

Smith, G. P. and Scott, J. K. (1993). Libraries of peptides and proteins displayed on filamentous phage. *Meth. Enzymol.* **217**, 228–257.

Tosser, G., Delaunay, T., Kohli, E., Grosclaude, J., Pothier, P. and Cohen, J. (1994). Topology of bovine rotavirus (RF strain) VP6 epitopes by real-time biospecific interaction analysis. *Virology* **204**, 8–16.

Van Cott, T. C., Loomis, L. D., Redfield, R. R. and Birx, D. L. (1992). Real-time biospecific interaction analysis of antibody reactivity to peptides from the envelope glycoprotein, gp160, of HIV-1. *J. Immunol. Meth.* **146**, 163–176.

Van Cott, T. C., Bethke, F. R., Polonis, V. R., Gorny, M. K., Zolla-Pazner, S., Redfield, R. R. and Birx, D. L. (1994). Dissociation rate of antibody–gp120 binding interactions is predictive of V3-mediated neutralization of HIV-1. *J. Immunol.* **153**, 449–459.

Van Regenmortel, M. H. V. (1966). Plant virus serology. *Adv. Virus Res.* **12**, 207–271.

Van Regenmortel, M. H. V. (1998). From absolute to exquisite specificity. Reflections on the fuzzy nature of species, specificity and antigenic sites. *J. Immunol. Meth.* **216**, 37–48.

Van Regenmortel, M. H. V. (1999). The antigenicity of tobacco mosaic virus. *Philos. Trans. R. Soc. London B Biol. Sci.* **354**, 559–568.

Van Regenmortel, M. H. V. (2001). Analysing structure–function relationships with biosensors. *Cell. Mol. Life Sci.* **58**, 794–800.

Van Regenmortel, M. H. V. (2002a). A paradigm shift is needed in proteomics: 'structure determines function' should be replaced by 'binding determines function'. *J. Mol. Recognit.* **15**, 349–351.

Van Regenmortel, M. H. V. (2002b). Reductionism and the search for structure–function relationships in antibody molecules. *J. Mol. Recognit.* **15**, 240–247.

Van Regenmortel, M. H. V., Altschuh, D., Pellequer, J.-L., Richalet-Sécordel, P. M., Saunal, H., Wiley, J. A. and Zeder-Lutz, G. (1994). Analysis of viral antigen using biosensor technology. *Meth. Companion Meth. Enzymol.* **6**, 177–187.

Weidenhaupt, M., Khalifa, M. B., Hugo, N., Choulier, L., Altschuh, D. and Vernet, T. (2002). Functional mapping of conserved, surface-exposed charges of antibody variable domains. *J. Mol. Recognit.* **15**, 94–103.

Wilson, T. M. A. (1984). Contranslational disassembly of tobacco mosaic virus in vitro. *Virology* **137**, 255–265.

Zeder-Lutz, G., Altschuh, D., Denery-Papini, S., Briand, J. P., Tribbick, G. and Van Regenmortel, M. H. V. (1993). Epitope analysis using kinetic measurements of antibody binding to synthetic peptides presenting single amino acid substitutions. *J. Mol. Recognit.* **6**, 71–79.

Zeder-Lutz, G., Benito, A. and Van Regenmortel, M. H. V. (1999). Active concentration measurements of recombinant biomolecules using biosensor technology. *J. Mol. Recognit.* **12**, 300–309.

Zeder-Lutz, G., Hoebeke, J. and Van Regenmortel, M. H. V. (2001). Differential recognition of epitopes present on monomeric and oligomeric forms of gp160 glycoprotein of human immunodeficiency virus type 1 by human monoclonal antibodies. *Eur. J. Biochem.* **268**, 2856–2866.

9 RT *In Situ* PCR: Protocols and Applications

Alcina Frederica Nicol[1] and Gerard J Nuovo[2]

[1] Laboratory of Immunology-IPEC/FIOCRUZ, Av. Brasil 4365, Manguinhos-RJ 21045-900, Brazil; [2] Ohio State University Medical Center, Department of Pathology, Columbus, OH 43210, USA

◆◆

Reverse transcriptase *in situ* polymerase chain reaction allows the direct cellular localization of any, including low-copy, RNA targets and requires several controls to document that nonspecific DNA synthesis pathways have been eliminated.

CONTENTS

Introductory statements
The theory behind RT *in situ* PCR
The positive control: nonspecific DNA synthesis
Inhibition of DNA repair
DNase digestion
Summation
RT *in situ* PCR for viral targets in cellular preparations
Importance of the specific localization of the signal with RT *in situ* PCR
The protocol for RT *in situ* PCR
Applications of RT *in situ* PCR

◆◆◆◆◆◆ INTRODUCTORY STATEMENTS

The *in situ* amplification of DNA and RNA has evolved a great deal since the first peer-review reports of each in 1990 and 1991, respectively (Haase *et al.*, 1990; Nuovo *et al.*, 1992a,b). Although, of course, the polymerase chain reaction (PCR) continues to be the mainstay of *in situ* amplification, other methods such as strand displacement amplification, have been shown to be effective (Nuovo, 1997). Whatever the method, the idea is the same—to increase the amount of target in the cell to an amount easily detectable *in situ*. This can be done for DNA or for RNA targets, where an initial reverse transcriptase (RT) step must be employed to convert the RNA to cDNA so that it can be amplified. It may surprise the reader to state that RT *in situ* PCR is, in some ways, technically easier to do than PCR *in situ* hybridization (for DNA targets). The process was made even easier about 10 years ago with the widespread use of the enzyme rTth and the "EZ" RT *in situ* PCR protocol (Nuovo *et al.*, 1995a). This, in combination with a solid understanding of the mechanisms of the nonspecific signal that may be evident when using *in situ* PCR in tissue sections, and the ability to reproducibly eliminate this nonspecific signal with overnight DNase digestion, makes it a relatively simple matter to detect *in situ* PCR-amplified RNAs (cDNAs) (Cioc and Nuovo, 2002; Cioc *et al.*, 2004; Morrison *et al.*, 2002; Nuovo *et al.*, 1992a,b; Nuovo *et al.*, 1993a; Nuovo *et al.*, 1993b; Nuovo *et al.*,1994; Nuovo *et al.*,1995a). After one determines the optimal protease digestion time for a given sample, in a matter of a few hours it is possible to detect a given mRNA via the direct incorporation of the labeled nucleotide. In order to appreciate the foundation of RT *in situ* PCR, we believe it is important to review the data that explores the mechanism of the nonspecific signal in tissue sections before discussing the actual protocols for the *in situ* detection of PCR-amplified cDNA. Finally, applications of RT *in situ* PCR will be discussed focusing on HIV-1.

◆◆◆◆◆◆ THE THEORY BEHIND RT *IN SITU* PCR

It must be emphasized that one cannot perform target specific direct incorporation of the reporter molecule (e.g. digoxigenin dUTP) for DNA targets in paraffin embedded tissue sections which is referred to as *in situ* PCR. The explanation is that nonspecific DNA synthesis, primarily in the form of DNA repair, invariably would cause a false positive signal with paraffin-embedded tissue sections (Nuovo *et al.*, 1991a; Nuovo *et al.*, 1991b). Although DNA repair can be blocked by pretreatment in a solution which contains a dideoxy nucleotide, the exacting requirements of correct protease digestion time and the preincubation time with the dideoxy nucleotide made this "dideoxy-assisted *in situ* PCR" prone to residual non specific false positive results (Nuovo, 1997). The nonspecific DNA synthesis pathways can be reliably and completely eliminated with an overnight digestion in RNase free DNase if (and only if) the sample was first optimally digested with a protease. Clearly, this would not be of use for the detection of either viral or native DNAs. However, these observations lend themselves very well to the *in situ* detection of RNA viruses or mRNAs using the direct incorporation of the reporter molecule. Specifically, one can first determine which protease digestion time gives the strongest signal in most cells for the particular tissue or cell sample. Protease digestion is followed by DNase digestion and, then, by the target specific synthesis of a cDNA molecule using the enzymes RT or rTth (Applied Biosystems). One can then perform *in situ* PCR with direct incorporation of the digoxigenin dUTP (or some other reporter nucleotide) using the target specific primers. To prove that the native DNA-derived signal was eliminated, one uses the tissue sample on the glass slide that was DNase digested and where the primers were omitted during the RT step; alternatively, irrelevant primers can be employed. Irrelevant primers are defined as not possibly having any targets in the tissue being analyzed; rabies primers are good examples as-of course-most tissues will not contain the rabies virus. Provided that protease conditions were optimal, this section should show no signal and would thus serve as the negative control. Similarly, one could demonstrate that protease digestion for the tissue was optimal and that the PCR and detection steps were working properly with the tissue sample that was not subject to DNase digestion. The relationship between protease digestion, length of fixation time in formalin, and the signal with the positive and negative control is discussed at length by Nuovo (1997). The salient points will now be reviewed;

◆◆◆◆◆◆ THE POSITIVE CONTROL: NONSPECIFIC DNA SYNTHESIS

The subheading of this section reflects two different perspectives of the results found in the non-DNased tissue section. On the one hand, the ability to invariably see an intense signal in tissue sections with optimal protease digestion pretreatment irrespective of the primers used indicates successful synthesis and detection of DNA within the nucleus—hence, the term the positive control. The other perspective views the DNA synthesis in the positive cells derived in this manner correctly as not being target specific, and the possible cause of a false positive reading when interpreting the RT *in situ* PCR results. Of course, both perspectives are valid.

The Sources of the Signal with the Positive Control

Three potential DNA synthesis pathways may operate inside the cell during *in situ* PCR. These are:

(1) target specific amplification;
(2) non target primer-dependent amplification (mispriming);
(3) DNA repair (primer independent pathway).

Note: A fourth pathway – primer oligomerization – that may dominate in solution phase PCR does not appear to occur inside intact cells, possibly due to the high protein concentration present in a cell (Chou *et al.*, 1992; Nuovo *et al.*, 1991a; Nuovo, 1997).

Of these, DNA repair is invariably present in paraffin embedded tissues or any other cell/tissue preparation that has been exposed to dry heat of at least 55°C for ≥1 h. Indeed, this primer independent signal may be evident even after several minutes of exposure to such dry heat. DNA repair will usually involve all of the different cell types in a tissue section. In comparison, the target specific cDNA based signal for RT *in situ* PCR will rarely involve all cell types unless the target chosen is a "housekeeper" transcript. Thus, the presence of a signal in all cell types after RT *in situ* PCR strongly suggests a nonspecific signal. Some basic knowledge of surgical pathology/histopathology is therefore invaluable to anyone undertaking RT *in situ* PCR. Alternatively, a signal that is restricted to only one or a few cell types after RT *in situ* PCR suggests a target specific signal. This simple but important point can be easily demonstrated with viral infections, given the fastidious tropism exhibited by many viruses (see below). Of course, this observation is also evident for human mRNAs where cytoplasmic localization is the expected result.

Mispriming is another source of DNA synthesis that is operative during *in situ* PCR if the hot start maneuver is not used. In comparison, the hot start method will not inhibit the DNA repair pathway. One can demonstrate the existence of mispriming by using frozen, fixed tissues that lack the primer independent pathway or by using primers that do not correspond to any target in the tissue cells. Of course, the hot start maneuver also reduces/eliminates mispriming in solution phase PCR, which increases both the sensitivity and specificity of the reaction.

As noted above, primer oligomerization does not appear to induce any signal with *in situ* PCR. This can be demonstrated by using paraffin embedded tissues in which DNA repair and mispriming have been blocked by DNase digestion after optimal protease digestion. In such cases, *in situ* PCR using primers that do not correspond to any target in the cells show no nuclear signal. However, it is important to stress that nonspecific *cytoplasmic* staining can be seen if a post PCR stringent wash is not performed. This represents nonspecific "stickiness" of the labeled primer oligomers formed in the *amplifying solution* on cytoplasmic proteins and it does not represent primer oligomerization formed inside the intact cell (Nuovo, 1997). This type of background is analogous to that encountered with standard *in situ* hybridization or immunohistochemistry.

Target specific DNA synthesis does, of course, operate during *in situ* PCR if the primers correspond to a cellular target in the sample. This can easily be demonstrated using cellular preparations in which two different cell types are mixed, one containing and the other lacking the target of interest, respectively. If hot start (to block mispriming) *in situ* PCR is performed on the non heated sample (to avoid DNA repair), then only the cell type that possesses the primer target will demonstrate a signal (Nuovo *et al.*, 1992a,b; Nuovo, 1997).

It is important to appreciate that the signal in tissue sections undergoing RT *in situ* PCR, due to either DNA repair or target specific DNA synthesis, is *highly* dependent on formalin fixation and protease digestion times. It is not surprising that formalin fixation is critical in obtaining results from RT *in situ* PCR. The extent of formalin fixation, which cross-links proteins and nucleic acids, has important effects on variables such as probe entry, intensity of the signal, and the requirement for protease digestion in standard *in situ* hybridization.

❖❖❖❖❖❖ INHIBITION OF DNA REPAIR

The primer independent signal serves as a useful function for *in situ* PCR with tissue sections; the signal generated from DNA repair under optimal protease digestion demonstrates successful DNA synthesis inside the nucleus and detection of the resultant labeled product. This observation is the basis of the start up protocol recommended for people just learning to do RT *in situ* PCR. However, because it masks the target specific signal, clearly one must be able to either eliminate or block the nonspecific signal to perform target specific *in situ* PCR in tissue sections for either RNA or DNA.

On theoretical grounds, one may envision several ways to either prevent or block DNA repair when performing *in situ* PCR. First, DNA could be digested with DNase to the point that it would no longer support Taq polymerase-mediated DNA synthesis. One would predict that prolonged DNase digestion may be needed to render all areas of DNA damage unavailable to synthesis from the activity of Taq polymerase. An alternative approach would be to modify the native DNA such that it could no longer serve as a template for Taq polymerase. One such mechanism uses dideoxy nucleotides which, when added to the growing DNA chain, block the addition of any additional nucleotides. This technique of preventing the phosphodiester linkage of additional nucleotides to a DNA fragment is commonly used in sequencing DNA. Thus, it would appear feasible that one might be able to block DNA repair during *in situ* PCR by first pretreating the slides in a solution which contains a dideoxy nucleotide.

We have been able to prevent DNA repair by both DNase digestion and blockage after the incorporation of a dideoxy nucleotide prior to the *in situ* PCR step (Nuovo *et al.*, 1991a; Nuovo *et al.*, 1993a; Nuovo, 1997). However, as is true for the nonspecific DNA repair signal itself, the results one obtains with either DNase digestion or dideoxy blockage are strongly dependent on the length of time the tissue has been fixed in 10% buffered formalin, as well as the length of time of protease digestion. Dideoxy blockage, however, is not useful for RT *in situ* PCR. One reason is that the window for optimal protease treatment of sections tends to be narrow, making false positive results from DNA repair more likely. More importantly, dideoxy blockage would not prevent target specific amplification of the DNA target that corresponds to the resultant mRNA (unless the primers where chosen to amplify across an intron. This would invariably lead to a false positive (albeit nuclear) signal for RT *in situ* PCR when one is analyzing for a cellular transcript. However, DNA repair, mispriming, and target specific DNA synthesis can all be eliminated by DNase *digestion after optimal protease digestion*. Thus, we will focus our attention on DNase digestion for RT *in situ* PCR. Some investigators have questioned whether there might not be an advantage to use frozen, fixed tissues for RT *in situ* PCR. The rationale is that the DNA repair system would not be operative and RNA degradation would be, perhaps, less likely as one would have strict control over tissue preparation and fixation. We do not recommend the use of frozen, fixed tissues for RT *in situ* PCR. First, DNase digestion will still be required for cellular genes, for the same reasons as was just discussed with dideoxy blockage. Second, it is technically more difficult to place three frozen tissue sections on a glass slide than to place three paraffin embedded tissue sections. Finally, we have had good success in detecting a variety of RNAs present in tissues obtained from routine surgical biopsies or autopsy material. Such tissues are placed in 10% buffered formalin and the resultant cross-linking of the RNAs with cellular proteins and nucleic acids likely protects them from RNase digestion until the protease digestion step. Developing a methodology based on formalin-fixation therefore has definitive advantages when clinical specimens are to be examined.

◆◆◆◆◆◆ DNASE DIGESTION

When considering the enormous amount of DNA present in the nucleus that may participate in nonspecific mispriming or DNA repair pathways, one would hypothesize that in order to render the native DNA nonamplifiable, it would have to be exposed for an extended period of time to a high concentration of DNase. It is important to stress that cross-linking fixative, such as buffered formalin, may protect the DNA from DNase digestion by steric hindrance from the resultant protein-DNA cross-links. This point suggests several possibilities about the relationship between the signal after DNase digestion and the length of protease digestion time. Specifically, if the tissue has been fixed for a prolonged time in formalin, extended digestion in protease may be required in order to allow the DNase access to the DNA. Of course, this strong relationship between the times of protease digestion and formalin fixation, and elimination of the signal with overnight DNase digestion have been noted above and is discussed in detail elsewhere (Nuovo *et al.*, 1993a; Nuovo, 1997). Note the following points:

(1) When the signal with the non-DNase treated tissue is optimal (ie, protease digestion is optimal), no signal is noted after DNase digestion;
(2) When the signal with the non-DNase treated tissue is suboptimal due to inadequate protease digestion, a signal is often present with the overnight DNase digestion and is usually stronger than the signal with the non-DNase treated tissue.

The first observation demonstrates that DNase digestion was adequate in rendering the native DNA non-amplifyable. This serves as the basis for the negative control and the test for RT *in situ* PCR. That is, if one is to use direct incorporation of the reporter molecule after the RT step, it is critical to be assured that any resulting signal must be derived from the cDNA because the native DNA has been rendered unavailable for amplification.

The second observation can be considered a "paradoxical" enhancement of signal with DNase digestion in the setting of inadequate protease digestion. The term enhanced is used because when the length of protease time is inadequate, not only is a signal evident in the DNased tissue, but it is often stronger than the signal evident in the serial section on the same glass slide that was not DNased. The fact that a signal is present after DNase digestion with suboptimal protease implies that something is preventing the enzyme from exerting its maximum activity. How is the DNase able to actually enhance the *in situ* PCR signal under suboptimal conditions of protease digestion? One possible explanation is that DNase digestion, although blocked from rendering the entire DNA template (with its gaps) incapable of DNA repair to the steric interference from persistent DNA-protein cross-links, may still create new gaps and/or extend existent gaps which can then be filled during the subsequent *in situ* PCR step. Although further testing of this model is needed, the important points for RT *in situ* PCR are:

(1) under optimal protease conditions, the non specific nuclear based signal should be present in many of the cells in a tissue section;
(2) under these same protease conditions, this signal can be completely eliminated with overnight DNase digestion;
(3) these two observations serve as the basis for the negative and positive controls for RT *in situ* PCR. Most importantly when performing RT *in situ* PCR, because the optimal protease digestion time varies with fixation time and tissue type, it is important to remember that it is essential to perform negative and positive controls on the same tissue sections as the test run.

A brief mention of two other points should be made. First, we have tried shorter times for DNase digestion. Under optimal protease digestion times, we were unable to reliably eradicate the nonspecific primer independent signal with *in situ* PCR in tissue sections when DNase digestion was performed up to 4 h. However, the nonspecific signal was eradicated after optimal protease digestion using DNase digestion times of 8 – 15 h. Second, a common misinterpretation among investigators is that the occurrence of a positive signal in the negative control (DNase digested overnight) is caused by DNase concentrations. Assuming that one uses 10 U of RNase free DNase per section, the persistence of a signal in the negative control represents too short a protease digestion time rather than an inadequate amount of DNase. Increasing the DNase concentration up to 5-fold will *not* eliminate the nonspecific signal under these conditions. However, increasing the protease digestion time will result in the elimination of the signal.

◆◆◆◆◆◆ SUMMATION

In summary, for RT *in situ* PCR, we strongly recommend that one uses DNase digestion to block the nonspecific pathways of mispriming and DNA repair, as well as target specific amplification of the genomic based sequence. Because each of these pathways is adequately blocked after optimal protease digestion, one may then perform a direct incorporation of the reporter nucleotide after *RT under cold start conditions.* That is, one adds the rTth to the amplifying solution before increasing the temperature of the thermal cycler. However, recall that under these conditions primer oligomerization may still occur in the amplifying solution. Even though the DNA labeled from primer oligomerization in the amplifying solution does not induce a nuclear signal with RT *in situ* PCR, it can serve as a source of nonspecific signal by "sticking" to cellular proteins, primarily the cell membrane and basement membrane. This is not surprising to those who utilize *in situ* hybridization, where one routinely adds large amounts of labeled nucleotide (from 75 to 150 base pairs) and detects signal only in cells that contain the target sequence. It is assumed that the probe enters all cells, but is washed out *after a wash of adequate stringency* in cells lacking the target. When one generates primer oligomers (or DNA synthesized from mispriming) in the amplifying solution, an analogous situation occurs. It follows that a non specific signal with RT *in situ* PCR may be generated by avoiding a stringent wash after the PCR step, due to the DNA synthesized from primer oligomerization in the over lying solution. Again, this background is due to nonspecific "sticking" of the labeled DNA in the solution and cellular proteins and should not be confused with background from DNA repair that is nuclear based and is not affected by DNA synthesis in the reaction solution. The practical implications for RT *in situ* PCR are clear. After RT *in situ* PCR, perform a stringent wash to eliminate nonspecific binding of labeled DNA in the overlying solution to cellular proteins.

◆◆◆◆◆◆ RT *IN SITU* PCR FOR VIRAL TARGETS IN CELLULAR PREPARATIONS

It is important to stress that the discussions listed above were based on work with paraffin embedded tissue sections. If one wishes to use formalin fixed cellular preparations that have not been embedded in paraffin (or exposed to dry heat), one should recall that it is possible to obtain direct incorporation of the reporter molecule with *in situ* PCR for DNA targets under defined conditions which include the hot start maneuver and defined protease digestion. It follows that one should be able to perform target specific hot start RT *in situ* PCR in cellular preparations (or frozen, formalin fixed tissues) for RNA viruses without the DNase digestion step. To test this hypothesis, we mixed measles infected HeLa cells and peripheral blood lymphocytes from a non-infected individual. The measles infected cells are easily recognized on microscopic examination due to their large size, intranuclear inclusions and multinucleation. Only the measles infected cells show a signal with RT *in situ* PCR if measles specific primers are used (Nuovo, 1997). It should be stressed that DNase digestion was not performed in these experiments. However, we recommend that the reader still performs DNase digestion for RT *in situ* PCR even if cell suspensions or frozen, fixed preparations are being used. In this way, one can perform "cold start" PCR and not be concerned about the possibility of erroneous signal due to DNA repair which can occur in cytospin preparations if they have been inadvertently exposed to dry heat prior to *in situ* PCR. Further, we believe that it is useful to induce the DNA repair pathway for RT *in situ* PCR as it serves as a control to demonstrate that the conditions are adequate to support DNA synthesis inside the cell. More specifically, the signal in the positive control (no DNase treatment using heated samples) demonstrates that one has reached the optimal protease digestion time for that tissue or cellular preparation. Further, the signal in the positive control demonstrates that the user is successfully synthesizing DNA in the nucleus.

◆◆◆◆◆◆ IMPORTANCE OF THE SPECIFIC LOCALIZATION OF THE SIGNAL WITH RT *IN SITU* PCR

As stressed in this chapter, one cannot interpret RT *in situ* PCR without the negative (DNase, no RT or RT with irrelevant primers) and positive (no DNase) controls. No signal with the negative control and an intense *nuclear* signal in the positive control are required before one can confidently analyze the test section. The nuclear localization of the positive control reflects the fact that the DNA repair pathway as well as mispriming and target specific genomic based DNA synthesis all occur in the nucleus. Another important principle in the interpretation of RT *in situ* PCR is the specific subcellular localization of the signal. For cellular mRNAs, the signal should localize to the cytoplasm. In comparison, one should see nuclear localization of the signal in the positive control (no DNase treatment). Another important point to make using this example is that the MMP-9 is not evident in the normal cervical epithelium that is only a few mm away from the cancer cells (Nuovo *et al.*, 1995a). Figure 9.1 illustrates this with another example—specifically, rabies cDNA localization to the cytoplasm of large neurons (compare this to the negative and positive PCR *in situ* controls); in contrast TNFα mRNA localized to astrocytes and microglial cells in the brain of cases of rabies encephalitis (Nuovo, unpublished data). These two examples remind us that localization to specific cell types is another hallmark of successful RT *in situ* PCR.

The specific subcellular distribution of RNAs can show much variation, especially for viral RNAs. HIV-1 RNA (for example, the *gag* transcript) typically localizes to the nucleus, as discussed below (Nuovo *et al.*, 1993b; Nuovo *et al.*, 1994). Most RNA viruses localize to the cytoplasm (Figure 9.1, that also illustrates rotaviral and coxsackie virus localization) but subcellular localization to, for example, the nuclear membrane is not unusual (for example, for hepatitis C RNA). Interestingly, an antigen of hepatitis C localizes to the cytoplasm in the area directly adjacent to the nuclear membrane (Nuovo, 1997). Measles RNA localizes to the nucleus or cytoplasm, depending on whether the RT step utilizes the sense or antisense strand (Nuovo, 1997). Figure 9.1 also demonstrates the detection of bacterial consensus rRNA by RT *in situ* PCR

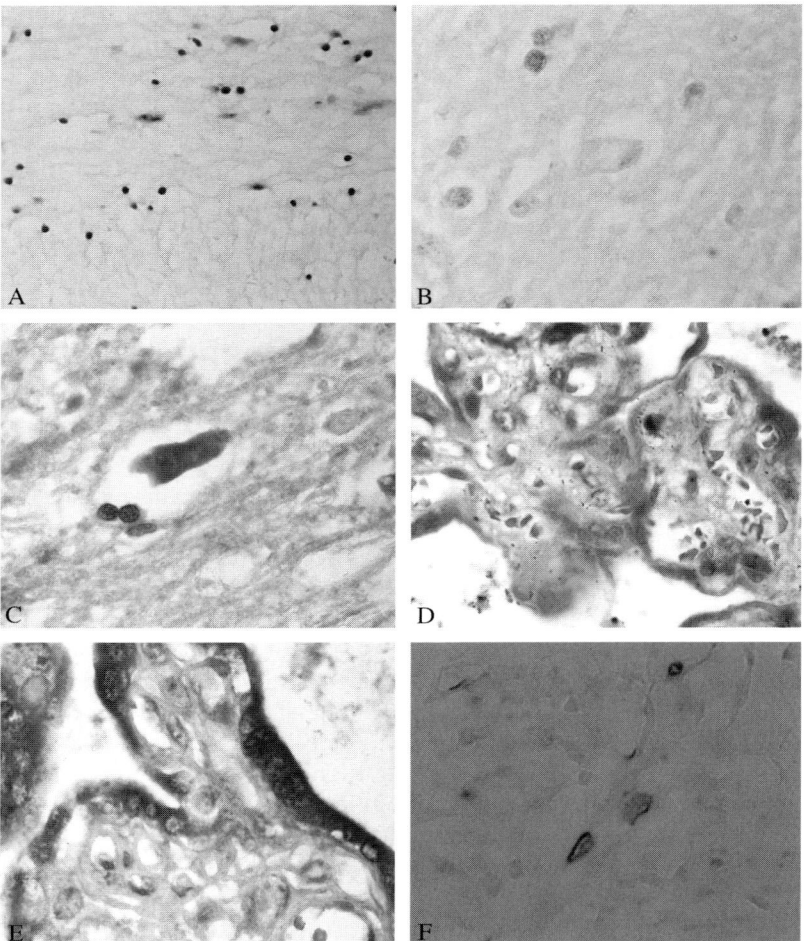

Figure 9.1. The importance of the exact localization of the signal with RT *in situ* PCR. Panels A and B show the expected results for the no DNase control and DNase, no RT (or irrelevant primers) control in a case of rabies encephalitis. Note the strong nuclear based signal for A and the lack of any signal for B. Panel C shows that rabies RNA localizes to the same brain tissue after optimal protease digestion to the cytoplasm of large neurons. Panels D, E and F show that the signal for RT *in situ* PCR for different targets after optimal protease and sufficient DNase digestion will show specific localization patterns. Panel D shows the detection of Klebsiella in the villi of a case of severe in utero bacterial infection; note the rod shaped forms. Panel E shows the cytoplasmic localization to trophoblasts of the placenta in a case of fatal in utero infection by coxsackie virus while panel F shows the cytoplasmic localization to endothelial cells in a case of fatal rotaviral sepsis. (Magnification: Panel A. 400 × . Panel B. C. D. E. F. 1000 × .). (See colour plate 26).

where, as expected, the bacterial forms are evident. Human mRNAs may also show different patterns. Using the megakaryocyte cell line DAMI and the fibrosacroma cell line HT1080, we noted either a diffuse cytoplasmic signal (for MMPs and TIMPs in HT1080 cells) or a reticulated cytoplasmic signal for gelsolin and glycoprotein IIB mRNA. Interestingly, the PCR amplified cDNA based glycoprotein IIB signal was also seen around the nucleoli of the nucleus, whereas another RNA in these same cells, amyloid precursor protein, localized to the nucleoli in these cells. It is presumed that these disparate localization patterns for different RNAs as determined by RT *in situ* PCR in the same cell types reflects the distribution of pre-mRNAs as they are processed from the nucleus to the cytoplasm (Nuovo *et al.*, 1992a,b).

In summary, two important variables with regard to the specificity of the signal with RT *in situ* PCR are:

(1) the signal should localize to the *entire* nucleus with the positive control and, in most cases, show a different pattern, usually cytoplasmic or sometimes part of the nucleus, for the test (cDNA) specimen;
(2) a relatively low percentage of the cells in a tissue section should be positive with RT *in situ* PCR and the positive cells should be restricted to specific cell types. By comparison, the positive control should label at least 50% of the cells in a tissue sample and various cell types should show a signal. An important exception to the latter statement applies to tissue that was fixed for a short period of time; cell types more resistant to protease (e.g. squamous cells and fibroblasts) may show signal with the positive control whereas other cell types sensitive to over digestion with protease (e.g. lymphocytes) may have no signal.

◆◆◆◆◆◆ THE PROTOCOL FOR RT *IN SITU* PCR

The Two-step Procedure

The most important part of the protocol for RT *in situ* PCR is that the negative and positive control should be performed on the same tissue section with the actual test. This is critical as it provides assurance that the native DNA repair and mispriming pathways have been eliminated and that the protease conditions are optimal to amplify DNA inside the cells.

If a signal is seen with the negative control (DNase, irrelevant primers), then inadequate protease digestion is by far the most likely problem. The test should be repeated after increasing the protease digestion time. Alternatively, an absent or weak signal in the no DNase treated control slide, in conjunction with poor tissue morphology, most likely implies repeating the test with a decreased protease digestion time.

The DNase solution can be removed with a short rinse in DEPC treated water and then 100% ethanol (which facilitates drying of the slide) using sterile technique. At times, it may not be possible to place three tissue sections on the same glass slide as the sections are too large. For example, brain sections from an autopsy are usually 15 – 20 mm in size, which permits only one section to be placed on a silane-coated slide. In such cases one can divide the tissue into three parts. One part can be DNased without RT (negative control), the other not DNased (positive control), and the last DNased and treated with RT. If one uses this method it is very useful to have a serial section stained with hematoxylin and eosin (H&E) to determine which part of the slide may be best suited for the DNase and RT step. This is particularly important when studying cancers as often one must use large resected sections containing a relatively small area afflicted by cancer. However, in such cases where large tissue sections are being analyzed, we recommend using three serial sections (each sections 4 μm apart, so that the same cells are being tested), and using one slide as the no DNase control, another as the DNase, irrelevant primers control, and the final slide as the DNase, test primers sample.

The following is the recommended protocol for RT *in situ* PCR.

Slide Preparation

1. Place three 4 μm thick sections or three cellular suspensions on a silane coated glass slide
2. Deparaffinize tissue section: 3 min fresh xylene and 3 min fresh 100% ethanol[1]; air dry
3. Digest with pepsin (2 – 6 mg/ml) at RT for 15 – 90 min depending on the length of time of fixation in formalin and tissue type
4. Inactivate pepsin with 1 min wash in DEPC water, 1 min wash in 100% ethanol[2]; air dry

Digestion with RNase Free DNase

1. Digest two of the three tissue sections overnight with RNase free DNase. Per each section add 1 μl of 10 × EZ rTth buffer, 1 μl of RNase-free DNase (Boerhinger Mannheim, 10 U/μl) and 8 μl of DEPC water
2. Cover the solution with the inside of the autoclaved polypropylene bag[3] to prevent drying
3. Place slides in humidity chamber at 37°C overnight
4. Remove coverslip, wash for 1 min in DEPC water and 100% ethanol, air dry.

[1] One does not need to take RNase precautions until after protease digestion.
[2] We remove the protease and DNase by simply flooding the slides while horizontal with 1 ml of DEPC water or ethanol that is stored in sterile 50 ml conical tubes; we use sterile 1 ml pipette tips to transfer the DEPC water and ethanol to the slides. These sterile precautions are not needed after the cDNA synthesis step.
[3] The polypropylene bags are best prepared by cutting in 15 cm sheets, placing in plastic container (e.g. empty pipette rack), then autoclaving although the latter is not essential since one can use the inner part of the plastic which has not been exposed to handling and the concomitant possibility of RNase contamination. The two layers of the plastic bag may be difficult to separate after autoclaving. They can be separated with a sterile toothpick or needle and then cut to size.

RT and PCR Steps

Although the enzyme Taq polymerase has some RT activity, the level of activity is far below that of conventional RTs, such as that obtained from MMV. Further, the reagent conditions that would support the synthesis of cDNA from RNA by Taq polymerase are not conducive to the PCR amplification activity of the enzyme. These problems have been resolved by the modified Taq polymerase rTth. Using this enzyme and the reagent conditions described below, one can synthesize cDNA from RNA and then PCR amplify it in one reaction mixture by varying the time and temperatures of the thermal cycler.

(1) Prepare the following solution: 10 μl of the EZ rTth buffer[4], 1.6 μl each of dATP, dCTP, dGTP, and dTTP, (10 mM stock[4]), 1.6 μl of 2% (w/v) bovine serum albumin 1.0 μl of RNasin[4], 3.0 μl of primer 1 and primer 2 (20 μM stock of each primer)[5], 13 μl DEPC water, 12.4 μl of 10 mM MnCl, 0.6 μl of digoxigenin dUTP (1 mM stock) and 2.0 μl of the rTth.
(2) cDNA synthesis: 65°C for 30 min
(3) Denaturation: 94 °C for 3 min
(4) PCR amplification: 20 cycles at 60°C for 1 min and 94°C for 30 s
(5) Do stringent wash (0.2 XSSC and 0.2% BSA at 60 °C for 10 min)
(6) Incubate with an alkaline phosphatase-antidigoxigenin conjugate (1:150 dilution in 0.1M Tris HCl, pH 7.4, and 0.1 M NaCl) for 30 min
(7) Chromagen: nitroblue tetrazolium and 5-bromo-4-chloro-3-indolyl-phosphate (NBT/BCIP) for 10 min, counterstain for 3 min in nuclear fast red, permount and coverslip.

◆◆◆◆◆◆ APPLICATIONS OF RT *IN SITU* PCR

HIV-1 Infection of the Cervix

HIV-1 has become an important risk factor for the human papillomavirus infection (HPV) and the development of HPV associated lesions in the female genital tract. HIV-1 also may increase the oncogenicity of high-risk HPV types. The Center for Disease Control, CDC 1993 declared invasive cervical cancer as an AIDS defining illness in HIV positive women, furthermore cervical cancer is the second most common female cancer worldwide.

The vagina and cervix represent the first line of physical and immunological defense against sexually transmitted diseases (STDs), moreover the cervical epithelial and antigen-presenting cells (Langerhans cells and endocervical macrophges) may play a critical role in HIV-1 infection (Nuovo *et al.*, 1993b).

In fact, women have assumed an important role in the epidemic of HIV-1 infection. In 1990 HIV-1 infection was the sixth most common

---

[4] Reagents available in the EZ RT PCR kit (Perkin Elmer).
[5] Omit the primers or use irrelevant primers for the negative control; substitute DEPC water or use irrelevant primers.

cause of death among 25–44 year old women in the United States. Currently, women represent 43% of all adults living with HIV and AIDS worldwide and, on a world-wide basis, heterosexual transmission is the most common method of spread of HIV-1, with women being at greater risk of infection than men. HIV and HPV are both sexually transmitted diseases. Prevalence of HPV infection among HIV-1 seropositive women is, therefore, high due to risk factors such as having sexual intercourse at an early age, having multiple partners and harbouring other STDs. Several studies have demonstrated high levels of HPV DNA in cervicovaginal washings (2 – 3 times more) and in anal swabs (15 times more) in HIV-1 seropositive than in seronegative women (Davis *et al.*, 2001; Fruchter *et al.*, 1996; Laga *et al.*, 1992; Luque *et al.*, 1999; Minkoff *et al.*, 1999; Ruche *et al.*, 1998). Furthermore HIV-1 infection has become an important risk factor in HPV infection, and in the development of HPV-associated lesions in the female genital tract. In fact, HIV seropositive women are about five times as likely as HIV seronegative women to have squamous intraepithelial lesions (SIL, equivalent to cervical intraepithelial neoplasia). Many authors have reported a high prevalence and severity of genital tract HPV infection in HIV-1 positive woman (Fruchter *et al.*, 1996; Laga *et al.*, 1992; Luque *et al.*, 1999; Minkoff *et al.*, 1999; Ruche *et al.*, 1998). Recent studies have correlated the plasma level of HIV-1 as a predictor for HPV infection (Davis *et al.*, 2001). Although high levels of HIV-1 RNA were found to be associated with cervical infection by oncogenic HPVs, most HPV infections are self limited in immuno-competent individuals, such that only 2 – 3% of the patients develop dysplasia, despite a much greater prevalence of assymptomatic HPV infection. The determining factors of disease progression are mainly the HPV genotype, the viral load and the persistence of infection (Nuovo, 1997).

Cell-mediated immune responses play an important role in controlling HPV-associated neoplasias. HPV-associated lesions are usually transient, and presumably regress as a result of a cellular immune response. HPV undergoes a period of clinical latency but frequently reappears, especially in HIV seropositive women (Fruchter *et al.*, 1996; Nuovo, 1997). Furthermore premalignant lesions can regress spontaneously, and if progression to cervical tumors occurs, a cellular infiltrate is seen consisting of $CD4^+$ and $CD8^+$ T cells, monocytes, macrophages and granulocytes (Nuovo *et al.*, 1993b).

We have performed several studies investigating the localization of HIV-1 and HPV in the cervices of women with HIV-1 infection. It is important to stress that, for HPV, standard *in situ* hybridization is usually adequate, as infection by this virus is associated with copy numbers that are often greater than 500 per cell (see chapter by Holm); standard *in situ* hybridization can detect a target when about 10 copies per cell are evident (Nuovo *et al.*, 1991a; Nuovo, 1997). However, HIV-1 DNA is usually found as one integrated copy per infected cell and, thus, one must use PCR *in situ* hybridization to detect the provirus. In cells latently infected by HIV-1, rare transcripts are made and, thus, RT *in situ* PCR is the preferable method. Productive HIV-1 infection is associated with a variety of transcripts, including spliced messages, often in copy numbers greater than 10 per infected cell. However, we still recommend using RT *in situ* PCR under such conditions as the RNA probes are more expensive and have a much shorter shelf-life than primers used in RT *in situ* PCR.

Our studies with HIV-1 and HPV localization in the female genital tract have documented the following (reviewed in Nuovo, 1997):

1. HPV localizes to the dysplastic squamous cells and is not found in any inflammatory cell or normal appearing glandular or squamous cell;
2. HPV copy number is lowest in the basal layer of the epithelium, and increases in the dysplastic cells towards the surface. This increase is marked in low grade SILs and less pronounced in high grade SILs;
3. The primary target for HIV-1 in the cervix is the endocervical macrophage;
4. The S-100 Langerhans cell (intraepithelial) can also be infected by HIV-1;
5. The squamous cell is very rarely infected by HIV-1—thus, co-infection by HIV-1 and HPV occurs rarely, if at all;
6. Surprisingly, the CD4 cell in the cervix (so-called chronic cervicitis) is rarely infected by HIV-1.

These data address certain important points regarding HIV-1 sexual transmission. First, since the endocervical macrophage is the primary target, it is clear why women are at increased risk for transmission by heterosexual sex. Macrophages are much less common at sites with thickened squamous epithelium, such as the penis, which also may serve as a physical barrier to viral transmission. Further, this data explains why men are at an increased risk through homosexual intercourse since the ano-rectal junction is histologically equivalent to the transformation zone of the cervix. These data also demonstrate that enhancement of HPV infection by HIV-1 is not caused as a result of direct co-infection, but rather by possibly altering T-cell responses and the associated cytokine milieu.

References

Chou, Q., Russell, M., Birch, D. E., Raymond, J. and Bloch, W. (1992). Prevention of pre-PCR mis-priming and primer dimerization improves low copy number amplifications. *Nucleic Acid. Res.* **20**, 1717–1723.

Cioc, A. and Nuovo, G. J. (2002). Correlation of viral detection with histology in cardiac tissue from patients with sudden, unexpected death. *Mod. Pathol.* **15**, 914–922.

Cioc, A., Nuovo, G. J., Allen, C., Kalmar, J. and Suster, S. (2004). Oral plasmablastic lymphomas in AIDS patients are associated with human herpes virus 8. *Am. J. Surg. Path* **28**, 41–45.

Davis, A. T., Chakraborty, H., Flowers, L. and Mosunjac, B. (2001). Cervical dysplasia in women infected with the human immunodeficiency virus (HIV): a correlation with HIV viral load and $CD4^+$ count. *Gynecol. Oncol.* **80**, 350–354.

Fruchter, R. G., Maiman, M., Sedlis, A., Bartley, L., Camilien, L. and Arrastia, C. D. (1996). Multiple recurences of cervical intraepithelial neoplasia in women with the human immunodeficiency virus. *Obstet. Gynecol.* **87**, 338–344.

Haase, A. T., Retzel, E. F. and Staskus, K. A. (1990). Amplification and detection of lentiviral DNA inside cells. *Proc. Natl Acad. Sci. USA* **87**, 4971–4975.

Hilders, C. G. J. M., Houbiers, J. G. A., Van Ravenswaay Claasen, H. H., Veldhuizen, R. W. and Fleuren, G. J. (1993). Association between HLA-expression and infiltration of immune cells in cervical carcinoma. *Lab Invest* **60**, 651–659.

Laga, M., Icenogle, J. P., Marselha, R., Manoka, A. T., Nzila, N., Ryder, R. W., Vermund, S. H., Heyward, W. L., Nelson, A. and Reeves, W. C. (1992). Genital papillomavirus infection and cervical dysplasia-opportunistic complications of HIV infection. *Int. J. Cancer* **50**, 45–48.

Luque, A. E., Demeter, L. M. and Reuchman, R. C. (1999). Association of human papillomavirus infection and disease with magnitude of human immunodeficiency virus type 1 (HIV-1) RNA plasma level among women with HIV-1 infection. *J. Infect. Dis.* **179**, 1405–1409.

Morrison, C., Eliezri, Y., Magro, C. and Nuovo, G. J. (2002). The histologic spectrum of epidermodysplasia verruciformis in transplant and AIDS patients. *J. Cutan. Pathol.* **29**, 480–489.

Minkoff, H. L., Eisenberger-Matityahu, D., Feldman, J., Burk, R. and Clark, L. (1999). Prevalence and incidence of cynecologic disorders among women infected with human immunodeficiency virus. *Am. J. Obstet. Gynecol.* **180**, 824–836.

Nuovo, G. J. (1997). *PCR In Situ Hybridization: Protocols And Applications*, 3rd edn. Lippincott, Williams & Wilkins, New York.

Nuovo, G. J. and Forde, A. (1995b). An improved system for reverse transcriptase in situ PCR. *J. Histotech.* **18**, 295–299.

Nuovo, G. J., Gallery, F., MacConnell, P., Becker, J. and Bloch, W. (1991a). An improved technique for the *in situ* detection of DNA after polymerase chain reaction amplification. *Am. J. Pathol.* **139**, 1239–1244.

Nuovo, G. J., MacConnell, P., Forde, A. and Delvenne, P. (1991b). Detection of human papillomavirus DNA in formalin fixed tissues by in situ hybridization after amplification by the polymerase chain reaction. *Am. J. Pathol.* **139**, 847–854.

Nuovo, G. J., Gorgone, G., MacConnell, P. and Goravic, P. (1992a). *In situ* localization of human and viral cDNAs after PCR-amplification. *PCR Meth. Appl.* **2**, 117–123.

Nuovo, G. J., Margiotta, M., MacConnell, P. and Becker, J. (1992b). Rapid *in situ* detection of PCR-amplified HIV-1 DNA. *Diagn. Mol. Pathol.* **1**, 98–102.

Nuovo, G. J., Gallery, F., Hom, R., MacConnell, P. and Bloch, W. (1993a). Importance of different variables for optimizing *in situ* detection of PCR-amplified DNA. *PCR Meth. Appl.* **2**, 305–312.

Nuovo, G. J., Forde, A., MacConnell, P. and Fahrenwald, R. (1993b). In situ detection of PCR-amplified HIV-1 nucleic acids and tumor necrosis factor cDNA in cervical tissues. *Am. J. Pathol.* **143**, 40–48.

Nuovo, G. J., Becker, J., Margiotta, M., Burke, M., Fuhrer, J. and Steigbigel, R. (1994). In situ detection of PCR-amplified HIV-1 nucleic acids in lymph nodes and peripheral blood in asymptomatic infection and advanced stage AIDS. *J. Acquired Immun. Def.* **7**, 916–923.

Nuovo, G. J., MacConnell, P., Valea, F. and French, D. L. (1995a). Correlation of the in situ detection of PCR-amplified metalloprotease cDNAs and their inhibitors with prognosis in cervical carcinoma. *Cancer Res.* **55**, 267–275.

Ruche, G. L., You, B., Mensah-Ado, I., Bergeron, C., Montcho, C., Ramon, R., Touré-Coulibaly, K., Welffens-Erra, C., Dabis, F. and Orth, G. (1998). Human papillomavirus and human immunodeficiency virus infections: relation with cervical dysplasia-neoplasia in African women. *Int. J. Cancer* **76**, 480–486.

10 Real-time Fluorescent PCR Techniques to Study Microbial–Host Interactions

Ian M Mackay[1,2], Katherine E Arden[1,2] and Andreas Nitsche[3]

[1] Clinical Virology Research Unit, Sir Albert Sakzewski Virus Research Centre, Royal Children's Hospital, Brisbane, Qld, Australia; [2] Clinical Medical Virology Centre, University of Queensland, Brisbane, Qld, Australia; [3] Robert Koch Institute, Berlin, Germany

◆◆

CONTENTS

Introduction
The ABC of PCR
The good with the bad
Generating a fluorescent signal
Real-time instrumentation
Detecting amplicon using fluorescence
Microbial genotyping
Nucleic acid quantitation
Multiplex real-time PCR
Application of real-time PCR to microbiology
Detecting agents of biowarfare
Host immunity: measuring the response to a microbe

◆◆◆◆◆◆ INTRODUCTION

The development and introduction of new investigative molecular technologies have begun to change the way we think about virulence and pathogenicity. One such technology is the real-time polymerase chain reaction (PCR). In the context of infectious disease, diagnostic microbiology can be divided into two broad sections: routine qualitative screening or detailed qualitative and quantitative investigations of the interplay between the microbe and host. The following review will describe how real-time PCR performs and how it may be used to detect microbial pathogens and elucidate the relationship they form with their host.

Research and diagnostic microbiology laboratories contain a mix of traditional and leading edge, in-house and commercial assays for the detection of microbes and the effects they impart upon target tissues, organs and systems. The wide variety of assays is the result of many factors including the perceived reliability of familiar techniques and technologies, the scope of a test to perform the required task, the existence and cost of commercial tests able to detect microbes of interest, the experience a laboratory has acquired with a particular technique and finally the degree of support offered by the assay manufacturer. Additionally, the cost of a new system may be over-estimated prior to its use in the laboratory, adding to slow uptake of new technologies. In the modern microbiology laboratory, an increasing role is played by molecular techniques and, in particular, the PCR (Freymuth et al., 1995; Mullis and Faloona, 1987). The PCR has undergone significant change over the last decade, to the extent that only a small proportion of scientists have been able or willing to keep abreast of the latest offerings. This chapter will review these changes, bringing the reader up to date with the second-generation of PCR technology: kinetic or real-time PCR, a tool gaining widespread acceptance in many scientific disciplines but especially in the microbiology laboratory (Ginzinger, 2002; Whelan and Persing, 1996).

♦♦♦♦♦♦ THE ABC OF PCR

Microbiology is in the midst of a new era. Technologies that detect nucleic acids have driven the evolution of a plethora of new experimental tools rapidly replacing some traditional methods, which depend on microbial phenotype rather than genotype. For example, it has become common to relegate viral culture to specialised virology laboratories rather than include it in the routine diagnostic laboratory, which can then focus on rapid result turnaround (Carman, 2001). The speed with which a negative result is provided is often as important as the return of a positive result and, when cultures can require weeks to complete, the PCR offers an attractive alternative (Carman, 2001). Of course, speed is the result of a number of factors, and they must all be taken into account when assessing the benefits of any new PCR-based assay. Faster assays must be described using the final optimised and validated test performed on real-world specimens instead of presenting preliminary data using idealised templates and conditions. Unfortunately, incomplete PCR assays are frequently rushed into publication resulting in problems reproducing data in other laboratories.

The PCR is the most commonly used molecular technique to detect and study microbes, appealing more widely than specific commercial template amplification technologies such as Abbott's ligase chain reaction (LCR) (Barany, 1991), Organon Teknika's nucleic acid sequence-based amplification (NASBA) (Compton, 1991; Kievits et al., 1991), Becton Dickinson's strand-displacement amplification (SDA) (Walker et al., 1992) and Gen-Probe's transcription-mediated amplification (TMA). PCR has also been applied more widely than the signal amplification methods such as Bayer's branched DNA (bDNA) and Digene's Hybrid Capture (Fredricks and Relman, 1999; Whelan and Persing, 1996; Persing, 1993). The PCR method utilises a pair of synthetic oligonucleotides called primers, each one hybridising in a 5' to 3' orientation to a single strand of a DNA target. The primer pair spans a region that is exponentially duplicated during the subsequent reaction. Each hybridised primer forms a starting point for the production of complementary DNA via the action of a DNA polymerase derived from thermophilic bacteria. The process of primer extension creates a strand that is the complement of the template through the sequential addition of deoxynucleotides. The PCR can be summarised in three steps: (i) double-stranded DNA (dsDNA) separation at temperatures above 90°C, (ii) primer annealing at a specific temperature commonly between 50 and 60°C and, (iii) optimal extension of the primed template at 70–78°C (Figure 10.1). A compact PCR format is recommended with some PCR assays which require only two steps: the denaturation and a combined annealing and extension step.

The oligonucleotide annealing temperature is usually referred to as the T_M (melting temperature) but is in fact 5–10°C below the T_M. The term T_M describes the temperature at which 50% of the oligonucleotide–target duplexes have formed. The T_M is dependent upon the concentration of the DNA, its length, nucleotide sequence and the composition of the solvent in which the DNA is suspended (Ririe *et al.*, 1997).

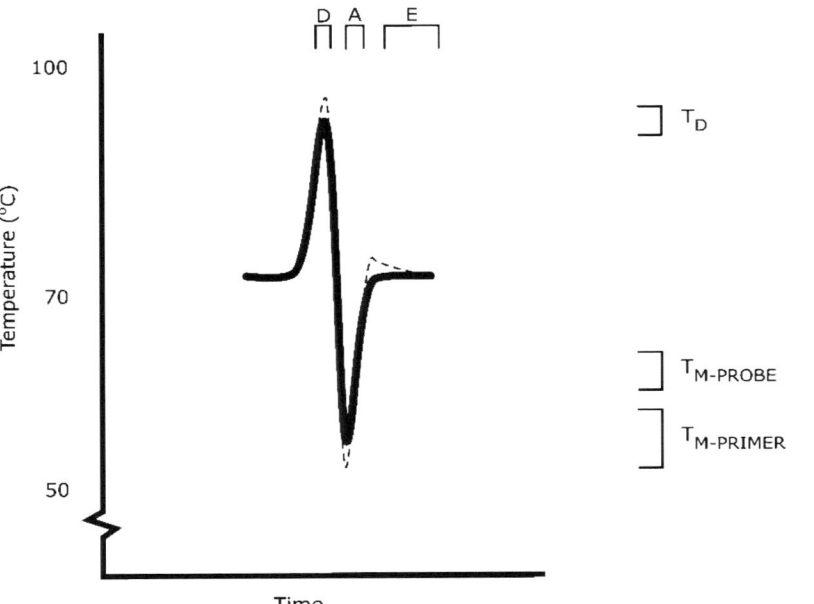

Figure 10.1. Time versus temperature plot during a single PCR Cycle. The denaturation (D), primer and probe annealing (A) and primer extension (E) steps are shown. At the indicated optimal temperature ranges, dsDNA denatures (T_D) then oligoprobes anneal ($T_{M\text{-PROBE}}$) followed by the primers ($T_{M\text{-PRIMER}}$) as a precursor to their extension. The actual thermal cycler incubation temperature (dashed line) may overshoot the desired temperature to varying degrees, depending on the quality of the thermal cycler employed.

The rate of temperature change in the reaction vessel, or ramp rate, the length of the incubation at each temperature and the number of times each cycle of temperatures is repeated, are all controlled by a programmable thermal cycler. Current technologies have significantly shortened the ramp rates and therefore total assay times, using electronically controlled heating blocks or fan-forced heated airflows.

Existing combinations of PCR and PCR product (amplicon) detection assays will be called "conventional PCR" throughout this chapter. These detection systems include analytical agarose gel electrophoresis (Kidd et al., 2000), Southern blot and ELISA-like systems such as ELAHA and ELOSA (van der Vliet et al., 1993; Chandelier et al., 2001; Mackay et al., 2001; Hyypiä et al., 1998). Traditional detection of amplified DNA relies upon its electrophoresis in the presence of ethidium bromide followed by visualisation or computer-assisted densitometric analysis of the bands during irradiation by ultraviolet light (Kidd et al., 2000). Whilst Southern blot detection of amplicon by hybridisation with a labelled oligonucleotide probe (oligoprobe) increases the specificity of amplicon detection, it is time consuming, frequently uses radioactive labels and requires multiple PCR product handling steps, increasing the risk of spreading amplicon throughout the laboratory (Holland et al., 1991). Alternatively, PCR–ELISA has been used to capture amplicon onto a solid phase via biotin or digoxigenin-labelled primers, oligoprobes or by direct capture after incorporation of the biotin or digoxigenin into the amplicon (van der Vliet et al., 1993; Keller et al., 1990; Kemp et al., 1990; Kox et al., 1996; Dekonenko et al., 1997; Watzinger et al., 2001). Once captured, amplicon is detected using an enzyme-labelled avidin or anti-digoxigenin reporter molecule in a manner similar to a standard ELISA format. PCR–ELISA has the added benefit of utilising hardware and techniques commonly available to the ELISA-enabled research and diagnostic microbiology laboratory.

The possibility that, in contrast to conventional PCR, real-time PCR could detect amplicon as it accumulated was welcomed by researchers and diagnostic scientists alike. This feat has expanded the view of PCR as a specialist tool to that of a versatile technology providing advanced and powerful systems for the research laboratory as well as permitting the development of contamination-resistant routine diagnostic applications for the clinical microbiology laboratory (Lomeli et al., 1989; Cockerill and Smith, 2002). Along the way, real-time PCR assays have provided insights into the nuts and bolts of the PCR as well as the performance of different nucleic acid extraction methods and the role some compounds play to inhibit amplification (Holland et al., 1991; Rosenstraus et al., 1998; Lee et al., 1993; Livak et al., 1995; Heid et al., 1996; Gibson et al., 1996; Niesters et al., 2000; Read, 2001; Biel et al., 2000; Petrik et al., 1997). Real-time PCR has allowed many more scientists to become familiar with the crucial factors contributing to successful amplification of nucleic acids and in its largest role to date, made the quantitation of gene transcription much simpler and faster. This has proven to be a valuable area of study since many cellular functions are regulated by changes in gene expression (Bustin, 2002; Balnaves et al., 1995). Today in microbiology, real-time PCR detects and quantifies nucleic acids from widely diverse targets including food, viral and non-viral vectors used in gene therapy protocols, genetically modified organisms and to study human and veterinary microbiology, oncology and immunology (Böhm et al., 1999; Kruse et al., 1997; Stordeur et al., 2002; Härtel et al., 1999; Fraaije et al., 2001; Nogva et al., 2000; Barzon et al., 2003; Klein, 2002; Rudi et al., 2002; Mackay et al., 2002; Ahmed, 2002; Mhlanga and Malmberg, 2001).

The monitoring of accumulating amplicon in real time has been made possible by the labelling of primers, oligoprobes or amplicon with molecules exhibiting fluorescent potential. The success of these assays revolves around a rapid and measurable signal change after the interaction of amplicon and fluorescent label (Morrison et al., 1989). The signal is related to the amount of amplicon present during each cycle, increasing as the DNA is replicated. The fluorescent chemistries have clear benefits over earlier radiogenic labels including an absence of radioactive emissions, easy disposal and an extended shelf life (Matthews and Kricka, 1988).

Although some of the oligoprobe systems have been given a specific nomenclature by their developer, we will use the term "fluorophore" to describe the fluorescent moieties, and their inclusion on an oligonucleotide will imply that the resulting oligoprobe has fluorogenic properties.

◆◆◆◆◆◆ THE GOOD WITH THE BAD

Traditional diagnostic microbiological assays include microscopy, microbial culture and ELISA, which aim to detect microbial antigens or the host response to microbial presence. The performance of these assays can be limited by poor sensitivity, slowly growing or poorly cytopathic microbes, reduced microbial viability, narrow detection windows, complex result interpretation, host immunosuppression, antimicrobial therapies, high levels of background signal and non-specific cross-reactions (Whelan and Persing, 1996; Carman et al., 2000). Nonetheless, microbial culture and rapid immunofluorescence assays are used to produce valuable epidemiological data, revealing new, uncharacterised or atypical microbes and yielding intact or infectious organisms for further study (Ogilvie, 2001). Although detection of microbial genomic nucleic acids is not identical to the detection of infectious particles, there is good correlation between infectivity and the viral genome as we have found for yellow fever virus (Bae et al., 2003).

Fluorescence microscopy still remains a popular and rapid screening tool for samples containing large microbial loads whilst ELISA and related methods are ideal for identifying detectable levels of immune or cell function modulators (Lipson, 2002). It is, therefore, clear that the role of the traditional assay, be it in screening or research, continues to be an important one (Biel and Madeley, 2001; Pfaller, 1999; Sintchenko et al., 1999; Ellis and Zambon, 2002; Clarke et al., 2002; Johnson, 2000).

The PCR does have some significant limitations. Our ability to design oligonucleotide primers only extends to our knowledge of the sequence of the template (usually the genome of a microorganism or its host) as well as the ability of publicly available sequence databases to suitably represent all variants of that target sequence. It is common for microbial genomes to contain unexpected "mutations", i.e. nucleotide changes compared to known microbial sequences, which reduce or abrogate the function of a PCR. Additionally, it is often the effects of these variations on colonisation and pathogenesis which warrant further study. Each PCR assay requires careful optimisation of several variables, ensuring that primer and magnesium concentrations are perfected as well as selecting the best polymerase for the assay (Kreuzer *et al.*, 2000; Wolffs *et al.*, 2004). Also, new batches of reagents must be carefully tested against previous batches (Burgos *et al.*, 2002). This testing is especially important for oligonucleotides since all manufacturers have "bad days" that equate to wasted time and added frustration for the assay developer. The use of standardised assay panels and the introduction of international units have begun to improve microbial diagnostics for the limited number of pathogens these encompass (Valentine-Thon, 2002).

False positive results due to amplicon carry-over contamination have always been a major concern for the routine implementation of PCR in the diagnostic laboratory and this has led to strict guidelines for laboratory design and work flow (McCreedy and Callaway, 1993). There are also occasions when the PCR is too sensitive, detecting a microbe that is present below pathogenic levels. Thus, care is required not only for assay design but also for result interpretation.

Real-time PCR assays also carry specific disadvantages compared to conventional PCR, including the inability of certain platforms to detect some fluorogenic chemistries and the relatively restricted multiplex capabilities of current systems. Also, the start-up expense of real-time PCR instruments is prohibitive for many low-throughput laboratories and ongoing costs are higher when compared to agarose gel amplicon detection although lower than many PCR–ELISA formats.

A significant improvement introduced by real-time PCR is the increased speed with which results can be produced. This is largely due to the removal of a separate post-PCR detection step, the use of sensitive fluorescence-detection equipment which permits earlier amplicon detection and the shortened cycle times possible on some instruments (Wittwer *et al.*, 1990, 1997b). A reduced amplicon size may also increase assay speed, however, we and others have shown that decreased product size does not strictly correlate with improved PCR efficiency but the distance between the primers and the oligoprobe does play a vital role (Nitsche *et al.*, 2000; Balnaves *et al.*, 1995; Lunge *et al.*, 2002).

◆◆◆◆◆◆ GENERATING A FLUORESCENT SIGNAL

Most of the popular real-time PCR chemistries involve hybridisation of one or more fluorescent oligoprobes to a complementary sequence on one of the amplicon strands. Therefore, the inclusion of more of the primer that creates the strand complementary to the oligoprobe – a process called asymmetric PCR – is often beneficial to the generation of an increased fluorescent signal (Gyllensten and Erlich, 1988; Barratt and Mackay, 2002). However, it is critical to determine which strand to target, particularly in the case of single stranded genomes such as those harboured by negative or positive sense viruses.

As mentioned earlier, post-amplification manipulation of the PCR product is not necessary to detect amplicon using real-time PCR because the fluorescent signals are directly measured as they exit the reaction vessel. Therefore, real-time PCR assays are often described as homogeneous or "closed" systems; combining template amplification and amplicon detection into a single reaction. Apart from the time saved by amplifying and detecting template in a single tube, there is minimal potential for amplicon carry-over contamination and the assay's performance can be closely scrutinised without introducing errors due to handling of the PCR product (Higuchi et al., 1993). An often overlooked benefit to the use of real-time PCR is its overall cost effectiveness on a per-run basis, when implemented in a high-throughput laboratory (Martell et al., 1999), particularly when replacing conventional PCR and gel or probe-based amplicon detection systems or microbial culture.

The most commonly used fluorogenic oligoprobes rely upon fluorescence resonance energy transfer (FRET; Figure 10.2) between fluorogenic labels or between one fluorophore and a dark, or black-hole, non-fluorescent quencher (NFQ), which disperses energy as heat rather than fluorescence (Didenko, 2001). FRET is a non-radiative process in which energy is passed between permissive molecules that are spatially separated by 10–100 Å and which have overlapping emission and absorption spectra (Stryer and Haugland, 1967; Heller and Morrison, 1985; Clegg, 1992). Förster primarily developed the theory behind this process (Förster, 1948). The energy transfer reduces the lifetime of the excited state of electrons in the original fluorophore by taking the emitted excess energy and expelling it either as heat or fluorescence. The efficiency of energy transfer is proportional to the inverse sixth power of the distance (R) between the donor and acceptor ($1/R^6$) fluorophores (Selvin, 1995; Didenko, 2001). FRET permits the determination of nucleic acid hybridisation, without the prior removal of unbound probe required for other hybridisation techniques (Cardullo et al., 1988).

Fluorescence data generated by real-time PCR assays are generally collected from PCR cycles that occur early in the reaction where amplification conditions are optimal and the fluorescence accumulates in proportion to the amplicon (Figure 10.3). Because the emissions from fluorescent chemistries are temperature-dependent, data is generally acquired only once per cycle, at the same temperature (Wittwer et al., 1997a). Signal detection occurring at the end point of the reaction is not ideal since amplicon accumulation may have been adversely affected by inhibitors, poorly optimised reaction conditions or saturation effects caused by excess double-stranded amplicon. In fact at the end point there may be no relationship between the initial template and final amplicon concentrations (Figure 10.3).

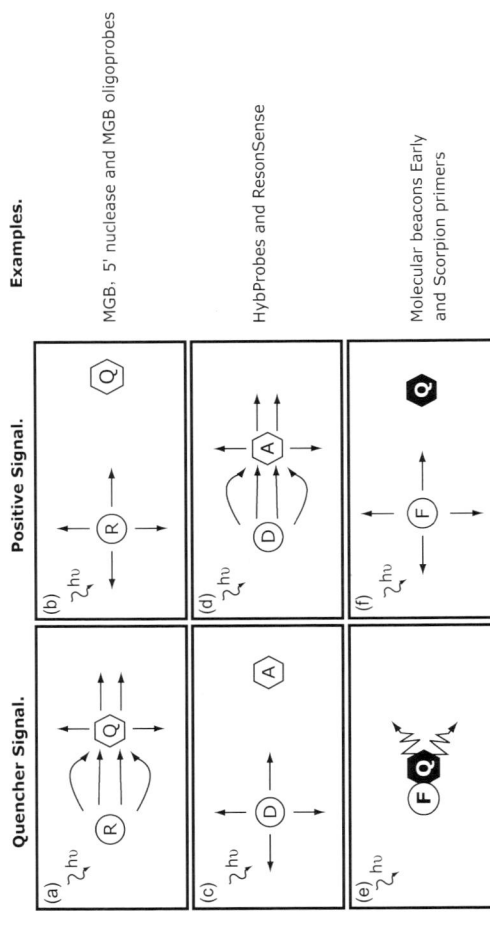

Figure 10.2. Mechanisms of fluorescence resonance energy transfer (FRET). When the reporter (R) and quencher (Q, unfilled) of a nuclease oligoprobe are in close proximity and illuminated by an instrument's light source (where h is Planck's constant and ν is the frequency of the electromagnetic radiation): (a) the quencher "hijacks" the emissions from excitation of the reporter. The quencher then emits this energy. When the fluorophores are separated, as occurs upon oligoprobe hydrolysis as depicted in (b), the quencher can no longer influence the reporter which now fluoresces at a distinctive wavelength recorded by the instrument. In the reverse process using adjacent oligoprobes (c), the fluorophores begin the cycle as separated entities. Whilst the emission of the donor (D) is monitored, it is the signal from the acceptor (A) produced when in close proximity to the donor that indicates a positive reaction (d). In (e), another form of quenching is shown, caused by the intimate contact of labels attached to hairpin oligoprobes. The fluorophore (F) and an NFQ (Q, filled) interact more by collision than FRET, disrupting each other's electronic structure and directly passing on the excitation energy which is dissipated as heat (jagged, arrows). When the labels are separated, as is the case in (f), the fluorophore is free to fluoresce.

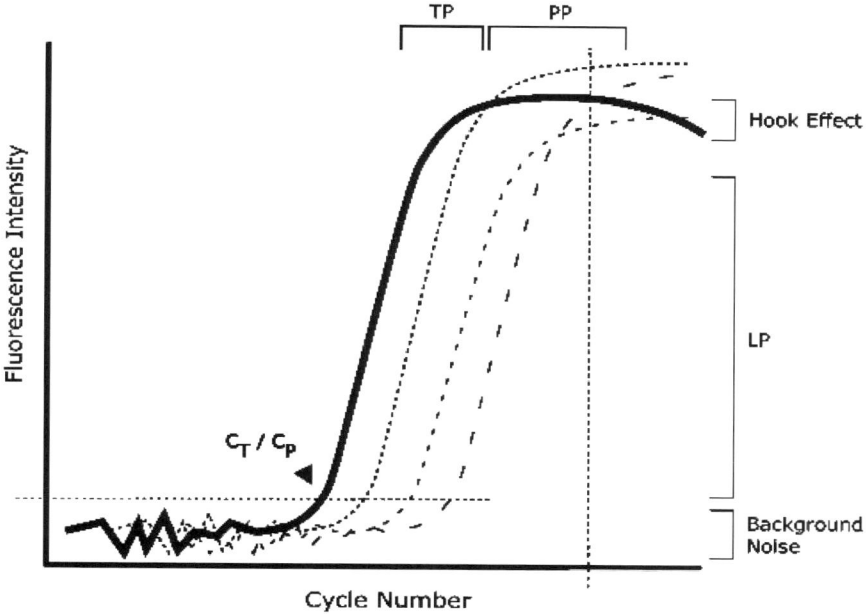

Figure 10.3. Kinetic analysis of amplicon accumulation. A chart of the amplification of a template by real-time PCR (solid line) ideally appears as a sigmoidal curve when plotted as cycle number versus fluorescence emission intensity. Early exponential amplification cannot be viewed because the signal is below the sensitivity of the detector. However, when enough amplicon is present, the assay's exponential progress can be monitored as the rate of amplification enters a log-linear phase (LP). Under ideal conditions, the amount of amplicon increases at a rate of one $\log_{10}$ every 3.32 cycles (i.e. it doubles, or increases by one $\log_2$ every cycle). As primers and enzymes become limiting and products accumulate that are inhibitory to the PCR or compete for hybridisation with an oligoprobe, the reaction slows, entering a transition phase (TP). Eventually a plateau phase (PP) is reached in which there is little or no increase in fluorescent signal although amplicon may continue to accumulate. Some fluorescent detection chemistries display an overall reduction in fluorescence intensity after the plateau phase (Hook). The point at which fluorescence surpasses a pre-defined background noise (dashed horizontal line) is called the threshold cycle or crossing point (C_T or C_P; indicated by an arrow). These data are used for the calculation of template quantity when constructing a standard curve. Traditional PCR data collection is performed at the end of the assay (dashed vertical line). Also shown are curves representing a serial titration of template (dashed curves), consisting of decreasing starting template concentrations, which produce increasing numerical C_T or C_P values.

The fractional cycle number at which the real-time fluorescence signal mirrors progression of the amplification reaction above the background noise level is used as an indicator of successful target amplification (Wilhelm et al., 2001a). Most commonly, this is called the threshold cycle (C_T) but the same value is described for use with the LightCycler where the fractional cycle is called the crossing point (C_P). The C_T is defined as the PCR cycle at which the gain in fluorescence generated by the accumulating amplicon exceeds 10 standard deviations of the mean baseline fluorescence, using data taken from cycles 3 to 15

(Jung *et al.*, 2000). The C_T and C_P are proportional to the number of template copies present in the sample (Gibson *et al.*, 1996). The values are assumed to represent equal amounts of amplicon in each vessel at the time of measurement, since the C_T and C_P values are acquired from a single fluorescence intensity value (Figure 10.3). In practice C_T and C_P are calculated after the definition of a noise band which excludes data from early PCR cycles that cannot be distinguished from background fluorescence. Therefore the final C_T and C_P value is the fractional cycle at which a specimen's plotted PCR curve intersects a single fluorescence value (usually at or close to the noise band; Figure 10.3, Wilhelm *et al.*, 2001a). The accuracy of the C_T or C_P depends upon the concentration and nature of the fluorescence-generating system, the amount of template initially present, the sensitivity of the fluorescence detector and its ability to discriminate specific fluorescence from background noise.

◆◆◆◆◆◆ REAL-TIME INSTRUMENTATION

Broadly speaking, real-time PCR instruments (Table 10.1) can be divided into two classes based on the method used to heat the reaction vessel: the block thermal cyclers and the air thermal cyclers. Block cyclers generally ramp at 1–10°C/s whereas heated air cyclers can ramp as quickly at 20°C/s. The faster rates permit 10–20 min assays (excluding hands-on time); however, these times are infrequently reported for microbial real-time PCR assays, probably due to technical difficulties when applied to clinical specimens under real-life conditions. An off-shoot of the block cyclers are miniaturised thermal cyclers capable of performing real-time PCR but constructed of sturdy solid-state and relatively low-cost materials permitting transport of the PCR laboratory onto the site of testing (Northrup, 1998; Ibrahim *et al.*, 1998).

The quality of the instrument has a significant role in the reproducibility of results, which subsequently influences how well a described protocol transports between laboratories around the world. A common discrepancy between systems is the temperature profile of a PCR cycle (Schoder *et al.*, 2003). Since temperature affects enzyme function, fluorescence and oligonucleotide binding, it is an important consideration during the purchase of a real-time PCR system.

The majority of PCR thermal cyclers use capped plastic reaction vessels to house the amplification reaction. However, the LightCycler (Roche Molecular Biochemicals, Germany) differs by using glass capillaries (Wittwer *et al.*, 1997a). The LightCycler's plastic and glass composite capillaries are optically clear and act as cuvettes for fluorescence analysis, as well as facilitating rapid heat transfer. However, they are fragile and require some experience to handle (Schalasta *et al.*, 2000a). For the Roche LightCycler and the Corbett Robotics RotorGene, the temperature is varied by rapidly heating and cooling air using a heating element and fan which produce ramp rates of 2.5 and 20°C/s, respectively.

Table 10.1. Popular real-time PCR instrumentation

Brand	Company	Light source	Capacity	Heating method	Real time
LightCycler	Roche	LED	32	Air	✓
iCycler	BioRad	Halogen	96–384	Block	✓
Mx4000	Stratagene	Halogen	96	Block	✓
RotorGene	Corbett	LEDs	36–72	Air	✓
7700 (discontinued)	Applied Biosystems	Laser	96	Block	✓
7900HT	Applied Biosystems	Laser	96	Block	✓
7000	Applied Biosystems		96	Block	✓
Chimaera	Hybaid	Halogen	96	Block	✓
DNA Engine Opticon 2	MJ Research	LEDs	96	Block	✓
Smartcycler	Cepheid	LEDs	16	Block	✓

Faster rates prolong polymerase survival and significantly shorten the assay's completion time but existing assays may require some fine tuning (Weis et al., 1992).

In the following sections, we will focus on real-time PCR performance especially the rapidly expanding range of real-time PCR chemistries. Also, we will review the application of real-time PCR to determine the interplay between microbe and host via microbial diagnosis and quantitation. We will also describe the limitations imparted upon real-time PCR by instrument design and the paucity of fluorophores. Although this review focuses on the application of fluorogenic chemistries to real-time PCR applications, they are also valuable as fluorogenic end-point amplicon detection assays. In these instances the ability to collect data as the amplicon accumulates, i.e. kinetic analysis, is lost but the systems retain a homogeneous format.

◆◆◆◆◆◆ DETECTING AMPLICON USING FLUORESCENCE

It is the amplicon detection processes that discriminate real-time PCR from conventional PCR assays. There are a range of chemistries currently in use and these can be broadly categorised as amplicon specific or non-specific (Whitcombe et al., 1999). To further clarify, the amplicon itself is produced from a PCR using sequence-specific primers. However, the specificity we refer to is that of amplicon detection. The difference arises from the use of a fluorophore that interacts with any and all dsDNA such as SYBR® green I, or the use of a sequence-specific, fluorogenic oligoprobe. In general, however, the specific and non-specific fluorogenic chemistries detect amplicon with equal sensitivity (Wittwer et al., 1997b) and have recently been reviewed in detail (Mackay et al., 2002). New fluorogenic systems continue to be described but few applications for the specific detection, quantitation and genotyping of microbes using these recent chemistries have been reported. Undoubtedly some of these new chemistries are the result of pure research projects; however, it is tempting to propose that at least some of these chemistries have been developed in order to circumvent patents rather than to significantly advance the technology.

DNA-associating Fluorophores

The DNA-associating fluorophore is the basis of the non-specific real-time PCR detection methods. Many of these molecules interact with dsDNA by associating with the minor groove of the DNA duplex. As a group, these fluorophores display minimal fluorescence when free in solution, but fluoresce strongly when associated with nucleic acids which occur with high affinity. The simplest real-time PCR reporter systems are included within the DNA-associating fluorophores. Ethidium bromide

(Higuchi *et al.*, 1992), YO-PRO-1 (Ishiguro *et al.*, 1995; Tseng *et al.*, 1997), SYBR® green I (Morrison *et al.*, 1998b), SYBR Gold (Tuma *et al.*, 1999), BEBO (Bengtsson *et al.*, 2003) and LCGreen (Vaisse *et al.*, 1996) are all molecules which fluoresce when associated with dsDNA which is exposed to a wavelength of light capable of exciting the fluorophore. This approach to detection is inexpensive and simpler than the design requirements of fluorogenic oligoprobes and is minimally affected by small changes in template sequence which may abrogate hybridisation of an oligoprobe (Komurian-Pradel *et al.*, 2001). However, recent evidence suggests that SYBR green I preferentially binds to amplicon species that melt at higher temperatures, indicating a preference for G + C-rich regions, which would prove detrimental to its use in multiplex assays (Giglio *et al.*, 2003). Possibly, SYBR Green I binds to high-temperature duplexes after melting off lower temperature duplexes, resulting in a hierarchy of melting peak heights expressing no relation to starting template concentration (Vaisse *et al.*, 1996). A recent addition to this chemistry, LCGreen, displays a higher sensitivity for lower temperature duplexes (Vaisse *et al.*, 1996).

Primer–dimer formation is common and is strongly associated with the entry of the PCR into a plateau phase of amplification (Figure 10.3; Chou *et al.*, 1992; Halford, 1999; Halford *et al.*, 1999). Association of a DNA-binding fluorophore with primer–dimer or with another non-specific amplification product can confuse interpretation of the PCR results. The problem of primer–dimer can be addressed using software capable of fluorescent melting curve analysis (FMCA). We will describe the use of FMCA for genotyping in a later section (Figure 10.18); however, in the context of SYBR green I, FMCA is completed in minutes, requires no amplicon manipulation and utilises the temperature at which a dsDNA amplicon denatures or "melts" (T_D; Figure 10.1). The shorter primer–dimer can theoretically be discriminated by its reduced T_D compared to the full-length amplicon. Practically, this discrimination is heavily reliant upon the G + C content of both the specific amplicon and any primer–dimer, as well as the length of the amplicon. Analysis of the melting character of an amplicon in the presence of SYBR green I has shown that the sensitivity of DNA-binding fluorophores is limited by non-specific amplification at low initial template concentrations (Wittwer *et al.*, 1997a). Unfortunately, limited template is a frequent occurrence when analysing patient specimens. The contribution of non-specific products to the accumulating fluorescence signal can be reduced in many instances by selecting a data collection temperature above the known T_D of the unwanted products (Morrison *et al.*, 1998a). However, there is no reliable real-time PCR solution to discriminate non-specific amplicon when it has a similar or higher T_D than the target amplicon species.

DNA binding fluorophores increase the T_D and broaden the melting transition, requiring substantial sequence change to produce a noticeable shift in the T_D compared to using the T_M to discriminate nucleotide polymorphisms. Oligoprobes permit the clear discrimination of single point mutation using the T_M (Wetmur, 1991). This is especially useful for discriminating related microbes.

Fluorogenic Oligoprobe and Primer Chemistries

The adoption of oligoprobes has added an additional layer of specificity to the PCR by confirming the sequence of the amplicon, in addition to the binding of a specific pair of primers. By using an excess of oligoprobe, the time required for it to hybridise with its target is significantly reduced (Wetmur, 1991). This is especially so when the amount of that target has been increased by PCR or some other molecular amplification process (Morrison et al., 1989).

While the most common oligoprobes are based on conventional nucleic acid chemistry, peptide nucleic acids (PNA) are becoming an increasingly popular choice for oligonucleotide backbones. PNA are a DNA analogue formed from neutral, repeated N-(2-aminoethyl) glycine units instead of the negatively charged sugar phosphates of DNA (Egholm et al., 1993). The PNA oligoprobe retains the same sequence recognition properties of DNA but it cannot be extended or hydrolysed by a DNA polymerase. A more recent family of DNA analogues are the locked nucleic acids (LNA; $2'$-O,$4'$-C-methylene-β-D-ribofuranosyl) (Singh et al., 1998; Koshkin et al., 1998; Obika et al., 1998; Kumar et al., 1998; Petersen and Wengel, 2003). LNA are modified nucleic acids in which the sugar has been conformationally "locked", imparting unprecedented hybridisation affinity towards DNA and RNA. LNA monomers can be added to a synthetic oligonucleotide as desired using conventional phosphoramidite chemistry and their addition increases the thermal stability of the oligo permitting the construction of shorter oligoprobes for real-time PCR applications (Simeonov and Nikiforov, 2002).

To be optimal, an oligonucleotide label must easily attach to DNA and be detectable at low concentrations. The label or labels should produce an altered signal upon specific hybridisation, remain biologically innocuous and be stable at elevated temperatures. Additionally, the label should not interfere with the activity of the polymerase (Matthews and Kricka, 1988; Holland et al., 1991).

General considerations for the design of a fluorogenic oligoprobe should include a length of 20–40 nt with a G + C content of 40–60%. The oligoprobe should neither contain clusters of a single nucleotide, particularly G, nor should it have repeated sequence patterns nor hybridise with the forward or reverse primers. A fluorescent label should efficiently absorb and emit energy and release its emissions at dissimilar wavelengths so that excitation and emission can occur concurrently. The relationship between a fluorescence signal indicating positive hybridisation and a signal from unwanted fluorescence is often referred to as the signal-to-noise ratio (signal:noise). A high signal:noise is preferred. Generally speaking, this can best be achieved by oligoprobe chemistries that utilise an NFQ and function through a unimolecular signalling configuration, i.e. only one molecule is required to directly indicate the presence of amplicon.

Deoxyguanosine nucleotides (G) naturally quench some fluorophores (e.g. FITC) in a position-dependent manner (Crockett and Wittwer, 2001). Natural quenching varies linearly with a defined concentration range of template. The level of quenching can be increased if more guanines are present or if a single guanine is located in the first non-hybridised position of the oligoprobe:amplicon duplex. In this position, the G will be located one nucleotide beyond the fluorophore-labelled terminus of the probe, where it will remain free once the oligoprobe has hybridised to the target amplicon. A hybridised G does not quench to the same degree as a free G. This knowledge has been applied to some amplicon detection approaches since a single labelled oligonucleotide is easy to design and use and relatively simple to synthesise. In addition, this approach does not require a DNA polymerase with nuclease activity (Crockett and Wittwer, 2001).

Ideally, an oligoprobe should have a T_M at least 5°C higher than that of the primers, to ensure the oligoprobe(s) hybridises with its template before extension of the primers begins (Landt, 2001). This caveat also applies to each component of a multi-oligonucleotide fluorogenic chemistry requiring two or more hybridisation events for effective signalling such as the tripartite molecular beacons (TMB), universal template (UT) oligoprobe or duplex scorpion primers.

All the non-incorporating, nucleotide-based, oligoprobe chemistries described in the following sections include a 3' phosphate or similar moiety, which blocks their extension by the DNA polymerase preventing the oligoprobe acting as a primer, but imparting no effect on the amplicon's yield. These fluorogenic chemistries can be divided into two classes: those which are destroyed to produce fluorescence and those which are not. Oligoprobes depending on a destructive process for signal generation are usually located close to the primer that hybridises to the same strand in order to maximise the chance that polymerase will make contact with the bound oligoprobe. Non-destructive oligoprobes are usually located as far as possible from the primer on the same strand to ensure signal is produced before the polymerase dislodges the duplex. The non-destructive chemistries include linear and hairpin oligoprobes, and incorporating primers. While reviewing these options one should note the number of hybridisation events required to generate a positive signal. Each fluorogenic oligonucleotide must meet and hybridise with its specific amplicon (a bimolecular system). If there is a second oligoprobe required, either to provide a quenching moiety or as a partner for FRET, the likelihood of a chance encounter between all three molecules decreases (a trimolecular system), especially as the amplicon concentration increases. Conversely, the fluorogenic primer systems are incorporated into the nascent amplicon. Therefore, a signal is generated for the remainder of the assay without the need for hybridisation to a new amplicon molecule each cycle (a unimolecluar system).

Destructive oligonucleotide systems

In 1991, Holland *et al.* described a technique that was to form the foundation for homogeneous PCR using fluorogenic oligoprobes. Radiolabelled amplicon was detected by monitoring the 5' to 3' nuclease activity of *Taq* DNA polymerase on specific oligoprobe and target DNA duplexes. The products were examined using thin layer chromatography and the presence or absence of hydrolysis was used as an indicator of specific duplex formation.

Lee *et al.* (1993) reported an innovative approach using nick-translation PCR in combination with a dual-fluorophore labelled oligoprobe. In the first truly homogeneous assay of its kind, one fluorophore was added to the 5' terminus and one to the middle of a sequence-specific oligoprobe. When in such close proximity, the 5' reporter fluorophore (6-carboxy-fluorescein, FAM) transferred laser-induced excitation energy by FRET to the 3' quencher fluorophore (6-carboxy-tetramethyl-rhodamine, TAMRA). TAMRA emitted the new energy at a wavelength that was monitored but not specifically utilised in the presentation of data. However, when the oligoprobe is hybridised to its template, the fluorophores were released upon hydrolysis of the oligoprobe component of the probe and target duplex due to the nuclease activity of the DNA polymerase. Once the labels were separated, the reporter's emissions were no longer quenched and the instrument could detect and present the resulting fluorescence data. These oligoprobes have been called 5' nuclease, hydrolysis or TaqMan® oligoprobes (Figure 10.4).

The use of 5' nuclease probes first required the development of a platform that could excite and detect fluorescence in addition to thermal cycling. A charge-coupled device (CCD) was described in 1992 for the quantification of conventional reverse transcription (RT)-PCR products (Nakayama *et al.*, 1992). In 1993, the CCD was combined with a thermal cycler resulting in the first real-time PCR fluorescence excitation and detection instrument (Higuchi *et al.*, 1993). To date, the commercial version of the platform, the ABI Prism® 7700 sequence detection system (Perkin Elmer Corporation/Applied Biosystems, USA), has been the most frequently reported instrument used with 5' nuclease oligoprobes.

A recent addition to the destructive chemistry is the UT oligoprobe (Figure 10.5; Zhang *et al.*, 2003). This system adds a generic or "universal" sequence to the 5' end of a PCR primer with which a common 5' nuclease oligoprobe can hybridise permitting the use of the same oligoprobe for different amplicons. The remainder of the primer provides assay specificity so that multiple primers can be used to amplify numerous targets all detected using the same oligoprobe. During the second PCR cycle, the nascent strand is copied and the polymerase encounters and hydrolyses the bound oligoprobe releasing the reporter and permitting fluorescence to be generated. Because the oligoprobe relies upon hybridisation with a primer, it is possible that this chemistry will also produce fluorescence from the formation of non-specific amplicon.

Another dual-labelled oligonucleotide sequence has been used as the signal-generating portion of the DzyNA-PCR system (Figure 10.6; Todd *et al.*, 2000; Applegate *et al.*, 2002). The reporter and quencher are attached to the termini of an oligonucleotide substrate. Cleavage of the oligosubstrate is performed by a DNAzyme, which is created during the PCR. This creation is the result of the PCR duplicating an antisense DNAzyme sequence included in the 5' tail of one of the primers. The duplicated sequence is the "functional" form of the DNAzyme.

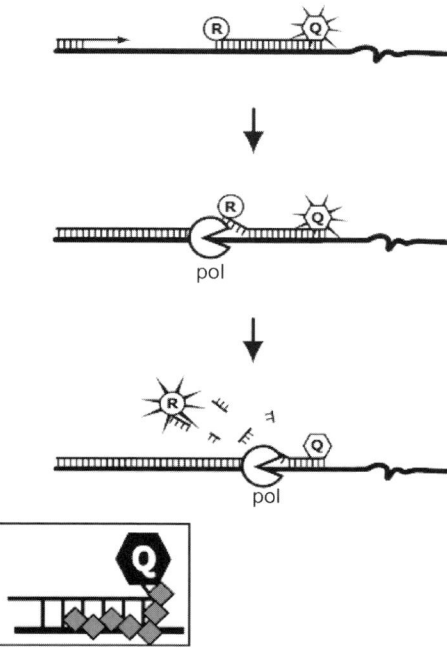

Figure 10.4. Function of 5' nuclease oligoprobes. Following primer hybridisation, the DNA polymerase (pol) progresses along the relevant strand during the extension step of the PCR, displacing and then hydrolysing the oligoprobe. Once the reporter (R) is removed from the extinguishing influence of the quencher (Q), it is able to release excitation energy at a wavelength that is monitored by the instrument and different from the emissions of the quencher. Inset shows the NFQ (Q; filled) and minor groove-binding molecule (grey diamonds) which make up the MGB nuclease oligoprobes. These bimolecular systems acquire data from the reporter's emissions: the opposite of the HybProbe chemistry. Data can be collected during the annealing or extension steps of the PCR.

Upon cleavage, the fluorophores are released allowing the production of fluorescence in an identical manner to a hydrolysed 5' nuclease oligoprobe.

Non-destructive oligonucleotide systems

Linear chemistries

The majority of fluorogenic oligoprobes fall into the class of linear oligoprobes. In fact, the use of a pair of adjacent, fluorogenic oligoprobes was first described in the mid-1980s, predating the popular 5' nuclease chemistry but failing to achieve the same degree of early commercial support. The pair of oligonucleotides were used to identify the distance between fluorophores on a complementary nucleic acid template and the system held promise for similar detection within living cells, hinting at further possibilities for diagnostic use (Heller and Morrison, 1985; Cardullo *et al.*, 1988). This trimolecular approach is now known

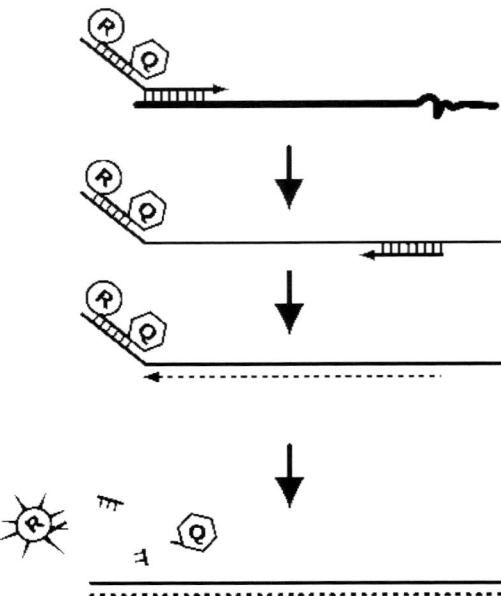

Figure 10.5. Function of the UT-oligoprobe. After the UT-primer is extended, the nascent strand acts as the template for the second primer. The polymerase encounters and hydrolyses the UT-oligoprobe whilst extending the second primer in the same fashion as a TaqMan oligoprobe. This trimolecular system can produce fluorescence data from the emissions of the released fluorophore during the annealing or extension steps.

commercially as hybridisation probes (HybProbes), having become the manufacturer's chemistry of choice for the LightCycler™ (Roche Molecular Biochemicals, Germany; Wittwer et al., 1997b). The upstream oligoprobe is labelled with a 3′ donor fluorophore (FITC) and the downstream probe is commonly labelled with either a proprietary LightCycler Red 640 or a Red 705 acceptor fluorophore at the 5′ terminus with a phosphate at the 3′ terminus to prevent extension by the DNA polymerase. When both oligoprobes are hybridised to the amplicon template, the two fluorophores are ideally located within 10 nt of each other (Figure 10.7). The ratio of acceptor to donor emissions can also be used as an internal reference signal rendering the results independent of absolute fluorescence (Huang et al., 2001).

The double-stranded oligoprobes function by displacement hybridisation (Figure 10.8; Li et al., 2002). In this process, a 5′ fluorophore-labelled oligonucleotide is, in its resting state, hybridised with a complementary, but shorter, quenching DNA strand that is 3′ end-labelled with an NFQ. When the full-length complementary sequence in the form of an amplicon is generated, the reporter strand preferentially hybridises to the longer target amplicon strand, disrupting the quenched oligoprobe duplex and permitting the fluorophore to emit its excitation energy directly.

The displacement hybridisation technique can also be used with a fluorophore-labelled primer and, due to added stringency provided by the longer complementary strand, the system performs its own

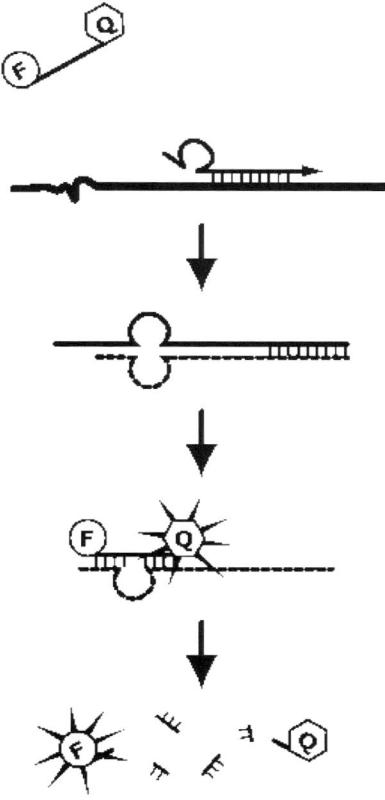

Figure 10.6. Function of the DzyNA primers. When the strand incorporating the primer is duplicated by a complementary strand (dashed line), a DNAzyme is created. A complementary, dual-labelled oligonucleotide substrate will be specifically cleaved by the DNAzyme releasing the fluorophore (F; circle) from its proximity to the quencher (Q; pentagon), releasing the labels and permitting fluorescence. Data can be collected during the annealing or extension steps of the PCR.

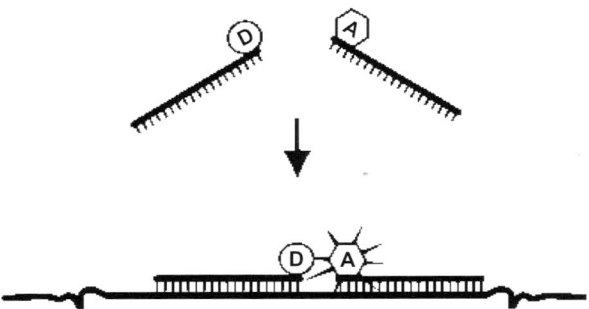

Figure 10.7. Function of HybProbes. Adjacent hybridisation results in a FRET signal due to interaction between the donor (D) and acceptor (A) spectra detected during the annealing step of the PCR. This trimolecular system (two oligoprobes and a target) acquires its data from the acceptor's emissions: the opposite of the 5' nuclease oligoprobe chemistry.

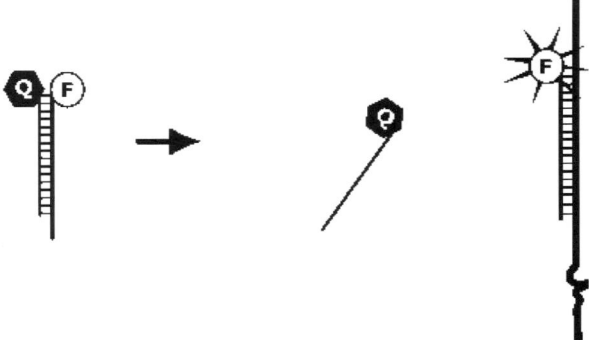

Figure 10.8. Function of displacement oligoprobes. The shorter NFQ-labelled strand (Q; hexagon) is displaced when the fluorophore-labelled (F; circle) strand preferentially hybridises to the longer specific amplicon strand. Data is collected from this trimolecular system during the annealing step of the PCR.

"hot-start" as was shown using an NFQ-labelled PNA strand (Q-PNA; Figure 10.9; Stender et al., 2002; Fiandaca et al., 2001). This and the previous chemistry are effectively trimolecular systems since the quenching strand must re-anneal after dissociating with amplicon in order to complete the signalling process. In the Q-PNA approach, a short quenching PNA probe is bound to an unincorporated fluorogenic primer such that the NFQ and fluorophore are adjacent, resulting in a quenched system. However, once the dsDNA amplicon is created by primer extension, the shorter Q-PNA is displaced in favour of the longer target amplicon, after which the fluorophore can fluoresce.

The first generation Light-up probe (Light-Up Technologies) is also a linear PNA, and is labelled with an asymmetric cyanine fluorophore, thiazole orange (Figure 10.10; Svanvik et al., 2001). When hybridised with

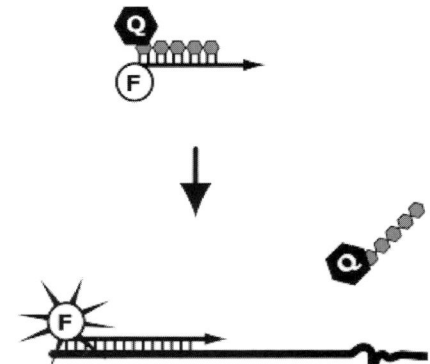

Figure 10.9. Function of the Q-PNA displacement primer. In the absence of the longer specific amplicon, quenching of the chemistry is achieved by a short NFQ-labelled PNA backbone (grey hexagons) designed to hybridise with the fluorophore-labelled primer (F; circle). Fluorescence data can be collected from this trimolecular system during the annealing and extension steps of the PCR.

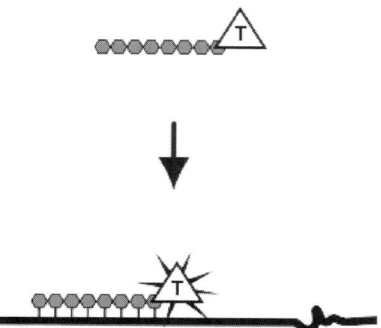

Figure 10.10. Function of the Light-Up Probe. These PNA (grey hexagons) oligoprobes fluoresce when their asymmetric thiazole orange fluorophore (T; open triangle) hybridises to the specific DNA strand. Data is collected from this bimolecular system during the annealing step of the PCR.

a nucleic acid target, either as a duplex or triplex, the fluorophore emits fluorescence due to association with amplicon. These probes do not interfere with the PCR, do not require conformational change and they are sensitive to single nucleotide mismatches, permitting further amplicon characterisation. The system is bimolecular and because a single reporter is used, a direct measurement of fluorescence can be made instead of the measurement of a change in fluorescence between two fluorophores (Svanvik et al., 2001; Isacsson et al., 2000). However, non-specific fluorescence has been reported during extended cycling (Svanvik et al., 2000).

The HyBeacon™ is a single linear oligonucleotide internally labelled with a fluorophore that emits an increased signal upon formation of a duplex between the target DNA strand and the HyBeacon (Figure 10.11; French et al., 2001, 2002). The HyBeacon and Light-up probe are relatively inexpensive and simple bimolecular systems to design and use.

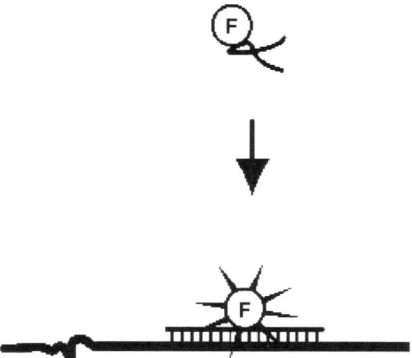

Figure 10.11. Function of the HyBeacon Oligoprobe. The fluorophore (F; circle) emits fluorescence when in close proximity to DNA as occurs upon hybridisation with the specific amplicon strand. Data is collected from this bimolecular system during the annealing step of the PCR.

The dual-labelled PNA was produced briefly as the Lightspeed probe or linear PNA beacon by Boston Probes Inc., USA (Stender et al., 2002). The oligoprobe is analogous to the TaqMan chemistry terminally labelled with a fluorophore and quencher, but it differs in its backbone, which is PNA (Figure 10.12). In an aqueous solution the PNA backbone brings the fluorophore and quencher into close proximity, quenching the system. When the probe hybridises to its specific target, it opens and fluorescence is possible.

A modification to the 5' nuclease chemistry resulted in the minor groove binding (MGB) oligoprobes (Figure 10.4, inset; Afonina et al., 2002a). This chemistry has a fluorescent reporter dye at the 5' end and a NFQ at the 3' end. In addition, the oligoprobe has an MGB molecule at the 3' end that hyper-stabilises each oligoprobe–target duplex by folding into the minor groove of the dsDNA (Kutyavin et al., 2000; Afonina et al., 2002b). In the unbound state the oligoprobe assumes a random coil configuration that is quenched. When hybridised, the stretched oligoprobe is able to fluoresce. However, as with many dual-labelled oligoprobe systems, this relaxed state can "leak" fluorescence at higher temperatures when the oligoprobe is prone to partial unfolding. When using 5' nuclease oligoprobes we found that this leakage can even produce a melting peak under controlled conditions (Mackay et al., 2003b).

The MGB chemistry permits the use of very short (12–17 nt) oligoprobes because of a 15–30°C rise in their T_M resulting from the interaction of the MGB and the DNA helix, in particular its stabilisation of A:T bonds. These short oligoprobes are theoretically ideal for detecting single nucleotide polymorphisms, since a short oligoprobe is more significantly destabilised by changes within the hybridisation site than a longer oligoprobe. However, in practice the degree of discrimination depends on the base to be determined, since some nucleotides can significantly influence fluorescence intensity. Originally this approach was designed

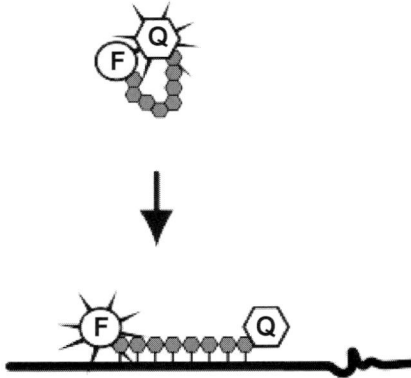

Figure 10.12. Function of the Lightspeed probe. In aqueous solution the PNA (grey hexagons) probe forms a random coil conformation that is quenched due to the proximity of the fluorophore (F; circle) and quencher (Q; hexagon). Upon hybridisation this bimolecular system is stretched open and the fluorophore can emit fluorescence which is acquired during the annealing step of the PCR.

for the allelic discrimination of genomic DNA where the ratio of heterozygotes to wild-type sequence was 1:1. In situations where this ratio is significantly skewed, the discriminatory power of polymorphism detection by oligoprobe may be reduced. A similar chemistry is the commercial MGB-Eclipse™ oligoprobe, which replaces the standard TAMRA or DABCYL quencher with a proprietary NFQ at the 5′ and a fluorophore at the 3′ end. The MGB molecule protects the oligoprobe from degradation by the polymerase and instead the signal is generated by stretching and relaxing of the oligoprobe during annealing and melting.

The result of combining a single sequence specific, Cy5-labelled linear oligoprobe with SYBR green I created the Bi-probe system. This functions via a variation of FRET recently termed Induced FRET (iFRET; Cardullo *et al.*, 1988; Howell *et al.*, 2002). Bi-probes are more specific than using SYBR green I alone and have an enhanced signal:noise (Brechtbuehl *et al.*, 2001; Walker *et al.*, 2001). Interestingly this approach functions using the Idaho Technologies LightCycler but not the closely related Roche LightCycler version 1.0. The disparity is due to the narrow bandpass filter sets employed by the latter instrument, which prohibited FRET between these two particular fluorophores. A similar technical problem rendered some proprietary LightCycler fluorophores unusable on the ABI PRISM 7700 (Nitsche *et al.*, 1999). The commercial form of the Bi-probe chemistry, called the ResonSense® probe (Defence Science and Technology Laboratory, United Kingdom; Lee *et al.*, 2002b), has overcome the platform-specific incompatibility possibly due to the substitution of SYBR Gold for SYBR Green I and the use of Cy5.5.

Single-fluorophore systems utilising a labelled oligoprobe or primer and FMCA to discriminate homozygous from heterozygous DNA without the need for FRET have also been described (Kurata *et al.*, 2001; Gundry *et al.*, 2003). The fluorophore is carefully chosen and positioned so that its emissions are quenched by proximity to a complementary guanine (Crockett and Wittwer, 2001). A commercial form, called the SimpleProbe, is used on the LightCycler and the LightTyper (Roche Diagnostics). This approach targets a single fluorescein-labelled hybridisation oligoprobe to polymorphism and produces an increased fluorescence when hybridised. Genotyping is performed by FMCA. The LightTyper is not a thermal cycler and as such this is not a real-time PCR system. The LightTyper is designed to rapidly characterise amplicon generated by PCR in a conventional thermal cycler. These systems suffer if used for the detection of nucleotide polymorphisms in a guanine-rich sequence.

Hairpin oligoprobes

Molecular beacons were the first fluorogenic hairpin oligoprobes described for real-time PCR applications. The hairpin oligoprobe's fluorogenic labels are called the fluorophore and quencher and are positioned at the termini of the oligoprobe (Figure 10.13). The most commonly used quencher, DABCYL (4-[4′-dimethylamino-phenylazo]-benzene), is an NFQ. The labels are held in close proximity by distal regions of homologous base pairing deliberately designed to create

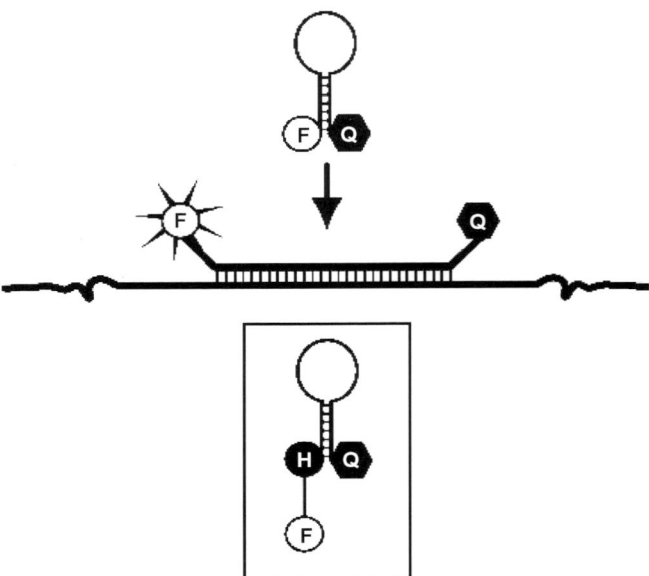

Figure 10.13. Function of the molecular beacon. Hybridisation of the loop section of the beacon to the target separates the fluorophore (F; open circle) and NFQ (Q; filled hexagon) allowing fluorescence. Data from this bimolecular system is collected during the annealing step of the PCR. Inset shows a wavelength-shifting hairpin oligoprobe incorporating a harvester molecule (H; filled circle).

a hairpin structure. The intimate proximity of the label molecules results in quenching either by FRET or direct energy transfer via a collisional mechanism (Tyagi *et al.*, 1998). In the presence of a sequence complementary to the molecular beacon's loop, the oligoprobe is shifted into an open configuration. The reporter is then removed from the quencher's influence and fluorescence can be detected (Tyagi and Kramer, 1996).

Wavelength-shifting hairpin oligoprobes are a recent improvement to the hairpin oligoprobe chemistry, making use of a second, "harvesting" fluorophore (Figure 10.13, inset). The harvester passes on excitation energy acquired from a blue light source as fluorescent energy in the far-red spectrum. A receptive "emitter" fluorophore can then be selected which uses the energy to produce light at characteristic wavelengths. This approach offers the potential for improved multiplex real-time PCR and nucleotide polymorphism analysis by increasing the number of emitters that can be excited using a single energising wavelength (Tyagi *et al.*, 2000). This is a useful workaround for instruments with a limited energising light source.

Recently, TMB have been added to the hairpin oligoprobe class of fluorogenic chemistry (Figure 10.14; Nutiu and Li, 2002). These highly complex oligoprobes combine a molecular beacon's hairpin with longer, unlabelled, single-stranded arms. Each arm is designed to hybridise to an oligonucleotide labelled with either a fluorophore or an NFQ. The system is quenched in the hairpin state due to the close proximity of the labels but

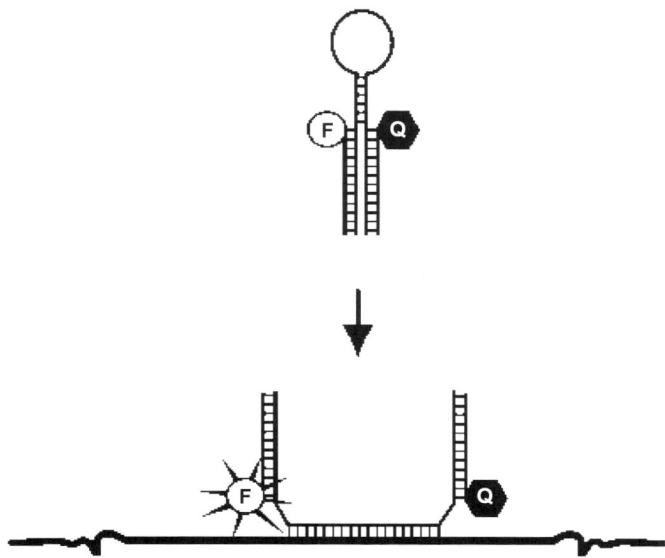

Figure 10.14. Function of the tripartite molecular beacon. The fluorophore (F; circle) is removed from the influence of the NFQ (Q; hexagon) upon binding to specific amplicon, which opens the hairpin and permits fluorescent emissions. Data are collected from this tetramolecular system during the annealing step of the PCR.

fluorescent when hybridised to the specific amplicon strand. However, for quenching to occur the arms must re-hybridise after the denaturation step. This is the only tetramolecular real-time PCR chemistry, so-called because it relies upon four intermolecular collisions to function correctly. The TMB chemistry may be subject to high background fluorescence in practical applications where genomic nucleic acids interfere with re-annealing.

Because the function of all the hairpin oligoprobes depends upon correct hybridisation of the stem, accurate design is crucial to their function and is considerably more challenging than for other oligoprobe chemistries.

Self-priming fluorogenic amplicon

The self-priming amplicon is similar in concept to the hairpin oligoprobe, except that the label(s) becomes irreversibly incorporated into the nascent amplicon. While these systems have only been described with a single primer, it is conceivable that both primers could be labelled in a similar way to provide a stronger, if more costly, fluorescent signal. The first contact between primer and template is a bimolecular event, but from then on the signalling is unimolecular. Intramolecular hybridisation is extremely fast and kinetically favourable since it does not rely upon the chance meeting of oligoprobe and amplicon in each cycle (Bustin, 2002). Fast cycling conditions appear to better suit the chemistries in this group (Thelwell *et al.*, 2000). Three approaches have been described: sunrise primers (commercially called Amplifluor™ hairpin primers), scorpion primers (Nazarenko *et al.*, 1997) and the recently described Light Upon eXtension primers (LUX; Whitcombe *et al.*, 1999).

Sunrise primers consist of a 5′ fluorophore and a 3′ DABCYL NFQ (Figure 10.15). The labels are brought together by complementary sequences that create a stem when the sunrise primer is closed. A target-specific primer sequence is located at the 3′ terminus downstream of the NFQ. The sunrise primer's sequence is intended to be duplicated by the nascent complementary strand and during subsequent annealing and extension steps, the stem is destabilised, the two fluorophores are forced approximately 20 nt (70 Å) apart and the fluorophore can emit excitation

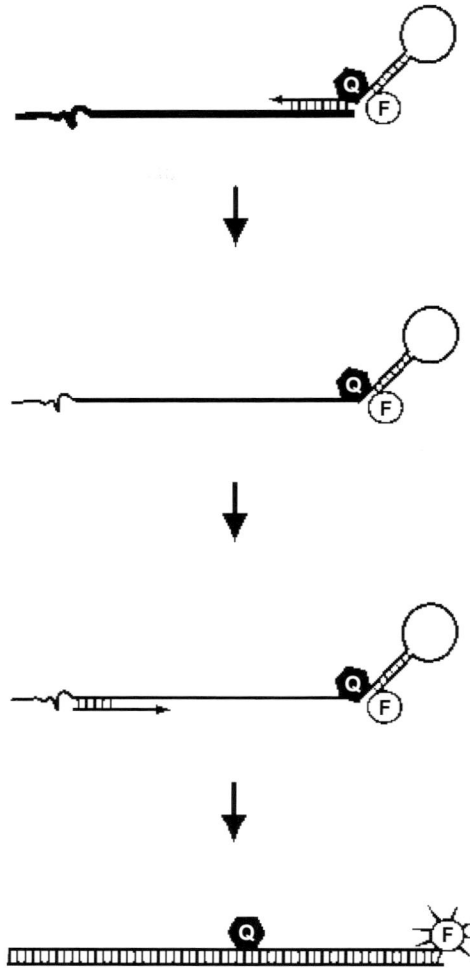

Figure 10.15. Function of the sunrise primer. The fluorophore (F; circle) is separated from the NFQ (Q; hexagon) during disruption of the sunrise primer's hairpin structure and free to fluoresce. This disruption occurs during extension of a nascent complementary DNA strand and whenever dsDNA duplexes form during re-annealing. Data from this unimolecular system can be collected during the annealing step of the PCR.

energy (Nazarenko *et al.*, 1997). It is possible, however, that this sequence duplication could also occur during the formation of non-specific amplicon.

The scorpion primer is almost identical in design to the sunrise primer except for a hexethylene glycol molecule that blocks duplication of the signalling portion of the scorpion (Figure 10.16). In addition to the difference in structure, the function of the scorpion primer differs in that the 5′ region of the oligonucleotide is designed to hybridise to a complementary region within the nascent amplicon strand creating an intramolecular signalling system. This approach to signal generation separates the labels disrupting the hairpin and permitting fluorescence in the same way that hairpin oligoprobes function (Whitcombe *et al.*, 1999). The duplex scorpion primer is initially a trimolecular system requiring the primer, quencher and target amplicon to interact. However, once the primer is incorporated into the nascent strand the system reverts to

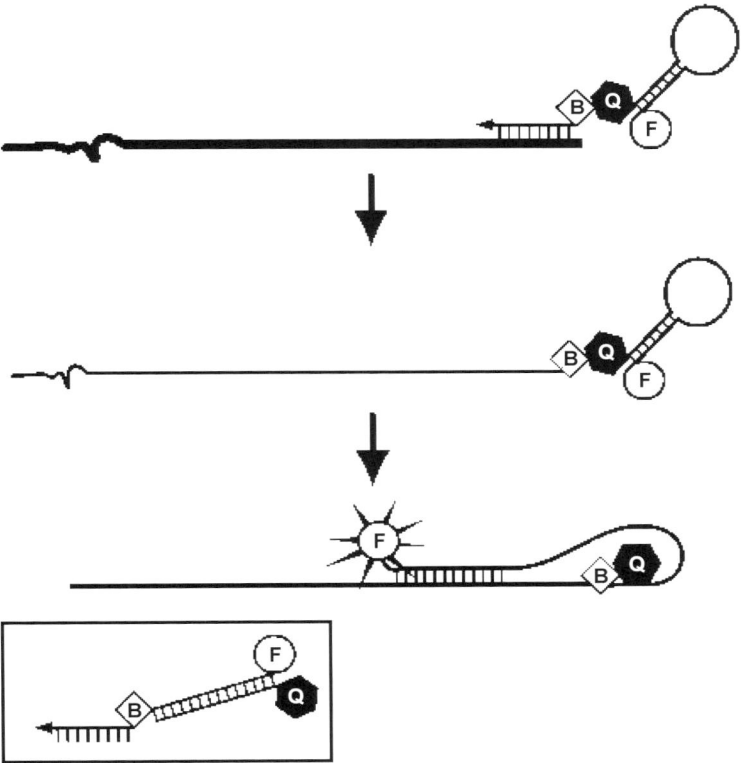

Figure 10.16. Function of the scorpion primer. The scorpion primer is blocked from being extended by a hexethylene glycol molecule (B; diamond) so that the hairpin can only be disrupted by specific hybridisation and not by the extension of a complementary amplicon strand as occurs for sunrise primers. The 5′ fluorophore (F; circle) is separated from a 3′ methyl red quencher (Q; hexagon) during self-hybridisation of the loop portion of the scorpion with a complementary region on the nascent amplicon strand. Inset shows a duplex scorpion. Data is collected from this unimolecular system during the annealing step of the PCR.

a bimolecular system which is simpler to synthesise and purify than the unimolecluar form. The stem-loop is exchanged for a separate, complementary oligonucleotide labelled with a quencher at the 5' terminus. The additional oligonucleotide interacts with the primer element which is terminally labelled with the fluorophore (Figure 10.16, inset; Solinas *et al.*, 2001). When it exists as a duplex, the chemistry is quenched but becomes fluorescent after hybridisation to the longer specific amplicon strand. Interestingly, because the quencher is not part of the same molecule, brighter fluorescence can be achieved than for the unimolecular version where the labels are neither absolutely quenched, nor freely fluorescent. Scorpion primers can be used for nucleotide polymorphism detection by designing the placement of either the primer component or the probe component to cover the polymorphism (Whitcombe *et al.*, 1999; Thelwell *et al.*, 2000).

A recently described variation of the hairpin chemistry is the self-quenching hairpin oligonucleotide primer which has been commercially labelled as the LUX fluorogenic primer (Figure 10.17; Nazarenko *et al.*, 2002). This chemistry is non-fluorescent in the absence of specific amplicon through the natural quenching ability of a carefully placed guanosine nucleotide. The natural quencher is brought into close proximity with the FAM or JOE fluorophore via a stretch of 5' and 3' complementary sequences. In the presence of the complementary target strand, a nascent strand is extended, which incorporates the LUX primer. The dsDNA opens the hairpin, permitting fluorescence from the fluorophore. This non-destructive chemistry is simple to design and use, relatively inexpensive and it does not require the inclusion of an additional oligoprobe. However, the presence of primer–dimer and

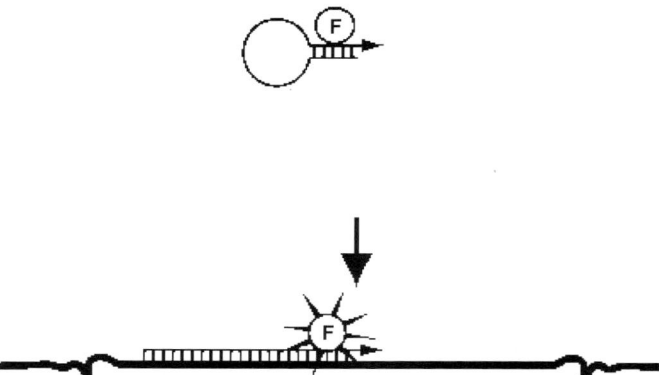

Figure 10.17. Function of the LUX primer. The LUX primer is labelled with a single fluorophore (F; circle) positioned next to a guanine nucleotide when the hairpin portion of the primer is intact. The G naturally quenches the fluorophore. In the presence of the specific target strand the primer hybridises, disrupts the hairpin and is extended. The fluorophore is now free to fluorescence and data can be collected from this unimolecular signalling system during the annealing or extension step of the PCR.

non-specific amplicon will, as for SYBR Green I and the sunrise primers, be displayed as a fluorogenic signal. The products can be discriminated using their differing T_D, however, there is no reliable way to account for their signal in quantitative real-time PCR.

A variant of these chemistries are the Angler® oligoprobes which closely resemble a Scorpion primer, except for the absence of a quenching moiety (Lee *et al.*, 2002b). Angler oligoprobes consist of a 3′ specific primer sequence linked via a hexethylene glycol molecule to a 5′, Cy5-labelled tail. The tail portion of the Angler is designed to self-hybridise to downstream sequences in the nascent strand producing a unimolecular signalling system. In contrast to the dual-labelled Scorpion, the Angler uses FRET between the terminal Cy5 and SYBR gold incorporated into the self-annealed duplex to produce a fluorescent signal.

Comparison of fluorogenic chemistries

When comparing the mechanism of signal generation by the different fluorescence systems, the most popular chemistries display unique quirks. The cyclical destruction of nuclease oligoprobes continues despite a plateau in amplicon accumulation whereas SYBR green I fluorescence generally increases non-specifically during later cycles even without template due to primer–dimer formation. HybProbe fluorescence begins to decrease as the rate of collision between the growing numbers of complementary amplicon strands increases resulting in a phenomenon called the "hook effect" (Figure 10.3). At this stage, the formation of dsDNA is favoured over the hybridisation of oligoprobe to its target DNA strand, adversely affecting total fluorescence. The possibility exists that some linear oligoprobes are consumed by sequence-related nuclease activity and it may also contribute to the hook effect as has been reported for HybProbes (Wilhelm *et al.*, 2001b; Wittwer *et al.*, 1997a; Lyamichev *et al.*, 1993).

◆◆◆◆◆◆ MICROBIAL GENOTYPING

Although nucleotide sequencing is the gold standard for in-depth characterisation of nucleic acids, it is a relatively lengthy process. The development of real-time PCR has partially addressed this by providing a tool capable of rapid detection of characterised nucleotide polymorphisms.

The non-destructive fluorescent chemistries are mostly intact at the end of the PCR and can be used to indirectly genotype a microbial template by characterising its amplicon (Bustin, 2000). Careful control of ionic strength, amplicon size and cooling and melting rates can improve the resolution of experimental data collected from FMCA. FMCA can also be used with a single labelled oligonucleotide (Gundry *et al.*, 2003). The SYBR green I and HybProbe chemistries are most commonly used to perform these analyses; however, the double-stranded and Light-up oligoprobes as well as the HyBeacons should be capable of functioning in this role. Other chemistries, such as the TaqMan and Eclipse oligoprobes and hairpin oligonucleotides and primers, require two sets of oligoprobes to differentiate a wild-type from an altered sequence. While this is a perfectly legitimate and functional approach for genotyping by real-time PCR, the extra fluorogenic oligonucleotides increase the overall cost and complexity of the assay. However, this approach does permit the creation of a dedicated mutation detection assay which will often perform more reliably than a single assay intended to detect both the wild-type and mutated sequences. Unfortunately, the use of two sets of fluorophores reduces the number of related but different microbes that can be discriminated in a single vessel, since two fluorophores must be assigned to analyse each nucleotide variant. The technique of FMCA has proven popular for the rapid diagnosis of human genetic disorders and has advanced the detection of multiple targets by real-time PCR. It should again be noted that genotyping two sequences by FMCA requires that the amount of one amplified sequence is less than 10-fold above the amount of the other sequence; otherwise the more common sequence will overpower detection of the more rare sequence.

The occurrence of a mismatch between a hairpin oligonucleotide and its target has a greater destabilising effect on the duplex than the introduction of an equivalent mismatch between the target and a linear oligoprobe. This is because the hairpin structure provides a highly stable alternate configuration. Therefore, hairpin oligonucleotides are more specific than the commonly used linear oligoprobes making hairpin oligonucleotides good candidates for detecting nucleotide changes, albeit more technically demanding to design (Tyagi et al., 1998). Despite the fact that hybridisation does not reach equilibrium using rapid ramp rates, the apparent T_M values deduced from FMCA are both reproducible and characteristic of a given probe and target duplex (Gundry et al., 1999). Importantly, different nucleotide changes destabilise hybridisation to differing degrees and this can be incorporated into the design of oligoprobes for genotyping assays to control the extent of discrimination between the melt peak temperatures. The least destabilising mismatches involve a change to a G on one of the strands (G:T, G:A, and G:G), whereas the most destabilising include a change to a C (C:C, C:A and C:T) (Bernard et al., 1998).

Because real-time PCR genotyping data is obtained at the completion of the PCR amplification phase, it is an end-point analysis, although the reaction vessel remains unopened throughout the process. The amplicon is denatured and rapidly cooled to encourage the formation of fluorophore and target strand complexes. The temperature is then gradually raised and the fluorescence from each vessel is continuously recorded. The detection of sequence variation using fluorescent chemistries relies upon duplex destabilisation incurred as a result of nucleotide changes. As mentioned earlier, the non-specific chemistries reflect these changes in the context of the entire dsDNA amplicon.

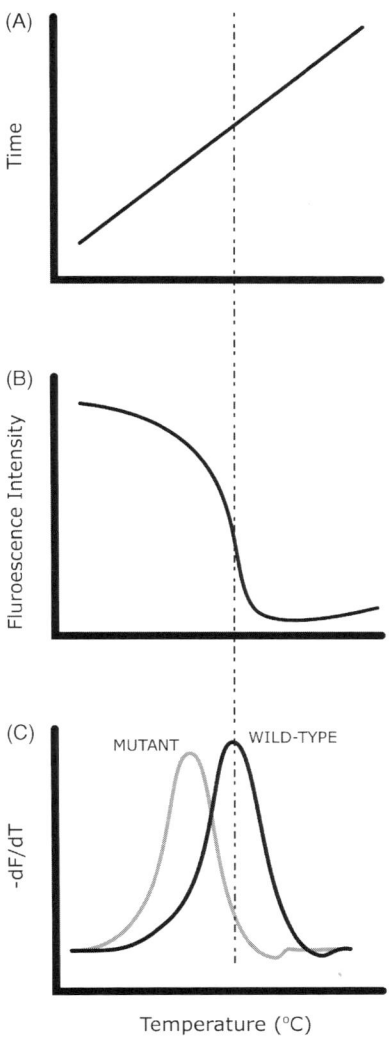

Figure 10.18. Fluorescence melting curve analysis (FMCA). At the completion of a real-time PCR using a fluorogenic chemistry, the reaction is rapidly heated and then cooled to a temperature below the expected T_D of the dsDNA or T_M of the oligoprobe(s). It is then heated to 85°C or more at a fraction of a degree per second (A). During heating, the raw data representing the emissions of a relevant fluorophore are constantly acquired (B). Software calculates the negative derivative of the fluorescence with temperature which is plotted against temperature to produce a melt peak indicative of the T_D of the dsDNA, or the T_M of the oligoprobe-target melting transition (dashed line and black peak; C). When sequence differences exist, the T_D or T_M is reduced (grey peak) due to the lower stability of heteroduplexes and the extent of the resulting temperature shift is used diagnostically to characterise an amplified template. Generally speaking, a ramp rate of 0.2°C/s permits clear discrimination between different genotypes. However, variation of the ramp rate can be helpful if unsatisfactory results are generated. Hybridisation can be enhanced by slowing the ramp rate whereas an accelerated ramp rate can help remove stable secondary structures at the oligoprobe hybridisation region resulting in sharper melt peaks.

The sequence changes have a different impact upon the specific fluorogenic chemistries, altering the expected T_M in a manner that reflects the particular nucleotide difference. The resulting rapid decrease in fluorescence using either approach can be presented as a "melt peak" using software capable of calculating the negative derivative of the fluorescence change with temperature and plotting that data against temperature (Figure 10.18).

◆◆◆◆◆◆ NUCLEIC ACID QUANTITATION

Although the terminology is often confused, real-time PCR does not inherently imply quantitative PCR. To determine the amount of template present in a sample, care must be taken to include the correct control system. External controls or "standards" are used for calculating the amount of template present in a patient sample and are amplified in a separate vessel to the unknown target (Figure 10.19). Internal controls are amplified in the same vessel as the unknown target, frequently indicating the occurrence of false negative reactions and examining the ability to amplify from a preparation of nucleic acids (Niesters, 2002). Certainly, the reliability of quantitative PCR methods is intimately associated with the choice and quality of the assay controls (Celi et al., 2000; Alexandre et al., 1998).

No matter what controls are used to generate quantitative data or monitor successful amplification, it is imperative to accurately determine their concentration and to ensure that internal controls are added at suitable levels in order to prevent extreme competition for reagents with the wild-type template (Zimmermann and Manhalter, 1996; Brightwell et al., 1998). A spectrometer alone is inadequate for quantifying a control molecule; however, in combination with experimental and statistical analysis, the reliability of the data is greatly enhanced (Glasel, 1995; Bagnarelli et al., 1995; Wang and Spadoro, 1998; Rodrigo et al., 1997; Taswell, 1981; Sykes et al., 1998). Finally, one must remember that the results of quantitation using a control need to be expressed relative to a suitable biological marker, e.g. in terms of the volume of plasma, the number of cells or the mass of tissue or genomic nucleic acid. Standardisation such as this will ensure comparability between assay results and testing locations (Niesters, 2001).

The ability to quantify nucleic acids is predominantly used in the field of gene expression or transcriptome analysis (PhorTech, 2003). For this application, primers are often carefully designed to span exon–intron boundaries or cross overly large expanses of intron to avoid co-amplification of genomic DNA (Yin et al., 2001). In our hands the enhanced specificity of an oligoprobe format is preferred to the use of non-specific chemistries for quantitation.

Another area suited to sensitive, specific quantitation with the ability to genotype a target is the detection of minimal residual disease in malignancy. The role of real-time PCR for the investigation of lymphoma

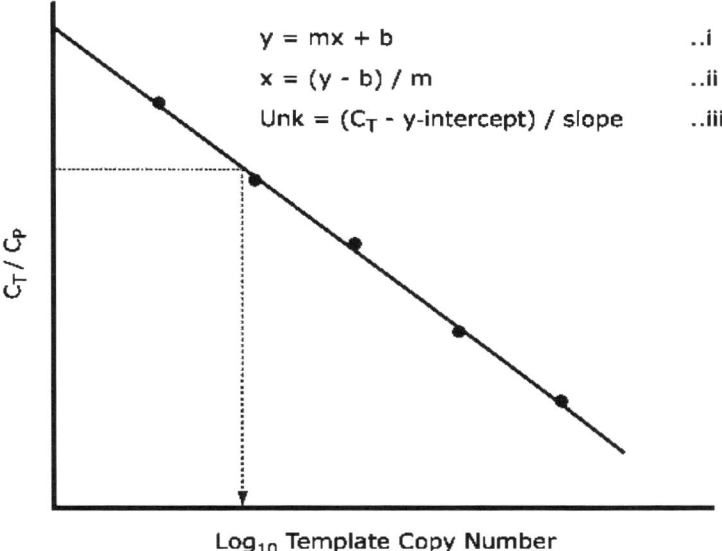

Figure 10.19. External standard curve for quantitation. Threshold cycle or crossing point data collected from an amplified titration of standard plotted against the concentration of each template. Through interpolation, C_T/C_P values for unknowns (Unk) permit the calculation of the starting template concentrations (dashed line). This is performed by rearranging the equation for the linear regression line (i), to solve for the x-axis value of interest, (ii). When the relevant values are substituted (iii), the concentration of template in each unknown can be calculated.

and leukaemia patients is expanding. This role examines and quantifies the number and genetic profile of malignant cells remaining in a clinically well patient as an indicator of response to therapy or to identify those patients at high risk of disease relapse (Sharp and Chan, 1999; Liu Yin, 2002).

External Control Templates

Standards used to monitor or enhance the power of PCR are most commonly created by cloning an amplicon, a portion of the target organism's genome or simply using the purified amplicon itself as the template (Borson et al., 1998). Because these molecules are amplified separately and because they are not inherently present within the sample matrix, they are called exogenous or external controls. These form the basis of external standard curves, which are created from the data produced by the individual amplification of a dilution series of control template (Figure 10.3). The concentration of an unknown, which is amplified in the same assay but in a separate vessel, can then be interpolated from the standard curve (Figure 10.19). While the external standard curve is the more commonly described approach for quantitative PCR, it frequently suffers from uncontrolled and unmonitored inter-vessel variations. Some real-time PCR systems have partially overcome this problem by including the ability to detect and correct for variation in the emissions of a non-participating, or "passive", internal reference fluorophore (6-carboxy-N,N,N',N'-tetramethylrhodamine, ROX). This reference is present in the reaction buffer and is used to indicate volume variations and non-specific fluorescence quenching. The corrected values, obtained from a ratio of the emission intensity of the fluorophore and ROX, are called RQ +. To further control amplification fluctuations, the fluorescence from a "no-template" control reaction (RQ −) is subtracted from RQ + for each sample, resulting in the ΔRQ value that indicates the magnitude of the reference signal generated for the given PCR (Gelmini et al., 1997).

The simplest mathematical representation of PCR amplification is described by the Equation:

$$N = N_0(E + 1)^C$$

where N is the number of amplified molecules, N_0 the initial number of template molecules, E the amplification efficiency and C the number of amplification cycles.

This equation is essential to obtaining accurate data from real-time PCR and is usually considered to apply to each reaction in an experimental run. However, the efficiency of the PCR is commonly calculated from the amplification of a standard curve, assuming that each unknown reaction will amplify with equivalent efficiency. Furthermore, the amount of amplicon in each reaction at the C_T is assumed to be equal because the fluorescence is the same (Figure 10.3). Nonetheless, by adopting these assumptions and deducing the linear regression line of best fit through the C_T values from a standard curve we can determine the reaction efficiency from the slope given by the following Equation:

$$E = 10^{-\text{Slope}} - 1$$

Internal Control Templates

The use of an internal control molecule has been described in the earliest of PCR experiments and is considered essential to increase assay quality (Reiss and Rutz, 1999) and as a competitive template for conventional quantitative PCR (Chehab et al., 1987; Rosenstraus et al., 1998). When such a control is added prior to template purification (extraction control) or amplification (amplification control), it is called an exogenous internal control since it does not occur naturally within the sample matrix, but it is co-amplified within the same reaction. Ideally the exogenous internal control template should hybridise to the same primers and have an identical amplification efficiency to the template under investigation (Zimmermann and Manhalter, 1996). The control should also contain a discriminating feature such as a change in its length (Orlando et al., 1998; Möller and Jansson, 1997; Brightwell et al., 1998; Hall et al., 1998, Celi et al., 2000) or more commonly in today's oligoprobe-based methods, a change in the sequence of the target (Alexandre *et al.*, 1998; Tarnuzzer *et al.*, 1996; Celi *et al.*, 2000; Natarajan *et al.*, 1994; Aberham *et al.*, 2001; Rosenstraus *et al.*, 1998; Kearns *et al.*, 2001b; Stöcher *et al.*, 2002; Gruber *et al.*, 2001). However, the internal control should not significantly interfere with the detection of small quantities of the template under investigation; therefore, its use must be carefully evaluated. Internal control templates that bind different primers or have different amplification efficiencies are still useful as standards for semi-quantitative PCR or relative quantitation.

An endogenous internal control occurs naturally within the specimen matrix and is therefore co-amplified with the specific target. Housekeeping genes are a common example that has been successfully used to quantify gene expression by RT-PCR and monitor template integrity after its purification (Chehab *et al.*, 1987). When housekeeping genes are used for the quantitation of RNA it is essential that they are minimally regulated and exhibit a constant and cell cycle-independent basal level of transcription (Selvey *et al.*, 2001). Studies have shown that an 18S rRNA target successfully meets the desired criteria for some applications (Selvey *et al.*, 2001; Thellin *et al.*, 1999). This is not the case for some commonly used genes including β-actin; unfortunately their use is widespread, especially for transcript analyses (Härtel *et al.*, 1999). To ensure the suitability of the housekeeping gene, a panel of candidate targets should be examined under the desired experimental conditions and the extent of change in the C_T used to indicate the stability of target expression (Löseke *et al.*, 2003).

We have found several promising candidate targets during a systematic quantitative study of the expression levels of 13 genes in 16 different tissues (Radonic *et al.*, 2004). As expected, no single gene was consistently expressed among all the tissues investigated. However, the RNA polymerase gene displayed the lowest variation over all tissues, even during mitogenic stimulation of a T-cell line. RNA polymerase mRNA encodes the principle enzyme used in mRNA transcription and is therefore part of a self-regulating cycle. It can be assumed that this gene is expressed steadily and independent of the cell state.

Real-time PCR assays that lack the capacity to correct for vessel-to-vessel variations such as those caused by variation in reverse transcription efficiency, cell number, nucleic acid isolation technique and PCR efficiency are more appropriately described as being semi-quantitative. To this day it is rare to find well-controlled real-time PCR assays among those in the literature. That will hopefully change as pressure to publish comprehensively optimised and validated assays is brought to bear by the editorial staff and peer reviewers of scientific journals (Hoorfar *et al.*, 2003).

Relative versus Absolute Quantitation

The amount of template in a sample can be described either relatively or absolutely. Relative quantitation is the simpler approach, which, as the name suggests, describes differences in the amount of a target sequence compared to its level in a related matrix, or within the same matrix by comparison to the signal from an endogenous or other reference control. This approach has proven particularly popular for gene expression studies and can be conveniently employed using the $2^{-\Delta\Delta C_T}$ method (Livak and Schmittgen, 2001). Absolute quantitation is more demanding to perform but states the exact number of nucleic acid targets present in the specimen in relation to a specific unit and it is therefore easier to compare data between different assays and laboratories (Freeman et al., 1999; Pfaffl et al., 2002). Absolute quantitation may be necessary when there are no sequential specimens to demonstrate a relative change in template load, or when no accurately standardised reference reagent or suitable endogenous control is available.

A highly accurate approach used for absolute quantitation by conventional PCR utilises competitive co-amplification of one or more dilutions of an internal control template of known concentration with a wild-type target nucleic acid of unknown concentration (Becker-Andre and Hahlbrock, 1989; Clementi et al., 1995; Gilliland et al., 1990; Siebert and Larrick, 1992). However, conventional competitive quantitation is technically demanding and time consuming, requiring significant development and optimisation compared to quantitation by real-time PCR. Thus, it is not suited to the quick decision making required within research and clinical environments (Locatelli et al., 2000; Wall and Edwards, 2002; Tanaka et al., 2000b). The existence of software with the ability to calculate the concentration of an unknown by comparing real-time PCR signals generated from a co-amplified target and internal control is rare but improving (Pfaffl et al., 2002). In addition, new or improved mathematical interpretations are appearing which aim to make quantitation more reliable and simpler (Liu and Saint, 2002; Rutledge and Côte, 2003).

Improved Quantitation Using Real-time PCR

Common microbiological methods to quantify an organism have traditionally required culture-based methods. While these methods are often considered the gold standard and are the only way to obtain data on microbial viability and infectivity, they are not without significant problems of their own. These problems are caused by variation in the method of inoculation, the choice of target cell, the type of culture medium, the sensitivity of the microbe to transport and storage conditions and the conditions of culture (Luria and Darnell, 1967). One must remember that titres obtained from culture only hold true under the specific conditions of that assay, making the results less portable than those obtained from nucleic acid quantitation by well-developed real-time PCR assays.

We have used conventional competitive PCR to obtain quantitative data with promising results (Mackay et al., 2001, 2003a). However, these approaches are laborious to develop and perform. The wide dynamic range of real-time PCR encompassing at least eight $\log_{10}$ copies of nucleic acid template is a significant benefit for microbial load determination (Ishiguro *et al.*, 1995; Kimura *et al.*, 1999; Najioullah *et al.*, 2001; Ryncarz *et al.*, 1999; Monopoeho *et al.*, 2000; Alexandersen *et al.*, 2001; Abe *et al.*, 1999; Locatelli *et al.*, 2000; Gruber *et al.*, 2001; Brechtbuehl *et al.*, 2001; Moody *et al.*, 2000). This broad dynamic range obviates the need to dilute an amplicon before detecting it or the repetition of an assay using a diluted template because a preliminary result falls above the upper limits of the detection assay. Both these problems occur when using conventional end-point PCR assays for quantitation because their detection systems cannot encompass the products of high template loads whilst maintaining adequate sensitivity (Weinberger *et al.*, 2000; Schaade *et al.*, 2000; Brechtbuehl *et al.*, 2001; Kawai *et al.*, 1999). The flexibility of real-time PCR is further demonstrated by its ability to detect one target in the presence of a vast excess of another target during duplexed assays or, with careful manipulation, to quantify expression from a single cell (Ryncarz *et al.*, 1999; Liss, 2002).

Real-time PCR is also a particularly attractive alternative to conventional PCR for the study of microbial load. We and others have found the consistently low inter-assay and intra-assay variability especially useful compared to other assays (Schutten *et al.*, 2000; Nitsche *et al.*, 2000; Locatelli *et al.*, 2000). Real-time PCR also has equivalent or improved sensitivity compared to microbial culture, or conventional single-round and nested PCR (Kearns *et al.*, 2001c; Clarke *et al.*, 2002; Kupferschmidt *et al.*, 2001; Kennedy *et al.*, 1997b; Locatelli *et al.*, 2000; Capone *et al.*, 2001; Leutenegger *et al.*, 1999; Smith *et al.*, 2001; Monopoeho *et al.*, 2000; van Elden *et al.*, 2001; Lanciotti *et al.*, 2000). It has been reported to be at least as sensitive as Southern blot, still considered by some as the gold standard for probe-based hybridisation assays (Capone *et al.*, 2001).

◆◆◆◆◆◆ MULTIPLEX REAL-TIME PCR

Conventional multiplex PCR applies a number of primer sets to potentially amplify multiple templates within a single reaction vessel (Chamberlain *et al.*, 1988; Burgart *et al.*, 1992). However, its adaptation to real-time PCR has clouded the traditional terminology. Multiplex real-time PCR more commonly describes the use of multiple fluorogenic oligoprobes for the discrimination of amplicons which have been produced by one or more primer pairs in a single vessel. The transfer of this technique from conventional PCR protocols has proven problematic because of the limited number of fluorophores available and the closed nature of real-time PCR which prohibits the monitoring of specificity made possible by size determination using agarose gel electrophoresis (Lee *et al.*, 1993). The commonly used monochromatic energising light source and the application of narrow bandpass filters on some real-time PCR instruments have limited their ability to perform multiplex PCR. Although excitation by a single wavelength produces bright emissions from an appropriate fluorophore, it restricts the number of fluorophores that can be successfully combined and discriminated in a single vessel (Tyagi *et al.*, 2000). The discovery and application of non-fluorescent quenching labels has made available some regions of the spectrum that were previously occupied by the emissions of early quenchers. This development has permitted a slight increase in the number of spectrally discernable oligoprobes per reaction.

Recent improvements to the design of hairpin primers and hairpin and nuclease oligoprobes, as well as novel combinations of fluorophores such as the Bi-probe and Light-up probes, have promised alternatives that should permit the discrimination of a greater variety of amplicons per reaction vessel.

Currently, the number of fluorophores that can be combined and clearly distinguished is extremely limited when compared to the capability of conventional multiplex PCR to discriminate amplicon. Some real-time PCR platforms have incorporated multiple light-emitting diodes to span the entire visible spectrum, or a tungsten light source, which emits light over a broad range of wavelengths permitting the excitation of a wide range of fluorophores. The inclusion of high-quality optical filters to clarify both the excitation and emission energies makes possible the use of a wider range of real-time PCR detection chemistries on one machine. Nonetheless, these improvements have to date permitted the multiplexing of only four different colours, of which one colour is ideally set aside for an internal control to monitor inhibition and one may be required for a passive reference. Some real-time PCR designs have astutely made use of single or multiple nucleotide changes between similar templates to discriminate the amplicon by T_M using FMCA and thus avoid the need for multiple fluorophores (Espy *et al.*, 2000a,b; Boivin *et al.*, 2004; Schalasta *et al.*, 2000a; Kearns *et al.*, 2001c; Loparev *et al.*, 2000; Read *et al.*, 2001; Whiley *et al.*, 2001). This approach has been common for the detection of human genetic diseases where as many as 27 possible nucleotide substitutions have been detected using only one or two fluorophores (Schütz *et al.*, 2000; Lay and Wittwer, 1997; Herrmann *et al.*, 2000; Lee *et al.*, 1999a; Gundry *et al.*, 1999; Bernard and Wittwer, 2000; Elenitoba *et al.*, 2001).

To date, only a handful of assays have been described in the literature that truly multiplex more than two fluorophores. Few of these duplexed assays have been applied to the diagnosis of infectious disease and those that have are frequently used to detect subtle changes within a single species rather than a number of different species. It is important to remember that the ultimate benefits to be gained from multiplex PCR derive from its use of minimal specimen to detect the maximum number of targets. Some approaches cannot technically be considered real-time homogeneous assays, because they require interruption of the procedure to transfer template, detect fluorescence using a separate end-point analysis or perform the assays in separate reaction vessels but within the same run (Stocher *et al.*, 2003). The best microbial multiplex real-time PCR can discriminate between four retroviral target sequences (Vet *et al.*, 1999); however, we and others have found conventional multiplex PCR using end-point detection capable of easily discriminating five or more different amplicons (Mackay *et al.*, 2003a; Quereda *et al.*, 2000). This discrepancy continues to highlight the greater flexibility of conventional PCR for true multiplex PCR assays (Kehl *et al.*, 2001; Echevarria *et al.*, 1998; Henegariu *et al.*, 1997; Weigl *et al.*, 2000; Stockton *et al.*, 1998).

Assays capable of detecting several viral genomes include those using the non-specific label, SYBR green I, to detect herpes simplex viruses (HSV), varicella zoster virus (VZV) or cytomegalovirus (CMV) in separate tubes (Nicoll *et al.*, 2001), or by adaptation of a conventional multiplex PCR, to identify HSV-1 and HSV-2, VZV and enteroviruses within a single capillary by applying FMCA (Read *et al.*, 2001).

Future development of novel chemistries and improved real-time PCR instrumentation and software will hopefully enhance our ability to perform truly multiplexed real-time PCR assays. Meanwhile, the promise of multiplex real-time PCR applications remains tantalisingly out of reach.

◆◆◆◆◆◆ APPLICATION OF REAL-TIME PCR TO MICROBIOLOGY

Microbiology has been a significant driving force for the development of real-time PCR technology. The resulting benefits to microbial studies have already been discussed in a number of elegant reviews (Niesters, 2001; Mackay *et al.*, 2002; Cockerill and Smith, 2002; Versalovic and Lupski, 2002; Niesters, 2002; Bretagne, 2003). The final sections of this chapter will update the reader on the use of real-time PCR for the study of viruses, bacteria, fungi, parasites and protozoans as well as some specialist areas that have been significantly enhanced by the technology.

Viruses

In microbiology, the application of real-time PCR has had the most dramatic impact upon virology where it has provided an alternative to assays which either use morbidity and mortality as an indicator of disease progression, or qualitatively investigate the role of viruses in a range of human diseases (Kato *et al.*, 2000). Also, epidemiological studies of co-infections have been improved by the use of duplex real-time PCR (Zerr *et al.*, 2000; Furuta *et al.*, 2001; Kearns *et al.*, 2001c).

Real-time PCR has been used as a tool to complement conventional research techniques when investigating direct and indirect links between viral infection and chronic conditions such as sarcoma (Kennedy *et al.*, 1997a,b, 1998a; O'Leary *et al.*, 2000a,b), carcinoma (Lo *et al.*, 1999; Capone *et al.*, 2001), cervical intraepithelial neoplasia (Josefsson *et al.*, 1999; Lefevre *et al.*, 2003; Swan *et al.*, 1997; Lanham *et al.*, 2001) and lymphoproliferative disorders (MacKenzie *et al.*, 2001; Jabs *et al.*, 2001).

More commonly, real-time PCR studies simply describe the presence of viruses in their host with a long-term view of improving clinical management of the affected patients. Viruses studied to date include the bunyaviruses (Stram *et al.*, 2004), caliciviruses (Song *et al.*, 1997), flaviviruses (Laue *et al.*, 1999; Kilpatrick *et al.*, 1996; White *et al.*, 2002; Callahan *et al.*, 2001; Ratge *et al.*, 2002; Ishiguro *et al.*, 1995; Komurian-Pradel *et al.*, 2001; Lanciotti *et al.*, 2000; Beames *et al.*, 2000), hepadnaviruses (Weinberger *et al.*, 2000; Chen *et al.*, 2001; Cane *et al.*, 1999; Brechtbuehl *et al.*, 2001), herpesviruses (Nitsche *et al.*, 2000; Takaya *et al.*, 1996; Song *et al.*, 2002; Whiley *et al.*, 2003b; Fan and Gulley, 2001; Schalasta *et al.*, 2000b; Kearns *et al.*, 2001a; Stevens *et al.*, 2002; Loparev *et al.*, 2000; Gallagher *et al.*, 1999; Kennedy *et al.*, 1997b; Fernandez *et al.*, 2002; Biggar *et al.*, 2000; Tanaka *et al.*, 2000b; Locatelli *et al.*, 2000; Ohyashiki *et al.*, 2000; Kearns *et al.*, 2001c; Lallemand *et al.*, 2000; White and Campbell, 2000; Najioullah *et al.*, 2001; Schaade *et al.*, 2000; Capone *et al.*, 2001; Lo *et al.*, 1999; Niesters *et al.*, 2000; Kimura *et al.*, 1999; Ryncarz *et al.*, 1999; Peter and Sevall, 2001; Hawrami and Breur, 1999; Furuta *et al.*, 2001), orthomyxoviruses (van Elden *et al.*, 2001; Ellis and Zambon, 2002), papovaviruses (Jordens *et al.*, 2000; Lefevre *et al.*, 2003; Biel *et al.*, 2000; Whiley *et al.*, 2001), paramyxoviruses (Smith *et al.*, 2001; Daniels *et al.*, 2001; Côte *et al.*, 2003; Mackay *et al.*, 2003b; Whiley *et al.*, 2002b; Aldous *et al.*, 2001), parvoviruses (Gruber *et al.*, 2001), pestiviruses (Vilcek and Paton, 2000), picornaviruses (Alexandersen *et al.*, 2001; Kares *et al.*, 2004; Lai *et al.*, 2003; Hu *et al.*, 2003; Reid *et al.*, 2004; Monpoeho *et al.*, 2002; Nijhuis *et al.*, 2002; Corless *et al.*, 2002; Watkins-Reidel *et al.*, 2002; Verstrepen *et al.*, 2001, 2002; Monopoeho *et al.*, 2000), polydnaviruses (Chen *et al.*, 2003), poxviruses (Espy *et al.*, 2002; Ibrahim *et al.*, 1997, 1998; Shaw *et al.*, 1995), reoviruses (Pang, 2004), retroviruses (Lewin *et al.*, 1999; Lew *et al.*, 2004a; Argaw *et al.*, 2002; Schutten *et al.*, 2000; Choo *et al.*, 2000; Klein *et al.*, 1999; Leutenegger *et al.*, 1999), rhabdoviruses (Smith *et al.*, 2002; Hughes *et al.*, 2004) and TT virus (Iriyama *et al.*, 1999).

Less commonly, studies of the pathogenic effects of viruses on the biology of specific cellular populations has provided a snapshot of the mechanism of disease progression as occurs during hepatitis C virus induced chronic liver disease (Shaw *et al.*, 1995).

Over the last decade, conventional quantitative PCR has repeatedly demonstrated that the determination of viral load is a useful marker of disease progression and a valuable way to monitor the efficacy of antiviral compounds (Holodniy *et al.*, 1991; Held *et al.*, 2000; Roberts *et al.*, 1998; Rollag *et al.*, 1998; Kaneko *et al.*, 1992; Clementi *et al.*, 1995; Clementi, 2000; Menzo *et al.*, 1992). Quantitative real-time PCR has similarly examined the interaction between virus and host and monitored changes in viral load resulting from antiviral therapy ultimately impacting on the treatment regimen selected (Nitsche *et al.*, 1999; Clementi, 2000; Limaye *et al.*, 2000). Because disease severity and viral load are linked, microbial quantitation by real-time PCR has proven beneficial when studying the impact of viral reactivation, persistence or isolated gene expression on the progression of disease (Ohyashiki *et al.*, 2000; Chen *et al.*, 2003; Kearns *et al.*, 2001c; Furuta *et al.*, 2001; Limaye *et al.*, 2001; Tanaka *et al.*, 2000a,b; Lallemand *et al.*,

2000; Laue *et al.*, 1999; Kimura *et al.*, 1999; Chang *et al.*, 1999; Lo *et al.*, 1999; Hawrami and Breur, 1999; Hoshino *et al.*, 2000; Nitsche *et al.*, 2000; Machida *et al.*, 2000; Limaye *et al.*, 2001; Najioullah *et al.*, 2001; Gault *et al.*, 2001). Altered microbial tropism or replication and the effect of these changes on the host cell can also be followed using real-time PCR (Kennedy *et al.*, 1998a, 1999a,b). Molecular assays are increasingly being used to augment or replace traditional assays. Two examples are the use of real-time PCR to calculate bacteriophage titer more accurately and over a broader dynamic range than plaque assays and the indirect determination of neutralising antibody titer to chicken anemia virus in chickens by quantifying viral load (Edelman and Barletta, 2003; van Santen *et al.*, 2004).

The speed and flexibility of real-time PCR has proven useful to commercial interests that require exquisite sensitivity to screen for microbial contamination within large-scale reagent preparations produced from eukaryotic expression systems or among livestock used for the production of food (Brorson *et al.*, 2001; Pinzani *et al.*, 2004; Stram *et al.*, 2004; Reid *et al.*, 2004; Lew *et al.*, 2004b; van Santen *et al.*, 2004; Rudi *et al.*, 2002; de Wit *et al.*, 2000). Additionally, highly sensitive assays are proving invaluable for the thorough assessment of viral gene therapy vectors prior to their use in clinical trials. Nuclease oligoprobes have been most commonly used for these studies, which assess the biodistribution, function and purity of the novel "drug" preparations (Gerard *et al.*, 1996; Suzuki *et al.*, 2003; Barzon *et al.*, 2003; Rohr *et al.*, 2002; Josefsson *et al.*, 2000; Hackett *et al.*, 2000; Sanburn and Cornetta, 1999; Choo *et al.*, 2000; Scherr *et al.*, 2001).

Likewise, the study of new and emerging viruses has embraced the use of homogeneous real-time PCR assays as tools to demonstrate and strengthen epidemiological links between unique viral sequences and the clinical signs and symptoms experienced by patients (Lanciotti and Kerst, 2001; Mackay *et al.*, 2003b; Smith *et al.*, 2001, 2002; Lanciotti *et al.*, 2000; Halpin *et al.*, 2000; Gibb *et al.*, 2001a,b).

The severe acute respiratory syndrome (SARS) arose in early 2003 in China and the causative agent was rapidly identified (Peiris *et al.*, 2003; Poutanen *et al.*, 2003; Tsang *et al.*, 2003; WHO, 2003). As soon as the first stretches of sequence were known, real-time PCR assays provided the potential for fast and reliable diagnoses and epidemiological studies (Drosten *et al.*, 2003; Emery *et al.*, 2004; Zhai *et al.*, 2004; Ruan *et al.*, 2003; Ng *et al.*, 2004). These assays are performed either as one-step or two-step RT-PCR assays. At the time of writing 48 complete SARS coronavirus (SARS-CoV) genomic sequences existed on GenBank. All these sequences are closely related. However, as with all RNA virus genomes, the coronaviruses display a tendency to vary with time and location (Wood, 2003). We recently described an approach to maintain the reliability of SARS coronavirus molecular assays. In this approach, three independent real-time PCR assays, based on the 5' nuclease format, were established to be performed under identical reaction conditions. All three assays are located in different, yet highly conserved regions of the known SARS-CoV genomes separated by approximately

9000 nucleotides. Because of this distance, under the chosen reaction conditions the primers used in one assay cannot react with primers of another assay performed in the same tube. The fluorescence signals produced by all three individual assays are indistinguishable; however, for diagnostic purposes it is not necessary to assign the signals to a certain amplicon. We were able to demonstrate that the sensitivity was maintained by the simultaneous amplification of three amplicons. The strength of this approach is that all three assays would need to be adversely affected by nucleotide change for the diagnostic system to fail, a virtually impossible occurrence. We believe that the amplification of more than one sequence stretch of the pathogen of interest is a useful technique to increase the reliability of diagnosis, especially for new and emerging microbes.

Nonetheless, the absolute reliance of PCR upon fully representative characterised genome sequences is highlighted in this area of microbial study. The early assay developer may suffer setbacks due to a paucity of sequence data representing all viral variants. We found this to be a problem when developing a 5' nuclease assay for human metapneumovirus (Mackay et al., 2003b). At the time of development the majority of viral sequences present on GenBank represented what was later to become known as the Type A virus – a bias which was reflected by the new assay.

Bacteria

Rapid real-time PCR assays provide significant benefit to the diagnosis of a bacterially infected host. The results can quickly inform the clinician, allowing a more specific and timely application of antibiotics, which are far more than antiviral therapies. This speed can limit the potential for toxicity caused by shotgun treatment regimens, reduce the duration of a hospital stay and prevent the improper use of antibiotics, minimising the potential for resistant bacterial strains to emerge.

Once again, elegant applications of real-time PCR can augment or even replace traditional culture and histochemical assays as was seen with the creation of a molecular assay capable of classifying bacteria in the same way as a Gram stain (Klaschik et al., 2002). However, particular bacterial species are the more frequent focus of real-time PCR assays especially when slow culture times can be replaced by rapid and specific gene detection. *Leptospira* genospecies, *Mycobacterium* and *Propionibacterium* species, *Chlamydia* species, *Legionella pneumophila* and *Listeria monocytogenes* have all been successfully detected and in some cases quantified using real-time PCR assays (Woo et al., 1997; DeGraves et al., 2003; Huang et al., 2001; Rudi et al., 2002; Nogva et al., 2000; Desjardin et al., 1998; Bassler et al., 1995; Miller et al., 2002; O'Mahony and Hill, 2002; Torres et al., 2000; Eishi et al., 2002; Kraus et al., 2001; de Viedma et al., 2002; Li et al., 2000; Creelan and McCullough, 2000; Hayden et al., 2001; Ballard et al., 2000; Wellinghausen et al., 2001; Reischl et al., 2002; Lunge et al., 2002). Real-time RT-PCR has proven useful for quantitation of bacterial transcripts in response to infection or the application of specific metabolic intermediates; however, these assays do not always agree with conventional methods of microbial detection, indicating the difference between detecting a live organism by culture and a dead organism by fluorescence or DNA detection (Goerke *et al.*, 2001; Corbella and Puyet, 2003; Desjardin *et al.*, 1998).

The detection of *Neisseria gonorrhoeae* has benefited from real-time PCR, particularly when used as a confirmatory test for commercial assays (Whiley *et al.*, 2002a). This example again demonstrates the need for care when choosing a PCR target, especially when that target exists on a plasmid which is exchanged among other bacteria, potentially providing confusing diagnostic results. *Neisseria meningitidis*, *Haemophilus influenzae* and *Streptococcus pneumoniae* are the major pathogens causing bacterial meningitis and the introduction of diagnostic real-time PCR has proven to be a powerful tool that we and others have quickly developed and deployed for the rapid discrimination of circulating pathogens (Corless *et al.*, 2001; Whiley *et al.*, 2003a; Probert *et al.*, 2002; Mothershed *et al.*, 2004).

The detection and monitoring of antibiotic resistance among clinical isolates of *Staphylococcus aureus*, *Staphylococcus epidermidis*, *Helicobacter pylori*, *Enterococcus faecalis* and *Enterococcus faecium* has benefited from the speed and reliability of real-time applications (Randegger and Hachler, 2001; Woodford *et al.*, 2002; Tan *et al.*, 2001; Hein *et al.*, 2001a; Shrestha *et al.*, 2002; Lindler *et al.*, 2001; Reischl *et al.*, 2000; Martineau *et al.*, 2000; Matsumura *et al.*, 2001; Gibson *et al.*, 1999; Chisholm *et al.*, 2001). These technologies have proven useful for determining the efficacy of antibiotic therapies when treating uncultivable organisms such as *Mycoplasma haemofelis* which infects cats (Tasker *et al.*, 2003, 2004). Meanwhile, the understanding and treatment of fulminant diseases such as meningitis, sepsis, inflammatory bowel disease and the sourcing of food poisoning outbreaks caused by characterised bacteria such as the group B *Streptococci*, *Mycobacterium* sp., *Escherichia coli*, *Bacteroides vulgatus* and *Salmonella* species have been enhanced by the speedy return of results from real-time and end-point fluorogenic assays (Taylor *et al.*, 2001; Bergeron *et al.*, 2000; Fujita *et al.*, 2002; Fortin *et al.*, 2001; Ibekwe *et al.*, 2002; Bellin *et al.*, 2001; Ke *et al.*, 2000; Chen *et al.*, 1997).

Real-time PCR has made possible the rapid quantitation and differentiation of some of the more exotic pathogenic bacteria such as the tick-borne spirochete *Borrelia burgdorferi* (Rauter *et al.*, 2002; Pietilä *et al.*, 2000; Pahl *et al.*, 1999), the methanotropic bio-remediating *Methylocystis* species (Kikuchi *et al.*, 2002) and bacteria capable of degrading agricultural herbicides such as the *Pseudomonas* sp. ADP and *Chelotobacter heintzii* (Devers *et al.*, 2004). The involvement of treponemes in the formation of periodontal disease has been studied using TaqMan chemistry, revealing a microbial role at every stage of disease (Asai *et al.*, 2002). In addition, measurement of the bacterial load of *Tropheryma whipplei* has permitted the discrimination of environmental contamination and low-level colonisation from active infection (Fenollar *et al.*, 2002).

Fungi, Parasites and Protozoans

A smaller but rapidly increasingly number of published applications have examined fungal, parasitic and protozoan pathogens of humans and plants. Real-time PCR assays have significantly contributed to the general diagnosis of invasive diseases, which are continually increasing with a rise in the population of immunocompromised patients. These pathogens include *Aspergillus fumigatus*, *Aspergillus flavus*, *Candida albicans*, *Candida parapsilosis*, *Candida tropicalis*, *Candida krusei*, *Candida kefyr* and *Candida glabrata* (Brandt *et al.*, 1998; O'Sullivan *et al.*, 2003; Guiver and Oppenheim, 2001; Costa *et al.*, 2002). Monitoring the transcriptional activity of certain *Aspergillus nidulans* transporter genes has provided important information about their role in multidrug resistance (Semighini *et al.*, 2002). The homogeneous nature of real-time PCR has proven useful when assays have been employed to investigate buildings for the presence of potentially harmful levels of toxigenic fungal spores, or conidia, such as those produced by *Stachybotrys chartarum* (Roe *et al.*, 2001; Haugland *et al.*, 1999; Cruz-Perez *et al.*, 2001). Quantifying *Septoria tritici*, *Stagonospora nodorum*, *Puccinia striifomis* and *Puccinia recondita*, which cause blotch and rust in wheat crops, permitted crops at risk of full-blown disease to be identified so that specific fungicide treatments could be quickly employed (Fraaije *et al.*, 2001). Similar benefits have resulted from the detection of *Glomus mosseae*, *Phytophthora infestans* and *Phytophthora citricola* which infect and damage a wide range of plants including those used as food crops (Böhm *et al.*, 1999). Rapid molecular assays are playing an increased role in the detection and monitoring of fungicide resistance, the presence of compatible mating types, pathogenic variation within fungal populations and for airborne monitoring of dispersed pathogens (Zhang *et al.*, 1996; McCartney *et al.*, 2003).

Cryptosporidium parvum oocysts and the spores from *Encephalitozoon* species have been successfully genotyped or speciated using real-time PCR significantly improving upon laboratory diagnosis using microscopy and histochemical staining, especially for low concentrations of excreted material (Tanriverdi *et al.*, 2002; Wolk *et al.*, 2002).

Rapid serological detection of *Toxoplasma gondii* is often hampered by the presence of the parasite in patients who are immunocompromised. Additionally, the length of time required for traditional culture or mouse inoculation is excessive. Therefore, rapid molecular methods have vastly improved detection of this microbe (Kupferschmidt *et al.*, 2001; Costa *et al.*, 2001). This technology is also useful to study how *Toxoplasma gondii* responds to antimicrobial therapies (Costa *et al.*, 2000).

Detection of malarial parasites using a mouse model in combination with real-time PCR has improved result turnaround and meant that parasite load data can be obtained (Bruña-Romero *et al.*, 2001; Witney *et al.*, 2001). Direct *in vivo* detection and quantitation of malarial parasites with a high level of sensitivity is also possible (Hermsen *et al.*, 2001; Lee *et al.*, 2002a) in addition to the indirect monitoring of stage-specific *Plasmodium falciparum* maturation by tracking the transcription of specific genes (Blair *et al.*, 2002).

Trichomonas vaginalis is the most common non-viral sexually transmitted organism in the world and the cost and time savings of real-time PCR assay have meant improvements to large-scale screening of patients at risk (Hardick *et al.*, 2003).

◆◆◆◆◆◆ DETECTING AGENTS OF BIOWARFARE

Perhaps no area of microbiology has impacted upon the community's psyche as much as the potential for microbes to be used as weapons of terror. While reported incidents of bioterrorism have to date been rare, further occurrences are considered by some to be inevitable and the potential for widespread disruption to communities is undoubtedly a serious threat to the population at large (Broussard, 2001). These disruptions need not take the form of widespread mortality, as small foci of morbidity would be sufficient to cause panic and disruption to essential services and financial markets. Hence, timely detection systems are an essential defence to minimise the effects of any such attack. Recently, there has been an explosion of literature indicating that real-time PCR is the tool of choice for rapidly detecting microbes used as agents of biological warfare. Many of these agents were, until recently, rarely encountered in the clinical laboratory and when they were, their detection relied upon relatively laborious and slow diagnostic techniques (Nulens and Voss, 2002). Nonetheless, this most recent area of application for real-time PCR technology enforces the role of real-time PCR as a new tool to complement the diagnostic arsenal rather than a complete replacement for traditional microbiological techniques.

There are three forms of human anthrax caused by *Bacillus anthracis*. These are the cutaneous, gastrointestinal and pulmonary forms. In contrast to self-limiting cutaneous anthrax, the ingestion or inhalation of endospores is generally fatal unless rapidly treated with antibiotics. The need for fast and reliable diagnostic tools became evident following the anthrax postal attacks late in 2001 in the United States. Real-time PCR assays permitted the rapid discrimination of weaponised pathogens from harmless laboratory-adapted or vaccine-related strains detecting *B. anthracis* spores and important plasmid or chromosomal markers (Makino *et al.*, 2001; Drago *et al.*, 2002; Uhl *et al.*, 2002; Oggioni *et al.*, 2002; Lee *et al.*, 1999b; Qi *et al.*, 2001). These assays could also discriminate pathogenic from mildly or apathogenic *Bacillus* spp. taken from colonies or enrichment broths (Ellerbrok *et al.*, 2002; Hurtle *et al.*, 2004). The addition of hand-held miniaturised real-time instruments and mobile laboratory systems has further enhanced the detection of bioweapons by providing the potential for on-site results in under 30 min during emergency situations (Higgins *et al.*, 2003a,b). Additionally *Francisella tularensis*, the cause of tularemia and *Yersinia pestis* the causative agent of plague have been detected using nuclease or HybProbe assays (Lindler *et al.*, 2001; Lindler and Fan, 2003). In some instances these assays can be modified for use in the field, permitting rapid and highly specific detection at the point of concern (Higgins *et al.*, 1998, 2000).

Smallpox is caused by infection with variola virus (VARV), a member of the family *Poxviridae*, genus *Orthopoxvirus*. VARV evokes the most serious concerns as a biowarfare agent because of low immunity among the population and the application of real-time PCR is attempting to address the diagnostic issues (Ibrahim *et al.*, 2003). The disease was declared eradicated in 1979 by a global vaccination program with the last naturally occurring VARV infection occurring in 1977 in Somalia (WHO, 1980; Behbehani, 1983). While humans are the only natural host for VARV, both vaccinia virus and cowpox virus have a much broader host spectrum. Although zoonotic infections of humans by monkeypox and other members of the genus *Orthopoxvirus* occur, potential VARV infections bear by far the most lethal risk for man (Centers for Disease Control and Prevention, 2003; Czerny *et al.*, 1991; Lewis-Jones, 2002; Stephenson, 2003). Due to this infectious potential, VARV has recently been flagged as a biological weapons threat (Henderson *et al.*, 1999; Whitley, 2003). The highest degree of sequence homology among characterised members of the genus *Orthopoxvirus* is found towards the centre of the genome while the terminal regions can exhibit considerable variability. Clearly, rapid and sensitive identification of variola virus and its discrimination from other members of the genus is fundamental to reliable diagnosis and risk evaluation of environmental samples. However, the consequences of a false positive or a false negative PCR result in smallpox diagnosis caused by unexpected sequence variation is considerable; therefore, identification of variola virus should also benefit from amplification of several independent targets by PCR.

Conventional VARV typing and sequencing assays may take more than 48 h to complete and, for obvious reasons, this is an unacceptable lag period. Even the use of end-point fluorescence detection decreases assay times drastically (Ibrahim *et al.*, 1997). Variola virus has also been used as a target to test new field-portable real-time PCR instruments in combination with rapid nucleic acid extraction techniques demonstrating their potential for detecting potential bioweapons on site (Ibrahim *et al.*, 1998). However, these applications have once again highlighted the importance of careful assay design when discriminating highly pathogenic microbes from innocuous, but closely related species (Espy *et al.*, 2002; Nitsche *et al.*, 2004, in press). This is an issue that clearly highlights technical difficulties, which remain in the area of diagnostic PCR design. In many cases gene sequencing is still the final stage for confirming a provisional identification of microbial material from a site or patient suspected of infection with a bioweapon.

◆◆◆◆◆◆ HOST IMMUNITY: MEASURING THE RESPONSE TO A MICROBE

The application of real-time PCR is beginning to provide evidence to support paradigm shifts in the way we conceptualise the interplay between microbe and host and in the way we describe these interactions

(Casadevall and Pirofski, 2003). Pathogen-centric views have dogmatically defined virulence as the result of microbial factors, when in fact the role of the host can be clearly shown to exert a significant influence (Casadevall and Pirofski, 2001; Mitchell, 1998). Increasingly, the old view is being challenged by our growing understanding that changing host immunity is as integral to the course of an infectious disease as the microbe causing the infection. This poses the question "what is a pathogen"? (Casadevall and Pirofski, 2002). Real-time PCR is helping formulate answers to this and other questions by defining the fitness of the host though quantitation of the immune response to infection and antimicrobial therapies and also as a tool to quantify the damage caused. While micro-array technologies provide a comprehensive snap-shot of the state of the host or invading microbe's transcriptome, arrays do not permit reliable quantitation of the target change over a broad dynamic range (Lockhart and Winzeler, 2000; Clewley, 2004; Lucchini et al., 2001). Real-time PCR has increasingly become the method of choice for validating and further characterising experimental data generated by micro-arrays and is the favoured tool for determining gene transcript abundance in basic research, molecular medicine and biotechnology. These determinations are essential markers of microbial–host interaction. While Northern blotting and RNase protection can determine both the size and amount of a transcript, the former is relatively insensitive, and both require large amounts of starting template (Wang and Brown, 1999). Considerably more has been published on the role of real-time PCR in the basic detection of cytokine transcripts from stored or *in vitro* stimulated blood cells than as an indicator of microbial activation (Kruse et al., 1997; Härtel et al., 1999; Blaschke et al., 2000; Stordeur et al., 2002). This is set to be a growth area for future real-time PCR applications.

Cytokines constitute the majority of mediators involved in the innate mammalian immune response recognising microbes, communicating with and recruiting leukocytes and removing invading microorganisms. However, it is the host's molecular pattern recognition systems operating via specialised receptors which trigger many of the events leading to cytokine induction post-infection (Strieter et al., 2003). Real-time PCR has identified critical mRNA from these receptors and from cytokine genes as a *de facto* indicator of protein production (Overbergh et al., 1999, 2003; Giulietti et al., 2001; Hein et al., 2001b). While this is not a perfect relationship insofar as protein levels are modified by more than simple transcript availability, studies have shown a good correlation between the transcriptome and the proteome for many genes (Balnaves et al., 1995). Additionally, it is often impossible to detect the expressed product in tissue samples due to the low expression levels and short half-lives of cytokine proteins (Broberg et al., 2003; Stordeur et al., 2002). Nonetheless, transcriptional studies using real-time PCR are enhanced by protein expression data.

Well-controlled quantitative real-time PCR experiments have permitted the differentiation and quantitation of IFN-α subclasses as a measure of the innate immune response to infectious or inactivated virus in humans (Löseke et al., 2003). The expression of other cytokines has also been monitored in murine models of microbial immunopathogenesis (Deng et al., 2003; Broberg et al., 2003). Microbial load studies in concert with cytokine protein detection have demonstrated relationships that may play a role in predicting the course and severity of post-transplant lymphoproliferative disease in EBV-positive patients (Muti et al., 2003). The direct impact of cytokines on viral gene expression and the effect of viral cytokine homologues on co-infecting viruses have potentially established new therapeutic strategies for diseases (Song et al., 2002).

Rapid advances in the performance and variety of fluorogenic chemistries and real-time PCR platforms together with the exponential increase in our understanding of the process have ensured real-time PCR is as important a technology for the diagnostic and research microbiology laboratory of tomorrow as agarose gel electrophoresis was to the laboratory of yesterday. Real-time PCR is no more or less than a diagnostic tool and therefore the data it generates are only as reliable as the design and implementation of each assay. Judicious application of the technology will both simplify and hasten the search for answers to many complex experimental and diagnostic questions. In order to more accurately understand the process of virulence, such reliable and robust tools will be essential for defining the interplay between the host and the invading microbe, and when necessary, to do so in a clinically relevant period of time.

References

Abe, A., Inoue, K., Tanaka, T., Kato, J., Kajiyama, N., Kawaguchi, R., Tanaka, S., Yoshiba, M. and Kohara, M. (1999). Quantitation of hepatitis B virus genomic DNA by real-time detection PCR. *J. Clin. Microbiol.* **37**, 2899–2903.

Aberham, C., Pendl, C., Gross, P., Zerlauth, G. and Gessner, M. (2001). A quantitative, internally controlled real-time PCR assay for the detection of parvovirus B19 DNA. *J. Virol. Meth.* **92**, 183–191.

Afonina, I. A., Reed, M. W., Lusby, E., Shishkina, I. G. and Belousov, Y. S. (2002a). Minor groove binder-conjugated DNA probes for quantitative DNA detection by hybridization-triggered fluorescence. *Biotechniques* **32**, 940–949.

Afonina, I. A., Sanders, S., Walburger, D. and Belousov, Y. S. (2002b). Accurate SNP typing by real-time PCR. *Pharmagenomics* Jan/Feb, 48–54.

Ahmed, F. E. (2002). Detection of genetically modified organisms in foods. *Trends Biotechnol.* **20**, 215–223.

Aldous, E. W., Collins, M. S., McGoldrick, A. and Alexander, D. J. (2001). Rapid pathotyping of Newcastle disease virus (NDV) using fluorogenic probes in a PCR assay. *Vet. Microbiol.* **80**, 201–212.

Alexandersen, S., Oleksiewicz, M. B. and Donaldson, A. I. (2001). The early pathogenesis of foot-and-mouth disease in pigs infected by contact: a quantitative time-course study using TaqMan RT-PCR. *J. Gen. Virol.* **82**, 747–755.

Alexandre, I., Zammatteo, N., Ernest, I., Ladriere, J.-M., Hamels, Le. S., Chandelier, N., Vipond, B. and Remacle, J. (1998). Quantitative determination of CMV DNA using a combination of competitive PCR amplification and sandwich hybridization. *Biotechniques* **25**, 676–683.

Applegate, T. L., Iland, H. J., Mokany, E. and Todd, A. V. (2002). Diagnosis and molecular monitoring of acute promyelocytic leukemia using DzyNA reverse transcription-PCR to quantify PML/RARα fusion transcripts. *Clin. Chem.* **48**, 1338–1343.

Argaw, T., Ritzhaupt, A. and Wilson, C. A. (2002). Development of a real time quantitative PCR assay for detection of porcine endogenous retrovirus. *J. Virol. Meth.* **106**, 97–106.

Asai, Y., Jinno, T., Igarashi, H., Ohyama, Y. and Ogawa, T. (2002). Detection and quantification of oral treponemes in subgingival plaque by real-time PCR. *J. Clin. Microbiol.* **40**, 3334–3340.

Bae, H. G., Nitsche, A., Teichmann, A., Biel, S. S. and Niedrig, M. (2003). Detection of yellow fever virus: a comparison of quantitative real-time PCR and plaque assay. *J. Virol. Meth.* **110**, 185–191.

Bagnarelli, P., Menzo, S., Valenza, A., Paolucci, S., Petroni, S., Scalise, G., Sampaolesi, R., Manzin, A. and Clementi, M. (1995). Quantitative molecular monitoring of human immunodeficiency virus type 1 activity during therapy with specific antiretroviral compounds. *J. Clin. Microbiol.* **33**, 16–23.

Ballard, A. L., Fry, N. K., Chan, L., Surman, S. B., Lee, J. V., Harrison, T. G. and Towner, K. J. (2000). Detection of *Legionella pneumophila* using real-time PCR hybridisation assay. *J. Clin. Microbiol.* **38**, 4215–4218.

Balnaves, M. E., Bonacquisto, L., Francis, I., Glazner, J. and Forrest, S. (1995). The impact of newborn screening on cystic fibrosis testing in Victoria. *Austr. J. Med. Genet.* **32**, 537–542.

Barany, F. (1991). The ligase chain reaction in a PCR world. *PCR Meth. Appl.* **1**, 5–16.

Barratt, K. and Mackay, J. F. (2002). Improving real-time PCR genotyping assays by asymmetric amplification. *J. Clin. Microbiol.* **40**, 1571–1572.

Barzon, L., Bonaguro, R., Castagliuolo, I., Chilosi, M., Franchin, E., Del Vecchio, C., Giaretta, I., Boscaro, M. and Palù, G. (2003). Gene therapy of thyroid cancer via retrovirally-driven combined expression of human interleukin-2 and herpes simplex virus thymidine kinase. *Eur. J. Endocrinol.* **148**, 73–80.

Bassler, H. A., Flood, S. J. A., Livak, K. J., Marmaro, J., Knorr, R. and Batt, C. A. (1995). Use of a fluorogenic probe in a PCR-based assay for the detection of *Listeria monocytogenes*. *Appl. Environ. Microbiol.* **61**, 3724–3728.

Beames, B., Chavez, D., Guerra, B., Notvall, L., Brasky, K. M. and Lanford, R. E. (2000). Development of a primary tamarin hepatocyte culture system for GB virus-B: a surrogate model for hepatitis C virus. *J. Virol.* **74**, 11764–11772.

Becker-Andre, M. and Hahlbrock, K. (1989). Absolute mRNA quantification using the polymerase chain reaction (PCR). A novel approach by a PCR aided transcript titration assay (PATTY). *Nucleic Acids Res.* **17**, 9437–9447.

Behbehani, A. M. (1983). The smallpox story: life and death of an old disease. *Microbiol. Rev.* **47**, 455–509.

Bellin, T., Pulz, M., Matussek, A., Hempen, H.-G. and Gunzer, F. (2001). Rapid detection of enterohemorrhagic *Escherichia coli* by real-time PCR with fluorescent hybridization probes. *J. Clin. Microbiol.* **39**, 370–374.

Bengtsson, M., Karlsson, H. J., Westman, G. and Kubista, M. (2003). A new minor groove binding asymmetric cyanine reporter dye for real-time PCR. *Nucleic Acids Res.* **31**, e45.

Bergeron, M. G., Ke, D., Ménard, C., Picard, F. J., Gagnon, M., Bernier, M., Ouellette, M., Roy, P. H., Marcoux, S. and Fraser, W. D. (2000). Rapid detection of group B streptococci in pregnant women at delivery. *N. Engl. J. Med.* **343**, 175–179.

Bernard, P. S. and Wittwer, C. T. (2000). Homogeneous amplification and variant detection by fluorescent hybridization probes. *Clin. Chem.* **46**, 147–148.

Bernard, P. S., Lay, M. J. and Wittwer, C. T. (1998). Integrated amplification and detection of the C677T point mutation in the methyltetrahydrofolate reductase gene by fluorescence resonance energy transfer and probe melting curves. *Anal. Biochem.* **255**, 101–107.

Biel, S. S. and Madeley, D. (2001). Diagnostic virology – the need for electron microscopy: a discussion paper. *J. Clin. Virol.* **22**, 1–9.

Biel, S. S., Held, T. K., Landt, O., Niedrig, M., Gelderblom, H. R., Siegert, W. and Nitsche, A. (2000). Rapid quantification and differentiation of human polyomavirus DNA in undiluted urine from patients after bone marrow transplantation. *J. Clin. Microbiol.* **38**, 3689–3695.

Biggar, R. J., Whitby, D., Marshall, V., Linhares, A. C. and Black, F. (2000). Human herpesvirus 8 in Brazilian Amerindians: a hyperendemic population with a new subtype. *J. Infect. Dis.* **181**, 1562–1568.

Blair, P. L., Witney, A., Haynes, J. D., Moch, J. K., Carucci, D. J. and Adams, J. H. (2002). Transcripts of developmentally regulated *Plasmodium falciparum* genes quantified by real-time RT-PCR. *Nucleic Acids Res.* **30**, 2224–2231.

Blaschke, V., Reich, K., Blaschke, S., Zipprich, S. and Neumann, C. (2000). Rapid quantitation of proinflammatory and chemoattractant cytokine expression in small tissue samples and monocyte-derived dendritic cells: validation of a new real-time RT-PCR technology. *J. Immunol. Meth.* **246**, 79–90.

Boivin, G., Côte, S., Déry, P., De Serres, G. and Bergeron, M. G. (2004). Multiplex real-time PCR assay for detection of influenza and human respiratory syncytial virus. *J. Clin. Microbiol.* **42**, 45–51.

Böhm, J., Hahn, A., Schubert, R., Bahnweg, G., Adler, N., Nechwatal, J., Oehlmann, R. and Oswald, W. (1999). Real-time quantitative PCR: DNA determination in isolated spores of the mycorrhizal fungus *Glomus mosseae* and monitoring of *Phytophthora infestans* and *Phytophthora citricola* in their respective host plants. *J. Phytopathol.* **147**, 409–416.

Borson, N. D., Strausbauch, M. A., Wettstein, P. J., Oda, R. P., Johnston, S. L. and Landers, J. P. (1998). Direct quantitation of RNA transcripts by competitive single-tube RT-PCR and capillary electrophoresis. *Biotechniques* **25**, 130–137.

Brandt, M. E., Padhye, A. A., Mayer, L. W. and Holloway, B. P. (1998). Utility of random amplified polymorphic DNA PCR and TaqMan automated detection in molecular identification of *Aspergillus fumigatus*. *J. Clin. Microbiol.* **36**, 2057–2062.

Brechtbuehl, K., Whalley, S. A., Dusheiko, G. M. and Saunders, N. A. (2001). A rapid real-time quantitative polymerase chain reaction for hepatitis B virus. *J. Virol. Meth.* **93**, 105–113.

Bretagne, S. (2003). Molecular diagnostics in clinical parasitology and mycology: limits of the current polymerase chain reaction (PCR) assays and interest of the real-time PCR assays. *Clin. Microbiol. Infect.* **9**, 505–511.

Brightwell, G., Pearce, M. and Leslie, D. (1998). Development of internal controls for PCR detection of *Bacillus anthracis*. *Mol. Cell. Probes* **12**, 367–377.

Broberg, E. K., Nygårdas, M., Salmi, A. A. and Hukkanen, V. (2003). Low copy number detection of herpes simplex virus type 1 mRNA and mouse Th1 type cytokine mRNAs by Light Cycler quantitative real-time PCR. *J. Virol. Meth.* **112**, 53–65.

Brorson, K., Swann, P. G., Lizzio, E., Maudru, T., Peden, K. and Stein, K. E. (2001). Use of a quantitative product-enhanced reverse transcriptase assay to monitor retrovirus levels in mAb cell-culture and downstream processing. *Biotechnol. Prog.* **17**, 188–196.

Broussard, L. A. (2001). Biological agents: weapons of warfare and bioterrorism. *Mol. Diagn.* **6**, 323–333.

Bruña-Romero, O., Hafalla, J. C. R., González-Aseguinolaza, G., Sano, G., Tsuji, M. and Zavala, F. (2001). Detection of malaria liver-stages in mice infected through the bite of a single *Anopheles* mosquito using a highly sensitive real-time PCR. *Int. J. Parasitol.* **31**, 1499–1502.

Burgart, L. J., Robinson, R. A., Heller, M. J., Wilke, W. W., Iakoubova, O. K. and Cheville, J. C. (1992). Multiplex polymerase chain reaction. *Mod. Pathol.* **5**, 320–323.

Burgos, J. S., Ramírez, C., Tenorio, R., Sastre, I. and Bullido, M. J. (2002). Influence of reagents formulation on real-time PCR parameters. *Mol. Cell. Probes* **16**, 257–260.

Bustin, S. A. (2000). Absolute quantification of mRNA using real-time reverse transcription polymerase chain reaction assays. *J. Mol. Endocrinol.* **25**, 169–193.

Bustin, S. A. (2002). Quantification of mRNA using real-time reverse transcription PCR (RT-PCR): trends and problems. *J. Mol. Endocrinol.* **29**, 23–39.

Callahan, J. D., Wu, S.-J. L., Dion-Schultz, A., Mangold, B. E., Peruski, L. F., Watts, D. M., Porter, K. R., Murphy, G. R., Suharyono, W., King, C.-C., Hayes, C. G. and Temenak, J. J. (2001). Development ad evaluation of serotype- and group-specific fluorogenic reverse transcriptase PCR (TaqMan) assays for dengue virus. *J. Clin. Microbiol.* **39**, 4119–4124.

Cane, P. A., Cook, P., Ratcliffe, D., Mutimer, D. and Pillay, D. (1999). Use of real-time PCR and fluorimetry to detect lamivudine resistance-associated mutations in hepatitis B virus. *Antimicrob. Agent Chem.* **43**, 1600–1608.

Capone, R. B., Pai, S. I., Koch, W. M. and Gillison, M. L. (2001). Detection and quantitation of human papillomavirus (HPV) DNA in the sera of patients with HPV-associated head and neck squamous cell carcinoma. *Clin. Cancer Res.* **6**, 4171–4175.

Cardullo, R. A., Agrawai, S., Flores, C., Zamecnik, P. C. and Wolf, D. E. (1988). Detection of nucleic acid hybridization by nonradiative fluorescence resonance energy transfer. *Proc. Natl Acad. Sci. USA* **85**, 8790–8794.

Carman, B. (2001). Molecular techniques should now replace cell culture in diagnostic virology laboratories. *Rev. Med. Virol.* **11**, 347–349.

Carman, W. F., Wallace, L. A., Walker, J., McIntyre, S., Noone, A., Christie, P., Millar, J. and Douglas, J. D. (2000). Rapid virological surveillance of community influenza infection in general practice. *Br. Med. J.* **321**, 736–737.

Casadevall, A. and Pirofski, L. (2001). Host pathogen interactions: the attributes of virulence. *J. Infect. Dis.* **184**, 337–344.

Casadevall, A. and Pirofski, L. (2002). What is a pathogen? *Ann. Med.* **34**, 2–4.

Casadevall, A. and Pirofski, L. (2003). Microbial virulence results from the interaction between host and microorganism. *Trends Microbiol.* **11**, 157–158.

Celi, F. S., Mentuccia, D., Proietti-Pannunzi, L., di Gioia, C. R. T. and Andreoli, M. (2000). Preparing poly-A-containing RNA internal standards for multiplex competitive RT-PCR. *Biotechniques* **29**, 454–458.

Centers for Disease Control and Prevention, (2003). Multistate outbreak of monkeypox – Illinois, Indiana and Wisconsin. *MMWR* **52**, 537–540.

Chamberlain, J. S., Gibbs, R. A., Ranier, J. E., Nguyen, P. N. and Caskey, C. T. (1988). Deletion screening of the Duchenne muscular dystrophy locus via multiplex DNA amplification. *Nucleic Acids Res.* **16**, 11141–11156.

Chandelier, A., Dubois, N., Baelen, F., De Leener, F., Warnon, S., Remacle, J. and Lepoivre, P. (2001). RT-PCR-ELOSA tests on pooled sample units for the detection of virus Y in potato tubers. *J. Virol. Meth.* **91**, 99–108.

Chang, L.-J., Urlacher, V., Iwakuma, T., Cui, Y. and Zucali, J. (1999). Efficacy and safety analyses of a recombinant human immunodeficiency virus type 1 derived vector system. *Gene Ther.* **6**, 715–728.

Chehab, F. F., Doherty, M., Cai, S., Kan, Y. W., Cooper, S. and Rubin, E. M. (1987). Detection of sickle cell anaemia and thalassaemias. *Nature* **329**, 293–294.

Chen, S., Yee, A., Griffiths, M., Larkin, C., Yamashiro, C. T., Behari, R., Paszko-Kolva, C., Rahn, K. and De Grandis, S. A. (1997). The evaluation of a fluorogenic polymerase chain reaction assay for the detection of *Salmonella* species in food commodities. *Int. J. Food Microbiol.* **35**, 239–250.

Chen, R. W., Piiparinen, H., Seppänen, M., Koskela, P., Sarna, S. and Lappalainen, M. (2001). Real-time PCR for detection and quantitation of hepatitis B virus DNA. *J. Med. Virol.* **65**, 250–256.

Chen, Y. P., Higgins, J. A. and Gundersen-Rindal, D. E. (2003). Quantitation of a *Glyptapanteles indiensis* polydnavirus gene expressed in parasitized host, *Lymantria dispar*, by real-time quantitative RT-PCR. *J. Virol. Meth.* **114**, 125–133.

Chisholm, S. A., Owen, R. J., Teare, E. L. and Saverymuttu, S. (2001). PCR-based diagnosis of *Helicobacter pylori* infection and real-time determination of clarithromycin resistance directly from human gastric biopsy samples. *J. Clin. Microbiol.* **39**, 1217–1220.

Choo, C. K., Ling, M. T., Suen, C. K. M., Chan, K. W. and Kwong, Y. L. (2000). Retrovirus-mediated delivery of HPV16 E7 antisense RNA inhibited tumorigenicity of CaSki cells. *Gynecol. Oncol.* **78**, 293–301.

Chou, Q., Russell, M., Birch, D. E., Raymond, J. and Bloch, W. (1992). Prevention of pre-PCR mis-priming and primer dimerization improves low-copy-number amplifications. *Nucleic Acids Res.* **20**, 1717–1723.

Clarke, S. C., Reid, J., Thom, L., Denham, B. C. and Edwards, G. F. S. (2002). Laboratory confirmation of meningococcal disease in Scotland, 1993–9. *J. Clin. Pathol.* **55**, 32–36.

Clegg, R. M. (1992). Fluorescence resonance energy transfer and nucleic acids. *Meth. Enzymol.* **211**, 353–388.

Clementi, M. (2000). Quantitative molecular analysis of virus expression and replication. *J. Clin. Microbiol.* **38**, 2030–2036.

Clementi, M., Menzo, S., Manzin, A. and Bagnarelli, P. (1995). Quantitative molecular methods in virology. *Arch. Virol.* **140**, 1523–1539.

Clewley, J. P. (2004). A role for arrays in clinical virology: fact or fiction? *J. Clin. Virol.* **29**, 2–12.

Cockerill, F. R. III and Smith, T. F. (2002). Rapid-cycle real-time PCR: a revolution for clinical microbiology. *ASM News* **68**, 77–83.

Compton, J. (1991). Nucleic acid sequence-based amplification. *Nature* **350**, 91–92.

Corbella, M. E. and Puyet, A. (2003). Real-time reverse transcription-PCR analysis of expression of halobenzoate and salicylate catabolism-associated operons in two strains of *Pseudomonas aeruginosa*. *Appl. Environ. Microbiol.* **69**, 2269–2275.

Corless, C. E., Guiver, M., Borrow, R., Edwards-Jones, V., Fox, A. J. and Kaczmarski, E. B. (2001). Simultaneous detection of *Neisseria meningitidis*, *Haemophilus influenzae*, and *Streptococcus pneumoniae* in suspected cases of meningitis and septicemia using real-time PCR. *J. Clin. Microbiol.* **39**, 1553–1558.

Corless, C. E., Guiver, M., Borrow, R., Edwards-Jones, V., Fox, A. J., Kaczmarski, E. B. and Mutton, K. J. (2002). Development and evaluation of a "real-time" RT-PCR for the detection of enterovirus and parechovirus RNA in CSF and throat swab samples. *J. Med. Virol.* **67**, 555–562.

Costa, J.-M., Pautas, C., Ernault, P., Foulet, F., Cordonnier, C. and Bretagne, S. (2000). Real-time PCR for diagnosis and follow-up of toxoplasma reactivation after allogeneic stem cell transplantation using fluorescence resonance energy transfer hybridization probes. *J. Clin. Microbiol.* **38**, 2929–2932.

Costa, J.-M., Ernault, P., Gautier, E. and Bretagne, S. (2001). Prenatal diagnosis of congenital toxoplasmosis by duplex real-time PCR using fluorescence resonance energy transfer hybridization probes. *Prenat. Diagn.* **21**, 85–88.

Costa, C., Costa, J.-M., Desterke, C., Botterel, F., Cordonnier, C. and Bretagne, S. (2002). Real-time PCR coupled with automated DNA extraction and detection of galactomannan antigen in serum by enzyme-linked immunosorbent assay for diagnosis of invasive aspergillosis. *J. Clin. Microbiol.* **40**, 2224–2227.

Côte, S., Abed, Y. and Boivin, G. (2003). Comparative evaluation of real-time PCR assays for detection of the human metapneumovirus. *J. Clin. Microbiol.* **41**, 3631–3635.

Creelan, J. L. and McCullough, S. J. (2000). Evaluation of strain-specific primer sequences from an abortifacient strain of ovine *Chlamydophila abortus* (*Chlamydia psittaci*) for the detection of EAE by PCR. *FEMS Microbiol. Lett.* **190**, 103–108.

Crockett, A. O. and Wittwer, C. T. (2001). Fluorescein-labelled oligonucleotides for real-time PCR: using the inherent quenching of deoxyguanosine nucleotides. *Anal. Biochem.* **290**, 89–97.

Cruz-Perez, P., Buttner, M. P. and Stetzenbach, L. D. (2001). Specific detection of *Stachybotrys chartarum* in pure culture using quantitative polymerase chain reaction. *Mol. Cell. Probes* **15**, 129–138.

Czerny, C. P., Eis-Hubinger, A. M., Mayr, A., Schneweis, K. E. and Pfeiff, B. (1991). Animal poxviruses transmitted from cat to man: current event with lethal end. *Zentralbl. Veterinarmed. B* **38**, 421–431.

Daniels, P., Ksiazek, T. and Eaton, B. T. (2001). Laboratory diagnosis of Nipah and Hendra virus infections. *Microbes Infect.* **3**, 289–295.

de Viedma, D. G., Diaz Infanntes, M., Lasala, F., Chaves, F., Alcalá, L. and Bouza, E. (2002). New real-time PCR able to detect in a single tube multiple rifamin resistance mutations and high-level isoniazid resistance mutations in *Mycobacterium tuberculosis*. *J. Clin. Microbiol.* **40**, 988–995.

de Wit, C., Fautz, C. and Xu, Y. (2000). Real-time quantitative PCR for retrovirus-like particle formation in CHO cell culture. *Biologicals* **28**, 137–148.

DeGraves, F. J., Gao, D. and Kaltenboeck, B. (2003). High-sensitivity quantitative PCR platform. *Biotechniques* **34**, 106–115.

Dekonenko, A., Ibrahim, M. S. and Schmaljohn, C. S. (1997). A colorimetric PCR–enzyme immunoassay to identify hantaviruses. *Clin. Diag. Virol.* **8**, 113–121.

Deng, X., Li, H. and Tang, Y.-W. (2003). Cytokine expression in respiratory syncytial virus-infected mice as measured by quantitative reverse-transcriptase PCR. *J. Virol. Meth.* **107**, 141–146.

Desjardin, L. E., Chen, Y., Perkins, M. D., Teixeira, L., Cave, M. D. and Eisenach, K. D. (1998). Comparison of the ABI 7700 system (TaqMan) and competitive PCR for quantification of IS6110 DNA in sputum during treatment of tuberculosis. *J. Clin. Microbiol.* **36**, 1964–1968.

Devers, M., Soulas, G. and Martin-Laurent, F. (2004). Real-time reverse transcription PCR analysis of expression of atrazine catabolism genes in two bacterial strains isolated from soil. *J. Microbiol. Meth.* **56**, 3–15.

Didenko, V. V. (2001). DNA probes using fluorescence resonance energy transfer (FRET): designs and applications. *Biotechniques* **31**, 1106–1121.

Drago, L., Lombardi, A., De Vecchi, E. and Gismondo, M. R. (2002). Real-time PCR assay for the rapid detection of *Bacillus anthracis* spores in clinical samples. *J. Clin. Microbiol.* **40**, 4399.

Drosten, C., Gunther, S., Preiser, W., van der, W. S., Brodt, H. R., Becker, S., Rabenau, H., Panning, M., Kolesnikova, L., Fouchier, R. A., Berger, A., Burguiere, A. M., Cinatl, J., Eickmann, M., Escriou, N., Grywna, K., Kramme, S., Manuguerra, J. C., Muller, S., Rickerts, V., Sturmer, M., Vieth, S., Klenk, H. D., Osterhaus, A. D., Schmitz, H. and Doerr, H. W. (2003). Identification of a novel coronavirus in patients with severe acute respiratory syndrome. *N. Engl. J. Med.* **348**, 1967–1976.

Echevarria, J. E., Erdman, D. D., Swierkosz, E. M., Holloway, B. P. and Anderson, L. J. (1998). Simultaneous detection and identification of human parainfluenza viruses 1, 2, and 3 from clinical samples by multiplex PCR. *J. Clin. Microbiol.* **36**, 1388–1391.

Edelman, D. C. and Barletta, J. (2003). Real-time PCR provides improved detection and titer determination of bacteriophage. *Biotechniques* **35**, 368–375.

Egholm, M., Buchardt, O., Christensen, L., Behrens, C., Freier, S. M., Driver, D. A., Berg, R. H., Kim, S. K., Norden, B. and Nielsen, P. E. (1993). PNA hybridizes to complementary oligonucleotides obeying the Watson–Crick hydrogen bonding rules. *Nature* **365**, 566–568.

Eishi, Y., Suga, M., Ishige, I., Kobayashi, D., Yamada, T., Takemura, T., Takizawa, T., Koike, M., Kudoh, S., Costabel, U., Guzman, J., Rizzato, G., Gambacorta, M., du Bois, R., Nicholson, A. G., Sharma, O. P. and Ando, M. (2002). Quantitative analysis of mycobacterial and propionibacterial DNA in lymph nodes of Japanese and European patients with sarcoidosis. *J. Clin. Microbiol.* **40**, 198–204.

Elenitoba, K. S. J., Bohling, S. D., Wittwer, C. T. and King, T. C. (2001). Multiplex PCR by multicolor fluorimetry and fluorescence melting curve analysis. *Nat. Med.* **7**, 249–253.

Ellerbrok, H., Nattermann, H., Özel, M., Beutin, L., Appel, B. and Pauli, G. (2002). Rapid and sensitive identification of pathogenic and apathogenic *Bacillus anthracis* by real-time PCR. *FEMS Microbiol. Lett.* **214**, 51–59.

Ellis, J. S. and Zambon, M. C. (2002). Molecular diagnosis of influenza. *Rev. Med. Virol.* **12**, 375–389.

Emery, S. L., Erdman, D. D., Bowen, M. D., Newton, B. R., Winchell, J. M., Meyer, R. F., Tong, S., Cook, B. T., Holloway, B. P., McCaustland, K. A., Rota, P. A., Bankamp, B., Lowe, L. E., Ksiazek, T. G., Bellini, W. J. and Anderson, L. J. (2004). Real-time reverse transcription-polymerase chain reaction assay for SARS-associated Coronavirus. *Emerg. Infect. Dis.* **10**, 311–316.

Espy, M. J., Ross, T. K., Teo, R., Svien, K. A., Wold, A. D., Uhl, J. R. and Smith, T. F. (2000a). Evaluation of LightCycler PCR for implementation of laboratory diagnosis of herpes simplex virus infections. *J. Clin. Microbiol.* **38**, 3116–3118.

Espy, M. J., Uhl, J. R., Mitchell, P. S., Thorvilson, J. N., Svien, K. A., Wold, A. D. and Smith, T. F. (2000b). Diagnosis of herpes simplex virus infections in the clinical laboratory by LightCycler PCR. *J. Clin. Microbiol.* **38**, 795–799.

Espy, M. J., Cockerill, F. R. III, Meyer, R. F., Bowen, M. D., Poland, G. A., Hadfield, T. L. and Smith, T. F. (2002). Detection of smallpox virus DNA by LightCycler PCR. *J. Clin. Microbiol.* **40**, 1985–1988.

Fan, H. and Gulley, M. L. (2001). Epstein-Barr viral load measurement as a marker of EBV-related disease. *Mol. Diagn.* **6**, 279–289.

Fenollar, F., Fournier, P.-E., Raoult, D., Gérolami, R., Lepidi, H. and Poyart, C. (2002). Quantitative detection of *Tropheryma whipplei* DNA by real-time PCR. *J. Clin. Microbiol.* **40**, 1119–1120.

Fernandez, C., Boutolleau, D., Manichanh, C., Mangeney, N., Agut, H. and Gautheret-Dejean, A. (2002). Quantitation of HHV-7 genome by real-time polymerase chain reaction assay using MGB probe technology. *J. Virol. Meth.* **106**, 11–16.

Fiandaca, M. J., Hyldig-Nielsen, J. J., Gildea, B. D. and Coull, J. M. (2001). Self-reporting PNA/DNA primers for PCR analysis. *Genome Res.* **11**, 609–611.

Förster, T. (1948). Zwischenmolekulare energiewanderung und fluoreszenz. *Ann. Phys.* **6**, 55–75.

Fortin, N. Y., Mulchandani, A. and Chen, W. (2001). Use of real-time polymerase chain reaction and molecular beacons for the detection of *Escherichia coli* 0157:H7. *Anal. Biochem.* **289**, 281–288.

Fraaije, B. A., Lovell, D. J., Coelho, J. M., Baldwin, S. and Hollomon, D. W. (2001). PCR-based assays to assess wheat varietal resistance to blotch (*Septoria tritici* and *Stagonospora nodorum*) and rust (*Puccinia striiformis* and *Puccinia recondita*) diseases. *Eur. J. Plant Pathol.* **107**, 905–917.

Fredricks, D. N. and Relman, D. A. (1999). Application of polymerase chain reaction to the diagnosis of infectious disease. *Clin. Infect. Dis.* **29**, 475–488.

Freeman, W. M., Walker, S. J. and Vrana, K. E. (1999). Quantitative RT-PCR: pitfalls and potential. *Biotechniques* **26**, 112–125.

French, D. J., Archard, C. L., Brown, T. and McDowell, D. G. (2001). HyBeacon™ probes: a new tool for DNA sequence detection and allele discrimination. *Mol. Cell. Probes* **15**, 363–374.

French, D. J., Archard, C. L., Andersen, M. T. and McDowell, D. G. (2002). Ultra-rapid analysis using HyBeacon™ probes and direct PCR amplification from saliva. *Mol. Cell. Probes* **16**, 319–326.

Freymuth, F., Eugene, G., Vabret, A., Petitjean, J., Gennetay, E., Brouard, J., Duhamel, J. F. and Guillois, B. (1995). Detection of respiratory syncytial virus by reverse transcription-PCR and hybridization with a DNA enzyme immunoassay. *J. Clin. Microbiol.* **33**, 3352–3355.

Fujita, H., Eishi, Y., Ishige, I., Saitoh, K. and Takizawa, T. (2002). Quantitative analysis of bacterial DNA from *Mycobacteria* spp. *Bacteroides vulgatus*, and *Escherichia coli* in tissue samples from patients with inflammatory bowel diseases. *J. Gastroenterol.* **37**, 509–516.

Furuta, Y., Ohtani, F., Sawa, H., Fukuda, S. and Inuyama, Y. (2001). Quantitation of varicella-zoster virus DNA in patients with Ramsay Hunt syndrome and zoster sine herpete. *J. Clin. Microbiol.* **39**, 2856–2859.

Gallagher, A., Armstrong, A. A., MacKenzie, J., Shield, L., Khan, G., Lake, A., Proctor, S., Taylor, P., Clements, G. B. and Jarrett, R. F. (1999). Detection of epstein-barr virus (EBV) genomes in the serum of patients with EBV-associated Hodgkin's disease. *Int. J. Can.* **84**, 442–448.

Gault, E., Michel, Y., Nicolas, J.-C., Belabani, C., Nicolas, J.-C. and Garbarg-Chenon, A. (2001). Quantification of human cytomegalovirus DNA by real-time PCR. *J. Clin. Microbiol.* **39**, 772–775.

Gelmini, S., Orlando, C., Sestini, R., Vona, G., Pinzani, P., Ruocco, L. and Pazzagli, M. (1997). Quantitative polymerase chain reaction-based homogeneous assay with fluorogenic probes to measure c-cerB-2 oncogene amplification. *Clin. Chem.* **43**, 752–758.

Gerard, C. J., Arboleda, M. J., Solar, G., Mule, J. J. and Kerr, W. G. (1996). A rapid and quantitative assay to estimate gene transfer into retrovirally transduced hematopoietic stem/progenitor cells using a 96-well format PCR and fluorescent detection system universal for MMLV-based proviruses. *Hum. Gen. Ther.* **7**, 343–354.

Gibb, T. R., Norwood, D. A. Jr., Woollen, N. and Henchal, E. A. (2001a). Development and evaluation of a fluorogenic 5′-nuclease assay to identify Marburg virus. *Mol. Cell. Probes* **15**, 259–266.

Gibb, T. R., Norwood, D. A. Jr., Woollen, N. and Henchal, E. A. (2001b). Development and evaluation of a fluorogenic 5′ nuclease assay to detect and differentiate between Ebola virus subtypes Zaire and Sudan. *J. Clin. Microbiol.* **39**, 4125–4130.

Gibson, U. E. M., Heid, C. A. and Williams, P. M. (1996). A novel method for real time quantitative RT-PCR. *Genome Res.* **6**, 995–1001.

Gibson, J. R., Saunders, N. A. and Owen, R. J. (1999). Novel method for rapid determination of clarithromycin sensitivity in *Helicobacter pylori*. *J. Clin. Microbiol.* **37**, 3746–3748.

Giglio, S., Monis, P. T. and Saint, C. P. (2003). Demonstration of preferential binding of SYBR Green I to specific DNA fragments in real-time multiplex PCR. *Nucleic Acids Res.* **31**, e136.

Gilliland, G., Perrin, S. and Bunn, H. F. (1990). *Competitive PCR for Quantitation of mRNA. PCR Protocols: A Guide to Methods and Applications*. Academic Press, New York, pp. 60–69.

Ginzinger, D. G. (2002). Gene quantification using real-time quantitative PCR: an emerging technology hits mainstream. *Exp. Hematol.* **30**, 503–512.

Giulietti, A., Overbergh, L., Valckx, D., Decallonne, B., Bouillon, R. and Mathieu, C. (2001). An overview of real-time quantitative PCR: applications to quantify cytokine gene expression. *Methods* **25**, 386–401.

Glasel, J. A. (1995). Validity of nucleic acid purities monitored by 260 nm/280 nm absorbance ratios. *Biotechniques* **18**, 62–63.

Goerke, C., Bayer, M. G. and Wolz, C. (2001). Quantification of bacterial transcripts during infection using competitive reverse transcription-PCR (RT-PCR) and LightCycler RT-PCR. *Clin. Diag. Lab. Immunol.* **8**, 279–282.

Gruber, F., Falkner, F. G., Dorner, F. and Hämmerle, T. (2001). Quantitation of viral DNA by real-time PCR applying duplex amplification, internal standardization, and two-colour fluorescence detection. *Appl. Environ. Microbiol.* **67**, 2837–2839.

Guiver, M. and Oppenheim, B. A. (2001). Rapid identification of *Candida* species by TaqMan PCR. *J. Clin. Pathol.* **54**, 362–366.

Gundry, C. N., Bernard, P. S., Herrmann, M. G., Reed, G. H. and Wittwer, C. T. (1999). Rapid *F508del* and *F508C* assay using fluorescent hybridization probes. *Genet. Test* **3**, 365–370.

Gundry, C. N., Vandersteen, J. G., Reed, G. H., Pryor, R. J., Chen, J. and Wittwer, C. T. (2003). Amplicon melting analysis with labeled primers: a closed-tube method for differentiating homozygotes and heterozygotes. *Clin. Chem.* **49**, 396–406.

Gyllensten, U. B. and Erlich, H. A. (1988). Generation of single-stranded DNA by the polymerase chain reaction and its application to direct sequencing of the HLA-DQA locus. *Proc. Natl Acad. Sci. USA* **85**, 7652–7656.

Hackett, N. R., El Sawy, T., Lee, L. Y., Silva, I., O'Leary, J., Rosengart, T. K. and Crystal, R. G. (2000). Use of quantitative TaqMan real-time PCR to track the time-dependent distribution of gene transfer vectors *in vivo*. *Mol. Ther.* **2**, 649–656.

Halford, W. P. (1999). The essential prerequisites for quantitative RT-PCR. *Nat. Biotechnol.* **17**, 835.

Halford, W. P., Falco, V. C., Gebhardt, B. M. and Carr, D. J. J. (1999). The inherent quantitative capacity of the reverse transcription-polymerase chain reaction. *Anal. Biochem.* **266**, 181–191.

Hall, L. L., Bicknell, G. R., Primrose, L., Pringle, J. H., Shaw, J. A. and Furness, P. N. (1998). Reproducibility in the quantification of mRNA levels by RT-PCR-ELISA and RT competitive-PCR-ELISA. *Biotechniques* **24**, 652–657.

Halpin, K., Young, P. L., Field, H. E. and Mackenzie, J. S. (2000). Isolation of Hendra virus from pteropid bats: a natural reservoir of Hendra virus. *J. Gen. Virol.* **81**, 1927–1932.

Hardick, J., Yang, S., Lin, S., Duncan, D. and Gaydos, C. (2003). Use of Roche LightCycler instrument in a real-time PCR for *Trichomonas vaginalis* in urine samples from females and males. *J. Clin. Microbiol.* **41**, 5619–5622.

Härtel, C., Bein, G., Kirchner, H. and Klüter, H. (1999). A human whole-blood assay for analysis of T-cell function by quantification of cytokine mRNA. *Scand. J. Immunol.* **49**, 649–654.

Haugland, R. A., Vesper, S. J. and Wymer, L. J. (1999). Quantitative measurement of *Stachybotrys chartarum conidia* using real time detection of PCR products with the TaqMan (TM) fluorogenic probe system. *Mol. Cell. Probes* **13**, 329–340.

Hawrami, K. and Breur, J. (1999). Development of a fluorogenic polymerase chain reaction assay (TaqMan) for the detection and quantitation of varicella zoster virus. *J. Virol. Meth.* **79**, 33–40.

Hayden, R. T., Uhl, J. R., Qian, X., Hopkins, M. K., Aubry, M. C., Limper, A. H., Lloyd, R. V. and Cockerill, F. R. (2001). Direct detection of *Legionella* species from bronchoalveolar lavage and open lung biopsy specimens: comparison of LightCycler PCR, in situ hybridization, direct fluorescence antigen detection, and culture. *J. Clin. Microbiol.* **37**, 2618–2626.

Heid, C. A., Stevens, J., Livak, K. J. and Williams, P. M. (1996). Real time quantitative PCR. *Gen. Meth.* **6**, 986–994.

Hein, I., Lehner, A., Rieck, P., Klein, K., Brandl, E. and Wagner, M. (2001a). Comparison of different approaches to quantify *Staphylococcus aureus* by real-time quantitative PCR and application of this techniques for examination of cheese. *Appl. Environ. Microbiol.* **67**, 3122–3126.

Hein, J., Schellenberg, U., Bein, G. and Hackstein, H. (2001b). Quantification of murine IFN-γ mRNA and protein expression: impact of real-time kinetic RT-PCR using SYBR Green I dye. *Scand. J. Immunol.* **54**, 285–291.

Held, T. K., Biel, S. S., Nitsche, A., Kurth, A., Chen, S., Gelderblom, H. R. and Siegert, W. (2000). Treatment of BK virus-associated hemorrhagic cystitis and simultaneous CMV reactivation with cidofovir. *Bone Marrow Transpl.* **26**, 347–350.

Heller, M. J. and Morrison, L. E. (1985). Chemiluminescent and fluorescent probes for DNA hybridization. In *Rapid Detection and Identification of Infectious Agents* (D. T. Kingsbury and S. Falkow, eds), pp. 245–256. Academic Press, New York.

Henderson, D. A., Inglesby, T. V., Bartlett, J. G., Ascher, M. S., Eitzen, E., Jahrling, P. B., Hauer, J., Layton, M., McDade, J., Osterholm, M. T., O'Toole, T., Parker, G., Perl, T., Russell, P. K. and Tonat, K. (1999). Smallpox as a biological weapon: medical and public health management. Working group on civilian biodefense. *JAMA* **281**, 2127–2137.

Henegariu, O., Heerema, N. A., Dloughy, S. R., Vance, G. H. and Vogt, P. H. (1997). Multiplex PCR critical parameters and step-by-step protocol. *Biotechniques* **23**, 504–511.

Hermsen, C. C., Telgt, D. S. C., Linders, E. H. P., van de Locht, L. A. T. F., Eling, W. M. C., Mensink, E. J. B. M. and Sauerwein, R. W. (2001). Detection of *Plasmodium falciparum* malaria parasites in vivo by real-time quantitative PCR. *Mol. Biochem. Parasitol.* **118**, 247–251.

Herrmann, M. G., Dobrowolski, S. F. and Wittwer, C. T. (2000). Rapid β-globin genotyping by multiplexing probe melting temperature and color. *Clin. Chem.* **46**, 425–429.

Higgins, J. A., Ezzell, J., Hinnebusch, B. J., Shipley, M., Henchal, E. A. and Ibrahim, M. S. (1998). 5′ Nuclease PCR assay to detect *Yersinia pestis*. *J. Clin. Microbiol.* **36**, 2284–2288.

Higgins, J. A., Hubalek, Z., Halouzka, J., Elkins, K. L., Sjostedt, A., Shipley, M. and Ibrahim, M. S. (2000). Detection of *Francisella tularensis* in infected mammals and vectors using a probe-based polymerase chain reaction. *Am. J. Trop. Med. Hyg.* **62**, 310–318.

Higgins, J. A., Cooper, M., Schroeder-Tucker, L., Black, S., Miller, D., Karns, K. S., Manthey, E., Breeze, R. and Perdue, M. L. (2003a). A field investigation of *Bacillus anthracis* contamination of U.S. department of agriculture and other Washington, DC, buildings during the anthrax attack of October 2001. *Appl. Environ. Microbiol.* **69**, 593–599.

Higgins, J. A., Nasarabadi, S., Karns, J. S., Shelton, D. R., Cooper, M., Gbakima, A. and Koopman, R. P. (2003b). A handheld real time thermal cycler for bacterial pathogen detection. *Biosens. Bioelectron.* **18**, 1115–1123.

Higuchi, R., Dollinger, G., Walsh, P. S. and Griffith, R. (1992). Simultaneous amplification and detection of specific DNA sequences. *Biotechnology (NY)* **10**, 413–417.

Higuchi, R., Fockler, C., Dollinger, G. and Watson, R. (1993). Kinetic PCR analysis: real-time monitoring of DNA amplification reactions. *Biotechnology (NY)* **11**, 1026–1030.

Holland, P. M., Abramson, R. D., Watson, R. and Gelfand, D. H. (1991). Detection of specific polymerase chain reaction product by utilizing the 5′–3′ exonuclease activity of *Thermus aquaticus*. *Proc. Natl Acad. Sci. USA* **88**, 7276–7280.

Holodniy, M., Katzenstein, D., Sengupta, S., Wang, A. M., Casipit, C., Schwartz, D. H., Konrad, M., Groves, E. and Merigan, T. C. (1991). Detection and quantification of human immunodeficiency virus RNA in patient serum by use of the polymerase chain reaction. *J. Infect. Dis.* **163**, 862–866.

Hoorfar, J., Cook, N., Malorny, B., Wagner, M., De Medici, D., Abdulmawjood, A. and Fach, P. (2003). Making an internal amplification control mandatory for diagnostic PCR. *J. Clin. Microbiol.* **41**, 5835.

Hoshino, Y., Kimura, H., Kuzushima, K., Tsurumi, T., Nemoto, K., Kikuta, A., Nishiyama, Y., Kojima, S., Matsuyama, T. and Morishima, T. (2000). Early intervention in post-transplant lymphoproliferative disorders based on Epstein-Barr viral load. *Bone Marrow Transpl.* **26**, 199–201.

Howell, W. M., Jobs, M. and Brookes, A. J. (2002). iFRET: an improved fluorescence system for DNA-melting analysis. *Genome Res.* **12**, 1401–1407.

Hu, A., Colella, M., Tam, J. S., Rappaport, R. and Cheng, S.-M. (2003). Simultaneous detection, subgrouping, and quantitation of respiratory syncytial virus A and B by real-time PCR. *J. Clin. Microbiol.* **41**, 149–154.

Huang, J., DeGraves, F. J., Gao, D., Feng, P., Schlapp, T. and Kaltenboeck, B. (2001). Quantitative detection of *Chlamydia* spp. by fluorescent PCR in the LightCycler. *Biotechniques* **30**, 150–157.

Hughes, G. J., Smith, J. A., Hanlon, C. A. and Rupprecht, C. E. (2004). Evaluation of a TaqMan PCR assay to detect rabies virus RNA: influence of sequence variation and application to quantification of viral loads. *J. Clin. Microbiol.* **41**, 299–306.

Hurtle, W., Bode, E., Kulesh, D. A., Kaplan, R. S., Garrison, J., Bridge, D., House, M., Frye, M. S., Loveless, B. and Norwood, D. (2004). Detection of *Bacillus anthracis* gyrA gene by using a minor groove binder probe. *J. Clin. Microbiol.* **42**, 179–185.

Hyypiä, T., Puhakka, T., Ruuskanen, O., Mäkelä, M., Arola, A. and Arstila, P. (1998). Molecular diagnosis of human rhinovirus infections: comparison with virus isolation. *J. Clin. Microbiol.* **36**, 2081–2083.

Ibekwe, A. M., Watt, P. M., Grieve, C. M., Sharma, V. K. and Lyons, S. R. (2002). Multiplex fluorogenic real-time PCR for detection and quantification of *Escherichia coli* O157:H7 in dairy wastewater wetlands. *Appl. Environ. Microbiol.* **68**, 4853–4862.

Ibrahim, M. S., Esposito, J. J., Jahrling, P. B. and Lofts, R. S. (1997). The potential of $5'$ nuclease PCR for detecting a single-base polymorphism in *Orthopoxvirus*. *Mol. Cell. Probes* **11**, 143–147.

Ibrahim, M. S., Lofts, R. S., Jahrling, P. B., Henchal, E. A., Weedn, V. W., Northrup, M. A. and Belgrader, P. (1998). Real-time microchip PCR for detecting single-base differences in viral and human DNA. *Anal. Chem.* **70**, 2013–2017.

Ibrahim, M. S., Kulesh, D. A., Saleh, S. S., Damon, I. K., Esposito, J. J., Schmaljohn, A. L. and Jahrling, P. B. (2003). Real-time PCR assay to detect smallpox virus. *J. Clin. Microbiol.* **41**, 3835–3839.

Iriyama, M., Kimura, H., Nishikawa, K., Yoshioka, K., Wakita, T., Nishimura, N., Shibata, M., Ozaki, T. and Morishima, T. (1999). The prevalence of TT virus (TTV) infection and its relationship to hepatitis in children. *Med. Microbiol. Immunol. (Berl.)* **188**, 83–89.

Isacsson, J., Cao, H., Ohlsson, L., Nordgren, S., Svanvik, N., Westman, G., Kubista, M., Sjöback, R. and Sehlstedt, U. (2000). Rapid and specific detection of PCR products using light-up probes. *Mol. Cell. Probes* **14**, 321–328.

Ishiguro, T., Saitoh, J., Yawata, H., Yamagishi, H., Iwasaki, S. and Mitoma, Y. (1995). Homogeneous quantitative assay of hepatitis C virus RNA by polymerase chain reaction in the presence of a fluorescent intercalater. *Anal. Biochem.* **229**, 207–213.

Jabs, W. J., Hennig, H., Kittel, M., Pethig, K., Smets, F., Bucsky, P., Kirchner, H. and Wagner, H. J. (2001). Normalized quantification by real-time PCR of Epstein-Barr virus load in patients at risk for posttransplant lymphoproliferative disorders. *J. Clin. Microbiol.* **39**, 564–569.

Johnson, J. R. (2000). Development of polymerase chain reaction-based assays for bacterial gene detection. *J. Microbiol. Meth.* **41**, 201–209.

Jordens, J. Z., Lanham, S., Pickett, M. A., Amarasekara, S., Aberywickrema, I. and Watt, P. J. (2000). Amplification with molecular beacon primers and reverse line blotting for the detection and typing of human papillomaviruses. *J. Virol. Meth.* **89**, 29–37.

Josefsson, A., Livak, K. and Gyllensten, U. (1999). Detection and quantitation of human papillomavirus by using the fluorescent $5'$ exonuclease assay. *J. Clin. Microbiol.* **37**, 490–496.

Josefsson, A. M., Magnusson, P. K. E., Ylitalo, N., Sørensen, P., Qwarforth-Tubbin, P., Andersen, P. K., Melbye, M., Adami, H.-O. and Gyllensten, U. (2000). Viral load of human papilloma virus 16 as a determinant for development of cervical carcinoma in situ: a nested case-control study. *Lancet* **355**, 2189–2193.

Jung, R., Soondrum, K. and Neumaier, M. (2000). Quantitative PCR. *Clin. Chem. Lab. Med.* **38**, 833–836.

Kaneko, S., Murakami, S., Unoura, M. and Kobayashi, K. (1992). Quantitation of hepatitis C virus RNA by competitive polymerase chain reaction. *J. Med. Virol.* **37**, 278–282.

Kares, S., Lönnrot, M., Vuorinen, P., Oikarinen, S., Taurianen, S. and Hyöty, H. (2004). Real-time PCR for rapid diagnosis of entero- and rhinovirus infections using LightCycler. *J. Clin. Virol.* **29**, 99–104.

Kato, T., Mizokami, M., Mukaide, M., Orito, E., Ohno, T., Nakano, T., Tanaka, Y., Kato, H., Sugauchi, F., Ueda, R., Hirashima, N., Shimamatsu, K., Kage, M. and Kojiro, M. (2000). Development of a TT virus DNA quantification system using real-time detection PCR. *J. Clin. Microbiol.* **38**, 94–98.

Kawai, S., Yokosuka, O., Kanda, T., Imazeki, F., Maru, Y. and Saisho, H. (1999). Quantification of hepatitis C virus by TaqMan PCR: comparison with HCV amplicor monitor assay. *J. Med. Virol.* **58**, 121–126.

Ke, D., Ménard, C., Picard, F. J., Boissinot, M., Ouellette, M., Roy, P. H. and Bergeron, M. G. (2000). Development of conventional and real-time PCR assays for the rapid detection of group B streptococci. *Clin. Chem.* **46**, 324–331.

Kearns, A. M., Draper, B., Wipat, W., Turner, A. J. L., Wheeler, J., Freeman, R., Harwood, J., Gould, F. K. and Dark, J. H. (2001a). LightCycler-based quantitative PCR for detection of cytomegalovirus in blood, urine and respiratory samples. *J. Clin. Microbiol.* **39**, 2364–2365.

Kearns, A. M., Guiver, M., James, V. and King, J. (2001b). Development and evaluation of a real-time quantitative PCR for the detection of human cytomegalovirus. *J. Virol. Meth.* **95**, 121–131.

Kearns, A. M., Turner, A. J. L., Taylor, C. E., George, P. W., Freeman, R. and Gennery, A. R. (2001c). LightCycler-based quantitative PCR for rapid detection of human herpesvirus 6 DNA in clinical material. *J. Clin. Microbiol.* **39**, 3020–3021.

Kehl, S. C., Henrickson, K. J., Hua, W. and Fan, J. (2001). Evaluation of the hexaplex assay for detection of respiratory viruses in children. *J. Clin. Microbiol.* **39**, 1696–1701.

Keller, G. H., Huang, D.-P., Shih, J. W.-K. and Manak, M. M. (1990). Detection of hepatitis B virus DNA in serum by polymerase chain reaction amplification and microtiter sandwich hybridization. *J. Clin. Microbiol.* **28**, 1411–1416.

Kemp, D. J., Churchill, M. J., Smith, D. B., Biggs, B. A., Foote, S. J., Peterson, M. G., Samaras, N., Deacon, N. J. and Doherty, R. (1990). Simplified colorimetric analysis of polymerase chain reactions: detection of HIV sequences in AIDS patients. *Gene* **94**, 223–228.

Kennedy, M. M., Lucas, S. B., Jones, R. R., Howells, D. D., Picton, S. J., Hanks, E. E., McGee, J. O. and O'Leary, J. (1997a). HHV8 and Kaposi's sarcoma: a time cohort study. *J. Clin. Pathol.* **50**, 96–100.

Kennedy, M. M., Lucas, S. B., Russell-Jones, R., Howells, D. D., Picton, S. J., Bardon, A., Comley, I. L., McGee, J. O. and O'Leary, J. J. (1997b). HHV8 and female Kaposi's sarcoma. *J. Pathol.* **183**, 447–452.

Kennedy, M. M., Cooper, K., Howells, D. D., Picton, S., Biddolph, S., Lucas, S. B., McGee, J. O. and O'Leary, J. J. (1998a). Identification of HHV8 in early Kaposi's sarcoma: implications for Kaposi's sarcoma pathogenesis. *J. Clin. Pathol.* **51**, 14–20.

Kennedy, M. M., O'Leary, J. J., Oates, J. L., Lucas, S. B., Howells, D. D., Picton, S. and McGee, J. O. (1998b). Human herpes virus 8 (HHV-8) in Kaposi's sarcoma: lack of association with Bcl-2 and p53 protein expression. *J. Clin. Pathol.* **51**, 155–159.

Kennedy, M. M., Biddolph, S., Lucas, S. B., Howells, D. D., Picton, S., McGee, J. O. and O'Leary, J. J. (1999a). CD40 upregulation is independent of HHV-8 in the pathogenesis of Kaposi's sarcoma. *J. Clin. Pathol.* **52**, 32–36.

Kennedy, M. M., Biddolph, S., Lucas, S. B., Howells, D. D., Picton, S., McGee, J. O., Silva, I., Uhlmann, V., Luttich, K. and O'Leary, J. (1999b). Cyclin D1 expression and HHV-8 in Kaposi sarcoma. *J. Clin. Pathol.* **52**, 569–573.

Kidd, I. M., Clark, D. A. and Emery, V. C. (2000). A non-radioisotopic quantitative competitive polymerase chain reaction method: application in measurement of human herpesvirus 7 load. *J. Virol. Meth.* **87**, 177–181.

Kievits, T., van Gemen, B., van Strijp, D., Schukkink, R., Dircks, M., Adriaanse, H., Malek, L., Sooknanan, R. and Lens, P. (1991). NASBA isothermal enzymatic in vitro nucleic acid amplification optimized for the diagnosis of HIV-1 infection. *J. Virol. Meth.* **35**, 273–286.

Kikuchi, T., Iwasaki, K., Nishihara, H., Takamura, Y. and Yagi, O. (2002). Quantitative and rapid detection of the trichloroethylene-degrading bacterium *Methylocystis* sp. in groundwater by real-time PCR. *Appl. Microbiol. Biotechnol.* **59**, 731–736.

Kilpatrick, D. R., Nottay, B., Yang, C. F., Yang, S. J., Mulders, M. N., Holloway, B. P., Pallansch, M. A. and Kew, O. M. (1996). Group-specific identification of polioviruses by PCR using primers containing mixed-base or deoxyinosine residue at positions of codon degeneracy. *J. Clin. Microbiol.* **34**, 2990–2996.

Kimura, H., Morita, M., Yabuta, Y., Kuzushima, K., Kato, K., Kojima, S., Matsuyama, T. and Morishima, T. (1999). Quantitative analysis of Epstein-Barr virus load by using a real-time PCR assay. *J. Clin. Microbiol.* **37**, 132–136.

Klaschik, S., Lehmann, L. E., Raadts, A., Book, M., Hoeft, A. and Stuber, F. (2002). Real-time PCR for the detection and differentiation of Gram-positive and Gram-negative bacteria. *J. Clin. Microbiol.* **40**, 4304–4307.

Klein, D. (2002). Quantification using real-time PCR technology: applications and limitations. *Trends Mol. Med.* **8**, 257–260.

Klein, D., Janda, P., Steinborn, R., Salmons, B. and Günzburg, W. H. (1999). Proviral load determination of different feline immunodeficiency virus isolates using real-time polymerase chain reaction: Influence of mismatches on quantification. *Electrophoresis* **20**, 291–299.

Komurian-Pradel, F., Paranhos-Baccalà, G., Sodoyer, M., Chevallier, P., Mandrand, B., Lotteau, V. and André, P. (2001). Quantitation of HCV RNA using real-time PCR and fluorimetry. *J. Virol. Meth.* **95**, 111–119.

Koshkin, A. A., Singh, S. K., Nielsen, P., Rajwanshi, V. K., Kumar, R., Meldgaard, M., Olsen, C. E. and Wengel, J. (1998). LNA (locked nuciec acids): synthesis of the adenine, cytosine, guanine, 5-methylcytosine, thymine and uracil bicyclo-nucleoside monomers, oligomerisation, and unprecedented nucleic acid recognition. *Tetrahedron* **54**, 3607–3630.

Kox, L. F. F., Noordhoek, G. T., Kunakorn, M., Mulder, S., Sterrenburg, M. and Kolk, A. H. J. (1996). Microwell hybridization assay for detection of PCR products from *Mycobacterium tuberculosis* complex and the recombinant *Mycobacterium smegmatis* strain 1008 used as an internal control. *J. Clin. Microbiol.* **34**, 2117–2120.

Kraus, G., Cleary, T., Miller, N., Seivright, R., Young, A. K., Spruill, G. and Hnatyszyn, H. J. (2001). Rapid and specific detection of the *Mycobacterium tuberculosis* complex using fluorogenic probes and real-time PCR. *Mol. Cell. Probes* **15**, 375–383.

Kreuzer, K.-A., Bohn, A., Lass, U., Peters, U. R. and Schmidt, C. A. (2000). Influence of DNA polymerases on quantitative PCR results using TaqMan probe format in the LightCycler instrument. *Mol. Cell. Probes* **14**, 57–60.

Kruse, N., Pette, M., Toyka, K. and Rieckmann, P. (1997). Quantification of cytokine mRNA expression by RT PCR in samples of previously frozen blood. *J. Immunol. Meth.* **210**, 195–203.

Kumar, R., Singh, S. K., Koshkin, A. A., Rajwanshi, V. K., Meldgaard, M. and Wengel, J. (1998). The first analogues of LNA (locked nucleic acids): phosphorothioate-LNA and 2′-thio-LNA. *Bioorg. Med. Chem. Lett.* **8**, 2219–2222.

Kupferschmidt, O., Krüger, D., Held, T. K., Ellerbrok, H., Siegert, W. and Janitschke, K. (2001). Quantitative detection of *Toxoplasma gondii* DNA in human

Kurata, S., Kanagawa, T., Yamada, K., Torimura, M., Yokomaku, T., Kamagata, Y. and Kurane, R. (2001). Fluorescent quenching-based quantitative detection of specific DNA/RNA using a BODIPY® FL-labeled probe or primer. *Nucleic Acids Res.* **29**, e34.

Kutyavin, I. V., Afonina, I. A., Mills, A., Gorn, V. V., Lukhtanov, E. A., Belousov, E. S., Singer, M. J., Walburger, D. K., Lokhov, S. G., Gall, A. A., Dempcy, R., Reed, M. W., Meyer, R. B. and Hedgpeth, J. (2000). 3′-Minor groove binder-DNA probes increase sequence specificity at PCR extension temperatures. *Nucleic Acids Res.* **28**, 655–661.

Lai, K. K.-Y., Cook, L., Wendt, S., Corey, L. and Jerome, K. R. (2003). Evaluation of real-time PCR versus PCR with liquid-phase hybridization for detection of enterovirus RNA in cerebrospinal fluid. *J. Clin. Microbiol.* **41**, 3133–3141.

Lallemand, F., Desire, N., Rozenbaum, W., Nicolas, J.-C. and Marechal, V. (2000). Quantitative analysis of human herpesvirus 8 viral load using a real-time PCR assay. *J. Clin. Microbiol.* **38**, 1404–1408.

Lanciotti, R. S. and Kerst, A. J. (2001). Nucleic acid sequence-based amplification assays for rapid detection of West Nile and St. Louis Encephalitis. *J. Clin. Microbiol.* **39**, 4506–4513.

Lanciotti, R. S., Kerst, A. J., Nasci, R. S., Godsey, M. S., Mitchell, C. J., Savage, H. M., Komar, N., Panella, N. A., Allen, B. C., Volpe, K. E., Davis, B. S. and Roehrig, J. T. (2000). Rapid detection of West Nile virus from human clinical specimens, field-collected mosquitoes, and avian samples by a TaqMan reverse transcriptase-PCR assay. *J. Clin. Microbiol.* **38**, 4066–4071.

Landt, O. (2001). Selection of hybridisation probes for real-time quantification and genetic analysis. In *Rapid Cycle Real-Time PCR: Methods and Applications* (S. Meuer, C. Wittwer and K. Nakagawara, eds), pp. 35–41. Springer, Germany.

Lanham, S., Herbert, A. and Watt, P. (2001). HPV detection and measurement of HPV-16, telomerase, and survivin transcripts in colposcopy clinic patients. *J. Clin. Pathol.* **54**, 304–308.

Laue, T., Emmerich, P. and Schmitz, H. (1999). Detection of dengue virus RNA inpatients after primary or secondary dengue infection by using the TaqMan automated amplification system. *J. Clin. Microbiol.* **37**, 2543–2547.

Lay, M. J. and Wittwer, C. T. (1997). Real-time fluorescence genotyping of factor V Leiden during rapid-cycle PCR. *Clin. Chem.* **43**, 2262–2267.

Lee, L. G., Connell, C. R. and Bloch, W. (1993). Allelic discrimination by nick-translation PCR with fluorogenic probes. *Nucleic Acids Res.* **21**, 3761–3766.

Lee, L. G., Livak, K. J., Mullah, B., Graham, R. J., Vinayak, R. S. and Woudenberg, T. M. (1999a). Seven-color, homogeneous detection of six PCR products. *Biotechniques* **27**, 342–349.

Lee, M. A., Brightwell, G., Leslie, D., Bird, H. and Hamilton, A. (1999b). Fluorescent detection techniques for real-time multiplex strand specific detection of *Bacillus anthracis* using rapid PCR. *J. Appl. Microbiol.* **87**, 218–223.

Lee, M.-A., Tan, C.-H., Aw, L.-T., Tang, C.-S., Singh, M., Lee, S.-H., Chia, H.-P. and Yap, E. P. H. (2002a). Real-time fluorescence-based PCR for detection of malaria parasites. *J. Clin. Microbiol.* **40**, 4343–4345.

Lee, M. A., Siddle, A. L. and Page, R. H. (2002b). ResonSense®: simple linear fluorescent probes for quantitative homogeneous rapid polymerase chain reaction. *Anal. Chim. Acta* **457**, 61–70.

Lefevre, J., Hankins, C., Pourreaux, K., Voyer, H. and Coutlée, F. (2003). Real-time PCR assays using internal controls for quantitation of HPV-16 and β-globin DNA in cervicovaginal lavages. *J. Virol. Meth.* **114**, 135–144.

Leutenegger, C. M., Klein, D., Hofmann-Lehmann, R., Mislin, C., Hummel, U., Boni, J., Boretti, F., Guenzburg, W. H. and Lutz, H. (1999). Rapid feline immunodeficiency virus provirus quantitation by polymerase chain reaction using the TaqMan fluorogenic real-time detection system. *J. Virol. Meth.* **78**, 105–116.

Lew, A. E., Bock, R. E., Miles, J., Cuttell, L. B., Steer, P. and Nadin-Davis, S. A. (2004a). Sensitive and specific detection of bovine immunodeficiency virus and bovine syncytial virus by 5′ Taq nuclease assays with fluorescent 3′ minor groove binder-DNA probes. *J. Virol. Meth.* **116**, 1–9.

Lew, A. E., Bock, R. E., Molloy, J. B., Minchin, C. M., Robinson, S. J. and Steer, P. (2004b). Sensitive and specific detection of proviral bovine leukemia virus by 5′ Taq nuclease PCR using a 3′ minor groove binder fluorogenic probe. *J. Virol. Meth.* **115**, 167–175.

Lewin, S. R., Vesanen, M., Kostrikis, L., Hurley, A., Duran, M., Zhang, L., Ho, D. D. and Markowitz, M. (1999). Use of real-time PCR and molecular beacons to detect virus replication in human immunodeficiency virus type 1-infected individuals on prolonged effective antiretroviral therapy. *J. Virol.* **73**, 6099–6103.

Lewis-Jones, S. (2002). The zoonotic poxviruses. *Dermatol. Nurs.* **14**, 79–82.

Li, Q.-G., Liang, J.-X., Luan, G.-Y., Zhang, Y. and Wang, K. (2000). Molecular beacon-based homogeneous fluorescence PCR assay for the diagnosis of infectious diseases. *Anal. Sci.* **16**, 245–249.

Li, Q., Luan, G., Guo, Q. and Liang, J. (2002). A new class of homogeneous nucleic acid probes based on specific displacement hybridization. *Nucleic Acids Res.* **30**, e5.

Limaye, A. P., Jerome, K. R., Kuhr, C. S., Ferrenberg, J., Huang, M.-L., Davis, C. L., Corey, L. and Marsh, C. L. (2000). Quantitation of BK virus load in serum for the diagnosis of BK virus-associated nephropathy in renal transplant recipients. *J. Infect. Dis.* **183**, 1669–1672.

Limaye, A. P., Huang, M.-L., Leisenring, W., Stensland, L., Corey, L. and Boeckh, M. (2001). Cytomegalovirus (CMV) DNA load in plasma for the diagnosis of CMV disease before engraftment in hematopoietic stem-cell transplant recipients. *J. Infect. Dis.* **183**, 377–382.

Lindler, L. E. and Fan, W. (2003). Development of a 5′ nuclease assay to detect ciprofloxacin resistant isolates of the biowarfare agent *Yersinia pestis*. *Mol. Cell. Probes* **17**, 41–47.

Lindler, L. E., Fan, W. and Jahan, N. (2001). Detection of ciprofloxacin-resistant *Yersinia pestis* by fluorogenic PCR using the LightCycler. *J. Clin. Microbiol.* **39**, 3649–3655.

Lipson, S. M. (2002). Rapid laboratory diagnostics during the winter respiratory virus season. *J. Clin. Microbiol.* **40**, 733–734.

Liss, B. (2002). Improved quantitative real-time RT-PCR for expression profiling of individual cells. *Nucleic Acids Res.* **30**, e89.

Liu Yin, J. A. (2002). Minimal residual disease in acute myeloid leukaemia. *Best Pract. Res. Clin. Haematol.* **15**, 119–135.

Liu, W. and Saint, D. A. (2002). A new quantitative method of real time reverse transcription polymerase chain reaction assay based on simulation of polymerase chain reaction kinetics. *Anal. Biochem.* **302**, 52–59.

Livak, K. J. and Schmittgen, T. D. (2001). Analysis of relative gene expression data using real-time quantitative PCR and the $2_t^{-\Delta\Delta C}$ method. *Methods* **25**, 402–408.

Livak, K. J., Flood, S. J. A., Marmaro, J., Giusti, W. and Deetz, K. (1995). Oligonucleotides with fluorescent dyes at opposite ends provide a quenched probe system useful for detecting PCR product and nucleic acid hybridization. *PCR Meth. Appl.* **4**, 357–362.

Lo, Y. M. D., Chan, L. Y. S., Lo, K.-W., Leung, S.-F., Zhang, J., Chan, A. T. C., Lee, J. C. K., Hjelm, N. M., Johnson, P. J. and Huang, D. P. (1999). Quantitative analysis of cell-free Epstein-Barr virus DNA in plasma of patients with nasopharyngeal carcinoma. *Cancer Res.* **59**, 1188–1191.

Locatelli, G., Santoro, F., Veglia, F., Gobbi, A., Lusso, P. and Malnati, M. S. (2000). Real-time quantitative PCR for human herpesvirus 6 DNA. *J. Clin. Microbiol.* **38**, 4042–4048.

Lockhart, D. J. and Winzeler, E. A. (2000). Genomics, gene expression and DNA arrays. *Nature* **405**, 827–836.

Lomeli, H., Tyagi, S., Pritchard, C. G., Lizardi, P. M. and Kramer, F. R. (1989). Quantitative assays based on the use of replicatable hybridization probes. *Clin. Chem.* **35**, 1826–1831.

Loparev, V. N., McCaustland, K., Holloway, B. P., Krause, P. R., Takayama, M. and Schmid, D. S. (2000). Rapid genotyping of varicella-zoster virus vaccine and wild-type strains with fluorophore-labeled hybridization probes. *J. Clin. Microbiol.* **38**, 4315–4319.

Löseke, S., Grage-Griebenow, E., Wagner, A., Gehlhar, K. and Bufe, A. (2003). Differential expression of IFN-α subtypes in human PBMC: evaluation of novel real-time PCR assays. *J. Immunol. Meth.* **276**, 207–222.

Lucchini, S., Thompson, A. and Hinton, J. C. D. (2001). Microarrays for microbiologists. *Microbiology* **147**, 1403–1414.

Lunge, V. R., Miller, B. J., Livak, K. J. and Batt, C. A. (2002). Factors affecting the performance of 5′ nuclease PCR assays for *Listeria monocytogenes* detection. *J. Microbiol. Meth.* **51**, 361–368.

Luria, S. E. and Darnell, J. E. (1967). Titration of viruses. In *General Virology* (S. E. Luria and J. E. Darnell, eds), 2nd edn., pp. 35–49. New York, Wiley.

Lyamichev, V., Brow, M. A. D. and Dahlberg, J. E. (1993). Structure-specific endonucleolytic cleavage of nucleic acids by eubacterial DNA polymerases. *Science* **260**, 778–783.

Machida, U., Kami, M., Fukui, T., Kazuyama, Y., Kinoshita, M., Tanaka, Y., Kanda, Y., Ogawa, S., Honda, H., Chiba, S., Mitani, K., Muto, Y., Osumi, K., Kimura, S. and Hirai, H. (2000). Real-time automated PCR for early diagnosis and monitoring of cytomegalovirus infection after bone marrow transplantation. *J. Clin. Microbiol.* **38**, 1536–1542.

Mackay, I. M., Metharom, P., Sloots, T. P. and Wei, M. Q. (2001). Quantitative PCR-ELAHA for the determination of retroviral vector transduction efficiency. *Mol. Ther.* **3**, 801–807.

Mackay, I. M., Arden, K. E. and Nitsche, A. (2002). Real-time PCR in virology. *Nucleic Acids Res.* **30**, 1292–1305.

Mackay, I. M., Gardam, T., Arden, K. E., McHardy, S., Whiley, D. M., Crisante, E. and Sloots, T. P. (2003a). Co-detection and discrimination of six human herpesviruses by multiplex PCR-ELAHA. *J. Clin. Virol.* **28**, 291–302.

Mackay, I. M., Jacob, K. C., Woolhouse, D., Waller, K., Syrmis, M. W., Whiley, D. M., Siebert, D. J., Nissen, M. N. and Sloots, T. P. (2003b). Molecular assays for the detection of human metapneumovirus. *J. Clin. Microbiol.* **41**, 100–105.

MacKenzie, J., Gallagher, A., Clayton, R. A., Perry, J., Eden, O. B., Ford, A. M., Greaves, M. F. and Jarrett, R. F. (2001). Screening for herpesvirus genomes in common acute lymphoblastic leukemia. *Leukemia* **15**, 415–421.

Makino, S.-I., Cheun, H. I., Watarai, M., Uchida, I. and Takeshi, K. (2001). Detection of anthrax spores from the air by real-time PCR. *Lett. Appl. Microbiol.* **33**, 237–240.

Martell, M., Gómez, J., Esteban, J. I., Sauleda, S., Quer, J., Cabot, B., Esteban, R. and Guardia, J. (1999). High-throughput real-time reverse transcription-PCR quantitation of hepatitis C virus RNA. *J. Clin. Microbiol.* **37**, 327–332.

Martineau, F., Picard, F. J., Lansac, N., Ménard, C., Roy, P. H., Ouellette, M. and Bergeron, M. G. (2000). Correlation between resistance genotype determined by multiplex PCR assays and the antibiotic susceptibility patterns of *Staphylococcus aureus* and *Staphylococcus epidermidis*. *Antimicrob. Agent Chem.* **44**, 231–238.

Matsumura, M., Hikiba, Y., Ogura, K., Togo, G., Tsukuda, I., Ushikawa, K., Shiratori, Y. and Omata, M. (2001). Rapid detection of mutations in the 23S rRNA gene of *Helicobacter pylori* that confers resistance to clarithromycin treatment to the bacterium. *J. Clin. Microbiol.* **39**, 691–695.

Matthews, J. A. and Kricka, L. J. (1988). Analytical strategies for the use of DNA probes. *Anal. Biochem.* **169**, 1–25.

McCartney, H. A., Foster, S. J., Fraaije, B. A. and Ward, E. (2003). Molecular diagnostics for fungal plant pathogens. *Pest Manag. Sci.* **59**, 129–142.

McCreedy, B. J. and Callaway, T. H. (1993). Laboratory design and work flow. In *Diagnostic Molecular Microbiology: Principles and Applications* (D. H. Persing, T. F. Smith, F. C. Tenover and T. J. White, eds), pp. 149–159. ASM, Washington, DC.

Menzo, S., Bagnarelli, P., Giacca, M., Manzin, A., Varaldo, P. E. and Clementi, M. (1992). Absolute quantitation of viremia in human immunodeficiency virus infection by competitive reverse transcription and polymerase chain reaction. *J. Clin. Microbiol.* **30**, 1752–1757.

Mhlanga, M. M. and Malmberg, L. (2001). Using molecular beacons to detect single-nucleotide polymorphisms with real-time PCR. *Methods* **25**, 463–471.

Miller, N., Cleary, T., Kraus, G., Young, A. K., Spruill, G. and Hnatyszyn, H. J. (2002). Rapid and specific detection of *Mycobacterium tuberculosis* from acid-fast bacillus smear-positive respiratory specimens and BacT/ALERT MP culture bottles by using fluorogenic probes and real time-PCR. *J. Clin. Microbiol.* **40**, 4143–4147.

Mitchell, T. J. (1998). Molecular basis of virulence. *Arch. Dis. Child.* **78**, 197–209.

Monopoeho, S., Mignotte, B., Schwartzbrod, L., Marechal, V., Nicolas, J.-C., Billaudel, S. and Férré, V. (2000). Quantification of enterovirus RNA in sludge samples using single tube real-time RT-PCR. *Biotechniques* **29**, 88–93.

Monpoeho, S., Coste-Burel, M., Costa-Mattioli, M., Besse, B., Chomel, J. J., Billaudel, S. and Ferré, V. (2002). Application of a real-time polymerase chain reaction with internal positive control for detection and quantification of enterovirus in cerebrospinal fluid. *Eur. J. Clin. Microbiol. Infect. Dis.* **21**, 532–536.

Moody, A., Sellers, S. and Bumstead, N. (2000). Measuring infectious bursal disease virus RNA in blood by multiplex real-time quantitative RT-PCR. *J. Virol. Meth.* **85**, 55–64.

Möller, A. and Jansson, J. K. (1997). Quantification of genetically tagged cyanobacteria in Baltic sea sediment by competitive PCR. *Biotechniques* **22**, 512–518.

Morrison, L. E., Halder, T. C. and Stols, L. M. (1989). Solution-phase detection of polynucleotides using interacting fluorescent labels and competitive hybridization. *Anal. Biochem.* **183**, 231–244.

Morrison, T. B., Weis, J. J. and Wittwer, C. T. (1998a). Quantification of low-copy transcripts by continuous SYBR® green I monitoring during amplification. *Biotechniques* **24**, 954–962.

Morrison, T. M., Weis, J. J. and Wittwer, C. T. (1998b). Quantification of low-copy transcripts by continuous SYBR green I monitoring during amplification. *Biotechniques* **24**, 954–962.

Mothershed, E. A., Sacchi, C. T., Whitney, A. M., Barnett, G. A., Ajello, G. W., Schmink, S., Mayer, L. W., Phelan, M., Taylor, T. H., Bernhardt, S. A., Rosenstein, N. E. and Popovic, T. (2004). Use of real-time PCR to resolve slide agglutination discrepancies in serogroup identification of *Neisseria meningitidis*. *J. Clin. Microbiol.* **42**, 320–328.

Mullis, K. B. and Faloona, F. (1987). Specific synthesis of DNA *in vitro* via a polymerase-catalysed chain reaction. *Meth. Enzymol.* **155**, 335–350.

Muti, G., Klersy, C., Baldanti, F., Granata, S., Oreste, P., Pezzetti, L., Gatti, M., Gargantini, L., Caramella, M., Mancini, V., Gerna, G. and Morra, E. (2003). Epstein-Barr virus (EBV) load and interleukin-10 in EBV-positive and EBV-negative post-transplant lymphoproliferative disorders. *Br. J. Haem.* **122**, 927–933.

Najioullah, F., Thouvenot, D. and Lina, B. (2001). Development of a real-time PCR procedure including an internal control for the measurement of HCMV viral load. *J. Virol. Meth.* **92**, 55–64.

Nakayama, H., Yokoi, H. and Fujita, J. (1992). Quantification of mRNA by non-radioactive RT-PCR and CCD imaging system. *Nucleic Acids Res.* **20**, 4939.

Natarajan, V., Plishka, R. J., Scott, E. W., Lane, H. C. and Salzman, N. P. (1994). An internally controlled virion PCR for the measurement of HIV-1 RNA in plasma. *PCR Meth. Appl.* **3**, 346–350.

Nazarenko, I. A., Bhatnager, S. K. and Hohman, R. J. (1997). A closed tube format for amplification and detection of DNA based on energy transfer. *Nucleic Acids Res.* **25**, 2516–2521.

Nazarenko, I., Lowe, B., Darfler, M., Ikonomi, P., Schuster, D. and Raschtchian, A. (2002). Multiplex quantitative PCR using self-quenched primers labeled with a single fluorophore. *Nucleic Acids Res.* **30**, e37.

Ng, L. F. P., Wong, M., Koh, S., Ooi, E.-E., Tang, K.-F., Leong, H.-N., Ling, A.-E., Agathe, L. V., Tan, J., Liu, E. T., Ren, E.-C., Ng, L.-C. and Hibberd, M. L. (2004). Detection of severe acute respiratory syndrome coronavirus in blood of infected patients. *J. Clin. Microbiol.* **42**, 347–350.

Nicoll, S., Brass, A. and Cubie, H. A. (2001). Detection of herpes viruses in clinical samples using real-time PCR. *J. Virol. Meth.* **96**, 25–31.

Niesters, H. G. M. (2001). Quantitation of viral load using real-time amplification techniques. *Methods* **25**, 419–429.

Niesters, H. G. M. (2002). Clinical virology in real time. *J. Clin. Virol.* **25**, S3–S12.

Niesters, H. G., van Esser, J., Fries, E., Wolthers, K. C., Cornelissen, J. and Osterhaus, A. D. (2000). Development of a real-time quantitative assay for detection of Epstein-Barr virus. *J. Clin. Microbiol.* **38**, 712–715.

Nijhuis, M., van Maarseveen, N., Schuurman, R., Verkuijlen, S., de Vos, M., Hendricksen, K. and van Loon, A. M. (2002). Rapid and sensitive routine detection of all members of the genus *Enterovirus* in different clinical specimens by real-time PCR. *J. Clin. Microbiol.* **40**, 3666–3670.

Nitsche, A., Steuer, N., Schmidt, C. A., Landt, O. and Siegert, W. (1999). Different real-time PCR formats compared for the quantitative detection of human cytomegalovirus DNA. *Clin. Chem.* **45**, 1932–1937.

Nitsche, A., Steuer, N., Schmidt, C. A., Landt, O., Ellerbrok, H., Pauli, G. and Siegert, W. (2000). Detection of human cytomegalovirus DNA by real-time quantitative PCR. *J. Clin. Microbiol.* **38**, 2734–2737.

Nitsche, A., Ellerbrok, H. and Pauli, G. (2004). Detection of orthopoxvirus DNA by real-time PCR and identification of variola virus DNA by melting analysis. *J. Clin. Microbiol.* **42**, 1207–1213.

Nogva, H. K., Rudi, K., Naterstad, K., Holck, A. and Lillehaug, D. (2000). Application of 5'-nuclease PCR for quantitative detection of *Listeria* monocytogenes in pure cultures, water, skim milk and unpasteurized whole milk. *Appl. Environ. Microbiol.* **66**, 4266–4271.

Northrup, M. A. (1998). A miniature analytical instrument for nucleic acids based on micromachined silicon reaction chambers. *Anal. Chem.* **70**, 918–922.

Nulens, E. and Voss, A. (2002). Laboratory diagnosis and biosafety issues of biological warfare agents. *Clin. Microbiol. Infect.* **8**, 455–466.

Nutiu, R. and Li, Y. (2002). Tripartite molecular beacons. *Nucleic Acids Res.* **30**, e94.

Obika, S., Nanbu, D., Hari, Y., Andoh, J.-I., Morio, K.-I., Doi, T. and Imanishi, T. (1998). Stability and structural features of the duplexes containing nucleoside analogues with a fixed N-type conformation, 2'-O,4'-C-methyleneribonucleosides. *Tetrahedron Lett.* **39**, 5401–5404.

Oggioni, M. R., Meacci, F., Carattoli, A., Ciervo, A., Orru, G., Cassone, A. and Pozzi, G. (2002). Protocol for real-time PCR inhibition of anthrax spores from nasal swabs after broth enrichment. *J. Clin. Microbiol.* **40**, 3956–3963.

Ogilvie, M. (2001). Molecular techniques should not now replace cell culture in diagnostic virology laboratories. *Rev. Med. Virol.* **11**, 351–354.

Ohyashiki, J. K., Suzuki, A., Aritaki, K., Nagate, A., Shoji, N., Ohyashiki, K., Ojima, T., Abe, K. and Yamamoto, K. (2000). Use of real-time PCR to monitor human herpesvirus 6 reactivation after allogeneic bone marrow transplantation. *Int. J. Mol. Med.* **6**, 427–432.

O'Leary, J., Kennedy, M., Howells, D., Silva, I., Uhlmann, V., Luttich, K., Biddolph, S., Lucas, S., Russell, J., Bermingham, N., O'Donovan, M., Ring, M., Kenny, C., Sweeney, M., Sheils, O., Martin, C., Picton, S. and Gatter, K. (2000a). Cellular localisation of HHV-8 in Castleman's disease: is there a link with lymph node vascularity? *J. Clin. Pathol.* **53**, 69–76.

O'Leary, J. J., Kennedy, M., Luttich, K., Uhlmann, V., Silva, I., Russell, J., Sheils, O., Ring, M., Sweeney, M., Kenny, C., Bermingham, N., Martin, C., O'Donovan, M., Howells, D., Picton, S. and Lucas, S. B. (2000b). Localisation of HHV-8 in AIDS related lymphadenopathy. *J. Clin. Pathol.* **53**, 43–47.

O'Mahony, J. and Hill, C. (2002). A real time PCR assay for the detection and quantitation of *Mycobacterium avium* subsp. *paratuberculosis* using SYBR green and the Light Cycler. *J. Microbiol. Meth.* **51**, 283–293.

Orlando, C., Pinzani, P. and Pazzagli, M. (1998). Developments in quantitative PCR. *Clin. Chem. Lab. Med.* **36**, 255–269.

O'Sullivan, C. E., Kasai, M., Francesconi, A., Petraitis, V., Petraitiene, R., Kelaher, A. M., Sarafandi, A. A. and Walsh, T. J. (2003). Development and validation of a quantitative real-time PCR assay using fluorescence resonance energy transfer technology for detection of *Aspergillus fumigatus* in experimental invasive pulmonary aspergillosis. *J. Clin. Microbiol.* **41**, 5676–5682.

Overbergh, L., Valckx, D., Waer, M. and Mathieu, C. (1999). Quantification of murine cytokine mRNAs using real-time quantitative reverse transcriptase PCR. *Cytokine* **11**, 305–312.

Overbergh, L., Giulietti, A., Valckx, D., Decallonne, B., Bouillon, R. and Mathieu, C. (2003). The use of real-time reverse transcriptase PCR for the quantification of cytokine gene expression. *J. Biomol. Tech.* **14**, 33–43.

Pahl, A., Kühlbrandt, U., Brune, K., Röllinghoff, M. and Gessner, A. (1999). Quantitative detection of *Borrelia burgdorferi* by real-time PCR. *J. Clin. Microbiol.* **37**, 1958–1963.

Pang, X. L. (2004). Increased detection of rotavirus using a real time reverse transcription-polymerase chain reaction (RT-PCR) assay in stool specimens from children with diarrhea. *J. Med. Virol.* **72**, 496–501.

Peiris, J. S., Lai, S. T., Poon, L. L., Guan, Y., Yam, L. Y., Lim, W., Nicholls, J., Yee, W. K., Yan, W. W., Cheung, M. T., Cheng, V. C., Chan, K. H., Tsang, D. N., Yung, R. W., Ng, T. K., Yuen, K. Y. and SARS study group (2003). Coronavirus as a possible cause of severe acute respiratory syndrome. *Lancet* **361**, 1319–1325.

Persing, D. H. (1993). *In vitro* nucleic acid amplification techniques. In *Diagnostic Molecular Microbiology: Principles and Applications* (D. H. Persing, T. F. Smith, F. C. Tenover and T. J. White, eds), pp. 51–87. ASM, Washington, DC.

Peter, J. B. and Sevall, J. S. (2001). Review of 3200 serially received CSF samples submitted for type-specific HSV detection by PCR in the reference laboratory setting. *Mol. Cell. Probes* **15**, 177–182.

Petersen, M. and Wengel, J. (2003). LNA: a versatile tool for therapeutics and genomics. *Trends Biotechnol.* **21**, 74–81.

Petrik, J., Pearson, G. J. M. and Allain, J.-P. (1997). High throughput PCR detection of HCV based on semiautomated multisample RNA capture. *J. Virol. Meth.* **64**, 147–159.

Pfaffl, M. W., Horgan, G. W. and Dempfle, L. (2002). Relative expression software tool (REST©) for group-wise comparison and statistical analysis of relative expression results in real-time PCR. *Nucleic Acids Res.* **30**, e36.

Pfaller, M. A. (1999). Molecular epidemiology in the care of patients. *Arch. Pathol. Lab. Med.* **123**, 1007–1010.

PhorTech. Rapid rise in platforms for realtime PCR transforms DNA amplification instrumentation market. http://www.phortech.com/02ampi.htm, 2003.

Pietilä, J., He, Q., Oksi, J. and Viljanen, M. K. (2000). Rapid differentiation of *Borrelia garinii* from *Borrelia afzelii* and *Borrelia burgdorferi* sensu stricto by LightCycler fluorescence melting curve analysis of a PCR product of the recA gene. *J. Clin. Microbiol.* **38**, 2756–2759.

Pinzani, P., Bonciani, L., Pazzagli, M., Orlando, C., Guerrini, S. and Granchi, L. (2004). Rapid detection of *Oenococcus oeni* in wine by real-time quantitative PCR. *Lett. Appl. Microbiol.* **38**, 118–124.

Poutanen, S. M., Low, D. E., Henry, B., Finkelstein, S., Rose, D., Green, K., Tellier, R., Draker, R., Adachi, D., Ayers, M., Chan, A. K., Skowronski, D. M., Salit, I., Simor, A. E., Slutsky, A. S., Doyle, P. W., Krajden, M., Petric, M., Brunham, R. C. and McGeer, A. J. (2003). Identification of severe acute respiratory syndrome in Canada. *N. Engl. J. Med.* **348**, 1995–2005.

Probert, W. S., Bystrom, S. L., Khashe, S., Schrader, K. N. and Wong, J. D. (2002). 5' Exonuclease assay for detection of serogroup Y *Neisseria meningitidis*. *J. Clin. Microbiol.* **40**, 4325–4328.

Qi, Y., Patra, G., Liang, X., Williams, L. E., Rose, S., Redkar, R. J. and del Vecchio, V. G. (2001). Utilization of the rpoB gene as a specific chromosomal marker for real-time PCR detection of *Bacillus anthracis*. *Appl. Environ. Microbiol.* **67**, 3720–3727.

Quereda, C., Corral, I., Laguna, F., Valencia, M. E., Tenorio, A., Echevarria, J. E., Navas, E., Martín-Davila, P., Moreno, A., Moreno, V., Gonzalez-Lahoz, J. M., Arribas, J. R. and Guerrero, A. (2000). Diagnostic utility of a multiplex herpesvirus PCR assay performed with cerebrospinal fluid from human immunodeficiency virus-infected patients with neurological disorders. *J. Clin. Microbiol.* **38**, 3061–3067.

Radonic, A., Thulke, S., Mackay, I. M., Landt, O., Siegert, W. and Nitsche, A. (2004). Guideline to reference gene selection for quantitative real-time PCR. *Biochem. Biophys. Res. Commun.* **313**, 856–862.

Randegger, C. C. and Hachler, H. (2001). Real-time PCR and melting curve analysis for reliable and rapid detection of SHV extended-spectrum β-lactamases. *Antimicrob. Agent Chem.* **45**, 1730–1736.

Ratge, D., Scheiblhuber, B., Landt, O., Berg, J. and Knabbe, C. (2002). Two-round rapid-cycle RT-PCR in single closed capillaries increases the sensitivity of HCV RNA detection and avoids amplicon carry-over. *J. Clin. Virol.* **24**, 161–172.

Rauter, C., Oehme, R., Diterich, I., Engele, M. and Hartung, T. (2002). Distribution of clinically relevant *Borrelia* genospecies in ticks assessed by a novel, single-run, real-time PCR. *J. Clin. Microbiol.* **40**, 36–43.

Read, S. J. (2001). Recovery efficiencies of nucleic acid extraction kits as measured by quantitative LightCycler™ PCR. *J. Clin. Pathol.* **54**, 86–90.

Read, S. J., Mitchell, J. L. and Fink, C. G. (2001). LightCycler multiplex PCR for the laboratory diagnosis of common viral infections of the central nervous system. *J. Clin. Microbiol.* **39**, 3056–3059.

Reid, S. M., Gerris, N. P., Hutchings, G. H., King, D. P. and Alexandersen, S. (2004). Evaluation of real-time reverse transcription polymerase chain reaction assays for the detection of swine vesicular disease virus. *J. Virol. Meth.* **116**, 169–176.

Reischl, U., Linde, H.-J., Metz, M., Leppmeier, B. and Lehn, N. (2000). Rapid identification of methicillin-resistant *Staphylococcus aureus* and simultaneous species confirmation using real-time fluorescence PCR. *J. Clin. Microbiol.* **38**, 2429–2433.

Reischl, U., Linde, H.-J., Lehn, N., Landt, O., Barratt, K. and Wellinghausen, N. (2002). Direct detection and differentiation of *Legionella* spp. and *Legionella pneumophila* in clinical specimens by dual-colour real-time PCR and melting curve analysis. *J. Clin. Microbiol.* **40**, 3814–3817.

Reiss, R. A. and Rutz, B. (1999). Quality control PCR: a method for detecting inhibitors of *Taq* DNA polymerase. *Biotechniques* **27**, 920–926.

Ririe, K. M., Rasmussen, R. P. and Wittwer, C. T. (1997). Product differentiation by analysis of DNA melting curves during the polymerase chain reaction. *Anal. Biochem.* **245**, 154–160.

Roberts, T. C., Brennan, D. C., Buller, R. S., Gaudreault-Keener, M., Schnitzler, M. A., Sternhell, K. E., Garlock, K. A., Singer, G. G. and Storch, G. A. (1998). Quantitative polymerase chain reaction to predict occurrence of symptomatic cytomegalovirus infection and assess response to ganciclovir therapy in renal transplant recipients. *J. Infect. Dis.* **178**, 626–636.

Rodrigo, A. G., Goracke, P. C., Rowhanian, K. and Mullins, J. I. (1997). Quantitation of target molecules from polymerase chain reaction-based limiting dilution assays. *AIDS Res. Hum. Retro.* **13**, 737–742.

Roe, J. D., Haugland, R. A., Vesper, S. J. and Wymer, L. J. (2001). Quantification of *Stachybotrys chartarum* conidia in indoor dust using real time, fluorescent probe-based detection of PCR products. *J. Exposure Anal. Environ. Epidemiol.* **11**, 12–20.

Rohr, U.-P., Wulf, M.-A., Stahn, S., Steidl, U., Haas, R. and Kronenwett, R. (2002). Fast and reliable titration of recombinant adeno-associated virus type-2 using quantitative real-time PCR. *J. Virol. Meth.* **106**, 81–88.

Rollag, H., Sagedal, S., Holter, E., Degre, M., Ariansen, S. and Nordal, K. P. (1998). Diagnosis of cytomegalovirus infection in kidney transplant recipients by a quantitative RNA-DNA hybrid capture assay for cytomegalovirus DNA in leukocytes. *Eur. J. Clin. Microbiol. Infect. Dis.* **17**, 124–127.

Rosenstraus, M., Wang, Z., Chang, S.-Y., DeBonville, D. and Spadoro, J. P. (1998). An internal control for routine diagnostic PCR: design, properties, and effect on clinical performance. *J. Clin. Microbiol.* **36**, 191–197.

Ruan, Y. J., Wei, C. L., Ee, A. L., Vega, V. B., Thoreau, H., Su, S. T., Chia, J. M., Ng, P., Chiu, K. P., Lim, L., Zhang, T., Peng, C. K., Lin, E. O., Lee, N. M., Yee, S. L., Ng, L. F., Chee, R. E., Stanton, L. W., Long, P. M. and Liu, E. T. (2003). Comparative full-length genome sequence analysis of 14 SARS coronavirus isolates and common mutations associated with putative origins of infection. *Lancet* **361**, 1779–1785.

Rudi, K., Nogva, H. K., Moen, B., Nissen, H., Bredholt, S., Møretrø, T., Naterstad, K. and Holck, A. (2002). Development and application of new nucleic acid-based technologies for microbial community analyses in food. *Int. J. Food Microbiol.* **78**, 171–180.

Rutledge, R. J. and Côte, C. (2003). Mathematics of quantitative kinetic PCR and the application of standard curves. *Nucleic Acids Res.* **31**, e93.

Ryncarz, A. J., Goddard, J., Wald, A., Huang, M.-L., Roizman, B. and Corey, L. (1999). Development of a high-throughput quantitative assay for detecting herpes simplex virus DNA in clinical samples. *J. Clin. Microbiol.* **37**, 1941–1947.

Sanburn, N. and Cornetta, K. (1999). Rapid titer determination using quantitative real-time PCR. *Gene Ther.* **6**, 1340–1345.

Schaade, L., Kockelkorn, P., Ritter, K. and Kleines, M. (2000). Detection of cytomegalovirus DNA in human specimens by LightCycler PCR. *J. Clin. Microbiol.* **38**, 4006–4009.

Schalasta, G., Arents, A., Schmid, M., Braun, R. W. and Enders, G. (2000a). Fast and type-specific analysis of herpes simplex virus types 1 and 2 by rapid PCR and fluorescence melting-curve-analysis. *Infection* **28**, 85–91.

Schalasta, G., Eggers, M., Schmid, M. and Enders, G. (2000b). Analysis of human cytomegalovirus DNA in urines of newborns and infants by means of a new ultrarapid real-time PCR-system. *J. Clin. Virol.* **19**, 175–185.

Scherr, M., Battmer, K., Blömer, U., Ganser, A. and Grez, M. (2001). Quantitative determination of lentiviral vector particle numbers by real-time PCR. *Biotechniques* **31**, 520–526.

Schoder, D., Schmalwieser, A., Schauberger, G., Kuhn, M., Hoorfar, J. and Wagner, M. (2003). Physical characteristics of six new thermocyclers. *Clin. Chem.* **49**, 960–963.

Schutten, M., van den Hoogen, B., van der Ende, M. E., Gruters, R. A., Osterhaus, A. D. M. E. and Niesters, H. G. M. (2000). Development of a real-time quantitative RT-PCR for the detection of HIV-2 RNA in plasma. *J. Virol. Meth.* **88**, 81–87.

Schütz, E., von Ashen, N. and Oellerich, M. (2000). Genotyping of eight thiopurine methyltransferase mutations: three-color multiplexing, two-color/shared anchor, and fluorescence-quenching hybridization probe assays based on thermodynamic nearest-neighbour probe design. *Clin. Chem.* **46**, 1728–1737.

Selvey, S., Thompson, E. W., Matthaei, K., Lea, R. A., Irving, M. G. and Griffiths, L. R. (2001). β-actin – an unsuitable internal control for RT-PCR. *Mol. Cell. Probes* **15**, 307–311.

Selvin, P. (1995). Fluorescence resonance energy transfer. *Meth. Enzymol.* **246**, 300–334.

Semighini, C. P., Marins, M., Goldman, M. H. S. and Goldman, G. H. (2002). Quantitative analysis of the relative transcript levels of ABC transporter Atr genes in *Aspergillus nidulans* by real-time reverse transcription-PCR assay. *Appl. Environ. Microbiol.* **68**, 1351–1357.

Sharp, J. G. and Chan, W. C. (1999). Detection and relevance of minimal disease in lymphomas. *Cancer Metastasis Rev.* **18**, 127–142.

Shaw, R. D., Hempson, S. J. and Mackow, E. R. (1995). Rotavirus diarrhea is caused by nonreplicating viral particles. *J. Virol.* **69**, 5946–5950, Erratum: *J. Virol.* 1996. **70** (8), 5740.

Shrestha, N. K., Tuohy, M. J., Hall, G. S., Isada, C. M. and Procop, G. W. (2002). Rapid identification of *Staphylococcus aureus* and the mecA gene from BacT/ALERT blood culture bottles by using the LightCycler system. *J. Clin. Microbiol.* **40**, 2659–2661.

Siebert, P. D. and Larrick, J. W. (1992). Competitive PCR. *Nature* **359**, 557–558.

Simeonov, A. and Nikiforov, T. T. (2002). Single nucleotide polymorphism genotyping using short, fluorescently labeled locked nucleic acid (LNA) probes and fluorescence polarization detection. *Nucleic Acids Res.* **30**, e91.

Singh, S. K., Nielsen, P., Koshkin, A. A. and Wengel, J. (1998). LNA (locked nucleic acids): synthesis and high-affinity nucleic acid recognition. *Chem. Commun.* 455–456.

Sintchenko, V., Iredell, J. R. and Gilbert, G. L. (1999). Is it time to replace the petri dish with PCR? Application of culture-independent nucleic acid amplification in diagnostic bacteriology: expectations and reality. *Pathology* **31**, 436–439.

Smith, I. L., Halpin, K., Warrilow, D. and Smith, G. A. (2001). Development of a fluorogenic RT-PCR assay (TaqMan) for the detection of Hendra virus. *J. Virol. Meth.* **98**, 33–40.

Smith, I. L., Northill, J. A., Harrower, B. J. and Smith, G. A. (2002). Detection of Australian bat lyssavirus using a fluorogenic probe. *J. Clin. Virol.* **25**, 285–291.

Solinas, A., Brown, L. J., McKeen, C., Mellor, J. M., Nicol, J. T. G., Thelwell, N. and Brown, T. (2001). Duplex scorpion primers in SNP analysis and FRET applications. *Nucleic Acids Res.* **29**, e96.

Song, E. S., Lee, V., Surh, C. D., Lynn, A., Brumm, D., Jolly, D. J., Warner, J. F. and Chada, S. (1997). Antigen presentation in retroviral vector-mediated gene transfer in vivo. *Proc. Natl Acad. Sci. USA* **94**, 1943–1948.

Song, J., Ohkura, T., Sugimoto, M., Mori, Y., Inagi, R., Yamanishi, K., Yoshizaki, K. and Nishimoto, N. (2002). Human interleukin-6 induces human herpesvirus-8 replication in a body cavity-based lymphoma cell line. *J. Med. Virol.* **68**, 404–411.

Stender, H., Fiandaca, M., Hyldig-Nielsen, J. J. and Coull, J. (2002). PNA for rapid microbiology. *J. Microbiol. Meth.* **48**, 1–17.

Stephenson, J. (2003). Monkeypox outbreak a reminder of emerging infections vulnerabilities. *JAMA* **290**, 23–24.

Stevens, S. J. C., Verkuijlen, S. A. W. M., van den Brule, A. J. C. and Middeldorp, J. M. (2002). Comparison of quantitative competitive PCR with LightCycler-based PCR for measuring Epstein-Barr virus DNA load in clinical specimens. *J. Clin. Microbiol.* **40**, 3986–3992.

Stöcher, M., Leb, V., Hölzl, G. and Berg, J. (2002). A simple approach to the generation of heterologous competitive internal controls for real-time PCR assays on the LightCycler. *J. Clin. Virol.* **25**, S47–S53.

Stocher, M., Leb, V., Bozic, M., Kessler, H. H., Halwachs-Baumann, G., Landt, O., Stekel, H. and Berg, J. (2003). Parallel detection of five human herpes virus DNAs by a set of real-time polymerase chain reactions in a single run. *J. Clin. Virol.* **26**, 85–93.

Stockton, J., Ellis, J. S., Saville, M., Clewley, J. P. and Zambon, M. C. (1998). Multiplex PCR for typing and subtyping influenza and respiratory syncytial viruses. *J. Clin. Microbiol.* **36**, 2990–2995.

Stordeur, P., Poulin, L. F., Craciun, L., Zhou, L., Schandené, L., de Lavareille, A., Goriely, S. and Goldman, M. (2002). Cytokine mRNA quantification by real-time PCR. *J. Immunol. Meth.* **259**, 55–64.

Stram, Y., Kuznetzova, L., Guini, M., Rogel, A., Meirom, R., Chai, D., Yadin, H. and Brenner, J. (2004). Detection and quantitation of Akabane and Aino viruses by multiplex real-time reverse-transcriptase PCR. *J. Virol. Meth.* **116**, 147–154.

Strieter, R. M., Belperio, J. A. and Keane, M. P. (2003). Host innate defenses in the lung: the role of cytokines. *Curr. Opin. Infect. Dis.* **16**, 193–198.

Stryer, L. and Haugland, R. P. (1967). Energy transfer: a spectroscopic ruler. *Proc. Natl Acad. Sci. USA* **58**, 719–726.

Suzuki, K., Aoki, K., Ohnami, S., Yoshida, K., Kazui, T., Kato, N., Inoue, K., Kohara, M. and Yoshida, T. (2003). Adenovirus-mediated gene transfer of interferon α inhibits hepatitis C virus replication in hepatocytes. *Biochem. Biophys. Res. Commun.* **307**, 814–819.

Svanvik, N., Sehlstedt, U., Sjöback, R. and Kubista, M. (2000). Detection of PCR products in real time using light-up probes. *Anal. Biochem.* **287**, 179–182.

Svanvik, N., Westman, G., Wang, D. and Kubista, M. (2001). Light-up probes: thiazole orange-conjugated peptide nucleic acid for the detection of target nucleic acid in homogeneous solution. *Anal. Biochem.* **281**, 26–35.

Swan, D. C., Tucker, R. A., Holloway, B. P. and Icenogle, J. P. (1997). A sensitive, type-specific, fluorogenic probe assay for detection of human papillomavirus DNA. *J. Clin. Microbiol.* **35**, 886–891.

Sykes, P. J., Brisco, M. J., Snell, L. E., Dolman, G., Neoh, S.-H., Peng, L.-M. and Toogood, I. (1998). Minimal residual disease in childhood acute lymphoblastic leukaemia quantified by aspirate and trephine: is the disease multifocal? *Br. J. Haematol.* **103**, 60–65.

Takaya, K., Ogawa, Y., Hiraoka, J., Hosoda, K., Yamori, Y. and Nakao, K. (1996). Nonesense mutation of leptin receptor in the obese spontaneously hypertensive Koletsky rat. *Nat. Gen.* **14**, 130–131.

Tan, T. Y., Corden, S., Barnes, R. and Cookson, B. (2001). Rapid identification of methicillin-resistant *Staphylococcus aureus* from positive blood cultures by real-time fluorescence PCR. *J. Clin. Microbiol.* **39**, 4529–4531.

Tanaka, N., Kimura, H., Hoshino, Y., Kato, K., Yoshikawa, T., Asano, Y., Horibe, K., Kojima, S. and Morishima, T. (2000a). Monitoring four herpesviruses in unrelated cord blood transplantation. *Bone Marrow Transpl.* **26**, 1193–1197.

Tanaka, N., Kimura, H., Iida, K., Saito, Y., Tsuge, I., Yoshimi, A., Matsuyama, T. and Morishima, T. (2000b). Quantitative analysis of cytomegalovirus load using a real-time PCR assay. *J. Med. Virol.* **60**, 455–462.

Tanriverdi, S., Tanyeli, A., Baslamisli, F., Köksal, F., Kilinç, Y., Feng, X., Batzer, G., Tzipori, S. and Widmer, G. (2002). Detection and genotyping of oocysts of *Cryptosporidium parvum* by real-time PCR and melting curve analysis. *J. Clin. Microbiol.* **40**, 3237–3244.

Tarnuzzer, R. W., Macauley, S. P., Farmerie, W. G., Caballero, S., Ghassemifar, M. R., Anderson, J. T., Robinson, C. P., Grant, M. B., Humphreys-Beher, M. G., Franzen, L., Peck, A. B. and Schultz, G. S. (1996). Competitive RNA templates for detection and quantitation of growth factors, cytokines, extracellular matrix components and matrix metalloproteinases by RT-PCR. *Biotechniques* **20**, 670–674.

Tasker, S., Helps, C. R., Day, M. J., Gruffydd-Jones, T. J. and Harbour, D. A. (2003). Use of real-time PCR to detect and quantify *Mycoplasma haemofelis* and *Candidatus Mycoplasma haemominutum* DNA. *J. Clin. Microbiol.* **41**, 439–441.

Tasker, S., Helps, C. R., Day, M. J., Harbour, D. A., Gruffydd-Jones, T. and Lappin, M. R. (2004). Use of Taqman PCR to determine the response of *Mycoplasma haemofelis* infection to antibiotic treatment. *J. Microbiol. Meth.* **56**, 63–71.

Taswell, C. (1981). Limiting dilution assays for the determination of immuno-competent cell frequencies. *J. Immunol.* **126**, 1614–1619.

Taylor, M. J., Hughes, M. S., Skuce, R. A. and Neill, S. D. (2001). Detection of *Mycobacterium bovis* in bovine clinical specimens using real-time fluorescence and fluorescence resonance energy transfer probe rapid-cycle PCR. *J. Clin. Microbiol.* **39**, 1272–1278.

Thellin, O., Zorzi, W., Lakaye, B., De Borman, B., Coumans, B., Hennen, G., Grisar, T., Igout, A. and Heinen, E. (1999). Housekeeping genes as internal standards: use and limits. *J. Biotech.* **75**, 291–295.

Thelwell, N., Millington, S., Solinas, A., Booth, J. and Brown, T. (2000). Mode of action and appllication of Scorpion primers to mutation detection. *Nucleic Acids Res.* **28**, 3752–3761.

Todd, A. V., Fuery, C. J., Impey, H. L., Applegate, T. L. and Haughton, M. A. (2000). DzyNA-PCR: use of DNAzymes to detect and quantify nucleic acid sequences in a real-time fluorescent format. *Clin. Chem.* **46**, 625–630.

Torres, M. J., Criado, A., Palomares, J. C. and Aznar, J. (2000). Use of real-time PCR and fluorimetry for rapid detection of rifampin and isoniazid resistance-associated mutations in *Mycobacterium tuberculosis*. *J. Clin. Microbiol.* **38**, 3194–3199.

Tsang, K. W., Ho, P. L., Ooi, G. C., Yee, W. K., Wang, T., Chan-Yeung, M., Lam, W. K., Seto, W. H., Yam, L. Y., Cheung, T. M., Wong, P. C., Lam, B., Ip, M. S., Chan, J., Yuen, K. Y. and Lai, K. N. (2003). A cluster of cases of severe acute respiratory syndrome in Hong Kong. *N. Engl. J. Med.* **348**, 1977–1985.

Tseng, S. Y., Macool, D., Elliott, V., Tice, G., Jackson, R., Barbour, M. and Amorese, D. (1997). An homogeneous fluorescence polymerase chain reaction assay to identify *Salmonella*. *Anal. Biochem.* **245**, 207–212.

Tuma, R. S., Beaudet, M. P., Jin, X., Jones, L. J., Cheung, C.-Y., Yue, S. and Singer, V. L. (1999). Characterization of SYBR gold nucleic acid gel stain: a dye optimised for use with 300-nm ultraviolet transilluminators. *Anal. Biochem.* **268**, 278–288.

Tyagi, S. and Kramer, F. R. (1996). Molecular beacons: probes that fluoresce upon hybridization. *Nat. Biotechnol.* **14**, 303–308.

Tyagi, S., Bratu, D. P. and Kramer, F. R. (1998). Multicolor molecular beacons for allele discrimination. *Nat. Biotechnol.* **16**, 49–53.

Tyagi, S., Marras, S. A. E. and Kramer, F. R. (2000). Wavelength-shifting molecular beacons. *Nat. Biotechnol.* **18**, 1191–1196.

Uhl, J. R., Bell, C. A., Sloan, L. M., Espy, M. J., Smith, T. F., Rosenblatt, J. E. and Cockerill, F. R. III (2002). Application of rapid-cycle real-time polymerase chain reaction for the detection of microbial pathogens: the Mayo-Roche rapid anthrax test. *Mayo Clin. Proc.* **77**, 673–680.

Vaisse, C., Halaas, J. L., Horvath, C. M., Darnell, J. E., Stoffel, M. and Friedman, J. M. (1996). Leptin activation of Stat3 in the hypothalamus of wild-type and ob/ob mice but not db/db mice. *Nat. Gen.* **14**, 95–97.

Valentine-Thon, E. (2002). Quality control in nucleic acid testing-where do we stand? *J. Clin. Virol.* **25**, S13–S21.

van der Vliet, G. M. E., Hermans, C. J. and Klatser, P. R. (1993). Simple colorimetric microtiter plate hybridization assay for detection of amplified *Mycobacterium leprae* DNA. *J. Clin. Microbiol.* **31**, 665–670.

van Elden, L. J. R., Nijhuis, M., Schipper, P., Schuurman, R. and van Loon, A. M. (2001). Simultaneous detection of influenza viruses A and B using real-time quantitative PCR. *J. Clin. Microbiol.* **39**, 196–200.

van Santen, V. L., Kaltenboeck, B., Joiner, K. S., Macklin, K. S. and Norton, R. A. (2004). Real-time quantitative PCR-based serum neutralization test for detection and titration of neutralizing antibodies to chicken anemia virus. *J. Virol. Meth.* **115**, 123–135.

Versalovic, J. and Lupski, J. R. (2002). Molecular detection and genotyping of pathogens: more accurate and rapid answers. *Trends Microbiol.* **10**, S15–S21.

Verstrepen, W. A., Kuhn, S., Kockx, M. M. and van de Vyvere, M. E. (2001). Rapid detection of enterovirus RNA in cerebrospinal fluid specimens with a novel single-tube real-time reverse transcription-PCR assay. *J. Clin. Microbiol.* **39**, 4093–4096.

Verstrepen, W. A., Bruynseels, P. and Mertens, A. H. (2002). Evaluation of a rapid real-time RT-PCR assay for the detection of enterovirus RNA in cerebrospinal fluid specimens. *J. Clin. Virol.* **25**, S39–S43.

Vet, J. A. M., Majithia, A. R., Marras, S. A. E., Tyagi, S., Dube, S., Poiesz, B. J. and Kramer, F. R. (1999). Multiplex detection of four pathogenic retroviruses using molecular beacons. *Proc. Natl Acad. Sci. USA* **96**, 6394–6399.

Vilcek, Š. and Paton, D. J. (2000). A RT-PCR assay for the rapid recognition of border disease virus. *Vet. Res.* **31**, 437–445.

Walker, G. T., Little, M. C., Nadeau, J. G. and Shank, D. D. (1992). Isothermal in vitro amplification of DNA by a restriction enzyme/DNA polymerase system. *Proc. Natl Acad. Sci. USA* **89**, 392–396.

Walker, R. A., Saunders, N., Lawson, A. J., Lindsay, E. A., Dassama, M., Ward, L. R., Woodward, M. J., Davies, R. H., Liebana, E. and Threlfall, E. J. (2001). Use of a LightCycler gyrA mutation assay for rapid identification of mutations conferring decreased susceptibility to ciprofloxacin in multiresistant *Salmonella enterica* serotype typhimurium DT104 isolates. *J. Clin. Microbiol.* **39**, 1443–1448.

Wall, S. J. and Edwards, D. R. (2002). Quantitative reverse transcription-polymerase chain reaction (RT-PCR): a comparison of primer-dropping, competitive, and real-time RT-PCRs. *Anal. Biochem.* **300**, 269–273.

Wang, T. and Brown, M. J. (1999). mRNA quantitation by real-time TaqMan polymerase chain reaction: validation and comparison with RNase protection. *Anal. Biochem.* **269**, 198–201.

Wang, Z. and Spadoro, J. (1998).Determination of target copy number of quantitative standards used in PCR based diagnostic assays In *Gene Quantification* (F. Ferré, ed.). Birkhauser, Boston.

Watkins-Reidel, T., Woegerbauer, M., Hollemann, D. and Hufnagl, P. (2002). Rapid diagnosis of enterovirus infections by real-time PCR on the LightCycler using the TaqMan format. *Diagn. Microbiol. Infect. Dis.* **42**, 99–105.

Watzinger, F., Hörth, E. and Lion, T. (2001). Quantitation of mRNA expression by competitive PCR using non-homologous competitors containing a shifted restriction site. *Nucleic Acids Res.* **29**, e52.

Weigl, J. A. I., Puppe, W., Gröndahl, B. and Schmitt, H.-J. (2000). Epidemiological investigation of nine respiratory pathogens in hospitalized children in Germany using multiplex reverse-transcriptase polymerase chain reaction. *Eur. J. Clin. Microbiol. Infect. Dis.* **19**, 336–343.

Weinberger, K. M., Wiedenmann, E., Böhm, S. and Jilg, W. (2000). Sensitive and accurate detection of hepatitis B virus DNA using a kinetic fluorescence detection system (TaqMan PCR). *J. Virol. Meth.* **85**, 75–82.

Weis, J. H., Tan, S. S., Martin, B. K. and Wittwer, C. T. (1992). Detection of rare mRNAs via quantitative RT-PCR. *TIG* **8**, 263–264.

Wellinghausen, N., Frost, C. and Marre, E. (2001). Detection of Legionellae in hospital water samples by quantitative real-time LightCycler PCR. *Appl. Environ. Microbiol.* **67**, 3985–3993.

Wetmur, J. G. (1991). DNA probes: applications of the principles of nucleic acid hybridization. *Crit. Rev. Biochem. Mol. Biol.* **26**, 227–259.

Whelan, A. C. and Persing, D. H. (1996). The role of nucleic acid amplification and detection in the clinical microbiology laboratory. *Annu. Rev. Gen.* **50**, 349–373.

Whiley, D. M., Mackay, I. M. and Sloots, T. P. (2001). Detection and differentiation of human polyomaviruses JC and BK by LightCycler PCR. *J. Clin. Microbiol.* **39**, 4357–4361.

Whiley, D. M., LeCornec, G. M., Mackay, I. M., Siebert, D. J. and Sloots, T. P. (2002a). A real-time PCR assay for the detection of *Neisseria gonorrhoeae* by LightCycler. *Diagn. Microbiol. Infect. Dis.* **42**, 85–89.

Whiley, D. M., Syrmis, M. W., Mackay, I. M. and Sloots, T. P. (2002b). Detection of human respiratory syncytial virus in respiratory samples by LightCycler reverse transcriptase PCR. *J. Clin. Microbiol.* **40**, 4418–4422.

Whiley, D. M., Crisante, M. E., Syrmis, M. W., Mackay, I. M. and Sloots, T. P. (2003a). Detection of *Neisseria meningitidis* by LightCycler PCR. *Pathology* **35**, 347–349.

Whiley, D. M., Syrmis, M. W., Mackay, I. M. and Sloots, T. P. (2003b). Preliminary Comparison of Three LightCycler PCR assays for the detection of herpes simplex virus in swab specimens. *Eur. J. Clin. Microbiol. Infect. Dis.* **22**, 764–767.

Whitcombe, D., Theaker, J., Guy, S. P., Brown, T. and Little, S. (1999). Detection of PCR products using self-probing amplicons and fluorescence. *Nat. Biotechnol.* **17**, 804–807.

White, I. and Campbell, T. C. (2000). Quantitation of cell-free and cell-associated Kaposi's sarcoma associated herpesvirus DNA by real-time PCR. *J. Clin. Microbiol.* **38**, 1992–1995.

White, P. A., Pan, Y., Freeman, A. J., Marinos, G., Ffrench, R. A., Lloyd, A. R. and Rawlinson, W. D. (2002). Quantification of hepatitis C virus in human liver and serum samples by using LightCycler reverse transcriptase PCR. *J. Clin. Microbiol.* **40**, 4346–4348.

Whitley, R. J. (2003). Smallpox: a potential agent of bioterrorism. *Antiviral Res.* **57**, 7–12.

WHO, (1980). Declaration of global eradication of smallpox. *Wkly Epidemiol. Rec.* **55**, 145–152.

WHO, (2003). Acute respiratory syndrome. China, Hong Kong Special Administrative Region of China, and Vietnam. *Wkly Epidemiol. Rec.* **78**, 73–74.

Wilhelm, J., Hahn, M. and Pingoud, A. (2001a). Influence of DNA target melting behavior on real-time PCR quantification. *Clin. Chem.* **46**, 1738–1743.

Wilhelm, J., Pingoud, A. and Hahn, M. (2001b). Comparison between Taq DNA polymerase and its Stoffel fragment for quantitative real-time PCR with hybridization probes. *Biotechniques* **30**, 1052–1062.

Witney, A. A., Doolan, D. L., Anthony, R. M., Weiss, W. R., Hoffman, S. L. and Carucci, D. J. (2001). Determining liver stage parasite burden by real time quantitative PCR as a method for evaluating pre-erythrocytic malaria vaccine efficacy. *Mol. Biochem. Parasitol.* **118**, 233–245.

Wittwer, C. T., Fillmore, G. C. and Garling, D. J. (1990). Minimizing the time required for DNA amplification by efficient heat transfer to small samples. *Anal. Biochem.* **186**, 328–331.

Wittwer, C. T., Herrmann, M. G., Moss, A. A. and Rasmussen, R. P. (1997a). Continuous fluorescence monitoring of rapid cycle DNA amplification. *Biotechniques* **22**, 130–138.

Wittwer, C. T., Ririe, K. M., Andrew, R. V., David, D. A., Gundry, R. A. and Balis, U. J. (1997b). The LightCycler™: a microvolume multisample fluorimeter with rapid temperature control. *Biotechniques* **22**, 176–181.

Wolffs, P., Grage, H., Hagberg, O. and Rådstrom, P. (2004). Impact of DNA polymerases and their buffer systems on quantitative real-time PCR. *J. Clin. Microbiol.* **42**, 408–411.

Wolk, D. M., Schneider, S. K., Wengenack, N. L., Sloan, L. M. and Rosenblatt, J. E. (2002). Real-time PCR method for detection of *Encephalitozoon intestinalis* from stool specimens. *J. Clin. Microbiol.* **40**, 3922–3928.

Woo, T. H. S., Patel, B. K. C., Smythe, L. D., Symonds, M. L., Norris, M. A. and Dohnt, M. F. (1997). Identification of pathogenic *Leptospira* genospecies by continuous monitoring of fluorogenic hybridization probes during rapid-cycle PCR. *J. Clin. Microbiol.* **35**, 3140–3146.

Wood, L. (2003). Questions about comparative genomics of SARS coronavirus isolates. *Lancet* **362**, 578–579.

Woodford, N., Tysall, L., Auckland, C., Stockdale, M. W., Lawson, A. J., Walker, R. A. and Livermore, D. M. (2002). Detection of oxazolidinone-resistant *Enterococcus faecalis* and *Enterococcus faecium* strains by real-time PCR and PCR-restriction fragment length polymorphism analysis. *J. Clin. Microbiol.* **40**, 4298–4300.

Yin, J. L., Shackel, N. A., Zekry, A., McGuinness, P. H., Richards, C., van der Putten, K., McCaughan, G. W., Eris, J. M. and Bishop, G. A. (2001). Real-time reverse transcriptase-polymerase chain reaction (RT-PCR) for measurement of cytokine and growth factor mRNA expression with fluorogenic probes or SYBR green I. *Immunol. Cell. Biol.* **79**, 213–221.

Zerr, D. M., Huang, M.-L., Corey, L., Erickson, M., Parker, H. L. and Frenkel, L. M. (2000). Sensitive method for detection of human herpesvirus 6 and 7 in saliva collected in field studies. *J. Clin. Microbiol.* **38**, 1981–1983.

Zhai, J., Briese, T., Dai, E., Wang, X., Pang, X., Du, Z., Liu, H., Wang, J., Wang, H., Guo, Z., Chen, Z., Jiang, L., Zhou, D., Han, Y., Jabado, O., Palacios, G., Lipkin, W. I. and Yang, R. (2004). Real-time polymerase chain reaction for detecting SARS coronavirus, Beijing, 2003. *Emerg. Infect. Dis.* **10**, 300–303.

Zhang, J., Yankaskas, J. R. and Engelhardt, J. F. (1996). In vivo analysis of fluid transport in cystic fibrosis airway epithelia of bronchial xenografts. *Am. J. Physiol.* **270**, C1326–C1335.

Zhang, Y., Zhang, D., Li, W., Chen, J., Peng, Y. and Cao, W. (2003). A novel real-time quantitative PCR method using attached universal template probe. *Nucleic Acids Res.* **31**, e123.

Zimmermann, K. and Manhalter, J. W. (1996). Technical aspects of quantitative competitive PCR. *Biotechniques* **21**, 268–279.

11 Design and Use of Functional Gene Microarrays (FGAs) for the Characterization of Microbial Communities

Christopher W Schadt[1], Jost Liebich[2], Song C Chong[1], Terry J Gentry[1], Zhili He[1], Hongbin Pan[1] and Jizhong Zhou[1]

[1] Environmental Sciences Division, Oak Ridge National Laboratory, P.O. Box 2008, Oak Ridge, TN 37831-6038, USA; [2] Institute of Chemistry and Dynamics of the Geosphere IV: Agrosphere, Forschungszentrum Jülich GmbH, 52425 Jülich, Germany

CONTENTS

Introduction and overview of FGAs
Functional gene diversity and data acquisition for probe design
Design of specific oligonucleotide probes for FGAs
Microarray construction, labeling, hybridization and image acquisition
Data analysis techniques
Evaluation and validation of FGA results

◆◆◆◆◆◆ INTRODUCTION AND OVERVIEW OF FGAs

The recent development of microarrays as powerful, high-throughput genomic technology has spurred investigators toward their use for the study of various biological processes. Although microarray technology has been used successfully to analyze global gene expression in pure cultures or tissue samples for many different organisms (Lockhart *et al.*, 1996; DeRisi *et al.*, 1997; Schena *et al.*, 1995, 1996; Ye *et al.*, 2000; Thompson *et al.*, 2002; Liu *et al.*, 2003a,b), adapting microarrays for use in environmental studies presents great challenges in terms of design, use and data analysis (Zhou and Thompson, 2002; Zhou, 2003). Recently, various formats of environmental microarrays have been proposed, developed and evaluated for species detection and microbial community analyses in complex environments as reviewed recently by Zhou (2003). These studies have indicated that microarray-based genomic technologies have great potential as specific, sensitive, quantitative, and high-throughput tools for microbial detection, identification and characterization in natural environments. This chapter will focus on and discuss recent work on the development and use of functional gene microarray

(FGA) technology and introduce readers to issues and methodology surrounding their design and use. As you will see in the following discussions this methodological debate is far from over, as this revolutionary technology is still very much in a state of continuing development.

The genes encoding functional enzymes involved in various biogeochemical cycling (e.g. nitrogen, carbon and sulfur) and bioremediation processes are very useful as signatures for monitoring the potential activities and physiological status of microbial populations and communities that drive these processes in the environment. Microarrays containing functional gene sequence information are often referred to as functional gene arrays (FGAs) because they are primarily used for analysis of microbial community activities in the environment (Zhou and Thompson, 2002; Wu et al., 2001). Similar to the microarrays used for monitoring gene expression, both oligonucleotides and DNA fragments derived from functional genes can be used for fabricating FGAs. To construct microarrays containing large DNA fragments as probes, the fragments are generally amplified by polymerase chain reaction (PCR) from environmental clones or from pure culture genomic DNAs (Wu et al., 2001). Obtaining all the diverse environmental clones and bacterial strains from various sources as templates for amplification can be an overwhelming obstacle. As a result, construction of comprehensive FGAs based on PCR gene fragments that adequately encompass diverse environmental sequences is a near impossibility.

To circumvent this problem, FGAs containing synthetic oligonucleotides (oligos) have been developed for use. The main advantage of oligo FGAs is that construction is much easier than DNA-based FGAs because the probes can be directly designed and synthesized based on sequence information from public databases. Therefore, comprehensive arrays representing the extreme diversity of environmental sequences can be constructed. Several studies have applied and evaluated the usefulness of this approach for select groups of microorganisms (Taroncher-Oldenburg et al., 2002; Koizumi et al., 2002; Bodrossy et al., 2003; Denef et al., 2003; Tiquia et al., 2004; Rhee et al., 2004). In one such recent study originating from our laboratory, a 50mer oligo FGA was constructed and evaluated encompassing 1033 genes involved in nitrogen transformations (*nirS*, *nirK*, *nifH* and *amoA*), methane consumption (*pmoA*) and dissimilatory sulfate reduction (*dsrA/B*) from sequences available in public databases and our own environmental sequence collections (Tiquia et al., 2004). Under the hybridization conditions of 50°C and 50% formamide, genes having <86–90% sequence identity could be clearly differentiated. This level of hybridization specificity is higher than those of PCR fragment-based FGAs (Wu et al., 2001). Based on our comparisons of sequences from pure cultures involved in nitrification, denitrification, nitrogen fixation, methane oxidation and sulfate reduction, the average percent similarity of such functional genes at species level is usually much less than 85%, suggesting that oligo-based FGAs can provide species-level resolution. Also the detection limits of approximately 8–10 ng for pure genomic DNA was 10 times lower than the PCR fragment-based FGAs (Zhou and

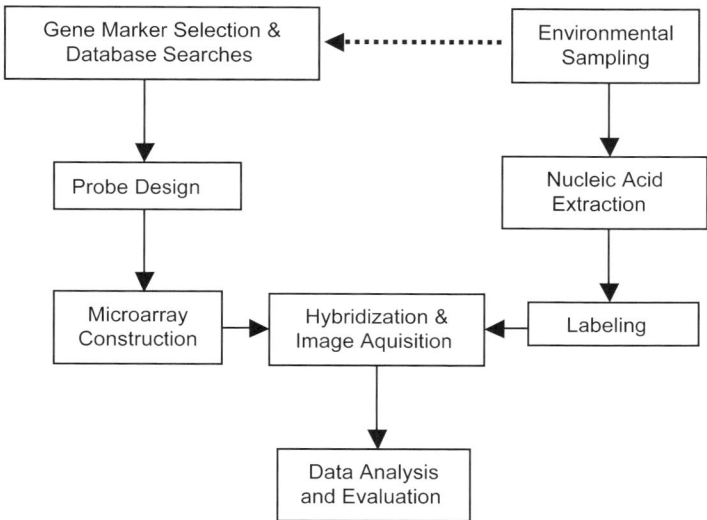

Figure 11.1. Schematic representation of the processes involved in designing and utilizing FGAs for the study of microbial communities.

Thompson, 2002; Wu *et al.*, 2001). In addition, similar to the DNA fragment-based FGAs, a strong linear relationship was observed between signal intensity and target DNA concentrations from 8 to 1000 ng for all six different functional gene groups ($r^2 = 0.96 - 0.98$). Furthermore, 5 μg of bulk community DNA from marine sediments was sufficient to obtain reasonably good hybridizations useful in profiling differences between communities. These results suggest that the developed 50mer FGA has potential as specific, sensitive, and potentially quantitative parallel tools for characterizing the composition, structure, activities and dynamics of microbial communities in natural environments. Based on such results and the methods presented below, a much more comprehensive 50mer FGA of several thousand gene probes is being designed and tested in our laboratory (Figure 11.1) (based on genes in Table 11.1 and others). The probes on the arrays represent very diverse groups of functional genes involved in nitrogen cycling, carbon cycling, sulfate reduction, phosphorus utilization, organic contaminant degradation and metal resistance.

◆◆◆◆◆◆ FUNCTIONAL GENE DIVERSITY AND DATA ACQUISITION FOR PROBE DESIGN

The first and one of the most important steps in design of oligo FGAs is to identify the optimal target genes that will be useful for tracking the microbial processes of interest (Figure 11.1). In this section, we have provided an overview of the diversity of publicly available functional gene sequence data for several microbially driven processes and suggested suitable genetic markers that could be used in probe design.

Table 11.1. Various microbial functional gene sequences available from public databases of potential use in FGAs. All categories exclude data from vascular plants and metazoans. The availability of environmental sequence data (*) and a representative reference of such is also provided. Categories including multiple genes or protein subunits are indicated as (all)

Gene/category	Sequences #	Example reference
Nitrogen cycling		
Nitrogenase-nifH	1784*	Hurek et al. (1997)
Nitrogenase-nifD	180*	Ueda et al. (1995)
Nitrogenase-nifK	89	
Ammonium monooxygenase-amoA	1158*	Nold et al. (2000)
Hydroxylamine oxidoreductase	15	
Nitrate reductase-napA	148	
Nitrate reductase-narB	50	
Nitrate reductase-narG	544*	Philippot et al. (2002)
Nitrate reductase-nasA	120*	Allen et al. (2001)
Nitrite reductase-nirK	264*	Liu et al. (2003a,b)
Nitrite reductase-nirS	411*	Liu et al. (2003a,b)
Nitrous oxide reductase-nosZ	273*	Scala and Kerkhof (1999)
Nitric oxide reductase-norB	68*	Braker and Tiedje (2003)
Urease (all)	1707	
Subtotal	6811	
Methane oxidation & reduction		
Soluble methane monooxygenase (all)	250*	Horz et al. (2001)
Particulate methane monooxygenase (all)	503*	Nold et al. (2000)
Methyl coenzyme M reductase	670*	Hallam et al. (2003)
Subtotal	1423	
Carbon polymer degradation		
Exoglucanase/cellobiohydrolase (all)	120	
Cellulase/endoglucanase (all)	920	
Chitinase (all)	1544*	Metcalfe et al. (2002)

(*continued*)

Table 11.1. *Continued*

Gene/category	Sequences #	Example reference
Laccase (all)	194*	Lyons et al. (2003)
Lignin peroxidase	15	
Mannanase (all)	400	
Polygalacturonase	156	
Subtotal	3349	
Carbon fixation		
Formyltetrahydrofolate synthetase (FTHFS)	206*	Leaphart and Lovell (2001)
Rubisco small subunit	747	
Rubisco large subunit	3474*	Elsaied and Naganuma (2001)
Rubisco small subunit (all euks)	112	
Rubisco large subunit (all euks)	787*	Elsaied and Naganuma (2001)
Carbon monoxide dehydrogenase (CODH)	14	
Subtotal	4441	
Sulphate reduction		
Sulfite reductase (*dsrA/B*)	924*	Wagner et al. (1998)
Adenosine phosphosulfate (*apsA*)	81*	Deplancke et al. (2000)
Subtotal	1006	
Phosphorus cycling		
Exopolyphosphatase	207	
Polyphosphate glucokinase	19	
Phytase	27	
Polyphosphate kinase	185*	McMahon et al. (2002)
Subtotal	438	
Organic remediation pathways		
Aniline (all)	36	
Atrazine (all)	113*	Martin-Laurent et al. (2003)
Benzene (all)	83	
Biphenyl (all)	237	
Dibenzothiophene (all)	101*	Duarte et al. (2001)

(continued)

Table 11.1. *Continued*

Gene/category	Sequences #	Example reference
2,4-Dichlorophenoxyacetic acid (all)	82*	Shaw and Burns, unpublished
Naphthalene (all)	392*	Baldwin et al. (2003)
Naphthalene dioxygenase α subunit	140*	Baldwin et al. (2003)
Naphthalene dioxygenase β subunit	21	
Naphthalene dioxygenase ferredoxin	23	
Naphthalene dioxygenase ferredoxin reductase	14	
Dihydrodiol naphthalene dehydrogenase	22	
Dihydroxynaphthalene dioxygenase	18	
Hydroxychromene carboxylate isomerase	42	
Hydroxybenzylidenepyruvate hydratase-aldolase	18	
Salicylaldehyde dehydrogenase	21	
Transcriptional regulator	35*	Park et al. (2002)
Nitrotoluene (all)	33	
n-Octane (all)	64	
Pentachlorophenol (all)	55*	Beaulieu et al. (2000)
Pyrene (all)	140	
Toluene-aerobic (all)	98*	Cassidy (2002)
Toluene-anaerobic (all)	32	
Trichloroethylene/perchloroethylene (all)	30	
Xylene (all)	77	
Other organic remediation genes	3786	
Subtotal	5359	
Metal resistance and efflux		
Aluminum resistance (all)	36	
Arsenic resistance (all)	407	
Cadmium resistance (all)	224*	Oger et al. (2003)
Chromium resistance (all)	125	
Cobalt resistance (all)	44	
Copper resistance (all)	311	
Lead resistance (all)	12	
Mercury resistance (all)	649	
Mercuric reductase-merA	149	
Organomercurial lyase-merB	33	
Mercury binding-merP	82	

(continued)

Table 11.1. *Continued*

Gene/category	Sequences #	Example reference
Mercury transport-merT	87	
Mercury transport-merC	45	
Mercury transport-merE	18	
Mercury transport-merF	7	
Phenylmercury resistance-merG	5	
Transcriptional regulator-merR	152	
Additional regulator-merD	50	
Nickel resistance (all)	52	
Selenium resistance (all)	4	
Silver resistance (all)	53	
Tellurium resistance (all)	321	
Vanadium resistance (all)	8	
Zinc resistance (all)	52	
Subtotal	2301	

We utilized a variety of combinatorial search strings within the GenBank ENTREZ interface in order to acquire these sequences, in combination with a locally executable program to extract and download pertinent information (database of origin, species name, gene descriptions, phylogenetic lineage, DNA coding sequence, protein sequence, etc.) starting only from a list of GI (gene identification) numbers in tab delimited format. These databases were then further screened to remove duplicate and non-homologous sequences. We found that such an approach greatly reduces the time and effort involved in sorting through and collecting the vast amounts of sequence data available for probe design. By conducting searches within the protein sequence database, many duplicated entries associated with the nucleotide database are avoided, and direct retrieval of coding DNA sequences is possible that avoids problems associated with intron containing sequences. Additionally, we have recently developed the program further so that once keywords are decided upon it automatically updates the databases (based on modification date) at user defined intervals and will also remove duplicate sequences. To illustrate the selection of genetic markers based on the data available from such searches, we discuss below some examples of this process for selected key functional genes. Table 11.1 additionally summarizes the results of many such searches and amounts of available sequence information.

Nitrogen Cycling

Microbial nitrogen transformations involve three major microbially driven processes; nitrogen fixation, nitrification and denitrification. These involve many different enzymes including nitrogenase, ammonium monooxygenase, hydroxylamine oxidoreductase, nitrite oxidase, nitrate reductase, nitrite reductase, nitric oxide reductase, nitrous oxide reductase, assimilatory nitrate reductase and urease (Table 11.1). Thus, it is essential to understand the diversity of functional genes contributing to the nitrogen cycle contained in microbial communities. Over 6000 nitrogen cycle related gene sequences are available from GenBank and other databases (Table 11.1). However, it is beyond the scope of this chapter to discuss all of these processes and enzymes, so we have focused on the genes involved in microbial nitrogen fixation to illustrate the selection of appropriate markers.

Nitrogen is one of the major components of living cells and about 79% of the earth atmosphere is nitrogen in the form of N_2 gas. However, nitrogen is often a limiting factor for growth and biomass production in both aquatic and terrestrial environments (Vitousek and Howarth, 1991). In order to use nitrogen for growth, N_2 must be fixed to ammonium (NH_4) or nitrate (NO_3) which occurs primarily through microbial nitrogen fixation. Nitrogen fixation is exclusively performed by prokaryotes. Some live independently of other organisms (free living nitrogen-fixing bacteria such as *Azotobacter, Beijerinckia, Desulfovibrio*, purple sulfur bacteria, purple non-sulfur bacteria, and green sulfur bacteria) while others live in symbiotic association with plants (*Rhizobium, Frankia, Azospirillum*) and some are found in both situations (e.g. *Bradyrhizobium*). Microorganisms perform nitrogen fixation with an evolutionarily conserved nitrogenase protein complex, and all consist of two major proteins found in both archaea and eubacteria (Howard and Rees, 1996). Component I consists of an $\alpha_2\beta_3$ tetramer molybdoferredoxin (MoFe protein) or dinitrogenase (EC 1.18.6.1), the active site for N_2 reduction, and is encoded by the genes *nifD* and *nifK*. Component II consists of a homodimer, azoferredoxin (Fe protein) or dinitrogenase reductase (EC 1.19.6.1), that couples ATP hydrolysis to interprotein electron transfer, and is encoded by *nifH* (Dean *et al.*, 1993). Dinitrogenase genes are further divided into four clusters (Normand and Bousquet, 1989; Normand *et al.*, 1992; Chien and Zinder, 1994). Cluster I contains most of eubacterial MoFe containing dinitrogenase. Cluster II consists of archaea, eubacterial alternative dinitrogenase with non-Mo, non-V. Cluster III, contains Mo dinitrogenase (*nif-1*) genes from the Gram positive eubacteria *Clostridium* and the *nif-2* genes from the archaea *Methanosarcia* and the sulfate reducer *Desulfovibrio*. Cluster IV, containing methanogens, utilizes a distantly related gene similar to bacteriochlorophyll. Numerous other genes are also known to be related to or interact with *nif* proteins such as *nifJ, nifF, nifM, nifB* and are often present in a complex extended *nif* operon.

However, of all these potential targets for a functional marker, the *nifH* gene is the most widely used. Dinitrogenase reductase (*nifH*) is especially evolutionarily conserved and has often been used as a genetic marker for detecting nitrogen fixing microorganisms in natural environments (Kirshtein *et al.*, 1991; Zehr *et al.*, 1995; Widmer *et al.*, 1999) and also as a basis for the phylogenetic analysis of *nifH* containing organisms (Zehr and McReynolds, 1989). Over 1700 *nifH* gene sequences are available from public DNA databases and this is expanding rapidly. However, *nifD* and *nifK* have also been used as additional information, especially for resolving differences between closely related sequences. For these reasons, the *nifH* gene makes an excellent functional marker, but the importance of *nifD* and *nifK* for resolving finer scale differences should not be discounted.

Carbon Cycling

Microbially driven aspects of the carbon cycle could play important roles in determining the amplitude of anthropogenic effects on climate change and their potential mitigation. Microorganisms play important roles through direct effects on such processes as methane oxidation and production, carbon fixation, and the breakdown and decomposition of organic substrates. While a discussion of all these aspects is not possible, we provide potential target genes for monitoring these processes in Table 11.1. To illustrate some of the issues surrounding selection of suitable genetic markers for these processes, we will proceed with an overview of aerobic methane oxidation. Methanotrophic bacteria oxidize methane for energy production and biosynthesis of organic compounds. These organisms are ubiquitous in environments such as oceanic and inland waters, wetlands, soils, groundwaters and even the deep subsurface and are of great interest to microbial ecologists because of the potentially important role they could play in mitigating global warming (Holmes *et al.*, 1995; Dunfield *et al.*, 1999). They are also of interest for use within certain industrial applications such as bioremediation or steps in the synthesis of certain organic compounds (Sullivan *et al.*, 1998).

The first step in the oxidation of methane is the conversion of methane to methanol by the enzyme methane monooxygenase (EC 1.14.13.25). After this step methanol is converted to formaldehyde via a non-specific alcohol dehydrogenase (e.g. EC 1.1.99.8), where it enters one of two different pathways (RuMP or serine-isocytrate) for the production of C_3 compounds. As a specific genetic marker for methanotrophy we are therefore left with methane monooxygenase. This key enzyme exists in two forms: the cytoplasmic, soluble, methane monooxygenase (sMMO) and the membrane-bound, particulate, methane monooxygenase (pMMO). Of the two forms of MMO, all known methanotrophs carry pMMO whereas only some select methanotrophs (Type II) carry both pMMO and sMMO (Murrell *et al.*, 2000). However, pMMO is also evolutionarily related to the important nitrogen cycling enzyme ammonium monooxygenase (AMO) and they share a high degree of sequence similarity in some regions of these genes (Holmes *et al.*, 1995). Because it is present in all known methanotrophs, pMMO carries distinct advantages as a specific functional marker for methanotrophy (Murrell *et al.*, 1998). Practically speaking, however, because of the high degree of similarity between the two genes (AMO and pMMO) any probe design effort must most likely consider both simultaneously. Both pMMO and AMO are made up of two different polypeptide subunits, the smaller of which (*pmoA* and *amoA*) contain the active sites and have been much more widely sequenced and studied as functional and phylogenetic markers.

While many organisms can utilize methanol and some other methyl compounds (methylotrophy), currently true methanotrophy is only known to occur within the beta and gamma-proteobacteria. However, studies of environmental libraries of *pmoA* suggest diversity within these bacteria is high, and many lineages contained within them are only known from such environmental libraries. Additionally, several studies have suggested that there may be divergent lineages related to either *pmoA* or *amoA* that could possibly fall outside the proteobacteria (Nold et al., 2000). Currently ~503 *pmoA* genes are listed in GenBank. However, because the closely related *amoA* can often be amplified with the same PCR primers (Nold et al., 2000), care should be taken to ensure that such gene annotations are correctly identified in sequences retrieved from databases.

Sulfur Cycling

The most widespread biochemical reaction of the sulfur cycle is probably the assimilatory sulfate reduction in which inorganic sulfate is reduced to become integrated in amino acids and proteins where they play an essential role in the formation of secondary structures. This process is followed by the release of reduced hydrogen sulfide by the degradation (or desulfurylation) of the organic sulfur compounds. This process is important for almost every life form on Earth and therefore can be found throughout all biological kingdoms. More specific to prokaryotes are sulfide/sulfur oxidation and dissimilatory sulfate reduction (sulfur respiration) for which we discuss probe selection. Sulfate respiration, using oxidized sulfur forms as a terminal electron acceptor, is widespread often leading to the formation of black metal sulfides (e.g. iron sulfide) and the toxic gas hydrogen sulfide (H_2S) (Rabus et al., 2000). Sulfate reducers occur in a wide variety of environments such as marine sediments, deep-sea hydrothermal vents, freshwater systems, anaerobic sludge and as endosymbionts (Jørgensen, 1982; Singleton, 1993; Finegold and Jousimies-Somer, 1997; Manz et al., 1998; Dubilier et al., 2001; Laue et al., 2001; Castro et al., 2002; Liu et al., 2003a,b; Nakagawa et al., 2004). Some sulfate reducing bacteria are also capable of oxidizing organic contaminants of anthropogenic origin (e.g. petroleum hydrocarbons) along with other naturally occurring substances (Lovley, 1997; Kleikemper et al., 2002). Others are able to reduce many different metals beyond sulfur species, for example, some aid in immobilizing potentially hazardous metals like uranium via such reductions (Chang et al., 2001). Both features make this group of interest for bioremediation processes in anoxic environments. In many cases, sulfate reducing bacteria may also compete with methanogens and denitrifiers for electron donors and sulfate reducing bacteria seem to often dominate if sufficient sulfate supply is available (Lovley and Klug, 1983).

Dissimilatory sulfate reduction is generally a three-enzyme transformation involving activation of sulfate by ATP sulfurylase (EC 2.7.7.4), reduction of the product adenosine-phosphosulfate (*aps*) to sulfite by adenylylsulfate reductase (EC 1.8.99.2) and subsequent further reduction to hydrogen sulfide by dissimilatory sulfite reductase (*dsr*, EC 1.8.99.3), the latter enzyme consists of at least three subunits. However, only APS reductase and *dsr*-genes are suitable indicators of dissimilatory sulfate reduction in the environment, since the first step is also involved in assimilatory sulfate reduction that is widely evolutionarily distributed. Furthermore, several organisms are known to use only sulfite as an electron acceptor but not sulfate, and they lack adenylylsulfate reductase genes (Huber *et al.*, 1997; Holliger *et al.*, 1998; Molitor *et al.*, 1998; Laue *et al.*, 2001). Because of this limitation, *aps* genes have only more recently been applied as a marker for this process, primarily to distinguish sulfate/sulfite reducers from those only capable of reducing sulfite (Deplancke *et al.*, 2000; Friedrich, 2002). As a result, in our recent searches, only 81 nucleotide sequences of APS genes from different species or subspecies were found in publicly available databases, whereas 924 sequences can be retrieved for *dsrA* and *dsrB* genes. Because of this better studied diversity, *dsr* genes are a more indicative marker for this overall function. *dsrA* and *dsrB* genes occur in all known sulfate or sulfite reducing bacteria and can be targeted by a single set of conserved primers (Karkhoff-Schweizer *et al.*, 1995; Wagner *et al.*, 1998) allowing probe design for both subunits. A third subunit for dissimilatory sulfite reductase, encoded by *dsrD*, is also known but available sequence information to date is mostly limited to cultured species (Karkhoff-Schweizer *et al.*, 1995). A large portion of the *dsrA* and *dsrB* sequence data were obtained from uncultivated organisms, after the discovery of the conserved nature of dissimilatory sulfite reductases over different phyla made their detection by PCR possible (Karkhoff-Schweizer *et al.*, 1995; Wagner *et al.*, 1998). Many of these sequences are phylogenetically divergent relative to those known sequences from cultured organisms (Liu *et al.*, 2003a,b; Nakagawa *et al.*, 2004).

According to sequence information deposited in GenBank and other publicly accessible sequence databases, 99 species have been identified to belong to this functional group spreading over five bacterial phyla (Chlorobi, Firmicutes, Nitrosospira, Proteobacteria, Thermodesulfobacteria) and two archaeal phyla (Crenarchaeota, Euryarchaeota), with members of the genus *Desulfovibrio* forming the largest group. However, more species existing in nature are able to perform sulfate respiration, some of which are likely to fall into novel to date yet unknown lineages as shown recently by Mori *et al.* (2003). Sequence comparison of these highly conserved genes over distant phylogenetic groups of archaeal and bacterial origin suggests that these genes may have been horizontally transferred in some lineages (Larsen *et al.*, 1999; Klein *et al.*, 2001; Friedrich, 2002). Thus, *dsr*-gene sequences can only partially be used as a phylogenetic marker for sulfate and sulfite reducing microorganisms, but they are very suitable for functional diversity studies.

Organic Contaminant Degradation

There are thousands of different gene sequences available in public databases encoding various enzymes that transform one or more of hundreds of complex organic chemicals. A review of all available contaminant degradation genes is beyond the scope of this chapter, so for the purposes of illustrating the diversity of available degradative genes and how they may be used in microarrays, we have focused on the naphthalene degradation pathway. Naphthalene is the most widely studied member of the polycyclic aromatic hydrocarbons (PAHs) which are composed of multiple, fused aromatic rings. In addition to naturally occurring sources, environmental PAH-contamination often results from anthropogenic deposition of various fossil fuel-derived chemicals such as the wood preservative creosote (Sun *et al.*, 2003).

In Table 11.1, we have listed the enzymes which sequentially convert naphthalene to salicylate (Bosch *et al.*, 1999; Takizawa *et al.*, 1999). The initial reaction in the pathway occurs via a multi-component naphthalene dioxygenase (NDO) (EC 1.14.12.12) (Kauppi *et al.*, 1998). The α subunit of the NDO iron sulfur protein is believed to confer the specificity of the enzyme and, as indicated in Table 11.1, is the most studied of the naphthalene genes (Parales *et al.*, 2000; Wackett, 2002). The other three NDO subunits probably have limited impact on enzymatic specificity (Parales *et al.*, 1998; Romine *et al.*, 1999). For these reasons, the α subunit is potentially the best candidate for use as a microarray marker gene. It has been identified in numerous eubacteria including *Burkholderia, Comamonas, Cycloclasticus, Marinobacter, Neptunomonas, Polaromonas, Pseudoalteromonas, Pseudomonas, Ralstonia* and *Rhodococcus* spp. (Kurkela *et al.*, 1988; Denome *et al.*, 1993; Takizawa *et al.*, 1994; Fuenmayor *et al.*, 1998; Hedlund *et al.*, 1999, 2001; Larkin *et al.*, 1999; Melcher *et al.*, 2002; Jeon *et al.*, 2003; Kasai *et al.*, 2003). The α subunit gene is referred to by various names, including *doxB, nagAc, nahAc, narAa, ndoB, pahAc* and *phnAc*, that were initially chosen in part based on the substrate on which the host bacterium was isolated (Habe and Omori, 2003). Most of the α subunit sequence information from isolated organisms is derived from Gram negative bacteria, primarily *Pseudomonas* spp. containing *nah*-like NDO genes (Habe and Omori, 2003). But, it has recently been demonstrated that other NDO genes such as the *phn* genes, of which much less is known, may be prevalent in environmental samples (Lloyd-Jones *et al.*, 1999; Wilson *et al.*, 2003). Information on naphthalene degradation genes in Gram positive bacteria is limited and even less information is available for the fungal genes (Andreoni *et al.*, 2000; Larkin *et al.*, 1999). Several different primers have been designed for the α subunit genes (Hamann *et al.*, 1999; Lloyd-Jones *et al.*, 1999; Wilson *et al.*, 1999; Baldwin *et al.*, 2003), but only a handful of studies have directly amplified and sequenced the genes from soil, sediment, or groundwater samples (Wilson *et al.*, 1999; Stach and Burns, 2002; Jeon *et al.*, 2003). To our knowledge, the diversity of the α subunit gene has not been thus assessed in marine environments even though the gene has been identified in several marine bacteria (Geiselbrecht *et al.*, 1998; Hedlund *et al.*, 1999, 2001).

Another potential application of environmental microarrays is in the determination of horizontal gene transfer events. Gene transfer has contributed to the evolution of metabolic pathways and has also played a role in the spread of NDO genes (Herrick *et al.*, 1997; McGowan *et al.*, 1998; Habe and Omori, 2003; Wilson *et al.*, 2003). Bacteria have been identified that contain multiple copies of the *nah* genes, and it also appears that some isolates have mosaics of *nah* operons from different organisms (Bosch *et al.*, 1999; Ferrero *et al.*, 2002). Microarray technology could be used to quickly assess if an isolate contains multiple copies of a given gene and if some of the genes in a pathway are similar to those in one organism while other genes are similar to another organism. This knowledge could help to determine the factors involved in microbial adaptation following environmental contamination with xenobiotics (Rensing *et al.*, 2002; Top and Springael, 2003). However, for successful application of microarray technology to the study of pathway evolution, probes would be required for all of the genes in a given pathway, which in the case of naphthalene would necessitate more sequence information from many organisms for the other genes besides the NDO α subunit.

Metals Resistance

A recent search of public databases revealed 2303 sequences either identified as, or similar to, genes encoding microbial resistance to 15 different metals and metalloids. Sequences for the various mercury resistance genes were the most numerous; therefore, we have focused on these genes for illustration. Mercury compounds are widely distributed around the Earth, and their presence in the environment can occur through natural or anthropogenic processes with the latter estimated to account for approximately two-thirds of the worldwide Hg input (Mason *et al.*, 1994). Not surprisingly, mercury resistance is among the most common phenotypes observed in bacteria (Barkay *et al.*, 2003). It has been proposed that the emergence of the basic mercury resistance genes (*mer*) predates the divergence of Gram negative and positive bacteria (Osborn *et al.*, 1997). Of the *mer* functional genes, *merA*, which encodes the mercuric ion reductase enzyme (EC 1.16.1.1) that converts Hg^{2+} to the volatile species Hg^0, is the most studied and has the most available sequence data (Table 11.1) (Barkay *et al.*, 2003; Nascimento and Chartone-Souza, 2003). The *merA* gene is therefore a good candidate for microarray fabrication and monitoring bacterial mercury resistance, although other genes may be more appropriate for specific samples such as *merB* encoding organomercurial lyase (EC 4.99.1.2) for samples contaminated with organomercury compounds.

The *merA* gene is widely spread among both Gram negative and Gram positive bacteria including *Alcaligenes*, *Bacillus*, *Delftia*, *Exiguobacterium*, *Pantoea*, *Pseudomonas*, *Shigella*, *Staphylococcus* and *Xanthomonas* spp.

(Laddaga *et al.*, 1987; Yurieva *et al.*, 1997; Bogdanova *et al.*, 1998; Reniero *et al.*, 1998; Kholodii *et al.*, 2000; Venkatesan *et al.*, 2001; Sota *et al.*, 2003). Whole genome sequences of archaea have also indicated the presence of *mer*-like sequences, and the first report of functional, archaeal *merA* and *merR* genes has recently been published for *Sulfolobus solfataricus* (Schelert *et al.*, 2004). There is, however, very limited sequence data in the literature and databases regarding the diversity of *merA* genes from uncultured microorganisms in environmental samples. In fact, the only culture-independent *merA* sequences we found were from plasmids isolated from sewage sludge and soil via an exogenous plasmid isolation method (Schluter *et al.*, 2003; Schneiker *et al.*, 2001). Researchers have amplified *merA* from environmental samples; however, the PCR products were not sequenced but analyzed with other procedures (Felske *et al.*, 2003; Hart *et al.*, 1998). Further information on the diversity of uncultured *merA* sequences would be useful prior to microarray construction since the sequences from isolated organisms may not comprehensively represent the genetic diversity in the environment (Bruce *et al.*, 1995; Marchesi and Weightman, 2003). Additionally, only limited information is available for mercury resistance genes in fungi, although several genes involved in mercury (metal) resistance in yeast have recently been identified which may help to expand this knowledge base in the future (Furuchi *et al.*, 2002; Nguyên-nhu and Knoops, 2002; Westwater *et al.*, 2002; Gueldry *et al.*, 2003).

◆◆◆◆◆◆ DESIGN OF SPECIFIC OLIGONUCLEOTIDE PROBES FOR FGAs

Oligonucleotide-based microarrays are becoming more popular because they offer a number of advantages over cDNA microarrays. First, as stated previously, only sequence information is required and PCR amplification can therefore be avoided. Secondly, more flexibility to control specificity of hybridization can be achieved in probe design by the ability to strictly delimit parameters such as melting temperature (T_m), overall similarity (% homology) and other factors. Thirdly, oligonucleotide synthesis costs have dropped considerably in the last few years (Relogio *et al.*, 2002). In addition, oligonucleotide arrays provide potential solutions to some of the more complicated problems involved in environmental studies. For example, short oligonucleotides may be used to avoid highly conserved regions of orthologous genes that would not be possible with PCR amplification using conserved primers. The challenge for probe design is how to identify the optimum probes for each gene or each group of genes.

There are a number of pre-existing programs available for automated selection of oligonucleotide probes for DNA microarrays (Table 11.2). OligoArraySelector (Zhu *et al.*, 2003) runs on Linux/Unix systems and uses a BLAST approach to search for sequence similarity and compute the thermodynamic properties for only the most probable non-specific

Table 11.2. Oligonucleotide probe design programs

Name	OS	Reference
ArrayOligoSelector	Linux	Zhu et al. (2003), http://sourceforge.net/projects/arrayoligosel/
OligoArray	Windows and Unix/Linux	Rouillard et al. (2002), http://berry.engin.umich.edu/oligoarray
OligoArray 2.0	Unix/Linux	Rouillard et al. (2003), http://berry.engin.umich.edu/oligoarray2
OligoPicker	Linux	Wang and Seed (2003), http://pga.mgh.harvard.edu/oligopicker/index.html
OligoWiz (Web-based)	Unix	Nielsen et al. (2003), http://cbs.dtu.dk/services/oligowiz/
PRIMEGENS	Unix/Linux	Xu et al. (2002), http://compbio.ornl.gov/structure/primegens/
PROBEmer (Web-based)	Linux	Emrich et al. (2003), http://probemer.cs.loyola.edu
ProbeSelect	Unix/Linux	Li and Stormo (2001)
ROSO (Web-based)	Windows and Unix	Reymond et al. (2004), http://pbil.univ-lyon1.fr/roso
ArrayDesigner (Commercial)	N/A	TeleChem International Inc., http://arrayit.com
Sarani Goldminer (Commercial)	N/A	Strand Genomics, http://mail.strandgenomics.com/index.html

hybridization. Oligopicker (Wang and Seed, 2003) runs on Linux platforms and relies on BLAST search and 15-base stretch filtering. Those two programs select 70mer oligonucleotides for whole genomes. OligoArray (Rouillard *et al.*, 2002) can run on Windows, Unix or Linux systems and the oligonucleotide specificity is checked using BLAST (Altschul *et al.*, 1997) and possible secondary structures are predicted by the Mfold server (Zuker *et al.*, 1999). Its sister program, OligoArray 2.0 (Rouillard *et al.*, 2003), runs on Linux or Unix systems and the probe specificity is based on a comparison of sequence similarity between the specific target and putative non-specific targets. PRIMEGENS (Xu *et al.*, 2002) uses BLAST search and sequence alignment to select gene-specific fragments and then feeds those fragments to the Primer3 program (Rozen and Skaletsky, 2000) to design PCR primer pairs or probes on a genome scale. The program runs on Linux or Unix platforms and can also be used from a web interface. ProbeSelect (Li and Storomo, 2001) runs on Linux or Unix and uses a suffix tree to search for sequence similarity and the Myersgrep program (Myers, 1998) to search for matching sequences with few mismatches. This program can choose short (20–25 bases) or long (50 or 70 bases) oligonucleotides. Recently, some web-based probe design programs have been developed. OligoWiz (Nielsen *et al.*, 2003) is implemented as a client-server application. The server is responsible for the calculation of scores and utilizes the BLAST program for homology search. The client is used to submit jobs to the server, to visualize the scores and to fine-tune the placement of oligonucleotides. PROBEmer (Emrich *et al.*, 2003) uses suffix tree-based algorithms to identify common substrings. The program can design oligonucleotide probes for a single sequence or a defined group of sequences (16S rRNA gene) or PCR primers. ROSO (Reymond *et al.*, 2004) separates the time-consuming BLAST search from the fast step of thermodynamic analysis. The program can be used to select oligonucleotide probes or PCR primers. Probe design parameters, such as oligonucleotide length, number of probes for each gene and target T_m can be changed by users for some of the programs described.

Some of the programs mentioned above were used to design probes for the whole genome of *Methanococcus maripaludis* and a separate group of sequences of *nirS* and *nirK* (nitrite reductase) based on those publicly available and our own sequence collections. The results are summarized in Table 11.3. Most programs worked well for the whole genome data. However, serious problems occurred when they were used to design 50mer oligonucleotides for the *nirSK* group of sequences. First, too few specific probes and too many non-specific probes were designed. For the purposes of this test we categorized probes as non-specific if they had >85% similarity, identical stretches >15 bp, or mismatch free energy < −30 kcal/mol. Secondly, a large majority of sequences did not have probes if only unique oligonucleotides were selected. A similar situation would be expected in probe selection for many other genes that might be used for FGAs because the nature of sequence data for these arrays is quite different from whole genome data. For example, sequences for FGAs are often highly homologous, and many sequences originating

Table 11.3. Number and specificity of designed probes (50mer) by different programs

Programs used	Whole genome sequences of *M. maripaludis* (1766 ORFs)					Group sequences of *nirSK* (842 gene sequences)					
	Total ORFs	ORFs rejected	Probes designed	Specific probes	Non-specific	Total ORFs	ORFs rejected	Probes designed	Specific probes	Non-specific
ArrayOligoSelector	1766	7	1759	1415	344	842	0	842	117	725
OligoArray	1766	68	1698	1654	44	842	35	807	70	737
OligoArray 2.0	1766	68	1698	1464	234	842	51	791	35	756
OligoPicker	1766	18	1748	1745	3	842	657	185	141	44
CommOligo	1766	21	1745	1745	0	842	695	147	147	0

from phylogenetic and environmental studies are incomplete. Additionally, as little is known about the diversity that might be encountered in any given environmental sample, probe design should focus on the conserved, well-known, regions of a sequence. For example, when full-length sequences and shorter environmental sequences are used together, most algorithms will design what are thought to be unique probes for the full-length sequences outside of the conserved region used for environmental data.

For these reasons, a new probe design software tool called CommOligo is currently being developed and tested in our laboratory that will select optimal oligonucleotide probes for whole genomes, meta-genomes or groups of orthologous sequences such as those involved in FGAs. A multiple sequence alignment (MSA) approach is used to pre-process sequence data, and then users can choose regions for designing probes by masking based on the MSA results. The program then uses a new global alignment algorithm to design single or multiple unique probes for each gene and designate allowable parameters such as maximal similarity (default = 85%), maximal number of continuous match stretches (15 bases) and free energy (-30 kcal/mol) all of which can be controlled independently and simultaneously. The program is also able to design single or multiple group-specific probes for related groups of genes if it is not possible to select unique probes for a sequence. This new algorithm selects probes which have maximal similarities within a group and minimal similarities outside groups. Using defaults, group-specific probes should have a minimal similarity of 96% within a group and the same parameters as unique probes outside a group. Other filters, such as self-binding, mismatch position and GC content can also be used. While the above parameters are used as defaults, users may adjust all these parameters and values to meet their own needs. The program was evaluated using both whole genome and orthologous gene sequence data and compared with other software. For example, 147 specific (unique) probes were designed for the *nirS* and *nirK* sequences, and most importantly, this program did not choose any non-specific probes (Table 11.3). A group-specific probe algorithm has been evaluated using small to medium size data sets with documented phylogenetic relationships. For example, for *nirS* and *nirK* sequences, the program automatically formed 59 groups based on the default settings and each group had 2–30 sequences. Single or multiple probes could be designed for each group. Those group-specific probes covered an additional 180 sequences for which no unique probes were possible for single sequences because of their close similarities. For this particular data set, only 40% of sequences had unique or group-specific probes under default conditions. Relaxing design parameters will produce more probes but it also could potentially jeopardize the probe quality.

In summary, the application of oligonucleotide arrays for environmental studies presents many problems for probe design. To achieve the optimal specificity of oligonucleotide arrays, probe design criteria need to be further investigated and better algorithms are needed to facilitate these intensive computations. We are currently still in the pre-release stage of testing solutions for this problem presented in the discussion of CommOligo. In the meantime, great care must be taken when applying tools originally designed for whole genome data to the design of FGA probes.

◆◆◆◆◆◆ MICROARRAY CONSTRUCTION, LABELING, HYBRIDIZATION AND IMAGE ACQUISITION

Reproducibility is one of the most critical requirements for microarray fabrication. For reliable and reproducible data, the uniformity of spots across the entire array is crucial to simplify image analysis and enhance the accuracy of signal detection. While array construction (oligo synthesis and printing) of FGAs does not significantly differ from that of other types of arrays, several recommendations can be made. Various factors will affect the uniformity of spots including array substrate, slide quality, printing pins, printing buffer and environmental controls. For instance, significant variations could be caused by pin characteristics due to the mechanical difference in pin geometry, pin age and sample solutions. Additionally, the printing buffer is critical for obtaining homogeneous spots. Using saline sodium citrate (SSC) buffer, the spot homogeneity as well as binding efficiency is often poor, largely because of high evaporation rates. We and others have found more uniform spots can be obtained with the printing buffer containing 50% DMSO (dimethyl sulfoxide) and between 50 and 100 pmol/ul probe concentration (Hegde *et al.*, 2000; Diehl *et al.*, 2001; Wu *et al.*, 2001; Tiquia *et al.*, 2004). We use this in combination with aminopropyl silane coated glass slides (e.g. UltraGAPS, Corning, Corning, NY) and UV cross-linking at 200 mJ. In general, more cross-linking time or energy may bind oligos more strongly to slides, but may also interfere with proper hybridization. However, the individual slide manufacturer's recommendations for printing, cross-linking and pre-hybridization should always be consulted, as slight variations in slide chemistry and preparation procedure greatly affect these processes.

Protocol 1: Microarray Printing

1. Prepare printing oligo probe solution in a 384-well, v-bottom, printing plate. Final concentration will be 50–100 pmol/μl probe and 50% DMSO (generally 5 μl probe and 5 μl DMSO).
2. Cover the plate with plastic lid and mix in an orbital shaker at 700 rpm for 3 min.
3. Spin the printing plate using a centrifuge equipped with a rotor for microtiter plates at 500 rpm for 5 min.
4. Setup the array printer and software (we use a PixSys 5500 printer; Cartesian technologies, Inc. Irvine, CA). Print slides according to the manufacturer's protocol. The ideal relative humidity should be

between 40 and 60% at room temperature (20–25°C). The spot size should be approximately 100–150 μm, with 200–500 μm spacing distance using split pins from Telechem.
5. Allow the slides to dry for at least 2 h before proceeding to UV cross-linking and post-processing (according to slide manufacturer's protocol).

Total genomic DNAs are generally used as targets for functional gene studies. Thus, effective and repeatable DNA extraction from the environment is therefore a key step for FGA studies. We suggest several criteria for evaluating extraction methods following Hurt et al. (2001): (1) The nucleic acid recovery efficiency should be high and not biased so that the final nucleic acids are representative of the total nucleic acids within the naturally occurring microbial community. (2) The DNA should be of sufficient purity for reliable labeling and hybridization. (3) The extraction and purification protocol should be robust and repeatable. The DNA extraction and purification protocol described by Zhou et al. (1996) and modified for simultaneous DNA and mRNA extraction by Hurt et al. (2001), fulfill the above criteria for soils and sediments. However, other methods may be suitable or superior depending upon the sample type of interest.

Direct labeling procedures and PCR labeling amplifications with Cy3 or Cy5 fluorescent dye modified deoxynucleotides (dNTPs) based on Schena et al. (1995) are the most common labeling methods for whole genome array studies and have also been used successfully in environmental samples (Wu et al., 2001; Rhee et al., 2004). For DNA samples, direct labeling with random primers and Klenow fragment DNA polymerase I is widely used. Given the current sensitivity limits for detection (Cho and Tiedje, 2002) and the diverse nature of microbial communities, the likelihood of detecting genes present in lower numbers will increase with the amount of DNA template used for hybridization. We are routinely able to efficiently label 2–5 μg of target DNA using the methods outlined below. Targets can also be labeled via PCR using gene-specific primers, however, this method introduces biases inherent in such procedures and is used most often for detection of specific targets that might be of low number in an environmental sample or for validation of probe specificity.

Protocol 2: Direct Community DNA Labeling Procedure

1. In a 0.2 ml PCR tube combine:
   (a) 2–5 μg[1] purified community DNA (in 10 μl nuclease-free water).
   (b) 20 μl (750 ng/μl) random octamer primers (Invitrogen # Y01393).
2. Mix well and denature at 99.9°C for 5 min.
3. Place immediately on ice.

[1] The DNA template amount will vary, but the higher the amount used, the higher the likelihood of detecting genes present in low numbers. Additional positive control templates may be added as well to the same labeling reactions.

4. In a 1.5 ml microcentrifuge tube, combine:
   (a) 2.5 µl dNTP's (5 mM dATP, dTTP, dGTP and 2.5 mM dCTP).
   (b) 1 µl (1 mM) Cy3 or Cy5 dCTP.
   (c) 1.5 µl (40 U/µl) Klenow fragment (Invitrogen # Y01396).
   (d) 1.25 µl DTT (Invitrogen # Y00147).
   (e) 13.75 µl DNase- and RNase-free water.
5. Add this mixture to the 0.2 ml PCR from step 1 (total volume = 50 µl).
6. Mix well and centrifuge the mixture briefly at maximum speed.
7. Incubate at 37°C for 6 h or overnight.
8. Purify labeled target DNA using QIAquick PCR purification columns according to the manufacturer's instructions (Qiagen, Valencia, CA).
9. Quantify labeling efficiency as below.

Protocol 3: Quantifying Labeling Efficiency of Cy-Labeled DNA Targets

1. Use a spectrophotometer to quantify the OD at 550 for Cy3 and OD 650 for Cy5. Also, measure OD at 230, 260 and 280 to assess purity. This can be done by using only 1 µl of the labeled DNA and a NanoDrop™ ND-1000 spectrophotometer (NanoDrop Technologies, Inc., Montchanin, DE) or equivalent.
2. Calculate the amount of DNA as well as the specific activity of the labeled DNA. The specific activity is calculated as follows:

$$\text{Specific activity} = \frac{\text{amount of target DNA} \times 1000}{\text{pmole of dye incorporated} \times 324.5}$$

3. Dry in vacuum centrifuge (45°C) for 1 h. Do not use higher heat levels or heat lamps to accelerate evaporation as the fluorescent dyes could be degraded.

Temperature, concentration of formamide and the volume of the hybridization mixture are critically important parameters for all microarray hybridizations. Temperature and formamide concentration together control the specificity of the resulting hybridization. While this is true of all microarray hybridizations, this can be especially important in FGAs as many orthologous and highly similar genes may be present in any given sample. Uneven hybridizations resulting from fluctuations in volume across the array can also result in spurious signal strengths and strong backgrounds. However, in FGA studies where detection limits are always an issue, it is highly desirable to minimize the volume of hybridization solution. We utilize a 22 × 22 mm glass LifterSlip cover slip (Erie Scientific, Portsmouth, NH) that allows even hybridizations with as little as 15 µl of hybridization solution. These methods and materials are not the only ones available; other procedures have been used successfully by different researchers. We present several protocols here for illustration of the steps involved based on those currently employed in our laboratory.

Protocol 4: Hybridization and Washing

Buffer	Volume (μl)	Final concentration
Nuclease-free water	3.3	
Formamide	7.5	50%
20× SSC	2.5	3.33 ×
10% SDS	0.5	0.33%
Herring sperm DNA (10 mg/ml)	1.2	10.2 μg

1. Preheat microarray slide in hybridization chamber (Corning #2551) for 20 min at 50°C.
2. Resuspend sample in hybridization solution, spin down and heat at 95°C for at least 5 min in a thermocycler.
3. Dispense 15 μl of 3× SSC solution into the chamber hydration wells.
4. Deposit the hybridization (15 μl) solution directly onto the immobilized DNA probes and place a cover slip over the array, avoid bubble formation.
5. Close the hybridization chamber and ensure a proper seal is formed.
6. Incubate the chamber in a 50°C water bath for 12–15 h (overnight).

Post-hybridization wash

1. Place the slides, with the coverslips still affixed, in a pre-warmed washing buffer I (2× SSC and 0.1% SDS) and allow the coverslips to fall from the slide.
2. Place the slides in a pre-warmed washing buffer I (2× SSC and 0.1% SDS) and wash for 5 min with gentle shaking. Repeat this wash once.
3. Place the slides into fresh buffer II (0.1× SSC and 0.1% SDS) at ambient temperature for 5 min. Repeat this wash once.
4. Place the slides in buffer III (0.1× SSC) at ambient temperature for 1 min. Repeat wash four times.
5. Transfer the slides to a slide rack and immediately spin the slides dry at 600 rpm for 5 min in a centrifuge with a horizontal rotor for microtiter plates. As evaporation can be quite rapid, it is suggested that the slide be placed in the centrifuge immediately upon removal from the jar to avoid residual salt deposition.
6. Slides should be stored in the dark until ready for scanning.

Notes. We found that non-specific hybridization could be significant when the microarray slides were not warmed or the hybridization mixture remained at the room temperature for several minutes after hybridization and prior to washing. To minimize potential non-specific hybridization, the slides should be pre-warmed and the hybridization mixture should be kept above the hybridization temperature through all hybridization steps prior to washing. The above post-washing procedure is based on that of Corning and may vary by slide manufacturer, but we have found it is critical to proceed immediately to the first wash step when slides are removed from the chamber.

Microarray image processing is critical to control signal variation due to high background and weak signals, and to remove false positive signals (Schuchhardt *et al.*, 2000). One of the critical steps in the analysis of microarray image data is filtering noise versus true signals. Problematic false signals introduced by impurities in the arrays can be identified and removed by visual examination of each spot. This is particularly a critical procedure for reducing misinterpretations in the final results. With the current sensitivity limits of environmental FGAs, it is most often required to use very high laser power and photomultiplier tube (PMT) settings for detection of microarray signals. Compared to whole genome microarrays, that are usually scanned at much lower settings, one may thus find very high levels of background for FGAs. Because of this, it is important to prevent impurities from being introduced to the process and to examine all spots critically to distinguish real signals and false signals (Figure 11.1). This can most easily be done after the image is imported to software such as ImaGene™ for spot identification and quantification, and before proceeding to data analysis. The image in Figure 11.2 shows typical hybridization results for an FGA of nitrogen and sulfur cycle genes (Laser power 100%, PMT gain 95%).

Protocol 5: Image Acquisition and Processing

1. Scan the slide initially at a low resolution of 50 μm to obtain a quick display image and then finally at 5–10 μm using for instance the ScanArray 5000 System (GSI Lumonics, Watertown, MA). The emitted fluorescent signal is detected by a photomultiplier tube (PMT) at 570 nm (Cy3) or 670 nm (Cy5). The percentages of laser power and PMT used should be appropriately selected based on hybridization signal intensity observed in the low resolution scan so that the signals for most of the spots are not saturated.
2. Save the scanned display as a 16-bit TIFF and BMP file and quantify the intensity of each spot using ImaGene™ (BioDiscovery, Los Angeles, CA) or equivalent.
3. Assess spot quality and reliability, and perform background subtraction of the microarray data.

Notes. Besides ImaGene software, there are other software packages available for image processing, spot identification, quantitation and normalization. These imaging include GenPix Pro (Axon Instruments, Union City, CA), Array Pro (Media Cybernetic, Carlsbad, CA), Quant Array (Packard Biosciences, Boston, MA) and TIGR Spot Finder (The Institute of Genomic Research TIGR, Rockville, MD).

◆◆◆◆◆◆ DATA ANALYSIS TECHNIQUES

Data analysis techniques for FGAs should allow for the detection of genes that are significantly different between samples. However, FGAs pose

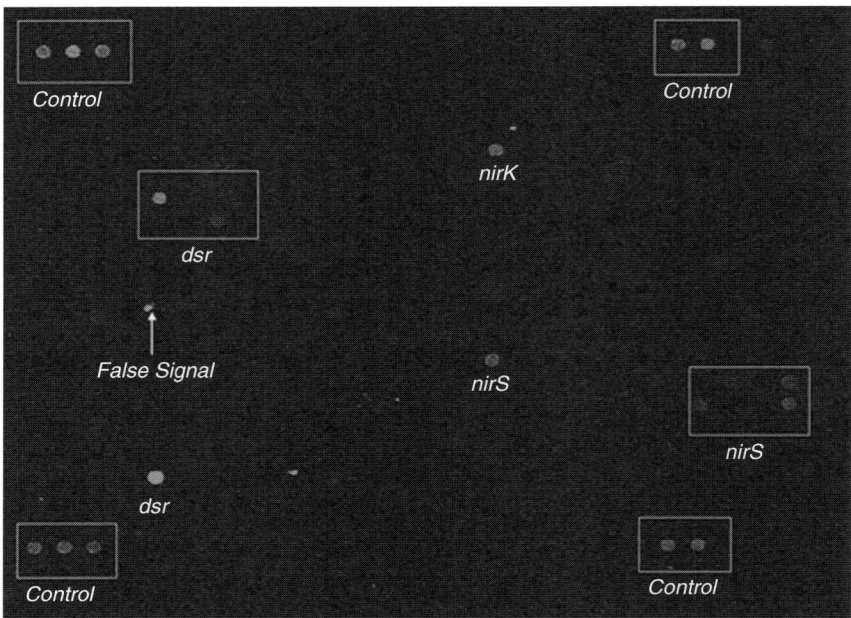

Figure 11.2. 50mer oligonucleotide microarray hybridization with a marine sediment genomic DNA sample. 3 ug of extracted marine sediment DNA was labeled Cy5 using methods outlined in the protocols outlined in this chapter. A small portion of an array image is shown to illustrate typical FGA images and true versus false signals. (See colour plate 27.)

a special challenge for environmental analysis since these samples often contain genomic DNA from a highly diverse range of organisms for which most of the genomic information is not known or knowable. Besides uncertain quantification of whole genome microarrays due to the inherently high variation associated with array fabrication, probe labeling, hybridization and image processing (Beißbarth et al., 2000; Zhou and Thompson, 2002), the targeted genes in FGAs are diluted in a complex pool of genomic DNA from both target and non-target organisms. Thus, potential target genes make up only an extremely small portion of the overall DNA in any given sample and thus sensitivity is a major issue in this kind of microarray analysis.

There are several software programs available, which deal with the task of data preparation and subsequently can be used for statistical analysis and evaluation (e.g. *ArrayStat*, Imaging Research, Inc., Ontario, Canada; *Cluster* 3.0, Human Genome Center, University of Tokyo, Japan; *GenePix Pro*, Axon, Foster City, CA; *GeneSpring*, Silicon Genetics, Redwood City, CA). In our lab, we routinely use ImaGene 5.5.4 (BioDiscovery, El Segundo, CA). Unlike the analysis of expression data using whole genome microarrays, FGAs focusing on environmental data are hampered by a low amount of positive spots for the reasons stated above. Therefore, a good allocation of positive and negative spots might be one of the first challenges for this type of microarray. If too few positive spots are present, placing the grid correctly can be very time consuming. We suggest users consider adding several positive control spots to each grid sector of a printed array. This allows for better spot locating ability as well as an easy way to implement localized quality control measures.

Because all raw microarray data, and especially FGAs, are affected by unspecific background binding of labeled DNA to the slides, correction for background signal intensity of the raw data is necessary. To perform this task, usually the mean or median value of the local background intensity of a spot is subtracted from the image intensities of the spot. Generally speaking, the application of median values is preferable because extreme outliers are not taken into calculations which could otherwise falsify the real value (Beißbarth *et al.*, 2000). However, both approaches have been used to process microarray data. The subtraction of local background data is especially imperative, whenever uneven background staining makes the usage of a global background values impossible. Other methods of background correction are also possible: Fluorescence intensity may also be corrected from empty spots (array positions that do not contain any DNA) and negative control spots (e.g. probes targeting human genes in soil or groundwater samples). To distinguish between background and real hybridizations, signal-to-noise ratios (SNR) are calculated and only those spots above a certain threshold (usually SNR ≥ 2 or 3) are considered as positive hybridizations. The SNR can be calculated as follows:

$$\text{SNR} = \frac{\text{signal mean} - \text{background mean}}{\text{background standard deviation}}.$$

However, the method of calculation sometimes varies between authors and thus the method must always be included when data are presented.

Meaningful interpretation and comparisons of microarray data require standardization of the measured raw signal intensities. Variability due to pre-hybridization handling and uneven labeling efficiency can be accounted for by the analysis of replicate microarray slides for each sample. However, if this does not lead to adequate information, raw data may be corrected by dividing their intensity values by negative control spots (Dennis *et al.*, 2003). If two or more samples are compared in two-color experiments, standardization of all data by calculating the ratio of cy5 labeled samples and a cy3 labeled control is always necessary. As mentioned above, in contrast to whole genome arrays, where overall expression levels between two differently labeled samples are compared, only a small fraction of genes can often be detected when environmental samples are hybridized to FGAs, thus quantitative comparisons (like generation of ratios and relative abundances of certain genes) in this type of microarray can only be generated for genes detectable in both samples (Talaat *et al.*, 2002). As a consequence, many standard normalization procedures are not applicable for this type of microarray analysis and the user should use great care in selecting such automated procedures.

In most cases, two or more replicates of each probe are printed on a microarray. While many software programs offer automatic calculation of the accordant means, data concerning the overall variability might be lost and thus should rather be handled separately (Beißbarth *et al.*, 2000). Microarray data is not in most cases normally distributed. Logarithmic transformation ($\log_{10}$ or $\log_2$) is needed and the transformed values will thus largely reflect the degree of abundance or expression increase of a treated sample versus the control (Dennis *et al.*, 2003). More information concerning normalization and data transformations that goes beyond the scope of this chapter may be found at Zar (1999). However, new approaches for normalization and interpretation of microarray data are constantly being considered (Piétu *et al.*, 1996; Chen *et al.*, 1996; Richmond *et al.*, 1999; Beißbarth *et al.*, 2000; Dudley *et al.*, 2002; Talaat *et al.*, 2002), but the user must be aware that most have not been fully evaluated for application to FGAs and their inherently different problems.

After processing the raw data, several statistical tests can be employed to discover prominent genes within specific samples or relationships between different samples. This can be done by describing similarities and distances of data sets by methods such as cluster analysis, principal component analysis (PCA) or with the aid of self-organizing maps (SOM). Similarity comparisons are the most obvious methods to compare microarray data. They can be calculated either from Euclidean distances, which calculate the absolute distances between two data points in space, or as Pearson correlations, which are insensitive to the amplitude of the signal intensity and as a result are most often used. In PCA, a multi-variant table with P columns in a P-dimensional (Euclidean) space is reduced to the two or three most representative dimensions. The first dimension explains as many of the differences in the data sets as possible and the second dimension as many as possible of what cannot be explained by the first one, and so on (Gilbert *et al.*, 2000). This approach has been widely employed to analyze microarray data for gene expression analysis (Hilsenbeck *et al.*, 1999; Thomas *et al.*, 2001; Spanakis and Brouty-Boyé, 1997).

Clustering analyses apply one of the above-mentioned similarity measurements to groups of genes with similar expression profiles, in order to detect clusters of genes that are presumably involved in a common process or that respond to a given treatment (Eisen *et al.*, 1998). A data set is divided in several subsets on the basis of their similarities. In hierarchical clustering, first the most closely related data are combined to form a cluster. Subsequently, the next cluster will be formed by a subset of two other data possibly including the already formed clusters. As a result, the data are combined to form a phylogenetic tree, the branch lengths of which represent the degree of similarity between the sets. However, several clustering methods exist, which might lead to slightly different interpretations (Gilbert *et al.*, 2000). Applied to FGAs, this type of analysis helps to identify similarities and differences in the genes present in the microbial communities at different sites as well as specific changes that might be due to a specific experimental treatment, etc. Such differences in clustering may indicate presumed differences in the function of interest, for example, nitrate reduction. SOM are a kind of artificial neural networks which can be considered as a type of mathematical cluster analysis. Data are iteratively relocated leading to adjoining clusters with high similarities and more distant clusters further apart. Similar patterns will occur as neighbors in SOM (Tamayo *et al.*, 1999).

◆◆◆◆◆◆ EVALUATION AND VALIDATION OF FGA RESULTS

Probes should be highly specific for the target gene in order to prevent cross-hybridization from similar environmental sequences which could lead to inaccurate results (Rhee *et al.*, 2004; Wu *et al.*, 2001). Specificity can be assessed by test hybridizations using pure culture genomic DNAs and PCR amplified genes, or by synthetic oligonucleotides. While templates may not always be available for testing with genomic DNA or PCR product, synthetic oligonucleotides can be synthesized to test the specificity of any probe. Additionally, more comprehensive testing can be achieved with oligos, as mismatches can be designed in any number and at any position along the probe template hybrid. Using such methods, 50mer FGAs tested in our laboratory have been shown to be specific when % homology is not $>85-88\%$ (Tiquia *et al.*, 2004; Rhee *et al.*, 2004). However, the free energy of potential probe target hybrids is possibly a better predictor of specificity, and with the techniques outlined here we have found that probe-hybrid combinations with ΔG values of >-30 kcal/mol were very specific (Rhee *et al.*, 2004; Liebich, unpublished data). However, as in sensitivity measurements, these values are dependent upon the specifics of the protocols in use, and have to be empirically determined for each study.

Evaluation of the sensitivity (e.g. lower detection limit) and the quantitative relationship between hybridization strength and DNA amount are critical for understanding the results of microarray-based approaches for detecting genes in environmental samples. Dilutions of pure culture genomic DNA hybridized against corresponding probes provide a rather straightforward approach for measuring sensitivity and evaluating the quantitative nature of FGAs. However, unlike whole genome arrays, detection limits for FGA and other environmental microarrays must account for the effects of heterogeneous non-target DNA sequences characteristic of environmental samples (Cho and Tiedje, 2002; Rhee *et al.*, 2004). Cho and Tiedje (2002) proposed that the detection limit of a PCR fragment of the denitrification gene *nirS* in an environmental sample was ~ 10 pg. This means only genes of organisms with a total DNA amount ~ 50 ng are detectable (assuming a 4 Mbp average genome size). In their experiments, 1 µg of total environmental DNA was analyzed, suggesting that a particular organism containing *nirS* must contribute at least 1/20 of the applied DNA amount to be detectable. However, increasing the amount of environmental DNA applied and other methods can improve these detection limitations. Using the same protocols suggested in this chapter, Rhee *et al.* (2004) estimated that several genes involved in biodegradation of naphthalene could be detected with 5 ng genomic DNA in the absence of background DNA. However, detection limits were 50–100 ng of pure culture genomic DNA when diluted in a heterogeneous background of *Shewanella oneidensis*. In this case, however, hybridizations were carried out with 5 µg of total sample DNA, suggesting the FGA could detect cells present at a level of about 1/50–1/100 of the total. After logarithmic transformation, the relationship between signal intensity and applied DNA amount is most often linear to at least 1 µg of target (Wu *et al.*, 2001; Cho and Tiedje, 2002; Tiquia *et al.*, 2004; Rhee *et al.*, 2004). However, the quantitative nature of the relationship is dependent upon the specifics of the protocols in use, and has to be empirically determined for each study. Ideally, quantitative controls should be introduced to every slide using a series of control spots and corresponding control DNA that is co-labeled with each sample (Chen *et al.*, 1996; Dudley *et al.*, 2002).

While FGAs have the potential to rapidly quantify thousands of different DNA/RNA sequences in environmental samples simultaneously (Cho and Tiedje, 2002; Dennis *et al.*, 2003; Rhee *et al.*, 2004; Tiquia *et al.*, 2004; Wu *et al.*, 2001), it may be useful to validate selected results using other techniques such as quantitative PCR (qPCR). For example, Rhee *et al.* (2004) used a real-time PCR-based qPCR approach to verify FGA data obtained from a PAH contaminated soil. The FGA analysis had indicated the presence of numerous organic contaminant degradation genes in the soil including several from the naphthalene catabolic pathway. The researchers designed primers for six of these naphthalene genes, four of which generated single PCR products from the soil DNA and were subsequently used for qPCR. The qPCR results corroborated the FGA data for each of the four genes by demonstrating significant correlations between the gene copy number and the FGA hybridization signals ($r^2 = 0.74$ for all genes and 0.96 for genes with SNR > 3). This approach for validation, however, is most useful when probes target relatively unique genes. When data originating from numerous orthologous gene sequences are used for probe design, the specific primers necessary for qPCR may not be possible to design.

Acknowledgements

The authors would like to thank Sonia Tiquia, Sung K. Rhee, and Liyou Wu for their pioneering work in the initial development of many of the protocols and ideas contained in this paper. The authors' efforts in preparing this chapter were supported by the US DOE Office of Science as part of its Biological and Environmental Research Programs in Natural and Accelerated Bioremediation Research, Genomes To Life, Biotechnology Investigations-Ocean Margins, and Carbon Sequestration (as part of the consortium on research to enhance Carbon Sequestration in Terrestrial Ecosystems-CSiTE). Jost Liebich's work was also supported by the postdoctoral program of the German Academic Exchange Service [DAAD]. Oak Ridge National Laboratory is managed by UT-Battelle, LLC, for the US Department of Energy under contract DE-AC05-00OR22725.

References

Allen, A., Booth, M. G., Frischer, M. E., Verity, P. G., Zehr, J. P. and Zani, S. (2001). Diversity and detection of nitrate assimilation genes in marine bacteria. *Appl. Environ. Microbiol.* **67**, 5343–5348.

Altschul, S., Madden, T., Schäffer, A., Zhang, J., Miller, W. and Lipman, D. (1997). Gapped BLAST and PSI-BLAST: a new generation of protein database search programs. *Nucleic Acids Res.* **25**, 3389–3402.

Andreoni, V., Bernasconi, S., Colombo, M., van Beilen, J. B. and Cavalca, L. (2000). Detection of genes for alkane and naphthalene catabolism in *Rhodococcus* sp. strain 1BN. *Environ. Microbiol.* **2**, 572–577.

Baldwin, B. R., Nakatsu, C. H. and Nies, L. (2003). Detection and enumeration of aromatic oxygenase genes by multiplex and real-time PCR. *Appl. Environ. Microbiol.* **69**, 3350–3358.

Barkay, T., Miller, S. M. and Summers, A. O. (2003). Bacterial mercury resistance from atoms to ecosystems. *FEMS Microbiol. Rev.* **27**, 355–384.

Beaulieu, M., Becaert, V., Deschenes, L. and Villemur, R. (2000). Evolution of bacterial diversity during enrichment of PCP-degrading activated soils. *Microb. Ecol.* **40**, 345–356.

Beißbarth, T., Fellenberg, K., Brors, B., Arribas-Prat, R., Boer, J. M., Hauser, N. C., Scheideler, M., Hoheisel, J. D., Schütz, G., Poustka, A. and Virgon, M. (2000). Processing and quality control of DNA array hybridization data. *Bioinformatics* **16**, 1014–1022.

Bodrossy, L., Stralis-Pavese, N., Jurrell, J. C., Radejewski, S., Weilharter, A. and Sessitsch, A. (2003). Development and validation of a diagnostic microbial microarray for methanotrophs. *Environ. Microbiol.* **5**, 566–582.

Bogdanova, E. S., Bass, I. A., Minakhin, L. S., Petrova, M. A., Mindlin, S. Z., Volodin, A. A., Kalyaeva, E. S., Tiedje, J. M., Hobman, J. L., Brown, N. L. and Nikiforov, V. G. (1998). Horizontal spread of *mer* operons among Gram-positive bacteria in natural environments. *Microbiology* **144**, 609–620.

Bosch, R., Garcia-Valdes, E. and Moore, E. R. B. (1999). Genetic characterization and evolutionary implications of a chromosomally encoded naphthalene-degradation upper pathway from *Pseudomonas stutzeri* AN10. *Gene* **236**, 149–157.

Braker, G. and Tiedje, J. M. (2003). Nitric oxide reductase (norB) genes from pure cultures and environmental samples. *Appl. Environ. Microbiol.* **69**, 3476–3483.

Bruce, K. D., Osborn, A. M., Pearson, A. J., Strike, P. and Ritchie, D. A. (1995). Genetic diversity within *mer* genes directly amplified from communities of noncultivated soil and sediment bacteria. *Mol. Ecol.* **4**, 605–612.

Cassidy, S. L. (2002). Microbial activity and biodiversity as indicators of hydrocarbon bioremediation. Thesis. University of London, United Kingdom.

Castro, H., Reddy, K. R. and Ogram, A. (2002). Composition and function of sulfate-reducing prokaryotes in eutrophic and pristine areas of the Florida Everglades. *Appl. Environ. Microbiol.* **68**, 6129–6137.

Chang, Y. J., Peacock, A. D., Long, P. E., Stephen, J. R., McKinley, J. P., Macnaughton, S. J., Hussain, A. K., Saxton, A. M. and White, D. C. (2001). Diversity and characterization of sulfate-reducing bacteria in groundwater at a uranium mill tailings site. *Appl. Environ. Microbiol.* **67**, 3149–3160.

Chen, Y., Dougherty, E. R. and Bittner, M. (1996). Ratio-based decisions and the quantitative analysis of cDNA microarray images. *J. Biomed. Opt.* **2**, 364–374.

Chien, Y. T. and Zinder, S. H. (1994). Cloning, DNA sequencing, and characterization of a nifD-homologous gene from the archaeon *Methanosarcina barkeri* 227 which resembles nifD from the eubacterium *Clostridium pasteurianum*. *J. Bacteriol.* **176**, 6590–6598.

Cho, J. C. and Tiedje, J. M. (2002). Quantitative detection of microbial genes by using DNA microarrays. *Appl. Environ. Microbiol.* **68**, 1425–1430.

Dean, D. R., Bolin, J. T. and Zheng, L. M. (1993). Nitrogenase metalloclusters: structures, organization, and synthesis. *J. Bacteriol.* **175**, 6737–6744.

Denef, V. J., Park, J., Rodrigues, J. L. M., Tsoi, T. V., Hashsham, S. A. and Tiedje, J. M. (2003). Validation of a more sensitive method for using spotted oligonucleotide DNA microarrays for functional genomics studies on bacterial communities. *Environ. Microbiol.* **5**, 933–943.

Dennis, P., Edwards, E. A., Liss, S. N. and Fulthorpe, R. (2003). Monitoring gene expression in mixed microbial communities by using DNA microarrays. *Appl. Environ. Microbiol.* **69**, 769–778.

Denome, S. A., Stanley, D. C., Olson, E. S. and Young, K. D. (1993). Metabolism of dibenzothiophene and naphthalene in *Pseudomonas* strains – complete DNA-sequence of an upper naphthalene catabolic pathway. *J. Bacteriol.* **175**, 6890–6901.

Deplancke, B., Hristova, K. R., Oakley, H. A., McCracken, V. J., Aminov, R., Mackie, R. I. and Gaskins, H. R. (2000). Molecular ecological analysis of the succession and diversity of sulfate-reducing bacteria in the mouse gastrointestinal tract. *Appl. Environ. Microbiol.* **66**, 2166–2174.

DeRisi, J. L., Iyer, V. R. and Brown, P. O. (1997). Exploring the metabolic and genetic control of gene expression on a genomic scale. *Science* **278**, 680–686.

Diehl, F., Grahlmann, S., Beier, M. and Hoheisel, J. D. (2001). Manufacturing DNA microarrays of high spot homogeneity and reduced background signal. *Nucleic Acids Res.* **29**, E38.

Duarte, G. F., Rosado, A. S., Seldin, L., de Araujo, W. and van Elsas, J. D. (2001). Analysis of bacterial community structure in sulfurous-oil-containing soils and detection of species carrying dibenzothiophene desulfurization (*dsz*) genes. *Appl. Environ. Microbiol.* **67**, 1052–1062.

Dubilier, N., Mulders, C., Ferdelman, T., de Beer, D., Pernthaler, A., Klein, M., Wagner, M., Erseus, C., Thiermann, F., Krieger, J., Giere, O. and Amann, R. (2001). Endosymbiotic sulphate-reducing and sulphide-oxidizing bacteria in an oligochaete worm. *Nature* **411**, 298–302.

Dudley, A. M., Aach, J., Steffen, M. A. and Church, G. M. (2002). Measuring absolute expression with microarrays with a calibrated reference sample and an extended signal intensity range. *Proc. Natl Acad. Sci.* **99**, 7554–7559.

Dunfield, P. F., Liesack, W., Henckel, T., Knowles, R. and Conrad, R. (1999). High-affinity methane oxidation by a soil enrichment culture containing a type II methanotroph. *Appl. Environ. Microbiol.* **65**, 1009–1014.

Eisen, M. B., Spellman, P. T., Brown, P. O. and Botstein, D. (1998). Cluster analysis and display of genome-wide expression patterns. *Proc. Natl Acad. Sci.* **95**, 14863–14868.

Elsaied, H. and Naganuma, T. (2001). Phylogenetic diversity of ribulose-1, 5-bisphosphate carboxylase/oxygenase large-subunit genes from deep-sea microorganisms. *Appl. Environ. Microbiol.* **67**, 1751–1765.

Emrich, S. J., Lowe, M. and Delcher, A. (2003). PROBEmer: a web-based software tool for selecting optimal DNA oligos. *Nucleic Acids Res.* **31**, 3746–3750.

Felske, A. D. M., Fehr, W., Pauling, B. V., von Canstein, H. and Wagner-Dobler, I. (2003). Functional profiling of mercuric reductase (*mer* A) genes in biofilm communities of a technical scale biocatalyzer. *BMC Microbiol.* **3**, 22, (published online- http://www.biomedcentral.com/1471-2180/3/22).

Ferrero, M., Llobet-Brossa, E., Lalucat, J., Garcia-Valdes, E., Rossello-Mora, R. and Bosch, R. (2002). Coexistence of two distinct copies of naphthalene degradation genes in *Pseudomonas* strains isolated from the western Mediterranean region. *Appl. Environ. Microbiol.* **68**, 957–962.

Finegold, S. M. and Jousimies-Somer, H. (1997). Recently described clinically important anaerobic bacteria: medical aspects. *Clin. Infect. Dis.* **25**(Suppl. 2), S88–S93.

Friedrich, M. W. (2002). Phylogenetic analysis reveals multiple lateral transfers of adenosine-5′-phosphosulfate reductase genes among sulfate-reducing microorganisms. *J. Bacteriol.* **184**, 278–289.

Fuenmayor, S. L., Wild, M., Boyes, A. L. and Williams, P. A. (1998). A gene cluster encoding steps in conversion of naphthalene to gentisate in *Pseudomonas* sp. strain U2. *J. Bacteriol.* **180**, 2522–2530.

Furuchi, T., Hwang, G. W. and Naganuma, A. (2002). Overexpression of the ubiquitin-conjugating enzyme Cdc34 confers resistance to methylmercury in *Saccharomyces cerevisiae*. *Mol. Pharmacol.* **61**, 738–741.

Geiselbrecht, A. D., Hedlund, B. P., Tichi, M. A. and Staley, J. T. (1998). Isolation of marine polycyclic aromatic hydrocarbon (PAH)-degrading *Cycloclasticus* strains from the Gulf of Mexico and comparison of their PAH degradation ability with that of Puget Sound *Cycloclasticus* strains. *Appl. Environ. Microbiol.* **64**, 4703–4710.

Gilbert, D. R., Schroeder, M. and van Helden, J. (2000). Interactive visualization and exploration of relationships between biological objects. *Trends Biotechnol.* **18**, 487–494.

Gueldry, O., Lazard, M., Delort, F., Dauplais, M., Grigoras, I., Blanquet, S. and Plateau, P. (2003). Ycf1p-dependent Hg(II) detoxification in *Saccharomyces cerevisiae*. *Eur. J. Biochem.* **270**, 2486–2496.

Habe, H. and Omori, T. (2003). Genetics of polycyclic aromatic hydrocarbon metabolism in diverse aerobic bacteria. *Biosci. Biotechnol. Biochem.* **67**, 225–243.

Hallam, S. J., Girguis, P. R., Preston, C. M., Richardson, P. M. and DeLong, E. F. (2003). Identification of methyl coenzyme M reductase A (mcrA) genes associated with methane-oxidizing archaea. *Appl. Environ. Microbiol.* **69**, 5483–5491.

Hamann, C., Hegemann, J. and Hildebrandt, A. (1999). Detection of polycyclic aromatic hydrocarbon degradation genes in different soil bacteria by polymerase chain reaction and DNA hybridization. *FEMS Microbiol. Lett.* **173**, 255–263.

Hart, M. C., Elliott, G. N., Osborn, A. M., Ritchie, D. A. and Strike, P. (1998). Diversity amongst *Bacillus mer A* genes amplified from mercury resistant isolates and directly from mercury polluted soil. *FEMS Microbiol. Ecol.* **27**, 73–84.

Hedlund, B. P., Geiselbrecht, A. D., Bair, T. J. and Staley, J. T. (1999). Polycyclic aromatic hydrocarbon degradation by a new marine bacterium, *Neptunomonas naphthovorans* gen. nov., sp. nov. *Appl. Environ. Microbiol.* **65**, 251–259.

Hedlund, B. P., Geiselbrecht, A. D. and Staley, J. T. (2001). *Marinobacter* strain NCE312 has a *Pseudomonas*-like naphthalene dioxygenase. *FEMS Microbiol. Lett.* **201**, 47–51.

Hegde, P., Qi, R., Abernathy, K., Gay, C., Dharap, S., Gaspard, R., Hughes, J. E., Snesrud, E., Lee, N. and Quackenbush, J. (2000). A concise guide to cDNA microarray analysis. *BioTechniques* **29**, 548–560.

Herrick, J. B., StuartKeil, K. G., Ghiorse, W. C. and Madsen, E. L. (1997). Natural horizontal transfer of a naphthalene dioxygenase gene between bacteria native to a coal tar-contaminated field site. *Appl. Environ. Microbiol.* **63**, 2330–2337.

Hilsenbeck, S. G., Friedrichs, W. E., Schiff, R., O'Connell, P., Hansen, R. K., Osborne, C. K. and Fuqua, S. A. (1999). Statistical analysis of array expression data as applied to the problem of tamoxifen resistance. *J. Natl Cancer Inst.* **91**, 53–459.

Holliger, C., Hahn, D., Harmsen, H., Ludwig, W., Schumacher, W., Tindall, B., Vazquez, F., Weiss, N. and Zehnder, A. J. (1998). *Dehalobacter restrictus* gen. nov. and sp. nov., a strictly anaerobic bacterium that reductively dechlorinates tetra- and trichloroethene in an anaerobic respiration. *Arch. Microbiol.* **169**, 313–321.

Holmes, A. J., Costello, A., Lidstrom, M. E. and Murrell, J. C. (1995). Evidence that participate methane monooxygenase and ammonia monooxygenase may be evolutionarily related. *FEMS Microbiol. Lett.* **132**, 203–208.

Horz, H. P., Yimga, M. T. and Liesack, W. (2001). Detection of methanotroph diversity on roots of submerged rice plants by molecular retrieval of *pmoA*, *mmoX*, *mxaF*, and 16S rRNA and ribosomal DNA including *pmoA*-based terminal restriction fragment length polymorphism profiling. *Appl. Environ. Microbiol.* **67**, 4177–4185.

Howard, J. B. and Rees, D. C. (1996). Structural basis of biological nitrogen fixation. *Chem. Rev.* **96**, 2965–2982.

Huber, H., Jannasch, H., Rachel, R., Fuchs, T. and Stetter, K. O. (1997). *Archaeoglobus veneficus* sp. nov., a novel facultative chemolithoautotrophic hyperthermophilic sulfite reducer, isolated from abyssal black smokers. *Syst. Appl. Microbiol.* **20**, 374–380.

Hurek, T., Egener, T. and Reinhold-Hurek, B. (1997). Divergence in nitrogenases of *Azoarcus* spp., Proteobacteria of the beta subclass. *J. Bacteriol.* **179**, 4172–4178.

Hurt, R. A., Qui, X., Wu, L., Roh, Y., Palumbo, A. V., Tiedje, J. M. and Zhou, J. (2001). Simultaneous recovery of RNA and DNA from soils and sediments. *Appl. Environ. Microbiol.* **67**, 4495–4503.

Jeon, C. O., Park, W., Padmanabhan, P., DeRito, C., Snape, J. R. and Madsen, E. L. (2003). Discovery of a bacterium, with distinctive dioxygenase, that is responsible for in situ biodegradation in contaminated sediment. *Proc. Natl Acad. Sci. USA* **100**, 13591–13596.

Jørgensen, B. B. (1982). Mineralization of organic-matter in the sea bed – the role of sulfate reduction. *Nature* **296**, 643–645.

Karkhoff-Schweizer, R. R., Huber, D. P. and Voordouw, G. (1995). Conservation of the genes for dissimilatory sulfite reductase from *Desulfovibrio vulgaris* and *Archaeoglobus fulgidus* allows their detection by PCR. *Appl. Environ. Microbiol.* **61**, 290–296.

Kasai, Y., Shindo, K., Harayama, S. and Misawa, N. (2003). Molecular characterization and substrate preference of a polycyclic aromatic hydrocarbon dioxygenase from *Cycloclasticus* sp. strain A5. *Appl. Environ. Microbiol.* **69**, 6688–6697.

Kauppi, B., Lee, K., Carredano, E., Parales, R. E., Gibson, D. T., Eklund, H. and Ramaswamy, S. (1998). Structure of an aromatic-ring-hydroxylating dioxygenase-naphthalene 1,2-dioxygenase. *Struct. Fold. Des.* **6**, 571–586.

Kholodii, G., Yurieva, O., Mindlin, S., Gorlenko, Z., Rybochkin, V. and Nikiforov, V. (2000). Tn*5044*, a novel Tn3 family transposon coding for temperature-sensitive mercury resistance. *Res. Microbiol.* **151**, 291–302.

Kirshtein, J. D., Paerl, H. W. and Zehy, J. (1991). Amplification, cloning, and sequencing of a nifH segment from aquatic microorganisms and natural communities. *Appl. Environ. Microbiol.* **57**, 2645–2650.

Kleikemper, J., Schroth, M. H., Sigler, W. V., Schmucki, M., Bernasconi, S. M. and Zeyer, J. (2002). Activity and diversity of sulfate-reducing bacteria in a petroleum hydrocarbon-contaminated aquifer. *Appl. Environ. Microbiol.* **68**, 1516–1523.

Klein, M., Friedrich, M., Roger, A. J., Hugenholtz, P., Fishbain, S., Abicht, H., Blackall, L. L., Stahl, D. A. and Wagner, M. (2001). Multiple lateral transfers of dissimilatory sulfite reductase genes between major lineages of sulfate-reducing prokaryotes. *J. Bacteriol.* **83**, 6028–6035.

Koizumi, Y., Kelly, J. J., Nakagawa, T., Urakawa, H., El-Fantroussi, S., Al-Muzaini, S., Fukui, M., Urushigawa, Y. and Stahl, D. A. (2002). Parallel characterization of anaerobic toluene- and ethylbenzene-degrading microbial consortia by PCR-denaturing gradient gel electrophoresis, RNA–DNA membrane hybridization, and DNA microarray technology. *Appl. Environ. Microbiol.* **68**, 3215–3225.

Kurkela, S., Lehvaslaiho, H., Palva, E. T. and Teeri, T. H. (1988). Cloning, nucleotide-sequence and characterization of genes encoding naphthalene dioxygenase of *Pseudomonas putida* strain NCIB9816. *Gene* **73**, 355–362.

Laddaga, R. A., Chu, L. E., Misra, T. K. and Silver, S. (1987). Nucleotide sequence and expression of the mercurial-resistance operon from *Staphylococcus aureus* plasmid pI258. *Proc. Natl Acad. Sci. USA* **84**, 5106–5110.

Larkin, M. J., Allen, C. C. R., Kulakov, L. A. and Lipscomb, D. A. (1999). Purification and characterization of a novel naphthalene dioxygenase from *Rhodococcus* sp. strain NCIMB12038. *J. Bacteriol.* **181**, 6200–6204.

Larsen, O., Lien, T. and Birkeland, N. K. (1999). Dissimilatory sulfite reductase from *Archaeoglobus profundus* and *Desulfotomaculum thermocisternum*: phylogenetic and structural implications from gene sequences. *Extremophiles* **3**, 63–70.

Laue, H., Friedrich, M., Ruff, J. and Cook, A. M. (2001). Dissimilatory sulfite reductase (desulfoviridin) of the taurine-degrading, non-sulfate-reducing bacterium *Bilophila wadsworthia* RZATAU contains a fused DsrB–DsrD subunit. *J. Bacteriol.* **183**, 1727–1733.

Leaphart, A. B. and Lovell, C. R. (2001). Recovery and analysis of formyltetrahydrofolate synthetase gene sequences from natural populations of acetogenic bacteria. *Appl. Environ. Microbiol.* **67**, 1392–1395.

Li, F. and Stormo, G. (2001). Selection of optimal DNA oligos for gene expression arrays. *Bioinformatics* **17**, 1067–1076.

Liu, X., Tiquia, S. M., Holguin, G., Wu, L., Nold, S. C., Devol, A. H., Luo, K., Palumbo, A., Tiedje, J. M. and Zhou, J. (2003a). Molecular diversity of denitrifying genes in continental margin sediments within the oxygen-deficient zone off the pacific coast of Mexico. *Appl. Environ. Microbiol.* **69**, 3549–3560.

Liu, Y., Zhou, J., Omelchenko, M., Beliaev, A., Venkateswaran, A., Stair, J., Wu, L., Thompson, D. K., Xu, D., Rogozin, I. B., Gaidamakova, E. K., Zhai, M., Makarova, K. S., Koonin, E. V. and Daly, M. J. (2003b). Transcriptome dynamics of *Deinococcus radiodurans* recovering from ionizing radiation. *Proc. Natl Acad. Sci. USA* **100**, 4191–4196.

Lloyd-Jones, G., Laurie, A. D., Hunter, D. W. F. and Fraser, R. (1999). Analysis of catabolic genes for naphthalene and phenanthrene degradation in contaminated New Zealand soils. *FEMS Microbiol. Ecol.* **29**, 69–79.

Lockhart, D. J., Dong, H., Byrne, M. C., Follettie, M. T., Gallo, M. V., Chee, M. S., Mittmann, M., Wang, C., Kobayashi, M., Horton, H. and Brown, E. L. (1996). Expression monitoring by hybridization to high-density oligonucleotide arrays. *Nat. Biotechnol.* **14**, 1675–1680.

Lovley, D. R. (1997). Potential for anaerobic bioremediation of BTEX in petroleum-contaminated aquifers. *J. Ind. Microbiol. Biotechnol.* **18**, 75–81.

Lovley, D. R. and Klug, M. J. (1983). Sulfate reducers can out-compete methanogens at fresh-water sulfate concentrations. *Appl. Environ. Microbiol.* **45**, 187–192.

Lyons, J. I., Newell, S. Y., Buchan, A. and Moran, M. A. (2003). Diversity of ascomycete laccase gene sequences in a southeastern U.S. salt marsh. *Microb. Ecol.* **45**, 270–281.

Manz, W., Eisenbrecher, M., Neu, T. R. and Szewzyk, U. (1998). Abundance and spatial organization of Gram-negative sulfate-reducing bacteria in activated sludge investigated by in situ probing with specific 16S rRNA targeted oligonucleotides. *FEMS Microbiol. Ecol.* **25**, 43–61.

Marchesi, J. R. and Weightman, A. J. (2003). Comparing the dehalogenase gene pool in cultivated α-halocarboxylic acid-degrading bacteria with the environmental metagene pool. *Appl. Environ. Microbiol.* **69**, 4375–4382.

Martin-Laurent, F., Piutti, S., Hallet, S., Wagschal, I., Philippot, L., Catroux, G. and Soulas, G. (2003). Monitoring of atrazine treatment on soil bacterial, fungal and atrazine-degrading communities by quantitative competitive PCR. *Pest. Manag. Sci.* **59**, 259–268.

Mason, R. P., Fitzgerald, W. F. and Morel, F. M. M. (1994). The biogeochemical cycling of elemental mercury – anthropogenic influences. *Geochim. Cosmochim. Acta* **58**, 3191–3198.

McGowan, C., Fulthorpe, R., Wright, A. and Tiedje, J. M. (1998). Evidence for interspecies gene transfer in the evolution of 2,4-dichlorophenoxyacetic acid degraders. *Appl. Environ. Microbiol.* **64**, 4089–4092.

McMahon, K. D., Dojka, M. A., Pace, N. R., Jenkins, D. and Keasling, J. D. (2002). Polyphosphate kinase from activated sludge performing enhanced biological phosphorus removal. *Appl. Environ. Microbiol.* **68**, 4971–4978.

Melcher, R. J., Apitz, S. E. and Hemmingsen, B. B. (2002). Impact of irradiation and polycyclic aromatic hydrocarbon spiking on microbial populations in marine sediment for future aging and biodegradability studies. *Appl. Environ. Microbiol.* **68**, 2858–2868.

Metcalfe, A. C., Krsek, M., Gooday, G. W., Prosser, J. I. and Wellington, E. M. H. (2002). Molecular analysis of a bacterial chitinolytic community in an upland pasture. *Appl. Environ. Microbiol.* **68**, 5042–5050.

Molitor, M., Dahl, C., Molitor, I., Schafer, U., Speich, N., Huber, R., Deutzmann, R. and Truper, H. G. (1998). A dissimilatory sirohaem-sulfite-reductase-type protein from the hyperthermophilic archaeon *Pyrobaculum islandicum*. *Microbiology* **144**, 529–541.

Mori, K., Kim, H., Kakegawa, T. and Hanada, S. (2003). A novel lineage of sulfate-reducing microorganisms: Thermodesulfobiaceae fam. nov., *Thermodesulfobium narugense*, gen. nov., sp. nov., a new thermophilic isolate from a hot spring. *Extremophiles* **7**, 283–290.

Murrell, J. C., Gilbert, B. and McDonald, I. R. (2000). Molecular biology and regulation of methane monooxygenase. *Arch. Microbiol.* **173**, 325–332.

Murrell, J. C., McDonald, I. R. and Bourne, D. G. (1998). Molecular methods for the study of methanotroph ecology. *FEMS Microbiol. Ecol.* **27**, 103–114.

Myers, E. W. (1998). A fast bit-vector algorithm for approximate string matching based on dynamic programming. *Ninth Combinatorial Pattern Matching Conference*. Springer, Piscataway, NJ, pp. 1–13.

Nakagawa, T., Nakagawa, S., Inagaki, F., Takai, K. and Horikoshi, K. (2004). Phylogenetic diversity of sulfate-reducing prokaryotes in active deep-sea hydrothermal vent chimney structures. *FEMS Microbiol. Lett.* **232**, 145–152.

Nascimento, A. M. and Chartone-Souza, E. (2003). Operon mer: bacterial resistance to mercury and potential for bioremediation of contaminated environments. *Genet. Mol. Res.* **2**(92), 101.

Nguyên-nhu, N. T. and Knoops, B. (2002). Alkyl hydroperoxide reductase 1 protects *Saccharomyces cerevisiae* against metal ion toxicity and glutathione depletion. *Toxicol. Lett.* **135**, 219–228.

Nielsen, H. B., Wernersson, R. and Knudsen, S. (2003). Design of oligonucleotides for microarrays and perspectives for design of multi-transcriptome arrays. *Nucleic Acids Res.* **31**, 3491–3496.

Nold, S. C., Zhou, J. Z., Devol, A. H. and Tiedje, J. M. (2000). Pacific northwest marine sediments contain ammonia-oxidizing bacteria in the beta subdivision of the Proteobacteria. *Appl. Environ. Microbiol.* **66**, 4532–4535.

Normand, P. and Bousquet, J. (1989). Phylogeny of nitrogenase sequences in Frankia and other nitrogen-fixing microorganisms. *J. Mol. Evol.* **29**, 436–447.

Normand, P., Gouy, M., Cournoyer, B. and Simonet, P. (1992). Nucleotide sequence of nifD from Frankia alni strain ARI3: phylogenetic inferences. *Mol. Biol. Evol.* **9**, 495–506.

Oger, C., Mahillon, J. and Petit, F. (2003). Distribution and diversity of a cadmium resistance (*cadA*) determinant and occurrence of IS257 insertion sequences in Staphylococcal bacteria isolated from a contaminated estuary (Seine, France). *FEMS Microbol. Ecol.* **43**, 173–183.

Osborn, A. M., Bruce, K. D., Strike, P. and Ritchie, D. A. (1997). Distribution, diversity and evolution of the bacterial mercury resistance (*mer*) operon. *FEMS Microbiol. Rev.* **19**, 239–262.

Parales, R. E., Emig, M. D., Lynch, N. A. and Gibson, D. T. (1998). Substrate specificities of hybrid naphthalene and 2,4-dinitrotoluene dioxygenase enzyme systems. *J. Bacteriol.* **180**, 2337–2344.

Parales, R. E., Lee, K., Resnick, S. M., Jiang, H. Y., Lessner, D. J. and Gibson, D. T. (2000). Substrate specificity of naphthalene dioxygenase: effect of specific amino acids at the active site of the enzyme. *J. Bacteriol.* **182**, 1641–1649.

Park, W., Padmanabhan, P., Padmanabhan, S., Zylstra, G. J. and Madsen, E. L. (2002). *nahR*, encoding a LysR-type transcriptional regulator, is highly conserved among naphthalene-degrading bacteria isolated from a coal tar waste-contaminated site and in extracted community DNA. *Microbiology* **148**, 2319–2329.

Philippot, L., Piutti, S., Martin-Laurent, F., Hallet, S. and Germon, J. C. (2002). Molecular analysis of the nitrate-reducing community from unplanted and maize-planted soils. *Appl. Environ. Microbiol.* **68**, 6121–6128.

Piétu, G., Alibert, O., Guichard, V., Lamy, B., Bois, F., Leroy, E., Mariage-Samson, R., Houlgatte, R., Soularue, P. and Auffray, C. (1996). Novel gene transcripts preferentially expressed in human muscles revealed by quantitative hybridization of a high density cDNA array. *Genome Res.* **6**, 492–503.

Rabus, R., Hansen, T. and Widdel, F. (2000). The dissimilatory sulfate- and sulfur reducing Prokaryotes. In *The Prokaryotes: An Evolving Electronic Resource for the Microbiological Community* (M. Dworkin et al., eds), 3rd edn. Springer, New York, release 3.3, http://link.springer-ny.com/link/service/books/10125/.

Relogio, A., Schwager, C., Richter, A., Ansorge, W. and Valcarcel, A. (2002). Optimization of oligonucleotide-based DNA microarrays. *Nucleic Acids Res.* **30**, e51.

Reniero, D., Mozzon, E., Galli, E. and Barbieri, P. (1998). Two aberrant mercury resistance transposons in the *Pseudomonas stutzeri* plasmid pPB. *Gene* **208**, 37–42.

Rensing, C., Newby, D. T. and Pepper, I. L. (2002). The role of selective pressure and selfish DNA in horizontal gene transfer and soil microbial community adaptation. *Soil Biol. Biochem.* **34**, 285–296.

Reymond, N., Charles, H., Duret, L., Calevro, F., Beslon, G. and Fayard, J.-M. (2004). ROSO: optimizing oligonucleotide probes for microarrays. *Bioinformatics* **20**, 271–273.

Rhee, S. K., Liu, X., Wu, L., Chong, S. C., Wan, X. and Zhou, J. (2004). Detection of genes involved in biodegradation and biotransformation in microbial communities by using 50-mer oligonucleotide microarrays. *Appl. Environ. Microbiol.* **70**, 4303–4317.

Richmond, C. S., Glasner, J. D., Mau, R., Jin, H. and Blattner, F. R. (1999). Genome-wide expression profiling in *Escherichia coli* K-12. *Nucleic Acids Res.* **27**, 3821–3835.

Romine, M. F., Stillwell, L. C., Wong, K. K., Thurston, S. J., Sisk, E. C., Sensen, C., Gaasterland, T., Fredrickson, J. K. and Saffer, J. D. (1999). Complete sequence of a 184-kilobase catabolic plasmid from *Sphingomonas aromaticivorans* F199. *J. Bacteriol.* **181**, 1585–1602.

Rouillard, J.-M., Herbert, C. and Zuker, M. (2002). OligoArray: genome-scale oligonucleotide design for microarrays. *Bioinformatics* **18**, 486–487.

Rouillard, J.-M., Zuker, M. and Gulari, E. (2003). OligoArray 2.0: design of oligonucleotide probes for DNA microarrays using thermodynamic approach. *Nucleic Acids Res.* **31**, 3057–3062.

Rozen, S. and Skaletsky, H. J. (2000). Primer3 on the WWW for general users and for biologist programmers. *Meth. Mol. Biol.* **13**, 365–386.

Scala, D. and Kerkhof, L. J. (1999). Diversity of nitrous oxide reductase (nosZ) genes in continental shelf sediments. *Appl. Environ. Microbiol.* **65**, 1681–1687.

Schelert, J., Dixit, V., Hoang, V., Simbahan, J., Drozda, M. and Blum, P. (2004). Occurrence and characterization of mercury resistance in the hyperthermophilic archaeon *Sulfolobus solfataricus* by use of gene disruption. *J. Bacteriol.* **2**, 427–437.

Schena, M., Shalon, D., Davis, R. W. and Brown, P. O. (1995). Quantitative monitoring of gene expression patterns with a complementary DNA microarray. *Science* **270**, 467–470.

Schena, M., Shalon, D., Heller, R., Chai, A., Brown, P. O. and Davis, R. W. (1996). Parallel human genome analysis: microarray-based expression monitoring of 1000 genes. *Proc. Natl Acad. Sci. USA* **93**, 10614–10619.

Schluter, A., Heuer, H., Szczepanowski, R., Forney, L. J., Thomas, C. M., Puhler, A. and Top, E. M. (2003). The 64 508 bp IncP-1β antibiotic multiresistance plasmid pB10 isolated from a waste-water treatment plant provides evidence for recombination between members of different branches of the IncP-1β group. *Microbiology* **149**, 3139–3153.

Schneiker, S., Keller, M., Droge, M., Lanka, E., Puhler, A. and Selbitschka, W. (2001). The genetic organization and evolution of the broad host range mercury resistance plasmid pSB102 isolated from a microbial population residing in the rhizosphere of alfalfa. *Nucleic Acids Res.* **29**, 5169–5181.

Schuchhardt, J., Beule, D., Malik, A., Wolski, E., Eickhoff, H., Lehrach, H. and Herzel, H. (2000). Normalization strategies for cDNA microarrays. *Nucleic Acids Res.* **28**, E47.

Singleton, R. Jr. (1993). The sulfate-reducing bacteria: an overview. In *The Sulfate-Reducing Bacteria: Contemporary Perspectives* (J. M. Odom and R. Singleton Jr., eds), *Brock/Springer Series in Contemporary Bioscience*, pp. 1–20. Springer, New York.

Sota, M., Kawasaki, H. and Tsuda, M. (2003). Structure of haloacetate-catabolic IncP-1β plasmid pUO1 and genetic mobility of its residing haloacetate-catabolic transposon. *J. Bacteriol.* **185**, 6741–6745.

Spanakis, E. and Brouty-Boyé, D. (1997). Discrimination of fibroblast subtypes by multivariate analysis of gene expression. *Int. J. Cancer* **71**, 402–409.

Stach, J. E. M. and Burns, R. G. (2002). Enrichment versus biofilm culture: a functional and phylogenetic comparison of polycyclic aromatic hydrocarbon-degrading microbial communities. *Environ. Microbiol.* **4**, 169–182.

Sullivan, J. P., Dickinson, D. and Chase, H. A. (1998). Methanotrophs, *Methylosinus trichosporium* OB3b, sMMO, and their application to bioremediation. *Crit. Rev. Microbiol.* **24**, 335–373.

Sun, C. G., Snape, C. E., McRae, C. and Fallick, A. E. (2003). Resolving coal and petroleum-derived polycyclic aromatic hydrocarbons (PAHs) in some contaminated land samples using compound-specific stable carbon isotope ratio measurements in conjunction with molecular fingerprints. *Fuel* **82**, 2017–2023.

Takizawa, N., Iida, T., Sawada, T., Yamauchi, K., Wang, Y. W., Fukuda, M. and Kiyohara, H. (1999). Nucleotide sequences and characterization of genes encoding naphthalene upper pathway of *Pseudomonas aeruginosa* PaK1 and *Pseudomonas putida* OUS82. *J. Biosci. Bioeng.* **87**, 721–731.

Takizawa, N., Kaida, N., Torigoe, S., Moritani, T., Sawada, T., Satoh, S. and Kiyohara, H. (1994). Identification and characterization of genes encoding polycyclic aromatic hydrocarbon dioxygenase and polycyclic aromatic hydrocarbon dihydrodiol dehydrogenase in *Pseudomonas putida* OUS82. *J. Bacteriol.* **176**, 2444–2449.

Talaat, A. M., Howard, S. T., Hale IV, W., Lyons, R., Garner, H. and Jonston, S. A. (2002). Genomic DNA standards for gene expression profiling in *Mycobacterium tuberculosis*. *Nucleic Acids Res.* **30**, e104.

Tamayo, P., Slonim, D., Mesirov, J., Zhu, Q., Kitareewan, S., Dmitrovsky, E., Lander, E. S. and Golub, T. R. (1999). Interpreting patterns of gene expression with self-organizing maps: methods and application to hematopoietic differentiation. *Proc. Natl Acad. Sci.* **96**, 2907–2912.

Taroncher-Oldenburg, G., Griner, E. M., Francis, C. A. and Ward, B. B. (2002). Oligonucleotide microarray for the study of functional gene diversity in the nitrogen cycle in the environment. *Appl. Environ. Microbiol.* **69**, 1159–1171.

Thomas, J. G., Olson, J. M., Tapscott, S. J. and Zhao, L. P. (2001). An efficient and robust statistical modeling approach to discover differentially expressed genes using genomic expression profiles. *Genome Res.* **11**, 1227–1236.

Thompson, D. K., Beliaev, A. S., Giometti, C. S., Tollaksen, S. L., Khare, T., Lies, D. P., Nealson, K. H., Lim, H., Yates III, J., Brandt, C. C., Tiedje, J. M. and Zhou, J. (2002). Transcriptional and proteomic analysis of a ferric uptake regulator (Fur) mutant of *Shewanella oneidensis*: possible involvement of Fur in energy metabolism, transcriptional regulation, and oxidative stress. *Appl. Environ. Microbiol.* **68**, 881–892.

Tiquia, S. M., Wu, L., Chong, S. C., Passovets, S., Xu, D., Xu, Y. and Zhou, J. (2004). Evaluation of 50-mer oligonucleotide arrays for detecting microbial populations in environmental samples. *BioTechniques* **36**, 664–675.

Top, E. M. and Springael, D. (2003). The role of mobile genetic elements in bacterial adaptation to xenobiotic organic compounds. *Curr. Opin. Biotechnol.* **14**, 262–269.

Ueda, T., Suga, Y., Yahiro, N. and Matsuguchi, T. (1995). Genetic diversity of N2-fixing bacteria associated with rice roots by molecular evolutionary analysis of a nifD library. *Can. J. Microbiol.* **41**, 235–240.

Venkatesan, M. M., Goldberg, M. B., Rose, D. J., Grotbeck, E. J., Burland, V. and Blattner, F. R. (2001). Complete DNA sequence and analysis of the large virulence plasmid of *Shigella flexneri*. *Infect. Immun.* **69**, 3271–3285.

Vitousek, P. M. and Howarth, R. W. (1991). Nitrogen limitation on land and in the sea: how can it occur? *Biogeochemistry* **13**, 87–115.

Wackett, L. P. (2002). Mechanism and applications of Rieske non-heme iron dioxygenases. *Enzyme Microb. Technol.* **31**, 577–587.

Wagner, M., Roger, A. J., Flax, J. L., Brusseau, G. A. and Stahl, D. A. (1998). Phylogeny of dissimilatory sulfite reductases supports an early origin of sulfate respiration. *J. Bacteriol.* **180**, 2975–2982.

Wang, X. and Seed, B. (2003). Selection of oligonucleotide probes for protein coding sequences. *Bioinformatics* **19**, 796–802.

Westwater, J., McLaren, N. F., Dormer, U. H. and Jamieson, D. J. (2002). The adaptive response of *Saccharomyces cerevisiae* to mercury exposure. *Yeast* **19**, 233–239.

Widmer, F., Shaffer, B. T., Porteous, L. A. and Seidler, R. J. (1999). Analysis of nifH gene pool complexity in soil and litter at a Douglas fir forest site in the Oregon Cascade mountain range. *Appl. Environ. Microbiol.* **65**, 374–380.

Wilson, M. S., Bakermans, C. and Madsen, E. L. (1999). In situ, real-time catabolic gene expression: extraction and characterization of naphthalene dioxygenase mRNA transcripts from groundwater. *Appl. Environ. Microbiol.* **65**, 80–87.

Wilson, M. S., Herrick, J. B., Jeon, C. O., Hinman, D. E. and Madsen, E. L. (2003). Horizontal transfer of *phnAc* dioxygenase genes within one of two phenotypically and genotypically distinctive naphthalene-degrading guilds from adjacent soil environments. *Appl. Environ. Microbiol.* **69**, 2172–2181.

Wu, L. Y., Thompson, D. K., Li, G., Hurt, R. A., Tiedje, J. M. and Zhou, J. (2001). Development and evaluation of functional gene arrays for detection of selected gene in the environment. *Appl. Environ. Microbiol.* **67**, 5780–5790.

Xu, D., Li, G., Wu, L., Zhou, J. and Xu, Y. (2002). PRIMEGENS: robust and efficient design of gene-specific probes for microarray analysis. *Bioinformatics* **18**(11), 1432–1437.

Ye, R. W., Tao, W., Beddzyk, L., Young, T., Chen, M. and Li, L. (2000). Global gene expression profiles of *Bacillus subtilis* grown under anaerobic conditions. *J. Bacteriol.* **182**, 4458–4465.

Yurieva, O., Kholodii, G., Minakhin, L., Gorlenko, Z., Kalyaeva, E., Mindlin, S. and Nikiforov, V. (1997). Intercontinental spread of promiscuous mercury-resistance transposons in environmental bacteria. *Mol. Microbiol.* **24**, 321–329.

Zar, J. H. (1999). *Biostatistical Analysis*. Prentice-Hall, Englewood Cliffs, NJ.

Zehr, J. P. and McReynolds, L. A. (1989). Use of degenerate oligonucleotides for amplification of the nifH gene from the marine cyanobacterium *Trichodesmium* spp. *Appl. Environ. Microbiol.* **55**, 2522–2526.

Zehr, J. P., Mellon, M., Braun, S., Litaker, W., Steppe, T. and Paerl, H. W. (1995). Diversity of heterotrophic nitrogen fixation genes in a marine cyanobacterial mat. *Appl. Environ. Microbiol.* **61**, 2527–2532.

Zhou, J. (2003). Microarrays for bacterial detection and microbial community analysis. *Curr. Opin. Microbiol.* **6**, 288–294.

Zhou, J. and Thompson, D. K. (2002). Challenges in applying microarrays to environmental studies. *Curr. Opin. Biotechnol.* **13**, 204–207.

Zhou, J., Bruns, M. A. and Tiedje, J. M. (1996). DNA recovery from soils of diverse composition. *Appl. Environ. Microbiol.* **62**, 461–468.

Zhu, J., Bozdech, Z. and DeRisi, J. (2003). *ArrayOligoSelector*. http://sourceforge.net/projects/arrayoligosel/.

Zuker, M., Mathews, D. H. and Turner, D. H. (1999). Algorithms and thermodynamics for RNA secondary structure prediction: a practical guide. In *RNA Biochemistry and Biotechnology* (J. Barciszewski and B. F. C. Clark, eds), pp. 11–43. Kluwer Academic Publishers, Dordrecht.

Index

acquired immunodeficiency syndrome (AIDS) 207–8
acridine orange (AO) 98
acrolein 140
acute osteomyelitis 200
adhesion forces 182
adsorption 167
AFM *see* atomic force microscope
AIDS *see* acquired immunodeficiency syndrome
aldehydes 140–2
algae, autofluorescence 98, 106
amino lipids 142
ammonium molybdate 150
amplicon
  contamination 260
  detection 266
  self-priming 279–83
anthrax 299
antibiotic resistance 297
antibody fragments 230–2
antigen–antibody interaction kinetics 228–30
antiviral therapy 294
AO *see* acridine orange
asymmetric PCR 261
atomic force microscope (AFM) 164–7, 190–1
  cell surface dynamics 178–9
  cell wall elasticity 179–80
  conformational changes 173–4
  high-resolution imaging 169
  membrane proteins 172–3
  molecular manipulation 183–7
  protein crystals 170–2
  receptor-ligand interactions 187–90
  sample preparation 167–8
  surface morphology 174–8
  surface properties 180–3
  tip artifacts 169–70
autofluorescence 98, 106–7
autolysins 176
autoradiography 5
avidin 8, 156
axial resolution 126

bacteria
  EM 138
  in situ hybridization 14
  membrane proteins 172
  methanotrophic 339
  nitrogen-fixing 338
  real-time fluorescent PCR 296–7
  sulfate reducing 340
bacteriophage titer 295
bacteriorhodopsin 184
BIACORE 215–17, 222
BiFC *see* bimolecular fluorescence complementation

bimolecular fluorescence complementation (BiFC) 50
binding activity 214
biofilm
  autofluorescence 106–7
  cell staining 107–9
  polymer staining 110–13
  reflection 106
  research 90
  samples 100–5
  surface morphology 174
biosensors 213
  antibody fragments 230–2
  antigen–antibody interaction kinetics 228–30
  BIACORE 215–17, 222
  proteins 215
  viral antigens 225
  viral diagnosis 224–6
  viral epitopes 217–24
  viral peptides 226
bioterrorism 299–300
biotin 5, 156
bone marrow 201
Borrelia burgdorferi 297
brain tumours 199
bud scars 180
buffer solutions 142

Ca^{2+} dyes 77
carbon cycling 339–40
carbon-coated plastic support films 148
carboxyfluorescein 124
CARS microscopy *see* coherent anti-stokes Raman scattering microscopy
cell attachment 168
cell cultures, EM 153
cell staining 107–9
cell surface dynamics 178–9
cell surface hydrophobicity 181
cell ultrastructure 139
cell viability dyes 78
cell wall dyes 75
cell wall elasticity 179–80, 190
cell–cell interactions 186
cells, surface morphology 174–8
cervix, HIV-1 251–3
chitin 179
chlorophyll A 98, 106
chronic osteomyelitis 200–3
CLSM *see* confocal laser scanning microscope
cluster analysis 356
CMV *see* cytomegalovirus
co-localization 106
coherent anti-stokes Raman scattering (CARS) microscopy 126
colloidal gold 154, 155

CommOligo 348
confocal laser scanning microscope (CLSM) 63, 90
confocal microscopy 38
conformational changes 173–4
continuous epitopes 217–18
contrasting 148–50
controls, fluorescent proteins 37
coronaviruses 295
cryo-electron microscopy 157–8
cryo-electron tomography 138, 157–8
cryo-fixation 140
cryo-sectioning 101, 138
cryo-ultramicrotomy 151–3
cryptotopes 218–19
cyanobacteria, autofluorescence 98, 106
cytokines, real-time fluorescent PCR 301
cytomegalovirus (CMV)
   in situ hybridization 15
   real-time PCR 293

deconvolution 122
degenerative disc disease 200
dehydration, EM 144
digital image analysis 113–22
digoxigenin 5
discontinuous epitopes 217–18
displacement hybridisation 272
DNA
   in situ amplification 239
   in situ hybridization 1–2
DNA microarrays 344
DNA probes 4–5
DNA repair 241–4
DNA-associating fluorophores 266–7
DNase digestion 244–5

EBV, *in situ* hybridization 15
EDL interaction *see* electrostatic double layer interaction
electron microscopy (EM) 137
   bacteria 138
   cell cultures 153
   cryo-ultramicrotomy 151–3
   dehydration 144
   embedding 144–6
   freeze-etching 153
   freeze-fracture 153
   freeze-substitution 146–7
   immunocytochemistry 154–7
   infiltration 144–5
   negative staining 148–50
   plastic support films 147–8
   three-dimensional imaging 157
   ultramicrotomy 150–1
electrostatic double layer (EDL) interaction 167
EM *see* electron microscopy
embedding 101, 140
   EM 145–6
endocytosis markers 64, 75–6
endoplasmic reticulum (ER), fluorescent proteins 43
enzymes 178

epitope tagging 37
epoxide resin 140
EPS *see* extracellular polymeric substances
ER *see* endoplasmic reticulum
extracellular polymeric substances (EPS) 110, 174

FCM *see* fluorescence correlation microscopy
FCS *see* fluorescence correlation spectroscopy
FDG *see* 2-^{18}fluoro-2-deoxy-glucose
ferritin 155
fever of unknown origin (FUO) 205–7
FGAs *see* functional gene microarrays
fixation
   electron microscopy 140, 143–4
   fluorescent proteins 36
   in situ hybridization 2–3
flat-embedding 153
FLIM *see* fluorescence lifetime imaging microscopy
flocculation 187
fluorescein 5
fluorescence correlation microscopy (FCM) 44, 125
fluorescence correlation spectroscopy (FCS) 125
fluorescence lifetime imaging microscopy (FLIM) 123–5
fluorescence microscopy 38, 259
fluorescence recovery after photobleaching (FRAP) 98
fluorescence resonance energy transfer (FRET) 29, 261
   biosensors 49–50
   protein–protein interactions
fluorescence speckle microscopy 42
fluorescent proteins 27–37
   gene expression 40
   immunolocalization 156–7
   plant pathology 45–8
   protein distribution 40–3
   subcellular compartments 43–4
2-^{18}fluoro-2-deoxy-glucose (FDG) 199
fluorobodies 37
fluorochromes 8, 94
fluorogenic oligoprobes 268–9
fluorophores
   DNA-associating 266–7
   real-time PCR 292
formaldehyde 141
formvar film 147–8
four-dimensional imaging 52
FRAP *see* fluorescence recovery after photobleaching
freeze-etching 153
freeze-fracture 153
freeze-substitution 146–7
FRET *see* fluorescence resonance energy transfer
fulminant diseases 297
functional gene microarrays (FGAs) 331–3
   carbon cycling 339–40
   data analysis 353–7
   evaluation 357–8
   metals resistance 343–4

microarray construction 349–53
nitrogen cycling 337–9
oligonucleotide probes 344–9
organic contaminant degradation 342–3
probe design 333–7
sulfur cycling 340–1
validation 357–8
fungi
　Ca^{2+} dyes 77
　cell wall dyes 75
　endocytosis marker dyes 75–6
　fluorescent protein probes 27–53
　live-cell imaging 63, 83
　membrane-selective dyes 64–70
　mitochondrial dyes 72–3
　multiple labelling 79
　nuclear dyes 70–2
　pH dyes 77
　real-time fluorescent PCR 298
　vacuolar dyes 73–5
FUO *see* fever of unknown origin

gallium scanning 205
gene diversity 333–44
gene expression
　β-glucuronidase 40
　nucleic acid quantitation 286
gene transfer 343
genetic markers 333–7
germination 178, 185
GFP *see* green fluorescent protein
beta-glucuronidase (GUS) 29, 40
glutaraldehyde 141
glycoconjugates 110
gold 155
green fluorescent protein (GFP) 30, 156
GUS *see* beta-glucuronidase

hairpin oligoprobes 277–9
Helicobacter pylori 14
heparin 189
herpes simplex virus (HSV) 293
high-resolution imaging 169
HIV *see* human immunodeficiency virus
HIV-1 251–3
host immunity 300–2
housekeeping genes 289
HPV *see* human papillomavirus
HSV *see* herpes simplex virus
human immunodeficiency virus (HIV) 207–8, 220
human papillomavirus (HPV)
　HIV-1 251–3
　in situ hybridization 15–17
hybridisation probes 272
hybridization reactions 7–8

imidoesters 140
immunocytochemistry 145, 154–7
immunolabellimg 154, 155
immunolocalizatiom 156–7
in situ hybridization 1–2
　controls 13–14
　methodology 2–14
　microorganisms 14–17
infiltration, EM 144–5
interferon-α, real-time fluorescent PCR 301
intramolecular hybridisation 279

laser capture microdissection (LCM) 50–2
laser scanning microscopy (LSM) 90–1
　biofilm imaging 106–13
　biofilm samples 100–5
　digital image analysis 113–23
　mounting 101
　one-photon excitation 91–3
　two-photon excitation 93–100
LCM *see* laser capture microdissection
lectins 110, 155–6
lenses 126–7
leucocyte scintigraphy (LS) 204
leukaemia 287
lifetime imaging 123
light microscopy
　cell ultrastructure 139
　resolution 126
light scattering 96
linear oligoprobes 271
live-cell imaging 28–30, 38–40, 191
　fungi 63, 83
　methodology 78–83
LNA *see* locked nucleic acids
locked nucleic acids (LNA) 268
LS *see* leucocyte scintigraphy
LSM *see* laser scanning microscopy
lymphoma 286–7

macromolecules 154
malignancy 286
mannoproteins 175
marine stromatolites 98
membrane proteins 172–3
membrane-selective dyes 64–70
mercury resistance 343
metals resistance 343–4
metatopes 218–19
microarray image processing 353
microarray technology 331
microbial biofilms 90
microbial contamination 295
microbial genotyping 283–6
microbial load 291, 302
microbial surfaces 163–4
microbial-host interactions 301
microinjection 28–9
microscope software 115
mispriming 241, 242
mitochondrial dyes 72–3
MMM *see* multifocoal multiphoton microscopy
molecular manipulation 183–7
molecular recognition 187–8
molecule elasticity 186
monoclonal antibodies 217, 226–7
mounting, biofilms 101
MSA *see* multiple sequence alignment
MTC *see* Mycobacterium tuberculosis complex

multidrug resistance 298
multifocoal multiphoton microscopy (MMM) 125
multiphoton microscopy 38–9
multiple sequence alignment (MSA) 348
multiplexd real-time PCR 291–3
Mycobacterium tuberculosis complex (MTC) 14

naphthalene 342
native cells 174–8
near-field scanning optical microscopy (NSOM) 127
negative staining 139, 148–50
Neisseria gonorrhoeae 297
neotopes 218–19
nitrogen cycling 337–9
nonspecific DNA synthesis 241–2
NSOM *see* near-field scanning optical microscopy
nuclear dyes 70–2
nuclease oligoprobes 295
nucleic acid quantitation 286–91
nucleic acid stains 108
nucleic acids 142

oligonucleotide probes 4, 344–9
oligonucleotides
    destructive 269–70
    non-destructive 271–7
oligoprobes
    fluorogenic 268–9
    hairpin 277–9
    linear 271
oncology 199
oncoproteins 16
one-photon LSM (1P-LSM) 90, 91–3
oral biofilms 98, 124
organic contaminant degradation 342–3
osmium tetroxide 140, 142–3
outer membrane protein 174
oxygen gradients 124

1P-LSM *see* one-photon LSM
2P-LSM *see* two-photon LSM
PAHs *see* polycyclic aromatic hydrocarbons
paraffin embedding 101
parasites, real-time fluorescent PCR 298
pathogenesis 188
PCA *see* principal component analysis
PCR *see* polymerase chain reaction
peptide nucleic acids (PNA) 268
peptide-antibody interactions 228
PET *see* positron emission tomography
pH dyes 77
phase contrast 28
phocus indicators 49
phospholipids 142
phosphotungstic acid 149
photostress 124
phycobilims 98, 106
plague 299
plant pathology, fluorescent proteins 45–8
plastic support films 147–8

PNA *see* peptide nucleic acids
polycyclic aromatic hydrocarbons (PAHs) 342
polymer staining 110–13
polymerase chain reaction (PCR) 239, 256–9
    absolute quantitation 290
    external control templates 287–8
    internal control templates 288–9
    multiplex real-time 291–3
    real-time 264–6
    relative quantitation 289–90
positron emission tomography (PET) 199–200, 208–9
    acute osteomyelitis 200
    AIDS 207–8
    chronic osteomyelitis 200–3
    fever of unknown origin 205–7
    oncology tracers 199
    prosthetic joint infections 203–5
    spondylodiscitis 200
primer-dimer formation 267
primers 286
principal component analysis (PCA) 356
probes
    detection 8–13
    in situ hybridization 3–5
    labeling 5–7
prosthetic joint infections 203–5
protein
    conformational changes 190
    crystals 170–2
    kinetics 42, 44–5
    structure–function relationship 213–15
protein A 156
protein–protein interactions, FRET 48–9
protocol 249–51
protozoans, real-time fluorescent PCR 298–9

QSAR *see* quantitative structure–activity relationship
quantification 119–21
quantitative structure–activity relationship (QSAR) 228

RCFPs *see* reef coral fluorescent proteins
real-time 178–9
real-time fluorescent PCR 255–6
    anthrax 299
    bacteria 296–7
    cytokines 301
    fungi 298
    host immunity 300–2
    interferon-α 301
    microbial genotyping 283–6
    parasites 298
    protozoans 298–9
    smallpox 300
    viruses 293–6
receptor–ligand interactions 187–90
recombinant antibodies 227–8
reef coral fluorescent proteins (RCFPs) 31
reflection imaging 106
refractive index 81–2
resolution 126

reverse transcriptase (RT) *in situ* polymerase chain reaction (PCR) 239–41
viral
rhodamine 123 72
RNA
  in situ amplication 239
  in situ hybridization 1–2
RNA polymerase 289
RNA probes 3–4
RNA viruses, RT *in situ* PCR 246
RT *see* reverse transcriptase

S-layers 170–1, 183, 190
SARS *see* severe acute respiratory syndrome
scorpion primers 280, 281
second harmonic generation (SHG) 126
self-organizing maps (SOM) 356
self-priming fluorogenic amplicon 279–83
severe acute respiratory syndrome (SARS) 295
sexually transmitted diseases (STDs) 251–2
SHG *see* second harmonic generation
signal localization 247–9
signal-to-noise ratio 79–81
slide preparation, *in situ* hybridization 3
smallpox 300
software
  digital image analysis 114–18
  FGAs 354
SOM *see* self-organizing maps
spatial versus temporal resolution 79–81
spondylodiscitis 200
SPR *see* surface plasmon resonance
STDs *see* sexually transmitted diseases
STED *see* stimulated emission depletion
stimulated emission depletion (STED) 127
streptavidin 8
subcellular compartments, fluorescent proteins 43–4
submolecular resolution 170–3
sulfur cycling 340–1
sunrise primers 280
surface morphology 174–8
surface plasmon resonance (SPR) 215
synthetic oligonucleotides 332
synthetic peptides 225

targets 246–7
TCSPC *see* time-correlated single-photon counting
TEM *see* transmission electron microscopy
time-correlated single-photon counting (TCSPC) 123
time-lapse imaging 83
tip artifacts 169–70
TIRFM *see* total internal reflection fluorescence microscopy
TMV *see* tobacco mosaic virus
TMVP *see* tobacco mosaic virus protein
tobacco mosaic virus protein (TMVP) 221–4
tobacco mosaic virus (TMV) 219
total internal reflection fluorescence microscopy (TIRFM) 39
transcriptome analysis 286
transmission electron microscopy (TEM) 140–4
treponemes 297
Trichomonas vaginalis 299
tuberculosis 14
two-dimensional protein crystals 170–2
two-photon LSM (2P-LSM) 93–100

ultramicrotomy 150–1
ultrastructure 139
uranyl acetate 140, 149

vacuolar dyes 73–5
varicella zoster virus (VZV) 293
variola virus (VARV) 300
VARV *see* variola virus
video-enhanced microscopy 29
viral antigens 225, 226
viral detection, DNA probes 4
viral diagnosis 224–6
viral epitopes 217–24
viral load 294
viral proteins 224
viruses
  in situ hybridization 14–15
  real-time fluorescent PCR 293–6
vitrification 138
VZV *see* varicella zoster virus